Probability and
Random Processes

Probability and Random Processes

With Applications to Signal Processing and Communications

Edition 2

Scott L. Miller

Professor
Department of Electrical and Computer Engineering
Texas A&M University

Donald Childers

Professor Emeritus
Department of Electrical and Computer Engineering
University of Florida

AMSTERDAM • BOSTON • HEIDELBERG • LONDON
NEW YORK • OXFORD • PARIS • SAN DIEGO
SAN FRANCISCO • SINGAPORE • SYDNEY • TOKYO

Academic Press is an imprint of Elsevier

Academic Press is an imprint of Elsevier
225 Wyman Street, Waltham, MA 02451, USA
525 B Street, Suite 1900, San Diego, CA 92101-4495, USA
The Boulevard, Langford Lane, Kidlington, Oxford OX5 1GB, UK
Radarweg 29, PO Box 211, 1000 AE Amsterdam, The Netherlands

Library of Congress Cataloging-in-Publication Data
Miller, Scott L.
    Probability and random processes : with applications to signal processing and communications / Scott L. Miller,
Donald Childers. — Ed. 2.
        p. cm.
    Includes bibliographical references and index.
    ISBN 978-0-12-386981-4 (alk. paper)
1. Signal processing—Mathematics. 2. Probabilities. 3. Stochastic processes. I. Childers, Donald G. II. Title.
    TK5102.9.M556 2012
    621.382'20151—dc23   2011053022

British Library Cataloguing in Publication Data
A catalogue record for this book is available from the British Library

ISBN: 978-0-12-386981-4

For all information on all Academic Press Publications
visit our Web site at www.elsevierdirect.com

Printed in the USA
Transfer to digital printing in 2014

Contents

Preface xi

Chapter 1: Introduction 1

   1.1 A Speech Recognition System 2
   1.2 A Radar System 4
   1.3 A Communication Network 4

Chapter 2: Introduction to Probability Theory 7

   2.1 Experiments, Sample Spaces, and Events 7
   2.2 Axioms of Probability 10
   2.3 Assigning Probabilities 13
   2.4 Joint and Conditional Probabilities 17
   2.5 Basic Combinatorics 20
   2.6 Bayes's Theorem 27
   2.7 Independence 29
   2.8 Discrete Random Variables 32
   2.9 Engineering Application—An Optical Communication System 38
       Exercises 43
       MATLAB Exercises 61

Chapter 3: Random Variables, Distributions, and Density Functions 63

   3.1 The Cumulative Distribution Function 64
   3.2 The Probability Density Function 69
   3.3 The Gaussian Random Variable 71
   3.4 Other Important Random Variables 78
       3.4.1 Uniform Random Variable 78
       3.4.2 Exponential Random Variable 79
       3.4.3 Laplace Random Variable 80
       3.4.4 Gamma Random Variable 81
       3.4.5 Erlang Random Variable 81
       3.4.6 Chi-Squared Random Variable 82
       3.4.7 Rayleigh Random Variable 82
       3.4.8 Rician Random Variable 83
       3.4.9 Cauchy Random Variable 84

3.5 Conditional Distribution and Density Functions 85
3.6 Engineering Application: Reliability and Failure Rates 91
       Exercises 97
       MATLAB Exercises 109

Chapter 4: Operations on a Single Random Variable *111*

4.1   Expected Value of a Random Variable 111
4.2   Expected Values of Functions of Random Variables 113
4.3   Moments 115
4.4   Central Moments 117
4.5   Conditional Expected Values 121
4.6   Transformations of Random Variables 122
       4.6.1 Monotonically Increasing Functions 122
       4.6.2 Monotonically Decreasing Functions 124
       4.6.3 Nonmonotonic Functions 125
4.7   Characteristic Functions 130
4.8   Probability-Generating Functions 136
4.9   Moment-Generating Functions 139
4.10 Evaluating Tail Probabilities 140
4.11 Engineering Application—Scalar Quantization 148
4.12 Engineering Application—Entropy and Source Coding 155
       Exercises 159
       MATLAB Exercises 174

Chapter 5: Pairs of Random Variables *177*

5.1   Joint Cumulative Distribution Functions 178
5.2   Joint Probability Density Functions 180
5.3   Joint Probability Mass Functions 186
5.4   Conditional Distribution, Density, and Mass Functions 188
5.5   Expected Values Involving Pairs of Random Variables 192
5.6   Independent Random Variables 197
5.7   Jointly Gaussian Random Variables 202
5.8   Joint Characteristic and Related Functions 206
5.9   Transformations of Pairs of Random Variables 210
5.10 Complex Random Variables 219
5.11 Engineering Application: Mutual Information, Channel
       Capacity, and Channel Coding 221
       Exercises 227
       MATLAB Exercises 242

Chapter 6: Multiple Random Variables *245*

6.1   Joint and Conditional PMFs, CDFs, and PDFs 245
6.2   Expectations Involving Multiple Random Variables 247
6.3   Gaussian Random Variables in Multiple Dimensions 249
6.4   Transformations Involving Multiple Random Variables 252

6.4.1 Linear Transformations 253
6.4.2 Quadratic Transformations of Gaussian Random Vectors 257
6.4.3 Order Statistics 260
6.4.4 Coordinate Systems in Three Dimensions 262
6.5 Estimation and Detection 264
6.5.1 Maximum a Posteriori Estimation 265
6.5.2 Maximum Likelihood Estimation 267
6.5.3 Minimum Mean Square Error Estimation 268
6.6 Engineering Application: Linear Prediction of Speech 272
Exercises 277
MATLAB Exercises 287

Chapter 7: Random Sums and Sequences **289**
7.1 Independent and Identically Distributed Random Variables 289
7.1.1 Estimating the Mean of IID Random Variables 290
7.1.2 Estimating the Variance of IID Random Variables 295
7.1.3 Estimating the CDF of IID Random Variables 297
7.2 Convergence Modes of Random Sequences 298
7.2.1 Convergence Everywhere 300
7.2.2 Convergence Almost Everywhere 301
7.2.3 Convergence in Probability 301
7.2.4 Convergence is the Mean Square Sense 302
7.2.5 Convergence in Distribution 302
7.3 The Law of Large Numbers 304
7.4 The Central Limit Theorem 306
7.5 Confidence Intervals 310
7.6 Random Sums of Random Variables 315
7.7 Engineering Application: A Radar System 317
Exercises 325
MATLAB Exercises 334

Chapter 8: Random Processes **335**
8.1 Definition and Classification of Processes 335
8.2 Mathematical Tools for Studying Random Processes 340
8.3 Stationary and Ergodic Random Processes 348
8.4 Properties of the Autocorrelation Function 356
8.5 Gaussian Random Processes 357
8.6 Poisson Processes 360
8.7 Engineering Application—Shot Noise in a p–n Junction Diode 365
Exercises 371
MATLAB Exercises 381

Chapter 9: Markov Processes **383**
9.1 Definition and Examples of Markov Processes 383
9.2 Calculating Transition and State Probabilities in Markov Chains 388
9.3 Characterization of Markov Chains 394
9.4 Continuous Time Markov Processes 401
9.5 Engineering Application: A Computer Communication Network 413

9.6 Engineering Application: A Telephone Exchange 416
    Exercises 419
    MATLAB Exercises 427

Chapter 10: Power Spectral Density ***429***
  10.1 Definition of PSD 430
  10.2 The Wiener–Khintchine–Einstein Theorem 433
  10.3 Bandwidth of a Random Process 439
  10.4 Spectral Estimation 441
      10.4.1 Non-parametric Spectral Estimation 441
      10.4.2 Parametric Spectral Estimation 448
  10.5 Thermal Noise 452
  10.6 Engineering Application: PSDs of Digital Modulation Formats 455
    Exercises 463
    MATLAB Exercises 469

Chapter 11: Random Processes in Linear Systems ***473***
  11.1 Continuous Time Linear Systems 473
  11.2 Discrete-Time Linear Systems 477
  11.3 Noise Equivalent Bandwidth 479
  11.4 Signal-to-Noise Ratios 480
  11.5 The Matched Filter 482
  11.6 The Wiener Filter 486
  11.7 Bandlimited and Narrowband Random Processes 494
  11.8 Complex Envelopes 499
  11.9 Engineering Application: An Analog Communication System 500
    Exercises 505
    MATLAB Exercises 514

Chapter 12: Simulation Techniques ***517***
  12.1 Computer Generation of Random Variables 517
      12.1.1 Binary Pseudorandom Number Generators 517
      12.1.2 Nonbinary Pseudorandom Number Generators 521
      12.1.3 Generation of Random Numbers from a Specified Distribution 523
      12.1.4 Generation of Correlated Random Variables 524
  12.2 Generation of Random Processes 525
      12.2.1 Frequency Domain Approach 525
      12.2.2 Time Domain Approach 529
      12.2.3 Generation of Gaussian White Noise 533
  12.3 Simulation of Rare Events 534
      12.3.1 Monte Carlo Simulations 535
      12.3.2 Importance Sampling 537
  12.4 Engineering Application: Simulation of a Coded Digital
      Communication System 539
    Exercises 543
    MATLAB Exercises 546

Appendices

A Review of Set Theory 547
B Review of Linear Algebra...551
C Review of Signals and Systems 559
D Summary of Common Random Variables 565
E Mathematical Tables 577
F Numerical Methods for Evaluating the Q-Function 587

Index *593*

Preface

This book is intended to be used as a text for either undergraduate level (junior/senior) courses in probability or introductory graduate level courses in random processes that are commonly found in Electrical Engineering curricula. While the subject matter is primarily mathematical, it is presented for engineers. Mathematics is much like a well crafted hammer. We can hang the tool on our wall and step back and admire the fine craftmanship used to construct the hammer, or we can pick it up and use it to pound a nail into the wall. Likewise, mathematics can be viewed as an art form or a tool. We can marvel at the elegance and rigor, or we can use it to solve problems. It is for this latter purpose that the mathematics is presented in this book. Instructors will note that there is no discussion of algebras, Borel fields or measure theory in this text. It is our belief that the vast majority of engineering problems regarding probability and random processes do not require this level of rigor. Rather, we focus on providing the student with the tools and skills needed to solve problems. Throughout the text we have gone to great effort to strike a balance between readability and sophistication. While the book provides enough depth to equip students with the necessary tools to study modern communication systems, control systems, signal processing techniques, and many other applications, concepts are explained in a clear and simple manner that makes the text accessible as well.

It has been our experience that most engineering students need to see how the mathematics they are learning relates to engineering practice. Towards that end, we have included numerous engineering application sections throughout the text to help the instructor tie the probability theory to engineering practice. Many of these application sections focus on various aspects of telecommunications since this community is one of the major users of probability theory, but there are applications to other fields as well. We feel that this aspect of the text can be very useful for accreditation purposes for many institutions. The Accreditation Board for Engineering and Technology (ABET) has stated that all electrical engineering programs should provide their graduates with a knowledge of probability and statistics including applications to electrical engineering. This text provides not only the probability theory, but also the applications to electrical engineering and a modest amount of statistics as applied to engineering.

A key feature of this text, not found in most texts on probability and random processes, is an entire chapter devoted to simulation techniques. With the advent of powerful, low-cost, computational facilities, simulations have become an integral part of both academic and industrial research and development. Yet, many students have major misconceptions about how to run simulations. Armed with the material presented in our chapter on simulation, we believe students can perform simulations with confidence.

It is assumed that the readers of this text have a background consistent with typical junior level electrical engineering curricula. In particular, the reader should have a knowledge of differential and integral calculus, differential equations, linear algebra, complex variables, discrete math (set theory), linear time-invariant systems, and Fourier transform theory. In addition, there are a few sections in the text that require the reader to have a background in analytic function theory (e.g., parts of Section 4.10), but these sections can be skipped without loss of continuity. While some appendices have been provided with a review of some of these topics, these presentations are intended to provide a refresher for those who need to ?rush up?and are not meant to be a substitute for a good course.

For undergraduate courses in probability and random variables, we recommend instructors cover the following sections:

Chapters 1-3: all sections,
Chapter 4: sections 1-6,
Chapter 5: sections 1-7 and 9,
Chapter 6: sections 1-3,
Chapter 7: sections 1-5.

These sections, along with selected application sections, could be covered in a one semester course with a comfortable pace. For those using this text in graduate courses in random processes, we recommend that instructors briefly review Chapters 1-7 focussing on those concepts not typically taught in an undergraduate course (e.g., 4.7-4.10, 5.8, 5.10, 6.4-6.5, and 7.6) and then cover selected topics of interest from Chapters 8-12.

We consider the contents of this text to be appropriate background material for such follow-on courses as Digital Communications, Information Theory, Coding Theory, Image Processing, Speech Analysis, Synthesis and Recognition, and similar courses that are commonly found in many undergraduate and graduate programs in Electrical Engineering. Where possible, we have included engineering application examples from some of these topics.

Introduction

The study of probability, random variables, and random processes is fundamental to a wide range of disciplines. For example, many concepts of basic probability can be motivated through the study of games of chance. Indeed, the foundations of probability theory were originally built by a mathematical study of games of chance. Today, a huge gambling industry is built on a foundation of probability. Casinos have carefully designed games that allow the players to win just enough to keep them hooked, while keeping the odds balanced slightly in favor of the "house." By nature, the outcomes of these games are random, but the casino owners fully understand that as long as the players keep playing, the theory of probability guarantees—with very high probability—that the casino will always come out ahead. Likewise, those playing the games may be able to increase their chances of winning by understanding and using probability.

In another application of probability theory, stock investors spend a great deal of time and effort trying to predict the random fluctuations in the market. Day traders try to take advantage of the random fluctuations that occur on a daily basis, while long-term investors try to benefit from the gradual trends that unfold over a much longer time period. These trends and fluctuations are random in nature and can only be described in a probabilistic fashion. Another business built on managing random occurrences is the insurance industry. Insurance premiums are calculated based on a careful study of the probabilities of various events happening. For example, the car insurance salesmen have carefully evaluated the inherent risk of various classes of drivers and will adjust the premiums of each class according to the probabilities that those drivers will have an accident. In yet another application of probability theory, a meteorologist tries to predict future weather events based on current and past meteorological conditions. Since these events are quite random, the weather forecast will often be presented in terms of probabilities (e.g., there is a 40% chance, or probability, of rain on Tuesday).

Since the theory of probability and random processes finds such a wide range of applications, students require various levels of understanding depending on the particular field they are preparing to enter. For those who wish to improve their proficiency at card games, a firm understanding of discrete probability may be sufficient. Those going into operations management need to understand queueing theory and therefore Markov and related random processes. A telecommunications engineer needs to have a firm understanding of models of noise and the design of systems to minimize the effects of noise.

Probability and Random Processes. DOI: 10.1016/B978-0-12-386981-4.00001-2

This book is not intended to serve the needs of all disciplines, but rather is focused on preparing the students entering the fields of electrical and computer engineering. One of the main goals of the text is to prepare the student to study random signals and systems. This material is fundamental to the study of digital signal processing (voice, image, video, etc.), communications systems and networks, radar systems, power systems, and many other applications within the engineering community. With this readership in mind, a background which is consistent with most electrical and computer engineering curricula is assumed. That is, in addition to fundamental mathematics including calculus, differential equations, linear algebra, and complex variables, the student is assumed to be familiar with the study of deterministic signals and systems. We understand that some readers may be very strong in these areas, while others may need to "brush up." Accordingly, we have included a few appendices which may help those that need a refresher and also provide a quick reference for significant results.

Throughout the text, the reader will find many examples and exercises which utilize MATLAB. MATLAB is a registered trademark of the MathWorks, Inc.; it is a technical software computing environment. Our purpose for introducing computer-based examples and problems is to expand our capabilities so that we may solve problems that might be too tedious or complex to do via hand calculations. Furthermore, MATLAB has nice plotting capabilities that can greatly assist the visualization of data. MATLAB is used extensively in practice throughout the engineering community; therefore, we feel it is useful for engineering students to gain exposure to this important software package. Examples in the text which use MATLAB are clearly marked with a small computer logo.

Before diving into the theory of discrete probability in the next chapter, we first provide a few illustrations of how the theory of probability and random processes is used in a few engineering applications. At the end of each subsequent chapter, the reader will find engineering application sections which illustrate how the material presented in that chapter is used in the real world. These sections can be skipped without losing any continuity, but we recommend that the reader at least skim through the material.

1.1 A Speech Recognition System

Many researchers are working on methods for computer recognition of speech. One application is to recognize commands spoken to a computer. Such systems are presently available from several vendors. A simple speech recognition system might use a procedure called template matching, which may be described as follows. We define a vocabulary, or a set of possible words for a computerized dictionary. This restricts the number of possible alternatives that must be recognized. Then a template for each word is obtained by digitizing the word as it is spoken. A simple dictionary of such templates is shown in Figure 1.1. The template may be the time waveform, the spectrum of the word, or a vector of selected features

Vocabulary   Template

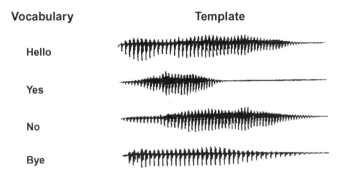

Figure 1.1
A simple dictionary of speech templates for speech recognition.

of the word. Common features might include the envelope of the time waveform, the energy, the number of zero crossings within a specified interval, and the like.

Speech recognition is a complicated task. Factors that make this task so difficult include interference from the surroundings, variability in the amplitude and duration of the spoken word, changes in other characteristics of the spoken word such as the speaker's pitch, and the size of the dictionary to name a few. In Figure 1.2, we have illustrated some of the variability that may occur when various talkers speak the same word. Here, we see that the waveform templates may vary considerably from speaker to speaker. This variability may be described

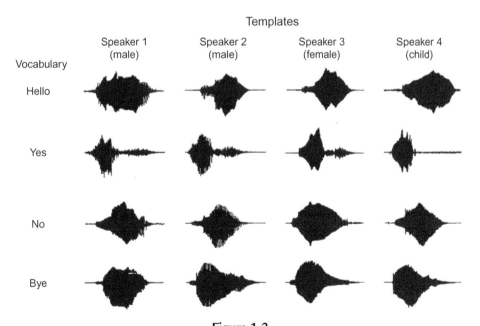

Figure 1.2
Variations in speech templates for different speakers.

by the theory of probability and random processes, which in turn may be used to develop models for speech production and recognition. Such models may then be used to design systems for speech recognition.

1.2 A Radar System

A classical problem drawing heavily on the theory of probability and random processes is that of signal detection and estimation. One example of such a problem is a simple radar system, such as might be used at an airport to track local air traffic. A known signal is converted to an electromagnetic wave and propagated via an antenna. This wave will reflect off an aircraft and return back to the antenna (as illustrated in Figure 1.3), where the signal is processed to gather information about the aircraft. In addition to being corrupted by a random noise and interference process, the returning signal itself may exhibit randomness as well. First, we must determine if there is a reflected signal present. Usually, we attempt to maximize the probability of correctly detecting an aircraft subject to a certain level of false alarms. Once we decide that the aircraft is there, we attempt to estimate various random parameters of the reflected signal to obtain information about the aircraft. From the time of arrival of the reflected signal, we can estimate the distance of the aircraft from the radar site. The frequency of the returned signal will indicate the speed of the aircraft. Since the desired signal is corrupted by noise and interference, we can never estimate these various parameters exactly. Given sufficiently accurate models for these random disturbances, however, we can devise procedures for providing the most accurate estimates possible. We can also use the theory of probability and random processes to analyze the performance of our system.

1.3 A Communication Network

Consider a node in a computer communication network, such as depicted in Figure 1.4, that receives packets of information from various sources and must forward them along toward their ultimate destinations. Typically, the node has a fixed, or at least a maximum, rate at

Figure 1.3
A radar system.

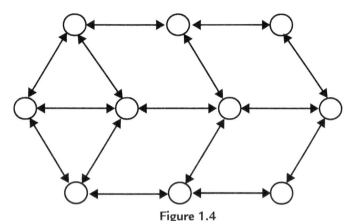

Figure 1.4
Nodes and links in a communications network.

which it can transmit data. Since the arrival of packets to a node will be quite random, the node will usually have some buffering capability, allowing the node to temporarily store packets which it cannot forward immediately. Given a random model of the arrival process of packets at a node, the theory of probability and random processes developed in this text will allow the network designer to determine how large a buffer is needed to insure a minimal probability of buffer overflow (and a resulting loss of information). Or, conversely, given a set buffer size, a limit on the amount of traffic (i.e., throughput) that the node can handle can be determined. Other random quantities such as the delay a packet encounters at the node can also be statistically characterized.

On a less local basis, when information is generated at one of the nodes with a specified destination, a route must be determined to get the packet from the source to the destination. Some nodes in the network may be more congested than others. Congestion throughout the network tends to be very dynamic and so the routing decision must be made using probability. Which route should the packet follow so that it is least likely to be dropped along the way? Or, maybe we want to find the path that will lead to the smallest average delay. Protocols for routing, flow control, and the likes are all based in the foundations of probability theory.

These few examples illustrate the diversity of problems that probability and random processes may model and thereby assist in the development of effective design solutions. By firmly understanding the concepts in this text, the reader will open up a vast world of engineering applications.

Introduction to Probability Theory

Many electrical engineering students have studied, analyzed, and designed systems from the point of view of steady-state and transient signals using time domain or frequency domain techniques. However, these techniques do not provide a method for accounting for variability in the signal nor for unwanted disturbances such as interference and noise. We will see that the theory of probability and random processes is useful for modeling the uncertainty of various events (e.g., the arrival of telephone calls and the failure of electronic components). We also know that the performance of many systems is adversely affected by noise, which may often be present in the form of an undesired signal that degrades the performance of the system. Thus, it becomes necessary to design systems that can discriminate against noise and enhance a desired signal.

How do we distinguish between a deterministic signal or function and a stochastic or random phenomenon such as noise? Usually, noise is defined to be any undesired signal, which often occurs in the presence of a desired signal. This definition includes deterministic as well as non-deterministic signals. A deterministic signal is one which may be represented by some parameter values, such as a sinusoid, which may be perfectly reconstructed given an amplitude, frequency, and phase. Stochastic signals, such as noise, do not have this property. While they may be approximately represented by several parameters, stochastic signals have an element of randomness which prevent them from being perfectly reconstructed from a past history. As we saw in Chapter 1 (Figure 1.2), even the same word spoken by different speakers is not deterministic; there is variability, which can be modeled as a random fluctuation. Likewise, the amplitude and/or phase of a stochastic signal cannot be calculated for any specified future time instant even though the entire past history of the signal may be known. However, the amplitude and/or phase of a stochastic signal can be predicted to occur with a specified probability, provided certain factors are known. The theory of probability provides a tool to model and analyze phenomena that occur in many diverse fields, such as communications, signal processing, control, and computers. Perhaps the major reason for studying probability and random processes is to be able to model complex systems and phenomena.

2.1 Experiments, Sample Spaces, and Events

The relationship between probability and gambling has been known for some time. Over the years, some famous scientists and mathematicians have devoted time to probability: Galileo wrote on dice games; Laplace worked out the probabilities of some gambling games; and

Probability and Random Processes. DOI: 10.1016/B978-0-12-386981-4.00002-4

7

Pascal and Bernoulli, while studying games of chance, contributed to the basic theory of probability. Since the time of this early work, the theory of probability has become a highly developed branch of mathematics. Throughout these beginning sections on basic probability theory, we will often use games of chance to illustrate basic ideas that will form the foundation for more advanced concepts. To start with, a few simple definitions are presented.

> **Definition 2.1:** An *experiment* is a procedure we perform (quite often hypothetical) that produces some result. Often the letter E is used to designate an experiment (e.g., the experiment E_5 might consist of tossing a coin five times).

> **Definition 2.2:** An *outcome* is a possible result of an experiment. The Greek letter xi (ξ) is often used to represent outcomes (e.g., the outcome ξ_1 of experiment E_5 might represent the sequence of tosses heads-heads-tails-heads-tails; however, the more concise HHTHT might also be used).

> **Definition 2.3:** An *event* is a certain set of outcomes of an experiment (e.g., the event C associated with experiment E_5 might be C = {all outcomes consisting of an even number of heads}).

> **Definition 2.4:** The *sample space* is the collection or set of "all possible" distinct (collectively exhaustive and mutually exclusive) outcomes of an experiment. The letter S is used to designate the sample space, which is the universal set of outcomes of an experiment. A sample space is called discrete if it is a finite or a countably infinite set. It is called continuous or a continuum otherwise.

The reason we have placed quotes about the words all possible in Definition 2.4 is explained by the following imaginary situation. Suppose we conduct the experiment of tossing a coin. While it is conceivable that the coin may land on edge, experience has shown us that such a result is highly unlikely to occur. Therefore, our sample space for such experiments typically excludes such unlikely outcomes. We also require, for the present, that all outcomes be distinct. Consequently, we are considering only the set of simple outcomes that are collectively exhaustive and mutually exclusive.

■ Example 2.1:

Consider the example of flipping a fair coin once, where fair means that the coin is not biased in weight to a particular side. There are two possible outcomes, namely, a head or a tail. Thus, the sample space, S, consists of two outcomes, ξ_1 = H to indicate that the outcome of the coin toss was heads, and ξ_2 = T to indicate that the outcome of the coin toss was tails.

■

■ Example 2.2:

A cubical die with numbered faces is rolled and the result observed. The sample space consists of six possible outcomes $\xi_1 = 1$, $\xi_2 = 2$, ..., $\xi_6 = 6$, indicating the possible faces of the cubical die that may be observed.

■

■ Example 2.3:

As a third example, consider the experiment of rolling two dice and observing the results. The sample space consists of 36 outcomes, which may be labeled by the ordered pairs $\xi_1 = (1, 1)$, $\xi_2 = (1, 2)$, $\xi_3 = (1, 3)$, ..., $\xi_6 = (1, 6)$, $\xi_7 = (2, 1)$, $\xi_8 = (2, 2)$, ..., $\xi_{36} = (6, 6)$; the first component in the ordered pair indicates the result of the toss of the first die and the second component indicates the result of the toss of the second die. Many events can be defined from this experiment, such as:

A = {the sum of the outcomes of the two rolls = 4},

B = {the outcomes of the two rolls are identical},

C = {the first roll was bigger than the second}.

An alternative way to consider this experiment is to imagine that we conduct two distinct experiments, with each consisting of rolling a single die. The sample spaces (S_1 and S_2) for each of the two experiments are identical, namely, the same as Example 2.2. We may now consider the sample space, S, of the original experiment to be the combination of the sample spaces, S_1 and S_2, which consist of all possible combinations of the elements of both S_1 and S_2. This is an example of a combined sample space.

■

■ Example 2.4:

For our fourth experiment, let us flip a coin until a tails occurs. The experiment is then terminated. The sample space consists of a collection of sequences of coin tosses. Label these outcomes as ξ_n, $n = 1, 2, 3, ...$. The final toss in any particular sequence is a tail and terminates the sequence. The preceding tosses prior to the occurrence of the tail must be heads. The possible outcomes that may occur are:

$$\xi_1 = (T), \quad \xi_2 = (H, T), \quad \xi_3 = (H, H, T), \quad ... .$$

Note that in this case, n can extend to infinity. This is another example of a combined sample space resulting from conducting independent, but identical experiments. In this example, the sample space is countably infinite, while the previous sample spaces were finite.

■

■ **Example 2.5:**

As a last example, consider a random number generator which selects a number in an arbitrary manner from the semi-closed interval $[0, 1)$. The sample space consists of all real numbers, x, for which $0 \leq x < 1$. This is an example of an experiment with a continuous sample space. We can define events on a continuous space as well, such as:

$$A = \{x < 1/2\},$$

$$B = \{|x - 1/2| < 1/4\},$$

$$C = \{x = 1/2\}.$$

Other examples of experiments with continuous sample spaces include the measurement of the voltage of thermal noise in a resistor or the measurement of the (x, y, z) position of an oxygen molecule in the atmosphere. Examples 2.1-2.4 illustrate discrete sample spaces.

■

There are also infinite sets that are uncountable and that are not continuous, but these sets are beyond the scope of this book. So for our purposes, we will consider only the preceding two types of sample spaces. It is also possible to have a sample space that is a mixture of discrete and continuous sample spaces. For the remainder of this chapter, we shall restrict ourselves to the study of discrete sample spaces.

A particular experiment can often be represented by more than one sample space. The choice of a particular sample space depends upon the questions that are to be answered concerning the experiment. This is perhaps best explained by recalling Example 2.3 in which a pair of dice was rolled. Suppose we were asked to record after each roll the sum of the numbers shown on the two faces. Then, the sample space could be represented by only eleven outcomes, $\xi_1 = 2$, $\xi_2 = 3$, $\xi_3 = 4, \ldots, \xi_{11} = 12$. However, the original sample space is in some way, more fundamental, since the sum of the die faces can be determined from the numbers on the die faces. If the second representation is used, it is not sufficient to specify the sequence of numbers that occurred from the sum of the numbers.

2.2 Axioms of Probability

Now that the concepts of experiments, outcomes, and events have been introduced, the next step is to assign probabilities to various outcomes and events. This requires a careful definition of probability. The words *probability* and *probable* are commonly used in everyday language. The meteorologist on the evening news may say that rain is probable for tomorrow or he may be more specific and state that the chance (or probability) of rain is 70%. Although this sounds like a precise statement, we could interpret it in several ways. Perhaps, it means that about 70% of the listening audience will experience rain. Or, maybe if tomorrow could be repeated many times, 70% of the tomorrows would have rain while the other 30% would not. Of course, tomorrow cannot be repeated and this experiment can only be run once. The outcome will be either rain or no rain. The meteorologist may like this interpretation since there is no way to repeat the experiment enough times to test the accuracy of the prediction. However, there is a similar interpretation that can be tested. We might say that any time a day with similar meteorological conditions presents itself, the following day will have rain 70% of the time. In fact, it may be his or her past experience with the given weather conditions that led the meteorologist to the prediction of a 70% chance of rain.

It should be clear from our everyday usage of the word *probability* that it is a measure of the likelihood of various events. So in general terms, probability is a function of an event that produces a numerical quantity that measures the likelihood of that event. There are many ways to define such a function, which we could then call probability. In fact, we will find several ways to assign probabilities to various events, depending on the situation. Before we do that, however, we start with three axioms that any method for assigning probabilities must satisfy:

Axiom 2.1: For any event A, $\Pr(A) \geq 0$ (a negative probability does not make sense).

Axiom 2.2: If S is the sample space for a given experiment, $\Pr(S) = 1$ (probabilities are normalized so that the maximum value is unity).

Axiom 2.3a: If $A \cap B = \varnothing$, then $\Pr(A \cup B) = \Pr(A) + \Pr(B)$.

As the word *axiom* implies, these statements are taken to be self-evident and require no proof. In fact, the first two axioms are really more of a self-imposed convention. We could have allowed for probabilities to be negative or we could have normalized the maximum probability to be something other than one. However, this would have greatly confused the subject and we do not consider these possibilities. From these axioms (plus one more to be presented shortly), the entire theory of probability can be developed. Before moving on to that task, a corollary to Axiom 2.3a is given.

Corollary 2.1: Consider M sets $A_1, A_2, ..., A_M$ which are mutually exclusive, $A_i \cap A_j = \varnothing$ for all $i \neq j$,

$$\Pr\left(\bigcup_{i=1}^{M} A_i\right) = \sum_{i=1}^{M} \Pr(A_i). \tag{2.1}$$

Proof: This statement can be proven using mathematical induction. For those students who are unfamiliar with this concept, the idea behind induction is to show that if the statement is true for $M = m$, then it must also hold for $M = m + 1$. Once this is established, it is noted that by Axiom 2.3a, the statement applies for $M = 2$, and hence it must be true for $M = 3$. Since it is true for $M = 3$, it must also be true for $M = 4$, and so on. In this way, we can prove that Corollary 2.1 is true for any finite M. The details of this proof are left as an exercise for the reader (see Exercise 2.7). $\square$

Unfortunately, the proof just outlined is not sufficient to show that Corollary 2.1 is true for the case of an infinite number of sets. That has to be accepted on faith and is listed here as the second part of Axiom 2.3.

Axiom 2.3b: For an infinite number of mutually exclusive sets, A_i, $i = 1, 2, 3...$, $A_i \cap A_j = \varnothing$ for all $i \neq j$,

$$\Pr\left(\bigcup_{i=1}^{\infty} A_i\right) = \sum_{i=1}^{\infty} \Pr(A_i). \tag{2.2}$$

It should be noted that Axiom 2.3a and Corollary 2.1 could be viewed as special cases of Axiom 2.3b. So, a more concise development could be obtained by starting with Axioms 2.1, 2.2, and 2.3b. This may be more pleasing to some, but we believe the approach given here is little easier to follow for the student learning the material for the first time.

The preceding axioms do not tell us directly how to deal with the probability of the union of two sets that are not mutually exclusive. This can be determined from these axioms as is now shown.

Theorem 2.1: For any sets A and B (not necessarily mutually exclusive),

$$\Pr(A \cup B) = \Pr(A) + \Pr(B) - \Pr(A \cap B). \tag{2.3}$$

Proof: We give a visual proof of this important result using the Venn diagram shown in Figure 2.1. To aid the student in the type of reasoning needed to complete proofs of this type, it is helpful to think of a pile of sand lying in the sample space shown in Figure 2.1. The probability of the event A would then be analogous to the mass of that subset of the sand pile that is above the region A and likewise for the probability of the event B. For the union of the two events, if we simply added the mass of the sand above A to the mass of the sand above B, we would double count that region which is common to both sets. Hence, it is necessary to subtract the probability of $A \cap B$. We freely admit that this proof is not very rigorous. It is possible to prove Theorem 2.1 without having to call on our sand analogy or even the use of Venn diagrams. The logic of the proof will closely follow what we have done here. The reader is led through that proof in Exercise 2.8.

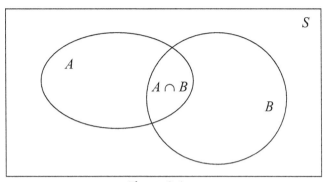

Figure 2.1
Venn diagram for proof of Theorem 2.1.

Many other fundamental results can also be obtained from the basic axioms of probability. A few simple ones are presented here. More will be developed later in this chapter and in subsequent chapters. As with Theorem 2.1, it might help the student to visualize these proofs by drawing a Venn diagram.

Theorem 2.2: $$\Pr(\overline{A}) = 1 - \Pr(A).$$

Proof: $$1 = \Pr(S) = \Pr(A \cup \overline{A}) \qquad \text{(by Axiom 2.2)}$$

$$= \Pr(A) + \Pr(\overline{A}) \qquad \text{(by Axiom 2.3a)}$$

$$\therefore \Pr(\overline{A}) = 1 - \Pr(A). \quad \square$$

Theorem 2.3: If $A \subset B$, then $\Pr(A) \le \Pr(B)$.

Proof: See Exercise 2.10. $\square$

2.3 Assigning Probabilities

In the previous section, probability was defined as a measure of the likelihood of an event or of events which satisfy Axioms 2.1-2.3. How probabilities are assigned to particular events was not specified. Mathematically, any assignment that satisfies the given axioms is acceptable. Practically speaking, we would like to assign probabilities to events in such a way that the probability assignment actually represents the likelihood of occurrence of that event. Two techniques are typically used for this purpose and are described in the following paragraphs.

In many experiments, it is possible to specify all of the outcomes of the experiment in terms of some fundamental outcomes which we refer to as *atomic outcomes*. These are the most basic events that cannot be decomposed into simpler events. From these atomic outcomes, we can build more complicated and more interesting events. Quite often we can justify assigning equal probabilities to all atomic outcomes in an experiment. In that case, if there are M mutually exclusive exhaustive atomic events, then each one should be assigned a probability of $1/M$. Not only does this make perfect common sense, it also satisfies the mathematical requirements of the three axioms which define probability. To see this, we label the M atomic outcomes of an experiment E as $\xi_1, \xi_2, \ldots, \xi_M$. These atomic events are taken to be mutually exclusive and exhaustive. That is, $\xi_i \cap \xi_j = \varnothing$ for all $i \neq j$, and $\xi_1 \cup \xi_2 \cup \ldots \cup \xi_M = S$. Then by Corollary 2.1 and Axiom 2.2,

$$\Pr(\xi_1 \cup \xi_2 \cup \ldots \cup \xi_M) = \Pr(\xi_1) + \Pr(\xi_2) + \ldots + \Pr(\xi_M) = \Pr(S) = 1. \quad (2.4)$$

If each atomic outcome is to be equally probable, then we must assign each a probability of $\Pr(\xi_i) = 1/M$ for there to be equality in the preceding equation. Once the probabilities of these outcomes are assigned, the probabilities of some more complicated events can be determined according to the rules set forth in Section 2.2. This approach to assigning probabilities is referred to as the *classical approach*.

■ Example 2.6:

The simplest example of this procedure is the coin flipping experiment of Example 2.1. In this case, there are only two atomic events, $\xi_1 = H$ and $\xi_2 = T$. Provided the coin is fair (again, not biased toward one side or the other), we have every reason to believe that these two events should be equally probable. These outcomes are mutually exclusive and collectively exhaustive (provided we rule out the possibility of the coin landing on end). According to our theory of probability, these events should be assigned probabilities of $\Pr(H) = \Pr(T) = 1/2$.   ■

■ Example 2.7:

Next consider the dice rolling experiment of Example 2.2. If the die is not loaded, the six possible faces of the cubicle die are reasonably taken to be equally likely to appear, in which case, the probability assignment is $\Pr(1) = \Pr(2) = \ldots = \Pr(6) = 1/6$. From this assignment, we can determine the probability of more complicated events, such as:

$$\Pr(\text{even number is rolled}) = \Pr(2 \cup 4 \cup 6)$$
$$= \Pr(2) + \Pr(4) + \Pr(6) \qquad \text{(by Corollary 2.1)}$$
$$= 1/6 + 1/6 + 1/6 \qquad \text{(by probability assignment)}$$
$$= 1/2.$$
■

■ Example 2.8:

In Example 2.3, a pair of dice were rolled. In this experiment, the most basic outcomes are the 36 different combinations of the six atomic outcomes of the previous example. Again, each of these atomic outcomes is assigned a probability of 1/36. Next, suppose we want to find the probability of the event $A = \{\text{sum of two dice} = 5\}$. Then,

$$\Pr(A) = \Pr((1, 4) \cup (2, 3) \cup (3, 2) \cup (4, 1))$$

$$= \Pr(1, 4) + \Pr(2, 3) + \Pr(3, 2) + \Pr(4, 1) \quad \text{(by Corollary 2.1)}$$

$$= 1/36 + 1/36 + 1/36 + 1/36 \quad \text{(by probability assignment)}$$

$$= 1/9 .$$

■

■ Example 2.9:

In this example, we will use the MATLAB command `rand` to simulate the flipping of coins and the rolling of dice. The command `rand (m,n)` creates a matrix of *m* rows and *n* columns where each element of the matrix is a randomly selected number equally likely to fall anywhere in the interval $(0, 1)$. By rounding this number to the nearest integer, we can create a randomly selected number equally likely to be 0 or 1. This can be used to simulate the flipping of a coin if we interpret 0 as "tails" and 1 as "heads" or vice versa. Similarly, if we multiply `rand(1)` by 6 and round up to the nearest integer, we will get one of the numbers 1,2,...,6 with equal probability. This can be used to simulate the rolling of a die. Try running the following script in MATLAB.

```
%  Simulation of coin flipping and die tossing.
coin_flip=round(rand(1))              % simulate flip of a coin.
die_toss=ceil(6*rand(1))              % simulate toss of one die.
dice_toss=ceil(6*rand(1,2))           % simulate toss of two dice.
```

You should find that each time you run this script, you get different (random) looking results. With any MATLAB command, if you want more information on what the command does, type `help` followed by the command name at the MATLAB prompt for detailed information on that command. For example, to get help on the `rand` command, type `help rand`.

■

Care must be taken when using the classical approach to assigning probabilities. If we define the set of atomic outcomes incorrectly, unsatisfactory results may occur. In Example 2.8, we may be tempted to define the set of atomic outcomes as the different sums that can occur on the two dice faces. If we assign equally likely probability to each of these outcomes, then we arrive at the assignment

$$\Pr(\text{sum}=2) = \Pr(\text{sum}=3) = \ldots = \Pr(\text{sum}=12) = 1/11 . \tag{2.5}$$

Anyone with experience in games involving dice knows that the likelihood of rolling a 2 is much less than the likelihood of rolling a 7. The problem here is that the atomic events we have assigned are not the most basic outcomes and can be decomposed into simpler outcomes as demonstrated in Example 2.8.

This is not the only problem encountered in the classical approach. Suppose we consider an experiment that consists of measuring the height of an arbitrarily chosen student in your class and rounding that measurement to the nearest inch. The atomic outcomes of this experiment would consist of all the heights of the students in your class. However, it would not be reasonable to assign an equal probability to each height. Those heights corresponding to very tall or very short students would be expected to be less probable than those heights corresponding to a medium height. So, how then do we assign probabilities to these events? The problems associated with the classical approach to assigning probabilities can be overcome by using the *relative frequency approach*.

The relative frequency approach requires that the experiment we are concerned with be repeatable, in which case, the probability of an event, A, can be assigned by repeating the experiment a large number of times and observing how many times the event A actually occurred. If we let n be the number of times the experiment is repeated and n_A the number of times the event A is observed, then the probability of the event A can be assigned according to

$$\Pr(A) = \lim_{n \to \infty} \frac{n_A}{n}. \tag{2.6}$$

This approach to assigning probability is based on experimental results and thus has a more practical flavor to it. It is left as an exercise for the reader (see Exercise 2.15) to confirm that this method does indeed satisfy the axioms of probability, and is thereby mathematically correct as well.

■ **Example 2.10:**

Consider the dice rolling experiment of Examples 2.3 and 2.8. We will use the relative frequency approach to assign the probability of the event $A = \{\text{sum of two dice} = 5\}$. We simulated the tossing of two dice using the following MATLAB code. The results of this dice tossing simulation are shown in Table 2.1.

Table 2.1: Simulation of Dice Tossing Experiment

n	1000	2000	3000	4000	5000	6000	7000	8000	9000	10,000
n_A	96	200	314	408	521	630	751	859	970	1095
n_A/n	0.096	0.100	0.105	0.102	0.104	0.105	0.107	0.107	0.108	0.110

```
% Simulation code for dice tossing experiment.
n=1000;                                 % number of times to toss the dice.
die1=ceil(6*rand(1,n));                 % Toss first die n times.
die2=ceil(6*rand(1,n));                 % Toss second die n times.
dice_sum=die1+die2;                     % Compute sum of two tosses.
nA=sum(dice_sum==5);                    % Count number of times sum = 5;
pA=nA/n                                 % Display relative frequency.
```

The next to last line of MATLAB code may need some explanation. The double equal sign asks MATLAB to compare the two quantities to see if they are equal. MATLAB responds with 1 for "yes" and 0 for "no." Hence, the expression `dice_sum==5` results in an *n* element vector where each element of the vector is either 0 or 1 depending on whether the corresponding element of `dice_sum` is equal to 5 or not. By summing all elements of this vector, we obtain the number of times the sum 5 occurs in *n* tosses of the dice. ■

To get an exact measure of the probability of an event, we must be able to repeat the event an infinite number of times—a serious drawback to this approach. In the dice rolling experiment of Example 2.8, even after rolling the dice 10,000 times, the probability of observing a 5 was measured to only two significant digits. Furthermore, many random phenomena in which we might be interested are not repeatable. The situation may occur only once, and therefore we cannot assign the probability according to the relative frequency approach.

2.4 Joint and Conditional Probabilities

Suppose that we have two events, A and B. We saw a few results in the previous section that dealt with how to calculate the probability of the union of two events, $A \cup B$. At least as frequently, we are interested in calculating the probability of the intersection of two events, $A \cap B$. This probability is referred to as the *joint probability* of the events A and B, $\Pr(A \cap B)$. Usually, we will use the simpler notation $\Pr(A, B)$. This definition and notation extends to an arbitrary number of sets. The joint probability of the events, $A_1, A_2, ..., A_M$, is $\Pr(A_1 \cap A_2 \cap ... \cap A_M)$ and we use the simpler notation $\Pr(A_1, A_2, ..., A_M)$ to represent the same quantity.

Now that we have established what a joint probability is, how does one compute it? To start with, by comparing Axiom 2.3a and Theorem 2.1, it is clear that if A and B are mutually exclusive, then their joint probability is zero. This is intuitively pleasing, since if A and B are mutually exclusive, then $\Pr(A, B) = \Pr(\varnothing)$, which we would expect to be zero. That is, an impossible event should never happen. Of course, this case is of rather limited interest, and we would be much more interested in calculating the joint probability of events that are not mutually exclusive.

In the general case when A and B are not necessarily mutually exclusive, how can we calculate the joint probability of A and B? From the general theory of probability, we can easily see two ways to accomplish this. First, we can use the classical approach. Both events A and B can be expressed in terms of atomic outcomes. We then write $A \cap B$ as the set of those atomic outcomes that are common to both and calculate the probabilities of each of these outcomes. Alternatively, we can use the relative frequency approach. Let $n_{A,B}$ be the number of times that A and B simultaneously occur in n trials. Then,

$$\Pr(A, B) = \lim_{n \to \infty} \frac{n_{A,B}}{n}. \tag{2.7}$$

■ Example 2.11:

 A standard deck of playing cards has 52 cards that can be divided in several manners. There are four suits (spades, hearts, diamonds and clubs) each of which has 13 cards (ace, 2, 3, 4,. . ., 10, jack, queen, king). There are two red suits (hearts and diamonds) and two black suits (spades and clubs). Also, the jacks, queens and kings are referred to as face cards, while the others are number cards. Suppose the cards are sufficiently shuffled (randomized) and one card is drawn from the deck. The experiment has 52 atomic outcomes corresponding to the 52 individual cards that could have been selected. Hence, each atomic outcome has a probability of $1/52$. Define the events: $A = \{\text{red card selected}\}$, $B = \{\text{number card selected}\}$, and $C = \{\text{heart selected}\}$. Since the event A consists of 26 atomic outcomes (there are 26 red cards), then $\Pr(A) = 26/52 = 1/2$. Likewise, $\Pr(B) = 40/52 = 10/13$ and $\Pr(C) = 13/52 = 1/4$. Events A and B have 20 outcomes in common, hence $\Pr(A, B) = 20/52 = 5/13$. Likewise, $\Pr(A, C) = 13/52 = 1/4$ and $\Pr(B, C) = 10/52 = 5/26$. It is interesting to note that in this example, $\Pr(A, C) = \Pr(C)$. This is because $C \subset A$ and as a result $A \cap C = C$.

■

Often the occurrence of one event may be dependent upon the occurrence of another. In the previous example, the event $A = \{\text{a red card is selected}\}$ had a probability of $\Pr(A) = 1/2$. If it is known that event $C = \{\text{a heart is selected}\}$ has occurred, then the event A is now certain (probability equal to 1), since all cards in the heart suit are red. Likewise, if it is known that the event C did not occur, then there are 39 cards remaining, 13 of which are red (all the diamonds). Hence, the probability of event A in that case becomes 1/3. Clearly, the probability of event A depends on the occurrence of event C. We say that the probability of A is conditional on C, and the probability of A given knowledge that the event C has occurred is referred to as the *conditional probability* of A given C. The shorthand notation $\Pr(A|C)$ is used to denote the probability of the event A given that the event C has occurred, or simply the probability of A given C.

Definition 2.5: For two events A and B, the probability of A conditioned on knowing that B has occurred is

$$\Pr(A|B) = \frac{\Pr(A, B)}{\Pr(B)}. \tag{2.8}$$

The reader should be able to verify that this definition of conditional probability does indeed satisfy the axioms of probability (see Exercise 2.21).

We may find in some cases that conditional probabilities are easier to compute than the corresponding joint probabilities and hence, this formula offers a convenient way to compute joint probabilities.

$$\Pr(A, B) = \Pr(B|A)Pr(A) = \Pr(A|B)\Pr(B). \tag{2.9}$$

This idea can be extended to more than two events. Consider finding the joint probability of three events, A, B, and C.

$$\Pr(A, B, C) = \Pr(C|A, B)\Pr(A, B) = \Pr(C|A, B)\Pr(B|A)\Pr(A). \tag{2.10}$$

In general, for M events, $A_1, A_2, ..., A_M$,

$$\Pr(A_1, A_2, ..., A_M) = \Pr(A_M|A_1, A_2, ..., A_{M-1})\Pr(A_{M-1}|A_1, ..., A_{M-2}) \\ ...\Pr(A_2|A_1)\Pr(A_1). \tag{2.11}$$

■ Example 2.12:

Return to the experiment of drawing cards from a deck as described in Example 2.11. Suppose now that we select two cards at random from the deck. When we select the second card, we do not return the first card to the deck. In this case, we say that we are selecting cards without replacement. As a result, the probabilities associated with selecting the second card are slightly different if we have knowledge of what card was drawn on the first selection. To illustrate this let $A = \{\text{first card was a spade}\}$ and $B = \{\text{second card was a spade}\}$. The probability of the event A can be calculated as in the previous example to be $\Pr(A) = 13/52 = 1/4$. Likewise, if we have no knowledge of what was drawn on the first selection, the probability of the event B is the same, $\Pr(B) = 1/4$. To calculate the joint probability of A and B, we have to do some counting.

To begin with, when we select the first card there are 52 possible outcomes. Since this card is not returned to the deck, there are only 51 possible outcomes for the second card. Hence, this experiment of selecting two cards from the deck has 52*51 possible outcomes each of which is equally likely and has a probability of $1/52*51$. Similarly,

(Continued)

there are 13*12 outcomes that belong to the joint event $A \cap B$. Therefore, the joint probability for A and B is $\Pr(A, B) = (13*12)/(52*51) = 1/17$. The conditional probability of the second card being a spade given that the first card is a spade is then $\Pr(B|A) = \Pr(A, B)/\Pr(A) = (1/17)/(1/4) = 4/17$. However, calculating this conditional probability directly is probably easier than calculating the joint probability. Given that we know the first card selected was a spade, there are now 51 cards left in the deck, 12 of which are spades, thus $\Pr(B|A) = 12/51 = 4/17$. Once this is established, then the joint probability can be calculated as $\Pr(A, B) = \Pr(B|A)\Pr(A) = (4/17)*(1/4) = 1/17$.   ∎

∎ Example 2.13:

In a game of poker you are dealt 5 cards from a standard 52-card deck. What is the probability you are dealt a flush in spades? (A flush is when you are dealt all five cards of the same suit.) What is the probability of a flush in any suit? To answer this requires a simple extension of the previous example. Let A_i be the event {ith card dealt to us is a spade}, $i = 1, 2, ..., 5$. Then,

$$\Pr(A_1) = 1/4,$$

$$\Pr(A_1, A_2) = \Pr(A_2|A_1)\Pr(A_1) = (12/51)*(1/4) = 1/17,$$

$$\Pr(A_1, A_2, A_3) = \Pr(A_3|A_1, A_2)\Pr(A_1, A_2) = (11/50)*(1/17) = 11/850,$$

$$\Pr(A_1, A_2, A_3, A_4) = \Pr(A_4|A_1, A_2, A_3)\Pr(A_1, A_2, A_3) = (10/49)*(11/850) = 11/4165,$$

$$\Pr(A_1, A_2, A_3, A_4, A_5) = \Pr(A_5|A_1, A_2, A_3, A_4)\Pr(A_1, A_2, A_3, A_4) = (9/48)*(11/4165) = 33/66640.$$

To find the probability of being dealt a flush in any suit, we proceed as follows:

$$\Pr(\text{flush}) = \Pr(\{\text{flush in spades}\} \cup \{\text{flush in hearts}\} \cup \{\text{flush in diamonds}\} \cup \{\text{flush in clubs}\})$$

$$= \Pr(\text{flush in spades}) + \Pr(\text{flush in hearts}) + \Pr(\text{flush in diamonds}) + \Pr(\text{flush in clubs}).$$

Since all four events in the preceding expression have equal probability, then

$$\Pr(\text{flush}) = 4*\Pr(\text{flush in spades}) = \frac{4*33}{66640} = \frac{33}{16660} = 0.001981.$$

So, we will be dealt a flush slightly less than two times in a thousand.   ∎

2.5 Basic Combinatorics

In many situations, the probability of each possible outcome of an experiment is taken to be equally likely. Often, problems encountered in games of chance fall into this category as was seen from the card drawing and dice rolling examples in the preceding sections. In these cases, finding the probability of a certain event, A, reduces to an exercise in counting,

$$\Pr(A) = \frac{\text{Number of outcomes in } A}{\text{Number of outcomes in entire sample space}}. \tag{2.12}$$

Sometimes, when the scope of the experiment is fairly small, it is straightforward to count the number of outcomes. On the other hand, for problems where the experiment is fairly complicated, the number of outcomes involved can quickly become astronomical, and the corresponding exercise in counting can be quite daunting. In this section, we present some fairly simple tools that are helpful for counting the number of outcomes in a variety of commonly encountered situations. Many students will have seen this material in their high school math courses or perhaps in a freshmen or sophomore level discrete mathematics course. For those who are familiar with combinatorics, this section can be skipped without any loss of continuity.

We start with a basic principle of counting from which the rest of our results will flow.

Principle of Counting: For a combined experiment, $E = E_1 \times E_2$ where experiment E_1 has n_1 possible outcomes and experiment E_2 has n_2 possible outcomes, the total number of possible outcomes in the combined experiment is $n = n_1 n_2$.

This can be seen through a simple example.

■ Example 2.14:

Suppose we form a two-digit word by selecting a letter from the set $\{A, B, C, D, E, F\}$ followed by a number from the set $\{0, 1, 2\}$. All possible combinations are enumerated in the following array.

$$\begin{bmatrix} A1 & B1 & C1 & D1 & E1 & F1 \\ A2 & B2 & C2 & D2 & E2 & F2 \\ A3 & B3 & C3 & D3 & E3 & F3 \end{bmatrix}$$

Since the first experiment (select a letter) had $n_1 = 6$ possible outcomes and the second experiment (select a number) had $n_2 = 3$ outcomes, there are a total of $n = n_1 n_2 = 6 \cdot 3 = 18$ possible outcomes in the combined experiment.

■

This result can easily be generalized to a combination of several experiments.

Theorem 2.4: A combined experiment, $E = E_1 \times E_2 \times E_3 \ldots \times E_m$, consisting of experiments E_i each with n_i outcomes, $i = 1, 2, 3, \ldots, m$, has a total number of possible outcomes given by

$$n = n_1 n_2 n_3 \ldots n_m = \prod_{i=1}^{m} n_i. \tag{2.13}$$

Proof: This can be proven through induction and is left as an exercise for the reader (see Exercise 2.29). □

■ Example 2.15:

In a certain state, automobile license plates consist of three letters (drawn from the 26-letter English alphabet) followed by three numbers (drawn from the decimal set 0,1,2, ... , 9). For example, one such possible license plate would be "ABC 123." This can be viewed as a combined experiment with six sub-experiments. The experiment "draw a letter" has 26 different outcomes and is repeated three times, while the experiment "draw a number" has 10 outcomes and is also repeated three times. The total number of possible license plates is then

$$n = 26 \cdot 26 \cdot 26 \cdot 10 \cdot 10 \cdot 10 = 26^3 10^3 = 17,576,000.$$

Once this state has registered more than approximately 17.5 million cars, it will have to adopt a new format for license plates.

■

In many problems of interest, we seek to find the number of different ways that we can rearrange or order several items. The orderings of various items are referred to as *permutations*. The number of permutations can easily be determined from the previous theorem and is given as follows:

Theorem 2.5 (Permutations): The number of *permutations* of n distinct elements is

$$n! = n(n-1)(n-2)...3 \cdot 2 \cdot 1 . \qquad (2.14)$$

Proof: We can view this as an experiment where we have items, numbered 1 through n which we randomly draw (without replacing any previously drawn item) until all n items have been drawn. This is a combined experiment with n sub-experiments, $E = E_1 \times E_2 \times E_3 ... E_n$. The first experiment is E_1 = "select one of the n items" and clearly has $n_1 = n$ possible outcomes. The second experiment is E_2 = "select one of the remaining items not chosen in E_1." This experiment has $n_2 = n-1$ outcomes. Continuing in this manner, E_3 has $n_3 = n-2$ outcomes, and so on, until we finally get to the last experiment where there is only one item remaining left unchosen and therefore the last experiment has only $n_n = 1$ outcome. Applying Theorem 2.4 to this situation results in Equation (2.14). □

■ Example 2.16:

A certain professor creates an exam for his course consisting of six questions. In order to discourage students from cheating off one another, he gives each student a different exam in such a way that all students get an exam with the same six questions, but each student is given the questions in a different order. In this manner, there are 6! = 720 different versions of the exam the professor could create.

■

A simple, but useful, extension to the number of permutations is the concept of *k-permutations*. In this scenario, there are n distinct elements and we would like to select a subset of k of these elements (without replacement).

> **Theorem 2.6 (k-permutations):** The number of k-permutations of n distinct elements is given by
>
> $$\frac{n!}{(n-k)!} = n(n-1)(n-2)\ldots(n-k+1). \qquad (2.15)$$
>
> **Proof:** The proof proceeds just as in the previous theorem.

■ Example 2.17:

A certain padlock manufacturer creates locks whose combinations consist of a sequence of three numbers from the set 0-39. The manufacturer creates combinations such that the same number is never repeated in a combination. For example, 2-13-27 is a valid combination while 27-13-27 is not (because the number 27 is repeated). How many distinct padlock combinations can be created? A straightforward application of Theorem 2.6 produces

$$\frac{n!}{(n-k)!} = \frac{40!}{37!} = 40 \cdot 39 \cdot 38 = 59,280 \ .$$

Note, the company can make more than 59,280 padlocks. It would just have to start re-using combinations at that point. There is no problem with having many padlocks with the same combination. The company just needs to create enough possibilities that it becomes too time consuming for someone to try to pick the lock by trying all possible combinations.

■

In many situations, we wish to select a subset of k out of n items, but we are not concerned about the order in which they are selected. In other words, we might be interested in the number of subsets there are consisting of k out of n items. A common example of this occurs in card games where there are 52 cards in a deck and we are dealt some subset of these 52 cards (e.g., 5 cards in a game of poker or 13 cards in a game of bridge). As far as the game is concerned, the order in which we are given these cards is irrelevant. The set of cards is the same regardless of what order they are given to us.

These subsets of k out of n items are referred to as *combinations*. The number of combinations can be determined by slightly modifying our formula for the number of k-permutations. When counting combinations, every subset that consists of the same

k elements is counted only once. Hence, the formula for the number of k-permutations has to be divided by the number of permutations of k elements. This leads us to the following result:

Theorem 2.7 (Combinations): The number of distinct subsets (regardless of order) consisting of k out of n distinct elements is

$$\binom{n}{k} = \frac{n!}{k!(n-k)!} = \frac{n(n-1)(n-2)\ldots(n-k+1)}{k(k-1)(k-2)\ldots 1}. \tag{2.16}$$

■ Example 2.18:

A certain lottery game requires players to select four numbers from the set 0-29. The numbers cannot be repeated and the order in which they are selected does not matter. The number of possible subsets of 4 numbers out of 30 is found from Theorem 2.7 as

$$\binom{30}{4} = \frac{30!}{26!4!} = \frac{30 \cdot 29 \cdot 28 \cdot 27}{4 \cdot 3 \cdot 2 \cdot 1} = 27,405 \ .$$

Thus, the probability of a player selecting the winning set of numbers is $1/27,405 = 3.649 \times 10^{-5}$.

■

The expression in Equation (2.16) for the number of combinations of k items out of n also goes by the name of the *binomial coefficient*. It is called this because the same number shows up in the power series representation of a binomial raised to a power. In particular, $\binom{n}{k}$ is the coefficient of the $x^k y^{n-k}$ term in the power series expansion of $(x + y)^n$ (see Equation E.13 in Appendix E). A number of properties of the binomial coefficient are studied in Exercise 2.31.

As a brief diversion, at this point it is worthwhile noting that sometimes the English usage of certain words is much different than their mathematical usage. In Example 2.17, we talked about padlocks and their combinations. We found that the formula for k-permutations was used to count the number padlock combinations (English usage), whereas the formula for the number of combinations (mathematical usage) actually applies to something quite different. We are not trying to intentionally introduce confusion here, but rather desire to warn the reader that sometimes the technical usage of a word may have a very specific meaning whereas the non-technical usage of the same word may have a much broader or even different meaning. In this case, it is probably pretty clear why the non-technical meaning of the word "combination" is more relaxed than its technical meaning. Try explaining to your first grader why he needs to know a 3-permutation in order to open his school locker.

The previous result dealing with combinations can be viewed in the context of partitioning of sets. Suppose we have a set of n distinct elements and we wish to partition this set into two groups. The first group will have k elements and the second group will then need to have the remaining $n-k$ elements. How many different ways can this partition be accomplished? The answer is that it is just the number of combinations of k out of n elements, $\binom{n}{k}$. To see this, we form the first group in the partition by choosing k out of the n elements. There are $\binom{n}{k}$ ways to do this. Since there are only two groups, all remaining elements must go in the second group.

This result can then be extended to partitions with more than two groups. Suppose, for example, we wanted to partition a set of n elements into three groups such that the first group has n_1 elements, the second group has n_2 elements, and the third group has n_3 elements, where $n_1 + n_2 + n_3 = n$. There are $\binom{n}{n_1}$ ways in which we could choose the members of the first group. After that, when choosing the members of the second group, we are selecting n_2 elements from the remaining $n - n_1$ elements that were not chosen to be in the first group. There are $\binom{n - n_1}{n_2}$ ways in which that can be accomplished. Finally, all remaining elements must be in the third and last group. Therefore, the total number of ways to partition n elements into three groups with n_1, n_2, n_3 elements, respectively, is

$$\binom{n}{n_1}\binom{n - n_1}{n_2} = \left(\frac{n!}{n_1!(n - n_1)!}\right)\left(\frac{(n - n_1)!}{n_2!(n - n_1 - n_2)!}\right) = \frac{n!}{n_1!n_2!(n - n_1 - n_2)!} = \frac{n!}{n_1!n_2!n_3!}.$$

(2.17)

The last step was made by employing the constraint that $n_1 + n_2 + n_3 = n$. Following the same sort of logic, we can extend this result to an arbitrary number of partitions.

Theorem 2.8 (Partitions): Given a set of n distinct elements, the number of ways to partition the set into m groups where the ith group has n_i elements is given by the *multinomial coefficient*,

$$\binom{n}{n_1, n_2, \ldots, n_m} = \frac{n!}{n_1!n_2!\ldots n_m!}.$$

(2.18)

Proof: We have already established this result for the cases when $m = 2, 3$. We leave it as an exercise for the reader (see Exercise 2.30) to complete the general proof. This can be accomplished using a relatively straightforward application of the concept of mathematical induction. □

▪ Example 2.19:

In the game of bridge, a standard deck of cards is divided amongst four players such that each player gets a hand of 13 cards. The number of different bridge games that could occur is then

$$\binom{52}{13, 13, 13, 13} = \frac{52!}{(13!)^4} = 5.365 \times 10^{28}.$$

(Continued)

Next, suppose I want to calculate the probability that when playing bridge, I get dealt a hand that is completely void of spades. Assuming all hands are equally likely to occur, we can calculate this as the ratio of the number of hands with no spades to the total number of hands. In this case, I am only concerned about my hand and not how the remaining cards are partitioned among the other players. There are $\binom{52}{13} = 6.35 \times 10^{11}$ different sets of 13 cards that I could be dealt. If we want to calculate how many of those have no spades, there are 39 cards in the deck that are not spades and we must be dealt a subset of 13 of those 39 cards to get a hand with no spades. Therefore, there are $\binom{39}{13} = 8.12 \times 10^{9}$ hands with no spades. Thus, the probability of receiving a hand with no spades is

$$\Pr(\text{no spades}) = \frac{\text{\# of hands with no spades}}{\text{total \# of hands}} = \frac{\binom{39}{13}}{\binom{52}{13}} = 0.0128 \,.$$

∎

Naturally, there are many more formulas of this nature for more complicated scenarios. We make no attempt to give any sort of exhaustive coverage here. Rather, we merely include some of the more commonly encountered situations in the field of combinatorics. Table 2.2 summarizes the results we have developed and provides a convenient place to find relevant formulas for future reference. In that table, we have also included the notation P_k^n to represent the number of k-permutations of n items and C_k^n to represent the number of combinations of k out of n items. This notation, or some variation of it, is commonly used in many textbooks as well as in many popularly used calculators.

Table 2.2: Summary of combinatorics formulas

	Situation	Notation	Formula
Combined Experiment	The number of outcomes in a combined experiment consisting of m sub-experiments each with n_i outcomes.		$\prod_{i=1}^{m} n_i$
Permutations	The number of ways to arrange n distinct objects.		$n!$
k-Permutations	The number ways to arrange k out of n distinct objects.	P_k^n	$\dfrac{n!}{(n-k)!}$
Combinations	The number of subsets of k out of n distinct objects,	$C_k^n, \binom{n}{k}$	$\dfrac{n!}{k!(n-k)!}$
Partitions	The number of ways to partition a set of n distinct objects into m groups where the ith group has n_i items.	$\binom{n}{n_1, n_2, \ldots, n_m}$	$\dfrac{n!}{\prod_{i=1}^{m}(n_i!)}$

2.6 Bayes's Theorem

In this section, we develop a few results related to the concept of conditional probability. While these results are fairly simple, they are so useful that we felt it was appropriate to devote an entire section to them. To start with, the following theorem was essentially proved in the previous section and is a direct result of the definition of conditional probability.

Theorem 2.9: For any events A and B such that $\Pr(B) \neq 0$,

$$\Pr(A|B) = \frac{\Pr(B|A)\Pr(A)}{\Pr(B)}. \tag{2.19}$$

Proof: From Definition 2.5,

$$\Pr(A, B) = \Pr(A|B)\Pr(B) = \Pr(B|A)\Pr(A). \tag{2.20}$$

Theorem 2.9 follows directly by dividing the preceding equations by $\Pr(B)$. $\square$

Theorem 2.9 is useful for calculating certain conditional probabilities since, in many problems, it may be quite difficult to compute $\Pr(A|B)$ directly, whereas calculating $\Pr(B|A)$ may be straightforward.

Theorem 2.10 (Theorem of Total Probability): Let $B_1, B_2, ..., B_n$ be a set of mutually exclusive and exhaustive events. That is, $B_i \cap B_j = \varnothing$ for all $i \neq j$ and

$$\bigcup_{i=1}^{n} B_i = S \Rightarrow \sum_{i=1}^{n} \Pr(B_i) = 1. \tag{2.21}$$

Then

$$\Pr(A) = \sum_{i=1}^{n} \Pr(A|B_i)\Pr(B_i) \tag{2.22}$$

Proof: As with Theorem 2.1, a Venn diagram (shown in Figure 2.2) is used here to aid in the visualization of our result. From the diagram, it can be seen that the event A can be written as

$$A = \{A \cap B_1\} \cup \{A \cap B_2\} \cup ... \cup \{A \cap B_n\} \tag{2.23}$$

$$\Rightarrow \Pr(A) = \Pr(\{A \cap B_1\} \cup \{A \cap B_2\} \cup ... \cup \{A \cap B_n\}) \tag{2.24}$$

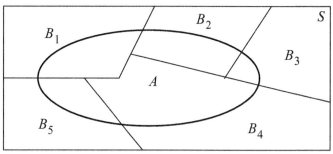

Figure 2.2
Venn diagram used to help prove the Theorem of Total Probability.

Also, since the B_i are all mutually exclusive, then the $\{A \cap B_i\}$ are also mutually exclusive so that

$$\Pr(A) = \sum_{i=1}^{n} \Pr(A, B_i) \qquad \text{(by Corollary 2.1)}, \qquad (2.25)$$

$$= \sum_{i=1}^{n} \Pr(A|B_i)\Pr(B_i) \qquad \text{(by Theorem 2.9).} \quad \square \qquad (2.26)$$

Finally by combining the results of Theorems 2.9 and 2.10, we get what has come to be know as Bayes's theorem.

Theorem 2.11 (Bayes's Theorem): Let $B_1, B_2, ..., B_n$ be a set of mutually exclusive and exhaustive events. Then,

$$\Pr(B_i|A) = \frac{\Pr(A|B_i)\Pr(B_i)}{\displaystyle\sum_{i=1}^{n} \Pr(A|B_i)\Pr(B_i)}. \qquad (2.27)$$

As a matter of nomenclature, $\Pr(B_i)$ is often referred to as the *a priori*[1] probability of event B_i, while $\Pr(B_i|A)$ is known as the *a posteriori*[2] probability of event B_i given A. Section 2.9 presents an engineering application showing how Bayes's theorem is used in the field of signal detection. We conclude here with an example showing how useful Bayes's theorem can be.

---

[1] The term *a priori* is Latin and is literally translated "from the former." In this context, it refers to probabilities that are formed from self-evident or presupposed models.

[2] The term *a posteriori* is also Latin and is literally translated "from the latter." In this context, it refers to probabilities that are derived or calculated after observing certain events.

■ Example 2.20:

A certain auditorium has 30 rows of seats. Row 1 has 11 seats, while Row 2 has 12 seats, Row 3 has 13 seats, and so on to the back of the auditorium where Row 30 has 40 seats. A door prize is to be given away by randomly selecting a row (with equal probability of selecting any of the 30 rows) and then randomly selecting a seat within that row (with each seat in the row equally likely to be selected). Find the probability that Seat 15 was selected given that Row 20 was selected and also find the probability that Row 20 was selected given that Seat 15 was selected. The first task is straightforward. Given that Row 20 was selected, there are 30 possible seats in Row 20 that are equally likely to be selected. Hence, $\Pr(\text{Seat 15}|\text{Row 20}) = 1/30$. Without the help of Bayes's Theorem, finding the probability that Row 20 was selected given that we know Seat 15 was selected would seem to be a formidable problem. Using Bayes's Theorem,

$$\Pr(\text{Row 20}|\text{Seat 15}) = \Pr(\text{Seat 15}|\text{Row 20})\Pr(\text{Row 20})/\Pr(\text{Seat 15}).$$

The two terms in the numerator on the right hand side are both equal to $1/30$. The term in the denominator is calculated using the help of the theorem of total probability.

$$\Pr(\text{Seat 15}) = \sum_{k=5}^{30} \frac{1}{k+10}\frac{1}{30} = 0.0342.$$

With this calculation completed, the a *posteriori* probability of Row 20 being selected given seat 15 was selected is given by

$$\Pr(\text{Row 20}|\text{Seat 15}) = \frac{\frac{1}{30}\frac{1}{30}}{0.0342} = 0.0325.$$

Note that the a *priori* probability that Row 20 was selected is $1/30 = 0.0333$. Therefore, the additional information that Seat 15 was selected makes the event that Row 20 was selected slightly less likely. In some sense, this may be counterintuitive, since we know that if Seat 15 was selected, there are certain rows that could not have been selected (i.e., Rows 1-4 have less that 15 seats) and, therefore, we might expect Row 20 to have a slightly higher probability of being selected compared to when we have no information about what seat was selected. To see why the probability actually goes down, try computing the probability that Row 5 was selected given the Seat 15 was selected. The event that Seat 15 was selected makes some rows much more probable while it makes others less probable and a few rows now impossible.   ■

2.7 Independence

In Example 2.20, it was seen that observing one event can change the probability of the occurrence of another event. In that particular case, the fact that it was known that Seat 15 was selected lowered the probability that Row 20 was selected. We say that the event A = {Row

20 was selected} is statistically dependent on the event B = {Seat 15 was selected}. If the description of the auditorium were changed so that each row had an equal number of seats (e.g., say all 30 rows had 20 seats each), then observing the event B ={Seat 15 was selected} would not give us any new information about the likelihood of the event A = {Row 20 was selected}. In that case, we say that the events A and B are statistically independent. Mathematically, two events A and B are independent if $\Pr(A|B)$ = $\Pr(A)$. That is, the a priori probability of event A is identical to the a posteriori probability of A given B. Note that if $\Pr(A|B)$ = $\Pr(A)$, then the following two conditions also hold (see Exercise 2.22)

$$\Pr(B|A) = \Pr(B), \tag{2.28}$$

$$\Pr(A, B) = \Pr(A)\Pr(B). \tag{2.29}$$

Furthermore, if $\Pr(A|B) \neq \Pr(A)$, then the other two conditions also do not hold. We can thereby conclude that any of these three conditions can be used as a test for independence and the other two forms must follow. We use the last form as a definition of independence since it is symmetric relative to the events A and B.

Definition 2.6: Two events are statistically independent if and only if

$$\Pr(A, B) = \Pr(A)\Pr(B). \tag{2.30}$$

■ Example 2.21:

Consider the experiment of tossing two numbered dice and observing the numbers that appear on the two upper faces. For convenience, let the dice be distinguished by color, with the first die tossed being red and the second being white. Let A = {number on the red die is less than or equal to 2}, B = {number on the white die is greater than or equal to 4}, and C = {the sum of the numbers on the two dice is 3}. As mentioned in the preceding text, there are several ways to establish independence (or lack thereof) of a pair of events. One possible way is to compare $\Pr(A, B)$ with $\Pr(A)\Pr(B)$. Note that for the events defined here, $\Pr(A) = 1/3$, $\Pr(B) = 1/2$, and $\Pr(C) = 1/18$. Also, of the 36 possible atomic outcomes of the experiment, 6 belong to the event $A \cap B$ and hence $\Pr(A, B) = 1/6$. Since $\Pr(A)\Pr(B) = 1/6$ as well, we conclude that the events A and B are independent. This agrees with intuition since we would not expect the outcome of the roll of one die to effect the outcome of the other. What about the events A and C? Of the 36 possible atomic outcomes of the experiment, two belong to the event $A \cap C$ and hence $\Pr(A, C) = 1/18$. Since $\Pr(A)\Pr(C) = 1/54$, the events A and C are not independent. Again, this is intuitive since whenever the event C occurs, the event A must also occur and so these two must be dependent. Finally, we look at the pair of events B and C. Clearly, B and C are mutually exclusive. If the white die shows a number greater than or equal to 4, there is no way the sum can be 3. Hence $\Pr(B, C) = 0$, and since $\Pr(B)\Pr(C) = 1/36$, these two events are also dependent.

■

The previous example brings out a point that is worth repeating. It is a common mistake to equate mutual exclusiveness with independence. Mutually exclusive events are not the same thing as independent events. In fact, for two events A and B for which $\Pr(A) \neq 0$ and $\Pr(B) \neq 0$, A and B can never be both independent and mutually exclusive. Thus, mutually exclusive events are necessarily statistically dependent.

A few generalizations of this basic idea of independence are in order. First, what does it mean for a set of three events to be independent? The following definition clarifies this and then we generalize the definition to any number of events.

Definition 2.7: The events A, B, and C are mutually independent if each pair of events is independent; that is

$$\Pr(A, B) = \Pr(A)\Pr(B), \tag{2.31a}$$

$$\Pr(A, C) = \Pr(A)\Pr(C), \tag{2.31b}$$

$$\Pr(B, C) = \Pr(B)\Pr(C), \tag{2.31c}$$

and in addition,

$$\Pr(A, B, C) = \Pr(A)\Pr(B)\Pr(C). \tag{2.31d}$$

Definition 2.8: The events $A_1, A_2, ..., A_n$ are independent if any subset of $k < n$ of these events are independent, and in addition

$$\Pr(A_1, A_2, ..., A_n) = \Pr(A_1)\Pr(A_2)...\Pr(A_n). \tag{2.32}$$

There are basically two ways in which we can use this idea of independence. As was shown in Example 2.21, we can compute joint or conditional probabilities and apply one of the definitions as a test for independence. Alternatively, we can assume independence and use the definitions to compute joint or conditional probabilities that otherwise may be difficult to find. This latter approach is used extensively in engineering applications. For example, certain types of noise signals can be modeled in this way. Suppose we have some time waveform $X(t)$ which represents a noisy signal which we wish to sample at various points in time, $t_1, t_2, ..., t_n$. Perhaps, we are interested in the probabilities that these samples might exceed some threshold, so we define the events $A_i = \Pr(X(t_i) > T)$, $i = 1, 2, ..., n$. How might we calculate the joint probability $\Pr(A_1, A_2, ..., A_n)$? In some cases, we have every reason to believe that the value of the noise at one point in time does not effect the value of the noise at another point in time. Hence, we assume that these events are independent and write $\Pr(A_1, A_2, ..., A_n) = \Pr(A_1)\Pr(A_2)...\Pr(A_n)$.

▪ Example 2.22:

Consider a communications network with nodes A, B, C, and D and links a_1, a_2, a_3, and a_4, as shown in the diagram. The probability of a link being available at any time is p. In order to send a messsage from node A to node D we must have a path of available links from A to D. Assuming independence of link availability, what is the probability of being able to send a message? Let L_k be the event that link a_k is available. Then

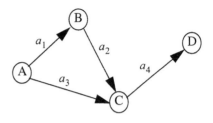

$$\Pr(A \to D) = \Pr((L_1 \cap L_2 \cap L_4) \cup (L_3 \cap L_4))$$
$$= \Pr(L_1 \cap L_2 \cap L_4) + \Pr(L_3 \cap L_4) - \Pr(L_1 \cap L_2 \cap L_3 \cap L_4)$$
$$= p^3 + p^2 - p^4.$$

▪

2.8 Discrete Random Variables

Suppose we conduct an experiment, E, which has some sample space, S. Furthermore, let ξ be some outcome defined on the sample space, S. It is useful to define functions of the outcome ξ, $X = f(\xi)$. That is, the function f has as its domain all possible outcomes associated with the experiment, E. The range of the function f will depend upon how it maps outcomes to numerical values but in general will be the set of real numbers or some part of the set of real numbers. Formally, we have the following definition.

> **Definition 2.9:** A *random variable* is a real valued function of the elements of a sample space, S. Given an experiment, E, with sample space, S, the random variable X maps each possible outcome, $\xi \in S$, to a real number $X(\xi)$ as specified by some rule. If the mapping $X(\xi)$ is such that the random variable X takes on a finite or countably infinite number of values, then we refer to X as a discrete random variable; whereas, if the range of $X(\xi)$ is an uncountably infinite number of points, we refer to X as a continuous random variable.

Since $X = f(\xi)$ is a random variable whose numerical value depends on the outcome of an experiment, we cannot describe the random variable by stating its value; rather, we must give it a probabilistic description by stating the probabilities that the variable X takes on a specific value or values (e.g., $\Pr(X=3)$ or $\Pr(X > 8)$). For now, we will focus on random variables which take on discrete values and will describe these random variables in terms of probabilities of the form $\Pr(X=x)$. In the next chapter when we study continuous random variables, we will find this description to be insufficient and will introduce other probabilistic descriptions as well.

Definition 2.10:   The *probability mass function* (PMF), $P_X(x)$, of a random variable, X, is a function that assigns a probability to each possible value of the random variable, X. The probability that the random variable X takes on the specific value x is the value of the probability mass function for x. That is, $P_X(x) = \Pr(X=x)$. We use the convention that upper case variables represent random variables while lower case variables represent fixed values that the random variable can assume.

Example 2.23:

A discrete random variable may be defined for the random experiment of flipping a coin. The sample space of outcomes is $S = \{H, T\}$. We could define the random variable X to be $X(H) = 0$ and $X(T) = 1$. That is, the sample space H, T is mapped to the set $\{0, 1\}$ by the random variable X. Assuming a fair coin, the resulting probability mass function is $P_X(0) = 1/2$ and $P_X(1) = 1/2$. Note that the mapping is not unique and we could have just as easily mapped the sample space $\{H, T\}$ to any other pair of real numbers (e.g., $\{1, 2\}$).

Example 2.24:

Suppose we repeat the experiment of flipping a fair coin n times and observe the sequence of heads and tails. A random variable, Y, could be defined to be the number of times tails occurs in n trials. It turns out that the probability mass function for this random variable is

$$P_Y(k) = \binom{n}{k}\left(\frac{1}{2}\right)^n, \quad k = 0, 1, ..., n \ .$$

The details of how this PMF is obtained will be deferred until later in this section.

Example 2.25:

Again, let the experiment be the flipping of a coin, and this time we will continue repeating the event until the first time a heads occurs. The random variable Z will represent the number of times until the first occurrence of a heads. In this case, the random variable Z can take on any positive integer value, $1 \le Z < \infty$. The probability mass function of the random variable Z can be worked out as follows:

$$\Pr(Z=n) = \Pr(n-1 \text{ tails followed by one heads}) = (\Pr(T))^{n-1}\Pr(H) = \left(\frac{1}{2}\right)^{n-1}\left(\frac{1}{2}\right) = 2^{-n}.$$

Hence,

$$P_Z(n) = 2^{-n}, \quad n = 1, 2, 3, ...$$

■ Example 2.26:

In this example, we will estimate the PMF in Example 2.24 via MATLAB simulation using the relative frequency approach. Suppose the experiment consists of tossing the coin $n = 10$ times and counting the number of tails. We then repeat this experiment a large number of times and count the relative frequency of each number of tails to estimate the PMF. The following MATLAB code can be used to accomplish this. Results of running this code are shown in Figure 2.3.

```
% Simulation code to estimate PMF of Example 2.17.
n=10;                           % Number of coin flips per experiment.
m=100;                          % Number of times to repeat experiment.
X=round(rand(n,m));             % Simulate coin flipping.
Y=sum(X);                       % Calculate number of tails per experiment.
Rel_Freq=hist(Y,[0:n])/m;       % Compute relative frequencies.
for k=0:n                       % Compute actual PMF.
  PMF(k+1)=nchoosek(n,k)*(2^(-n));
end
% Plot Results
plot([0:n],Rel_Freq,'o',[0:n],PMF,'*')
legend('Relative Frequency','True PMF')
xlabel('k')
ylabel('P_X(k)')
title('Comparison of estimated and true PMF for Example 2.26')
```

Try running this code using a larger value for m. You should see more accurate relative frequency estimates as you increase m.   ■

Figure 2.3
MATLAB Simulation results from Example 2.26.

From the preceding examples, it should be clear that the probability mass function associated with a random variable, X, must obey certain properties. First, since $P_X(x)$ is a probability it must be non-negative and no greater than 1. Second, if we sum $P_X(x)$ over all x, then this is the same as the sum of the probabilities of all outcomes in the sample space, which must be equal to 1. Stated mathematically, we may conclude that

$$0 \leq P_X(x) \leq 1, \tag{2.33a}$$

$$\sum_x P_X(x) = 1. \tag{2.33b}$$

When developing the probability mass function for a random variable, it is useful to check that the PMF satisfies these properties.

In the paragraphs that follow, we list some commonly used discrete random variables, along with their probability mass functions, and some real-world applications in which each might typically be used.

A. Bernoulli Random Variable This is the simplest possible random variable and is used to represent experiments which have two possible outcomes. These experiments are called Bernoulli trials and the resulting random variable is called a *Bernoulli random variable*. It is most common to associate the values $\{0,1\}$ with the two outcomes of the experiment. If X is a Bernoulli random variable, its probability mass function is of the form

$$P_X(0) = 1 - p, \ P_X(1) = p. \tag{2.34}$$

The coin tossing experiment would produce a Bernoulli random variable. In that case, we may map the outcome H to the value $X = 1$ and T to $X = 0$. Also, we would use the value $p = 1/2$ assuming that the coin is fair. Examples of engineering applications might include radar systems where the random variable could indicate the presence ($X = 1$) or absence ($X = 0$) of a target, or a digital communication system where $X = 1$ might indicate a bit was transmitted in error while $X = 0$ would indicate that the bit was received correctly. In these examples, we would probably expect that the value of p would be much smaller than 1/2.

B. Binomial Random Variable Consider repeating a Bernoulli trial n times, where the outcome of each trial is independent of all others. The Bernoulli trial has a sample space of $S = \{0, 1\}$ and we say that the repeated experiment has a sample space of $S_n = \{0, 1\}^n$, which is referred to as a *Cartesian space*. That is, outcomes of the repeated trials are represented as n element vectors whose elements are taken from S. Consider, for example, the outcome

$$\xi_k = \overbrace{(1, 1, ..., 1}^{k \text{ times}}, \overbrace{0, 0, ..., 0)}^{n - k \text{ times}} . \tag{2.35}$$

The probability of this outcome occurring is

$$\Pr(\xi_k) \ = \ \Pr(1, 1, ..., 1, 0, 0, ..., 0) \ = \ \Pr(1)\Pr(1)...\Pr(1)\Pr(0)\Pr(0)...\Pr(0)$$

$$= \ (\Pr(1))^k(\Pr(0))^{n-k} \ = \ p^k(1-p)^{n-k}. \tag{2.36}$$

In fact, the order of the 1's and 0's in the sequence is irrelevant. Any outcome with exactly k 1's and $n - k$ 0's would have the same probability. Now let the random variable X represent the number of times the outcome 1 occurred in the sequence of n trials. This is known as a *binomial random variable* and takes on integer values from 0 to n. To find the probability mass function of the binomial random variable, let A_k be the set of all outcomes which have exactly k 1's and $n - k$ 0's. Note that all outcomes in this event occur with the same probability. Furthermore, all outcomes in this event are mutually exclusive. Then,

$$P_X(k) \ = \ Pr(A_k) \ = \ (\text{\# of outcomes in } A_k)*(\text{probability of each outcome in } A_k)$$

$$= \ \binom{n}{k}p^k(1-p)^{n-k}, \quad k \ = \ 0, 1, 2, ..., n . \tag{2.37}$$

The number of outcomes in the event A_k is just the number of combinations of n objects taken k at a time. Referring to Theorem 2.7, this is the binomial coefficient,

$$\binom{n}{k} \ = \ \frac{n!}{k!(n-k)!}. \tag{2.38}$$

As a check, we verify that this probability mass function is properly normalized:

$$\sum_{k=0}^{n} \binom{n}{k}p^k(1-p)^{n-k} \ = \ (p + 1 - p)^n \ = \ 1^n \ = \ 1. \tag{2.39}$$

In the above calculation, we have used the binomial expansion

$$(a + b)^n \ = \ \sum_{k=0}^{n} \binom{n}{k}a^k b^{n-k}. \tag{2.40}$$

Binomial random variables occur, in practice, any time Bernoulli trials are repeated. For example, in a digital communication system, a packet of n bits may be transmitted and we might be interested in the number of bits in the packet that are received in error. Or, perhaps a bank manager might be interested in the number of tellers that are serving customers at a given point in time. Similarly, a medical technician might want to know how many cells from a blood sample are white and how many are red. In Example 2.23, the coin tossing experiment was repeated n times and the random variable Y represented the number of times tails occurred in the sequence of n tosses. This is a repetition of a Bernoulli trial, and hence the random variable Y should be a binomial random variable with $p = 1/2$ (assuming the coin is fair).

C. Poisson Random Variable Consider a binomial random variable, X, where the number of repeated trials, n, is very large. In that case, evaluating the binomial coefficients can pose numerical problems. If the probability of success in each individual trial, p, is very small, then the binomial random variable can be well approximated by a *Poisson random variable*. That is, the Poisson random variable is a limiting case of the binomial random variable. Formally, let n approach infinity and p approach 0 in such a way that $\lim_{n \to \infty} np = \alpha$. Then, the binomial probability mass function converges to the form

$$P_X(m) = \frac{\alpha^m}{m!}e^{-\alpha}, \quad m = 0, 1, 2, \dots , \tag{2.41}$$

which is the probability mass function of a Poisson random variable. We see that the Poisson random variable is properly normalized by noting that

$$\sum_{m=0}^{\infty} \frac{\alpha^m}{m!}e^{-\alpha} = e^{-\alpha}e^{\alpha} = 1, \tag{2.42}$$

(see Equation E.14 in Appendix E). The Poisson random variable is extremely important as it describes the behavior of many physical phenomena. It is commonly used in queuing theory and in communication networks. The number of customers arriving at a cashier in a store during some time interval may be well modeled as a Poisson random variable as may the number of data packets arriving at a node in a computer network. We will see increasingly in later chapters that the Poisson random variable plays a fundamental role in our development of a probabilistic description of noise.

D. Geometric Random Variable Consider repeating a Bernoulli trial until the first occurrence of the outcome ξ_0. If X represents the number of times the outcome ξ_1 occurs before the first occurrence of ξ_0, then X is a *geometric random variable* whose probability mass function is

$$P_X(k) = (1 - p)p^k, \quad k = 0, 1, 2, \dots . \tag{2.43}$$

We might also formulate the geometric random variable in a slightly different way. Suppose X counted the number of trials that were performed until the first occurrence of ξ_0. Then, the probability mass function would take on the form,

$$P_X(k) = (1 - p)p^{k-1}, \quad k = 1, 2, 3, \dots . \tag{2.44}$$

The geometric random variable can also be generalized to the case where the outcome ξ_0 must occur exactly m times. That is, the generalized geometric random variable counts the number of Bernoulli trials that must be repeated until the mth occurrence of the outcome ξ_0. We can derive the form of the probability mass function for the generalized geometric random variable from what we know about binomial random variables. For the mth occurrence of ξ_0 to occur on the kth trial, the first $k-1$ trials must have had $m-1$ occurrences of ξ_0 and $k-m$ occurrences of ξ_1. Then

$$P_X(k) = \Pr(\{((m-1) \text{ occurrences of } \xi_0 \text{ in } k-1) \text{ trials}\} \cap \{\xi_0 \text{ occurs on the } k\text{th trial}\})$$

$$= \binom{k-1}{m-1}p^{k-m}(1-p)^{m-1}(1-p) = \binom{k-1}{m-1}p^{k-m}(1-p)^n, \quad k = m, m+1, m+2, \dots$$

$$(2.45)$$

This generalized geometric random variable sometimes goes by the name of a *Pascal random variable* or the *negative binomial random variable*.

Of course, one can define many other random variables and develop the associated probability mass functions. We have chosen to introduce some of the more important discrete random variables here. In the next chapter, we will introduce some continuous random variables and the appropriate probabilistic descriptions of these random variables. However, to close out this chapter, we provide a section showing how some of the material covered herein can be used in at least one engineering application.

2.9 Engineering Application—An Optical Communication System

Figure 2.4 shows a simplified block diagram of an optical communication system. Binary data are transmitted by pulsing a laser or a light emitting diode (LED) that is coupled to an optical fiber. To transmit a binary 1, we turn on the light source for T seconds, while a binary 0 is represented by turning the source off for the same time period. Hence, the signal transmitted down the optical fiber is a series of pulses (or absence of pulses) of duration T seconds which represents the string of binary data to be transmitted. The receiver must convert this optical

Figure 2.4
Block diagram of an optical communication system.

signal back into a string of binary numbers; it does this using a photodetector. The received light wave strikes a photoemissive surface, which emits electrons in a random manner. While the number of electrons emitted during a T second interval is random and thus needs to be described by a random variable, the probability mass function of that random variable changes according to the intensity of the light incident on the photoemissive surface during the T second interval. Therefore, we define a random variable X to be the number of electrons counted during a T second interval, and we describe this random variable in terms of two conditional probability mass functions, $P_{X|0}(k) = \Pr(X=k|0 \text{ sent})$ and $P_{X|1}(k) = \Pr(X=k|1 \text{ sent})$. It can be shown through a quantum mechanical argument that these two probability mass functions should be those of Poisson random variables. When a binary 0 is sent, a relatively low number of electrons are typically observed; whereas, when a 1 is sent, a higher number of electrons is typically counted. In particular, suppose the two probability mass functions are given by

$$P_{X|0}(k) = \frac{R_0^k}{k!}e^{-R_0}, \quad k = 0, 1, 2, \dots , \tag{2.46a}$$

$$P_{X|1}(k) = \frac{R_1^k}{k!}e^{-R_1}, \quad k = 0, 1, 2, \dots . \tag{2.46b}$$

In these two PMFs, the parameters R_0 and R_1 are interpreted as the "average" number of electrons observed when a 0 is sent and when a 1 is sent, respectively. Also, it is assumed that $R_0 < R_1$, so when a 0 is sent we tend to observe fewer electrons than when a 1 is sent.

At the receiver, we count the number of electrons emitted during each T second interval and then must decide whether a "0" or "1" was sent during each interval. Suppose that during a certain bit interval it is observed that k electrons are emitted. A logical decision rule would be to calculate $\Pr(0 \text{ sent}|X=k)$ and $\Pr(1 \text{ sent}|X=k)$ and choose according to whichever is larger. That is, we calculate the a posteriori probabilities of each bit being sent, given the observation of the number of electrons emitted and choose the data bit which maximizes the a posteriori probability. This is referred to as a *maximum a posteriori* (MAP) decision rule and we decide that a binary 1 was sent if

$$\Pr(1 \text{ sent}|X=k) > \Pr(0 \text{ sent}|X=k); \tag{2.47}$$

otherwise, we decide a 0 was sent. Note that these desired a posteriori probabilities are backward relative to how the photodetector was statistically described. That is, we know the probabilities of the form $\Pr(X=k|1 \text{ sent})$ but we want to know $\Pr(1 \text{ sent}|X=k)$. We call upon Bayes's theorem to help us convert what we know into what we desire to know. Using the theorem of total probability,

$$P_X(k) = \Pr(X=k) = P_{X|0}(k)\Pr(0 \text{ sent}) + P_{X|1}(k)\Pr(1 \text{ sent}) . \tag{2.48}$$

The *a priori* probabilities $\Pr(0 \text{ sent})$ and $\Pr(1 \text{ sent})$ are taken to be equal (to 1/2), so that

$$P_X(k) = \frac{1}{2}\frac{R_0^k}{k!}e^{-R_0} + \frac{1}{2}\frac{R_1^k}{k!}e^{-R_1}. \tag{2.49}$$

Therefore, applying Bayes's theorem,

$$\Pr(0 \text{ sent}|X=k) = \frac{P_{X|0}(k)\Pr(0 \text{ sent})}{P_X(k)} = \frac{\dfrac{1}{2}\dfrac{R_0^k}{k!}e^{-R_0}}{\dfrac{1}{2}\dfrac{R_0^k}{k!}e^{-R_0} + \dfrac{1}{2}\dfrac{R_1^k}{k!}e^{-R_1}}, \tag{2.50}$$

and

$$\Pr(1 \text{ sent}|X=k) = \frac{P_{X|1}(k)\Pr(1 \text{ sent})}{P_X(k)} = \frac{\dfrac{1}{2}\dfrac{R_1^k}{k!}e^{-R_1}}{\dfrac{1}{2}\dfrac{R_0^k}{k!}e^{-R_0} + \dfrac{1}{2}\dfrac{R_1^k}{k!}e^{-R_1}}. \tag{2.51}$$

Since the denominators of both a posteriori probabilities are the same, we decide that a 1 was sent if

$$\frac{1}{2}\frac{R_1^k}{k!}e^{-R_1} > \frac{1}{2}\frac{R_0^k}{k!}e^{-R_0}. \tag{2.52}$$

After a little algebraic manipulation, this reduces down to choosing in favor of a 1 if

$$k > \frac{R_1 - R_0}{\ln(R_1/R_0)} \; ; \tag{2.53}$$

otherwise, we choose in favor of 0. That is, the receiver for our optical communication system counts the number of electrons emitted and compares that number with a threshold. If the number of electrons emitted is above the threshold, we decide that a 1 was sent; otherwise, we decide a 0 was sent.

We might also be interested in evaluating how often our receiver makes a wrong decision. Ideally, the answer is that errors are very rare, but still we would like to quantify this. Toward that end, we note that errors can occur in two manners. First a 0 could be sent and the number of electrons observed could fall above the threshold, causing us to decide that a 1 was sent. Likewise, if a 1 is actually sent and the number of electrons observed is low, we would mistakenly decide that a 0 was sent. Again, invoking concepts of conditional probability, we see that

$$\Pr(\text{error}) = \Pr(\text{error}|0 \text{ sent})\Pr(0 \text{ sent}) + \Pr(\text{error}|1 \text{ sent})\Pr(1 \text{ sent}). \tag{2.54}$$

Let x_0 be the threshold with which we compare X to decide which data bit was sent. Specifically, let[3] $x_0 = \lfloor (R_1 - R_0)/\ln(R_1/R_0) \rfloor$ so that we decide a 1 was sent if $X > x_0$, and we decide a 0 was sent if $X \leq x_0$. Then

$$\Pr(\text{error}|0 \text{ sent}) = \Pr(X > x_0|0 \text{ sent}) = \sum_{k = x_0 + 1}^{\infty} P_{X|0}(k)$$

$$= \sum_{k = x_0 + 1}^{\infty} \frac{R_0^k}{k!} e^{-R_0} = 1 - \sum_{k = 0}^{x_0} \frac{R_0^k}{k!} e^{-R_0}. \qquad (2.55)$$

Likewise,

$$\Pr(\text{error}|1 \text{ sent}) = \sum_{k = 0}^{x_0} P_{X|0}(k) = \sum_{k = 0}^{x_0} \frac{R_1^k}{k!} e^{-R_1}. \qquad (2.56)$$

Hence, the probability of error for our optical communication system is

$$\Pr(\text{error}) = \frac{1}{2} - \frac{1}{2} \sum_{k = 0}^{x_0} \frac{R_0^k e^{-R_0} - R_1^k e^{-R_1}}{k!}. \qquad (2.57)$$

Figure 2.5 shows a plot of the probability of error as a function of R_1 with R_0 as a parameter. The parameter R_0 is a characteristic of the photodetector used. We will see in later chapters that R_0 can be interpreted as the "average" number of electrons emitted during a bit interval when there is no signal incident on the photodetector. This is sometimes referred to as the "dark current." The parameter R_1 is controlled by the intensity of the incident light. Given a certain photodetector, the value of the parameter R_0 can be measured. The required value of R_1 needed to achieve a desired probability of error can be found from Figure 2.5 (or from Equation 2.57 which generated that figure). The intensity of the laser or LED can then be adjusted to produce the required value for the parameter R_1.

---

[3] The notation $\lfloor x \rfloor$ represents the integer part of x or the floor function. Mathematically, $\lfloor x \rfloor$ is the largest integer less than or equal to x.

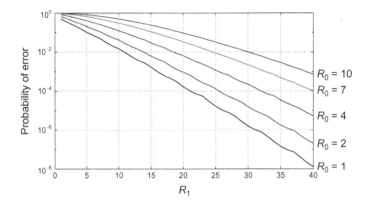

Figure 2.5

Probability of error curves for an optical communication system; curves are parameterized from bottom to top with R_0 =1, 2, 4, 7, 10.

Exercises

Section 2.1: Experiments, Sample Spaces, and Events

2.1 An experiment consists of rolling n (six-sided) dice and recording the sum of the n rolls. How many outcomes are there to this experiment?

2.2

    (a) An experiment consists of rolling a die and flipping a coin. If the coin flip is heads, the value of the die is multiplied by -1, otherwise it is left as is. What are the possible outcomes of this experiment?

    (b) Now, suppose we want to repeat the experiment in part (a) n times and record the sum of the results of each experiment. How many outcomes are there in this experiment and what are they?

2.3 An experiment consists of selecting a number x from the interval $[0, 1)$ and a number y from the interval $[0, 2)$ to form a point (x, y).

    (a) Describe the sample space of all points.

    (b) What fraction of the points in the space satisfy $x > y$?

    (c) What fraction of points in the sample space satisfy $x = y$?

2.4 An experiment consists of selecting a point (x, y) from the interior of the unit circle, $x^2 + y^2 < 1$.

    (a) What fraction of the points in the space satisfy $x > \dfrac{1}{2}$?

    (b) What fraction of the points in the space satisfy $x^2 + y^2 > \dfrac{1}{2}$?

    (c) What fraction of the points in the space satisfy $x + y > \dfrac{1}{2}$?

    (d) What fraction of the points in the space satisfy $x + y = \dfrac{1}{2}$?

2.5 An experiment consists of selecting two integers (n, k) such that $0 \le n < 5$ and $0 \le k < 10$.

(a) How many outcomes are in the sample space?

(b) What fraction of the outcomes in the sample space satisfy $n > k$?

(c) What fraction of the outcomes in the sample space satisfy $n < k$?

(d) What fraction of the outcomes in the sample space satisfy $n = k$?

2.6

(a) For each of the experiments described in Examples 2.1-2.5, state whether or not it is reasonable to expect that each outcome of the experiment would occur equally often.

(b) Create (a hypothethical) experiment of your own with a finite number of outcomes where you would not expect the outcomes to occur equally often.

Section 2.2: Axioms of Probability

2.7 Using mathematical induction, prove Corollary 2.1. Recall, Corollary 2.1 states that for M events A_1, A_2, ..., A_M which are mutually exclusive (i.e., $A_i \cap A_j = \varnothing$ for all $i \ne j$),

$$\Pr\left(\bigcup_{i=1}^{M} A_i\right) = \sum_{i=1}^{M} \Pr(A_i).$$

2.8 Develop a careful proof of Theorem 2.1 which states that for any events A and B,

$$\Pr(A \cup B) = \Pr(A) + \Pr(B) - \Pr(A \cap B).$$

One way to approach this proof is to start by showing that the set $A \cup B$ can be written as the union of three mutually exclusive sets,

$$A \cup B = \{A \cap (\overline{A \cap B})\} \cup \{A \cap B\} \cup \{B \cap (\overline{A \cap B})\}$$

and hence by Corollary 2.1,

$$\Pr(A \cup B) = \Pr(A \cap (\overline{A \cap B})) + \Pr(A \cap B) + \Pr(B \cap (\overline{A \cap B})).$$

Next, show that

$$\Pr(A \cap (\overline{A \cap B})) = \Pr(A) - \Pr(A \cap B)$$

and likewise

$$\Pr(B \cap (\overline{A \cap B})) = \Pr(B) - \Pr(A \cap B).$$

(*Hint:* recall DeMorgan's law) Put these results together to complete the desired proof.

2.9 Show that the above formula for the probability of the union of two events can be generalized to three events as follows:

$$Pr(A \cup B \cup C) = Pr(A) + Pr(B) + Pr(C)$$
$$- Pr(A \cap B) - Pr(A \cap C) - Pr(B \cap C) + Pr(A \cap B \cap C).$$

2.10 Prove Theorem 2.3 which states that if $A \subset B$ then $Pr(A) \le Pr(B)$.

2.11 Formally prove the *union bound* which states that for any events $A_1, A_2, ..., A_M$ (not necessarily mutually exclusive),

$$Pr\left(\bigcup_{i=1}^{M} A_i\right) \le \sum_{i=1}^{M} Pr(A_i).$$

2.12 An experiment consists of tossing a coin twice and observing the sequence of coin tosses. The sample space consists of four outcomes $\xi_1 = (H, H)$, $\xi_2 = (H, T)$, $\xi_3 = (T, H)$, and $\xi_4 = (T, T)$. Suppose the coin is not evenly weighted such that we expect a heads to occur more often than tails and as a result, we assign the following probabilities to each of the four outcomes:

$$Pr(H, H) = \frac{3}{8}, \quad Pr(H, T) = \frac{1}{4}, \quad Pr(T, H) = \frac{1}{4}, \quad Pr(T, T) = \frac{1}{8}.$$

(a) Does this probability assignment satisfy the three axioms of probability?
(b) Given this probability assignment, what is $Pr($first toss is heads$)$?
(c) Given this probability assignment, what is $Pr($second toss is heads$)$?

2.13 Repeat Exercise 2.12 if the probability assignment is changed to:

$$Pr(H, H) = \frac{25}{64}, \quad Pr(H, T) = \frac{15}{64}, \quad Pr(T, H) = \frac{15}{64}, \quad Pr(T, T) = \frac{9}{64}.$$

2.14 Consider the experiment of tossing a six-sided die as described in Example 2.2. Suppose the die is loaded and as such we assign the following probabilities to each of the six outcomes:

$$Pr(1) = \frac{1}{21}, Pr(2) = \frac{2}{21}, \quad Pr(3) = \frac{3}{21}, Pr(4) = \frac{4}{21}, \quad Pr(5) = \frac{5}{21}, Pr(6) = \frac{6}{21}.$$

Is this assignment consistent with the three axioms of probability?

Section 2.3: Assigning Probabilities

2.15 Demonstrate that the relative frequency approach to assigning probabilities satisfies the three axioms of probability.

2.16 We are given a number of darts. Suppose it is known that each time we throw a dart at a target, we have a probability of $1/4$ of hitting the target. An experiment consists of throwing three darts at the target and observing the sequence of hits and misses (e.g., one possible outcome might be (H,M,M)).
  (a) Find a probabilty assignment for the eight outcomes of this experiment that leads to a probabilty of 1/4 of hitting the target on any toss. Note, your assignment must satisfy the axioms of probability.
  (b) Is the probability assigment you chose unique? That is, are there other possible probability assignments that lead to a probabilty of 1/4 of hitting the target on any toss, or is yours the only valid assignment? Carefully prove your answer.

2.17 A spinner as shown in the figure selects a number from the set $\{1, 2, 3, 4\}$. If we select the outcomes $\{\xi_1 = 1, \xi_2 = 2, \xi_3 = 3, \xi_4 = 4\}$ as atomic outcomes and use the classical approach to assigning probabilities, then we would conclude that the spinner selects all four numbers each with equal probability, 1/4. Show how to describe this experiment in terms of more fundamental outcomes so that the classical approach leads to a more reasonable probability assignment.

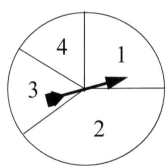

2.18 Consider a modified version of the experiment in Example 2.4 where we flip a coin until the first occurence of tails or until we flip the coin four times, whichever comes first.
  (a) List the possible outcomes of this experiment. How many are there? Would you expect each outcome to be equally probable?
  (b) Can you develop a method to use the classical approach to assign probabilities to this experiment? *Hint:* You are going to have to define an atomic outcome which is more fundamental than the outcomes listed in part (a).

    (c)   Given the probability assigment you developed in part (b), find the probabilities of each outcome listed in part (a).

2.19 Extend your solution to Exercise 2.18 to find the probability of each of the outcomes of Example 2.4.

2.20 If we roll two dice and observe the sum, the most common outcome is 7 and occurs with probability 1/6. But what if we roll more than 2 dice?
    (a)   Suppose we roll three dice and observe the sum. What is the most likely sum and what is the probability of its occurrence?
    (b)   Repeat part (a) for the case of four dice?

Section 2.4: Joint and Conditional Probabilities

2.21 Demonstrate that the definition of conditional probability $\Pr(A|B) = \Pr(A, B)\Pr(B)$ satisfies the three axioms of probability.

2.22 Prove that if $\Pr(B|A) = \Pr(B)$, then it follows that
    (a) $\Pr(A, B) = \Pr(A)\Pr(B)$ and
    (b) $\Pr(A|B) = \Pr(A)$.

Furthermore, show that if $\Pr(B|A) \neq \Pr(B)$, then the two conditions above do not hold as well.

2.23 A box of 30 diodes is known to contain five defective ones. If two diodes are selected at random without replacement, what is the probability that at least one of these diodes is defective?

2.24 Two balls are selected sequentially (without replacement) from an urn containing three red, four white, and five blue balls.
    (a)   What is the probability that the first is red and the second blue?
    (b)   What is the probability of selecting a white ball on the second draw if the first ball is replaced before the second is selected?
    (c)   What is the probability of selecting a white ball on the second draw if the first ball is not replaced before the second is selected?

2.25 Two six-sided (balanced) dice are thrown. Find the probabilities of each of the following events:
    (a)   a 5 does not occur on either throw;
    (b)   the sum is 7;

    (c)   a 5 and a 3 occur in any order;

    (d)   the first throw is a 5 and the second throw is a 5 or a 4;

    (e)   both throws are 5;

    (f)   either throw is a 6.

2.26 Two six-sided (balanced) dice are thrown. Find the probabilities of each of the following events:

    (a)   only 2, 3, or 4 appear on both dice;

    (b)   the value of the second roll subtracted from the value of the first roll is 2;

    (c)   the sum is 10 given that one roll is 6;

    (d)   the sum is 7 or 8 given one roll is 5;

    (e)   one roll is a 4 given the sum is 7.

2.27 Consider two events A and B such that $\Pr(A) > \Pr(B)$. Determine if $\Pr(A|B) > \Pr(B|A)$ is always true, sometimes true, or never true.

2.28 Prove that for any two events A and B, $\Pr(A \cap B) \leq \Pr(A) \leq \Pr(A \cup B)$.

Section 2.5: Basic Combinatorics

2.29 Use mathematical induction to prove Theorem 2.4. Recall that Theorem 2.4 states that a combined experiment, $E = E_1 \times E_2 \times E_3 \times \ldots \times E_m$, consisting of experiments E_i each with n_i outcomes, $i = 1, 2, 3, \ldots, m$, has a total number of possible outcomes given by

$$n = n_1 n_2 n_3 \ldots n_m = \prod_{i=1}^{m} n_i.$$

2.30 Use mathematical induction to prove Theorem 2.8. Recall that Theorem 2.8 states that given a set of n distinct elements, the number of ways to partition the set into m groups where the ith group has n_i elements is given by the *multinomial coefficient*,

$$\binom{n}{n_1, n_2, \ldots, n_m} = \frac{n!}{n_1! n_2! \ldots n_m!}.$$

2.31 Prove the following identities involving the binomial coefficient $\binom{n}{k} = \dfrac{n!}{k!(n-k)!}$.

    (a) $\displaystyle\binom{n}{k} = \binom{n}{n-k}$ (b) $\displaystyle\binom{n}{k} + \binom{n}{k+1} = \binom{n+1}{k+1}$

(c) $\displaystyle\sum_{k=0}^{n}\binom{n}{k} = 2^n$

(d) $\displaystyle\sum_{k=0}^{n}\binom{n}{k}(-1)^k = 0$

(e) $\displaystyle\sum_{k=1}^{n}\binom{n}{k}k = n2^{n-1}$

(f) $\displaystyle\sum_{k=0}^{n}\binom{n}{k}k(-1)^k = 0$

2.32 I deal myself 3 cards from a standard 52-card deck. Find the probabilities of each of the following events:
   (a)   2 of a kind (e.g., 2 fives or 2 kings);
   (b)   3 of a kind;
   (c)   3 of the same suit (a.k.a a flush, e.g., 3 hearts or 3 clubs);
   (d)   3 cards in consecutive order (a.k.a. a straight, e.g., 2-3-4 or 10-J-Q).

2.33 I deal myself 13 cards for a standard 52-card deck. Find the probabilities of each of the following events:
   (a)   exactly one heart appears in my hand (of 13 cards);
   (b)   at least 7 cards from a single suit appear in my hand;
   (c)   my hand is void (0 cards) of at least one suit.

2.34 For an alphabet consisting of 26 characters, find the following:
   (a)   The number of possible 4-letter sequences (words) where the same letter can appear in the word more than once.
   (b)   The number of possible 4-letter sequences (words) where the same letter cannot appear in the word more than once.
   (c)   The number of different ways the 26 letters can be ordered.

2.35 In pulse-code modulation (PCM), a PCM word consists of a sequence of binary digits (bits) of 1s and 0s.
   (a)   Suppose the PCM word length is n bits long. How many distinct words are there?
   (b)   If each PCM word, three bits long, is equally likely to occur, what is the probability of a word with exactly two 1s to occur?

2.36 In pulse-amplitude modulation (PAM), a PAM word consists of a sequence of pulses, where each pulse may take on a given number of amplitude levels. Suppose a PAM word is n pulses long and each pulse may take on m different levels.
   (a)   How many distinct PAM words are there?
   (b)   If each PAM word, 4 pulses long, is equally likely to occur and each pulse can have one of three levels, $\{0, 1, 2\}$, what is the probability of a PAM word occurring with exactly two pulses of level 2?

2.37 A certain communication system transmits text messages by representing each character with an n-bit binary codeword. Suppose it is necessary for this communication system to operate in such a way that there are always an equal number of 0s and 1s transmitted. Toward that end, the communication system uses a codebook consisting only of those n-bit words that have exactly $n/2$ 0s and $n/2$ 1s (where n is an even integer). For example, in the case of $n = 4$, there are exactly 6 four-bit codewords consisting of exactly two 1s and two 0s resulting in the codebook $\{(1100), (1010), (1001), (0110), (0101), (0011)\}$. Thus, with four bit codewords, we could represent an alphabet of only six characters.

(a) Find an expression for the number of codewords with half 1s and half 0s for an arbitrary even integer n.

(b) What is the minimum length of codeword we would need if the codebook needs to represent at least 100 different characters?

2.38 Phone numbers in the United States consist of 10 digits, a three-digit area code followed by a seven-digit number. Suppose there are several constraints that must be satisfied for a phone number to be valid, such as:

(i) Neither the first digit of any area code nor the first digit of a phone number can be 0, (since 0 is reserved to connenct to an operator).

(ii) The second digit of any valid area code must be either 0 or 1.

(iii) The second digit of any valid phone number must not be either 0 or 1 (the second and third constraints are no longer in place but were once used so that the phone company could easily determine whether you were making a local or long distance call).

(iv) The second and third digits of any valid area code cannot both be 1s.

(v) The second and third digits of any valid phone number cannot both be 1s. (three-digit numbers ending in 11 are reserved for special purposes, e.g., emergency, 911, or information, 411).

(a) Given the five constraints listed above, how many valid three-digit area codes are there?

(b) Given the five constraints listed above, how many valid seven-digit phone numbers are there?

(c) How many different 10-digit phone numbers can be constructed under these constraints?

2.39 A balanced coin is tossed nine times. Find the probabilities of each of the following events:

(a) exactly 3 heads occurred;

(b) at least 3 heads occurred;

(c) at least 3 heads and at least 2 tails occurred.

2.40 A blackjack hand consists of 2 cards drawn from a 52-card deck. The order in which the cards are drawn does not matter.
(a)   How many different blackjack hands are there?
(b)   In blackjack, the suit of the card is irrelevant to the game. If we do not make a distinction between the different suits, now how many different blackjack hands are there?
(c)   In blackjack, 10s, Jacks, Queens, and Kings are all worth 10 points and thus have an identical function in the game. If we do not make a distinction between 10s, Jacks, Queens, and Kings, nor between the different suits, now how many blackjack hands are there?

2.41 In the game of blackjack, Aces are worth 11 points (or they can also be counted as 1 point, but for the sake of this problem, we will only count them as 11 points), 10s, Jacks, Queens, and Kings are all worth 10 points and the other numbered cards are worth their face values (i.e., 2s are worth two points, 3s are worth three points, etc.). The suit of the card does not matter.
(a)   Find the probability of being dealt a two-card blackjack hand worth a total of 18 or more points?
(b)   Find the probability of being dealt a two-card blackjack hand worth a total of no less than 12 points and no more than 17 points?
Hint: Consider carefully the results of Problem 2.40 and think about what consists of an atomic outcome in this experiment.

2.42 In a game of blackjack, the player and the dealer are both dealt two cards. One of the dealer's cards is dealt face up so that the player gets to see it. Suppose you (as a player) are dealt a 10 and a 6 and you observe that one of the dealer's cards is a 7. Given the three observed cards, what is the probability that the dealers cards total more points than yours (before any additional cards are drawn)? Refer to Exercise 2.41 for a description of the point values of various cards in blackjack.

2.43 A poker hand consists of 5 cards drawn from a 52-card deck.
(a)   How many different poker hands are there?
(b)   How many different poker hands are there that contain all four aces? What then is the probability of being dealt four aces? What is the probability of being dealt four of a kind (i.e., four aces, or four kings, or four queens, etc.)?
(c)   How many different poker hands are there that contain exactly three aces? What then is the probability of being dealt three aces? What is the probability of being dealt three of a kind (i.e., three aces, or three kings, or three queens, etc.)?

2.44 A certain gym teacher has a class of 20 students. He wants to divide them into four teams of five students each in order to have a class basketball tournament.

(a) How many different ways can he divide the class into four teams?

(b) Tommy and Bobby are two of the students in the class and are best friends. Assuming the gym teacher assigns students in a completely random fashion, what is the probability that they get selected to be on the same team?

(c) Neither boy wants to be on a team with the class bully, Frank. What is the probability that neither Tommy nor Bobby are on the same team as Frank?

Section 2.6: Bayes's Theorem

2.45 Manufacturer X produces personal computers (PCs) at two different locations in the world. Fifteen percent of the PCs produced at location A are delivered to a retail outlet defective, while five percent of the PCs produced at location B are delivered defective to the same retail store. If the manufacturing plant at A produces 1,000,000 PCs per year and the plant at B produces 150,000 PCs per year, find the probability of purchasing a defective PC.

2.46 A communication system sends binary data {0 or 1} which is then detected at the receiver. The receiver occasionally makes mistakes and sometimes a 0 is sent and is detected as a 1 or a 1 can be sent and detected as a 0. Suppose the communication system is described by the following set of conditional probabilities:

$\Pr(0 \text{ received}|0 \text{ transmitted}) = 0.95$, $\Pr(1 \text{ received}|0 \text{ transmitted}) = 0.05$,

$\Pr(0 \text{ received}|1 \text{ transmitted}) = 0.10$, $\Pr(1 \text{ received}|1 \text{ transmitted}) = 0.90$.

(a) Assuming 0s and 1s are equally likely to be transmitted (i.e., $\Pr(0 \text{ transmitted}) = 1/2$ and $\Pr(1 \text{ transmitted}) = 1/2$), find $\Pr(0 \text{ received})$ and $\Pr(1 \text{ received})$.

(b) Suppose a 0 is detected at the receiver. What is the probability that the transmitted bit was actually a 1? Also, if a 1 was detected at the receiver, what is the probability that the transmitted bit was actually a 0?

(c) What is the probability that the detected bit is not equal to the transmitted bit. This is the overall probability of error of the receiver.

2.47 In this problem, we will modify the communication system described in Exercise 2.46 so that the detector at the receiver is allowed to make one of three possible decisions:

"0" the detector decides the received signal was a 0,

"1"   the detector decides the received signal was a 1,
"E"   the detector is not sure and declares the received signal an erasure
        (i.e., the receiver chooses not to choose).

The operation of the detector is described by the following set of conditional probabilities:

$$\Pr(0 \text{ received}|0 \text{ transmitted}) = 0.90\,, \Pr(0 \text{ received}|1 \text{ transmitted}) = 0.04\,,$$

$$\Pr(1 \text{ received}|0 \text{ transmitted}) = 0.01\,, \Pr(1 \text{ received}|1 \text{ transmitted}) = 0.80\,,$$

$$\Pr(E \text{ received}|0 \text{ transmitted}) = 0.09\,, \Pr(E \text{ received}|1 \text{ transmitted}) = 0.16\,.$$

Again, assume that 0s and 1s are equally likely to be transmitted.
(a)   What is the probability that a symbol is erased at the receiver?
(b)   Given that a received symbol is declared an erasure, what is the probability that a 0 was actually transmitted?
(c)   What is the probability of error of this receiver? That is, what is the probability that a 0 was transmitted and it is detected as a 1 or a 1 was transmitted and it is detected as a 0?

2.48 We are in possession of two coins, one which is fairly balanced and turns up heads with probability 1/2, the other is weighted such that heads shows up with probability 3/4 and tails with probability 1/4. The two coins are identical looking and feeling so we cannot tell which is which. In order to determine which is the biased coin we toss the coin 10 times and observe the number of heads that occurred.
(a)   If 7 heads were observed, what is the probability that the coin flipped was the fair coin?
(b)   If 3 heads were observed, what is the probability that the coin flipped was the fair coin?

2.49 Researchers are investigating the physical development of children over time. In the study, children are given a physical aptitude test at several stages in their development. Let P be the event that the child passes the physical aptitude test. Furthermore, let B be the event that the child taking the test was a boy, and G be the event that the child taking the test was a girl. The computer storing the data from this project experienced hard drive failure and only some of the data was recovered resulting in the partial database shown in the table. Use your knowledge of probability theory to fill in the missing items in the table.

Data for Exercise 2.49

Year in School	$\Pr(P)$	$\Pr(B)$	$\Pr(G)$	$\Pr(P\|B)$	$\Pr(P\|G)$	$\Pr(B\|P)$	$\Pr(G\|P)$
2nd grade		0.45		0.25	0.30		
4th grade	0.37			0.35	0.40		
6th grade				0.50	0.50	0.52	
8th grade	0.74			0.75			0.35

Section 2.7: Independence

2.50 Compare the two probability assignments in Exercises 2.12 and 2.13. Which of these two assignments corresponds to independent coin tosses?

2.51 Cards are drawn from a standard 52-card deck until an ace is drawn. After each card is drawn, it is put back in the deck and the cards are reshuffled so that each card drawn is independent of all others.
   (a) Find the probability that the first ace is drawn on the 5th selection.
   (b) Find the probability that *at least* 5 cards are drawn *before* the first ace appears.
   (c) Repeat parts (a) and (b) if the cards are drawn without replacement. That is, after each card is drawn, the card is set aside and not replaced in the deck.

2.52 Cards are drawn from a standard 52-card deck until the third club is drawn. After each card is drawn, it is put back in the deck and the cards are reshuffled so that each card drawn is independent of all others.
   (a) Find the probability that the 3rd club is drawn on the 8th selection.
   (b) Find the probability that *at least* 8 cards are drawn *before* the 3rd club appears.
   (c) Repeat parts (a) and (b) if the cards are drawn without replacement. That is, after each card is drawn, the card is set aside and not replaced in the deck.

2.53 A computer memory has the capability of storing 10^6 words. Due to outside forces, portions of the memory are often erased. Therefore, words are stored redundantly in various areas of the memory. If a particular word is stored in n different places in the memory, what is the probability that this word cannot be recalled if one-half of the memory is erased by electromagnetic radiation? *Hint:* Consider each word to be stored in a particular cell (or box).

These cells (boxes) may be located anywhere, geometrically speaking, in memory. The contents of each cell may be either erased or not erased. Assume n is small compared to the memory capacity.

2.54 If two events A and B are such that $\Pr(A)$ is not zero and $\Pr(B)$ is not zero, what combinations of independent (I), not independent (NI), mutually exclusive (M), and not mutually exclusive (NM) are permissible? In other words, which of the four combinations (I, M), (NI, M), (I, NM), and (NI, NM) are permissible? Construct an example for those combinations that are permissible.

2.55 Suppose two events A and B are independent.

(a)   Is it true that A is independent of $\bar{B}$? If yes, give a convincing proof, otherwise, give a counterexample.

(b)   Is it true that $\bar{A}$ is independent of $\bar{B}$? If yes, give a convincing proof, otherwise, give a counterexample.

2.56 Suppose we modify the communications network of Example 2.22 as shown in the diagram by adding a link from node B to node D. Assuming each link is available with probability p independent of any other link, what is the probability of being able to send a message from node A to D?

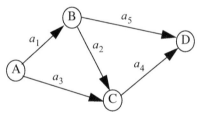

Section 2.8: Discrete Random Variables

2.57 A possible outcome of an experiment is the event A. The probability of this event is p. The experiment is performed n times, the outcome of any trial is not affected by the results of the previous trials. Define a random variable X to be the number of times the event A occurs in n trials.

(a)   What is the PMF $\Pr(X=x)$?

(b)   Show that the sum of the PMF over all x is 1.

(c)   What is the name of this PMF?

2.58 For each of the following probability mass functions, find the value of the constant c:

(a) $P_X(k) = c(0.37)^k, \ k = 0, 1, 2, \dots$

(b) $P_X(k) = c(0.82)^k, \ k = 1, 2, 3, \dots$

(c) $P_X(k) = c(0.41)^k, \ k = 0, 1, 2, \dots, 24$.

(d) $P_X(k) = c(0.91)^k, \ k = 1, 2, 3, \dots, 15$.

(e) $P_X(k) = c(0.41)^k, \ k = 0, 2, 4, \dots, 12$.

2.59 Consider a Bernoulli trial where $\Pr(1) = p$ and $\Pr(0) = 1 - p$. Suppose this Bernoulli trial is repeated n times.

(a) Plot the probability mass function for a Binomial random variable, X, with $p = 1/5$ and $n = 10$.

(b) Plot the corresponding probability mass function for a Poisson random variable X with $\alpha = np = 2$.

(c) Compare $\Pr(X \geq 5)$ as computed by both the Binomial and Poisson random variables. Is the Poisson random variable a good approximation for the Binomial random variable for this example?

2.60 Suppose the arrival of telephone calls at a switch can be modeled with a Poisson PMF. That is, if X is the number of calls that arrive in t minutes, then

$$\Pr(X=k) = \frac{(\lambda t)^k}{k!} e^{-\lambda t}, \quad k = 0, 1, 2, \dots,$$

where λ is the average arrival rate in calls/minute. Suppose that the average rate of calls is 10 min.

(a) What is the probability that fewer than three calls will be received in the first 6 s?

(b) What is the probability that fewer than three calls will be received in the first 6 min?

2.61 In a certain lottery, six numbers are randomly chosen from the set $\{0, 1, 2, \dots, 49\}$ (without replacement). To win the lottery, a player must guess correctly all six numbers but it is not necessary to specify in which order the numbers are selected.

(a) What is the probability of winning the lottery with only one ticket?

(b) Suppose in a given week, 6 million lottery tickets are sold. Suppose further that each player is equally likely to choose any of the possible number combinations and does so independent of the selections of all other players. What is the probability that exactly four players correctly select the winning combination?

(c) Again assuming 6 million tickets sold, what is the most probable number of winning tickets?

(d) Repeat parts (b) and (c) using the Poisson approximation to the binomial probability distribution. Is the Poisson distribution an accurate approximation in this example?

2.62 Imagine an audio amplifier contains six transistors. Harry has determined that two transistors are defective, but he does not know which two. Harry removes three transistors at random and inspects them. Let X be the number of defective transistors that Harry finds, where X may be 0, 1, or 2. Find the PMF for x.

2.63 A software manufacturer knows that 1 out of 10 software games that the company markets will be a financial success. The manufacturer selects 10 new games to market. What is the probability that exactly one game will be a financial success? What is the probability that at least two games will be a success?

2.64 In a digital communication system, a block of k data bits is mapped into an n-bit codeword that typically contains the k information bits as well as $n - k$ redundant bits. This is known as an (n, k) block code. The redundant bits are included to provide error correction capability. Suppose that each transmitted bit in our digital communication system is received in error with probability p. Furthermore, assume that the decoder is capable of correcting any pattern of t or fewer errors in an n bit block. That is, if t or less bits in an n bit block are received in error, then the codeword will be decoded correctly, whereas if more than t errors occur, the decoder will decode the received word incorrectly. Assuming each bit is received in error with probability $p = 0.03$, find the probability of decoder error for each of the following codes.

(a) $(n, k) = (7, 4)$, $t = 1$;

(b) $(n, k) = (15, 7)$, $t = 2$;

(c) $(n, k) = (31, 16)$, $t = 3$.

2.65 A roulette wheel consists of 38 numbers (18 are red, 18 are black, and 2 are green). Assume that with each spin of the wheel, each number is equally likely to appear.

(a) What is the probability of a gambler winning if he bets on a red number showing up?

(b) Suppose the gambler keeps betting on red until he finally wins. Let N be the number of times he plays/bets. Specify the probability mass function of the random variable N. That is, find $P_N(k) = \Pr(N=k)$.

(c) Now, suppose the gambler keeps betting on red until he wins twice. Let M be the number of times he plays/bets. Specify the probability mass function of the random variable M. That is, find $P_M(k) = \Pr(M=k)$.

2.66 Cards are drawn from a standard 52-card deck. After each card is drawn, it is put back in the deck and the cards are reshuffled so that each card drawn is independent of all others. Let N be the random variable that represents the number of cards that are drawn before the second appearance of an ace. For example, if the sequence of cards drawn was {2, 5, K, 7, A, 5, 3, J, A, ...}, then N would take on a value of 8. Find the probability mass function of N. That is, find $P_N(n) = \Pr(N=n)$.

2.67 Highway A and Highway B merge to form Highway C as shown in the figure. Engineers have studied the traffic patterns on the two merging roads and found that the number of cars per minute that travel each road can be well modeled as Poisson random variables as described below

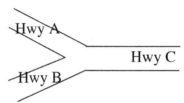

Highway A: N=# cars per minute, $\Pr(N=n) = \dfrac{(\lambda_A)^n e^{-\lambda_A}}{n!}$, $n = 0, 1, 2, 3, \ldots$,

Highway B: M=# cars per minute, $\Pr(M=m) = \dfrac{(\lambda_B)^m e^{-\lambda_B}}{m!}$, $m = 0, 1, 2, 3, \ldots$.

Let $K = M + N$ be the number of cars per minute on Highway C. Find the PMF of K, $P_K(k) = \Pr(K=k)$. Is K also a Poisson random varialbe or does it follow some other distribution? You may assume that in any interval of time, the number of cars on Highway A and the number of cars on Highway B are independent of each other.

2.68 An experiment consists of rolling a pair of (six-sided) dice and observing the sum. This experiment is repeated until the sum of 7 is observed at which point the experiment stops. Let N be the random variable which represents the number of times the experiment is repeated. That is, if the first occurrence of {sum=7} happens on the 5th roll of the dice, then $N = 5$.

(a) Find the probability mass function for the random variable N. That is, find $P_N(k) = \Pr(N=k)$ for all k.

(b) What is the probability that the experiment proceeds for at least 4 rolls? That is, find $\Pr(N \geq 4)$.

Miscellaneous Problems

2.69 I deal myself 5 cards from a standard 52-card deck. Find the probabilities of each of the following events:
(a) 2 of a kind;
(b) 3 of a kind;
(c) 2 pair (e.g., 2 eights and 2 queens);
(d) a flush (5 cards all of the same suit);
(e) a full house (3 of one kind and 2 of another kind);
(f) a straight (5 cards in consecutive order).

2.70 In the game of RISK, two players compete in a game of dice rolling for conquest of the world. One player is on "offense" while the other is on "defense." For this problem, the player on offense is allowed to roll multiple dice while the player on defense rolls a single die. Whoever rolls the higher number wins (i.e., the highest number rolled by the offense is compared with the number rolled by the defense). In case of a tie, the defense is declared the winner. The loser must remove one army from the board. Find the probability of the offense winning and the probability of the defense winning in each of the following scenarios:
(a) Both players roll only a single die.
(b) Offense rolls two dice while defense rolls one die.
(c) Offense rolls three dice while defense rolls one die.

2.71 Now consider a modified version of Problem 2.70 where the defense is also allowed to roll multiple dice. Each player's highest roll is compared with the other player's highest roll, their second highest roll is compared with the other player's second highest roll, etc. As before, any ties go to the defense.
(a) Suppose both players roll two dice. In this case, there are two armies to be lost since there are two dice comparisons (highest vs. highest and lowest vs. lowest). Find each of the following probabilities:

   (i)   Offense wins both comparisons (and thus defense loses two armies).
   (ii)   Offense wins one comparison and defense wins the other (and thus each lose one army).
   (iii) Defense wins both comparisons (and thus offense loses two armies).

(b) Repeat all the calculations in part (a) for the scenario where the offense rolls three dice and the defense rolls two dice. As before, there are two comparisons to be made in this scenario (highest vs. highest and second highest vs. second highest).

2.72 (Adapted from the minigame "Hide and Go Boom" from Nintendo's "Mario Party 4") In this game, 1 player competes against 3 other players. Players 2, 3, and 4 independently select one of four slots (labeled A, B, X, and Y) in which to hide. More than one player can hide in the same slot and none of these three "hiders" can co-ordinate their actions. Once the three are hidden, player 1 gets to select 3 of the 4 slots in which to "find" the other three players. If all three hidden players were located in the three slots chosen by player 1, then player 1 wins. If any of the three hidden players are located in the slot not selected by player 1, then the group of players 2, 3, and 4 win.

(a) What is the probability that player 1 wins?

(b) Suppose the three-player team was allowed to coordinate their hiding efforts and they decided to all hide in the same slot. What then would be the probability that player 1 wins?

(c) Now, suppose the three-player team was allowed to coordinate their hiding efforts and they decided to all hide in different slots. Now what is the probability that player 1 wins?

2.73 (Adapted from the battle game "Bowser's Bigger Blast" from Nintendo's "Mario Party 4") In this game, 4 players compete in a deadly game of chance against each other. On a stage, there are 5 detonators. One is connected to a bomb while the other four are disconnected and will have no effect. Player 1 must go first and push one of the five detonators. If he is unlucky enough to choose the live detonator, he will set off the bomb and lose the game (and his life). If he chooses one of the four "duds," he is safe and will remove that detonator from the game. At that point, player 2 must choose one of the four remaining detonators, and so on. If all four players are lucky enough to survive their choices, then the stage is reset (with the five detonators and one randomly selected to be live) and the procedure is repeated, until a player eventually loses. Is this a fair game? Or, is one player more likely to lose than the others? That is, find the probability of each player losing the game.

2.74 A certain light bulb manufacturer makes two types of bulbs, a low-cost short-life (S-type) bulb and a more expensive long-life (L-type) bulb. The two types of bulbs look identical and so the company must be sure to carefully label the boxes of bulbs. A box of bulbs is found on the floor of the manufacturing plant that (you guessed it) has not been labeled. In order to determine which types of bulbs are in the box, a promising young engineer suggested that they take one bulb from the box and run it until it burns out. After observing how long the bulb remains lit, they should be able to make a good guess as to which type of bulbs are in the box. It is known that the length of time (in hours), X, that a bulb lasts can be described by a geometric random variable

$$P_X(k) = (1-a)a^k, \ k = 0, 1, 2, \ldots.$$

The parameter a that appears in the above expression is $a = 0.99$ for the S-type bulbs and $a = 0.999$ for the L-type bulbs. It is known that of all the light bulbs the company manufactures 75% are S-type and 25% are L-type. Hence, before the experiment is run, the box in question has a 75% chance of being S-type and 25% chance of being L-type.

(a) If, after running the proposed experiment, it is observed that the bulb burned out after 200 hours, which type of bulb is most likely in the unmarked box? Mathematically justify your answer.

(b) What is the probability that your decision in part (a) turns out to be wrong? That is, if you decided that the box most likely contained L-type bulbs, what is the probability that the box actually contains S-type bulbs (or if you decided the box most likely contained S-type bulbs, what is the probability that the box actually contains L-type bulbs)?

2.75 This classic problem was given widespread pop culture attention in 2008 by the movie "21." A form of this problem known as the Monty Hall problem appeared in the "Ask Marilyn" column in *Parade Magazine* in 1990. Although the column's author, Marilyn vos Savant, provided the correct solution, many readers wrote the magazine insisting that her solution was wrong. Can you get this tricky problem correct?

> The Monty Hall Problem—You are on a game show where you are asked to select one of three doors and you get to keep the prize behind the door. Behind one of the three doors is a new car, while there are goats behind the other two. After you select your door, the host, who knows where the car is, will open one of the doors that you did not select, which he knows to contain a goat. After the goat has been revealed, the host asks if you would like to change your selection and choose instead the other unopened door. Are you better off keeping your original selection, changing to the other unopend door, or does it not matter?

MATLAB Exercises

2.76 Write MATLAB code to produce a randomly generated number which is equally likely to produce any number from the set $\{0, 1, 2, ..., 9\}$.

2.77 Write MATLAB code to produce a randomly generated number that follows the Bernoulli distribution for an arbitrary parameter, p.

2.78 Modify the MATLAB code in Example 2.26 to produce a random variable which follows a Binomial distribution for arbitrary parameters n, p.

2.79 Write MATLAB code to simulate a random variable, Z, whose PMF is given by $P_z(k) = 2^{-k}$, $k = 1, 2, 3, \ldots$. *Hint:* See Example 2.25 for specifics on how this random variable arises and then follow the lead of Example 2.26.

2.80

(a) Write and execute a MATLAB program to calculate $n!$ for an arbitrary n. Use your program to calculate $64!$.

(b) What is the largest integer n for which your program gives a finite answer?

(c) Sterling's approximation for the factorial function is given by

$$n! \approx \sqrt{2\pi}\left(n^{n+\frac{1}{2}}\right)e^{-n}\left(1 - \frac{1}{12n}\right).$$

Use your program to compare the true value of $n!$ with Sterling's approximation. For what ranges of n is the approximation within 1% of the true value?

2.81 Write your own program to evaluate the binomial coefficient $\binom{n}{k} = \dfrac{n!}{k!(n-k)!}$. Create your program in such a way so that it does not have to directly evaluate $n!$. That way the program will not crash if you use it to evaluate a binomial coefficient with n greater than the value you found in Exercise 2.80b. Use your program to evaluate $\binom{384}{15}$.

2.82 Write a MATLAB program to simulate the dice rolling game described in Exercise 2.70. Run your program a large number of times and use the relative frequency to verify that your answers to Exercise 2.70 are correct.

2.83 Write a MATLAB program to simulate the dice rolling game described in Exercise 2.71. Run your program a large number of times and use the relative frequency to verify that your answers to Exercise 2.71 are correct.

2.84 Write a MATLAB program to simulate the Monty Hall problem described in Exercise 2.75. Run your program a large number of times and use the relative frequency to verify that your answer to Exercise 2.75 is correct.

Random Variables, Distributions, and Density Functions

At the end of the last chapter, we introduced the concept of a random variable and gave several examples of common discrete random variables. These random variables were described by their probability mass functions. While this description works fine for discrete random variables, it is inadequate to describe random variables that take on a continuum of values. We will illustrate this through an example shortly. In this chapter, we introduce the cumulative distribution function (CDF) as an alternative description of random variables that is appropriate for describing continuous as well as discrete random variables. A related function, the probability density function (PDF), is also covered. With these tools in hand, the concepts of random variables can be fully developed. Several examples of commonly used continuous random variables are also discussed in this chapter.

To show the need for an alternative to the probability mass function, consider a discrete random variable, X, that takes on values from the set $\{0, 1/N, 2/N, \ldots, (N-1)/N\}$ with equal probability. That is, the probability mass function of X is

$$P_X\left(\frac{k}{N}\right) = \frac{1}{N}, \quad k = 0, 1, 2, \ldots, N-1. \tag{3.1}$$

This is the type of random variable that is produced by "random" number generators in high level languages such as Fortran and C, as well as math packages such as MATLAB, MathCAD, and Mathematica. In these cases, N is taken to be a fairly large number so that it appears that the random number can be anything in the continuous range $[0, 1)$. The reader is referred to Chapter 12, for more details on how computer-generated random numbers work. For now, consider the limiting case as $N \to \infty$ so that the random variable can truly fall anywhere in the interval $[0, 1)$. One curious result of passing to the limit is that now

$$P_X\left(\frac{k}{N}\right) = \lim_{N \to \infty} \frac{1}{N} = 0. \tag{3.2}$$

That is, each point has zero probability of occurring. Yet, something has to occur! This problem is common to continuous random variables, and it is clear that the probability mass

Probability and Random Processes. DOI: 10.1016/B978-0-12-386981-4.00003-6

function is not a suitable description for such a random variable. The next sections develop two alternative descriptions for continuous random variables, which will be used extensively throughout the rest of the text.

3.1 The Cumulative Distribution Function

Since a continuous random variable will typically have a zero probability of taking on a specific value, we avoid talking about such probabilities. Instead, events of the form $\{X \leq x\}$ can be considered.

Definition 3.1: The CDF of a random variable, X, is

$$F_X(x) = \Pr(X \leq x). \tag{3.3}$$

From this definition, several properties of the CDF can be inferred. First, since the CDF is a probability, it must take on values between 0 and 1. Since random variables are real valued, it is easy to conclude that $F_X(-\infty) = 0$ and $F_X(\infty) = 1$. That is, a real number cannot be less than $-\infty$ and must be less than ∞. Next, if we consider two fixed values, x_1 and x_2, such that $x_1 < x_2$, then the event $\{X \leq x_1\}$ is a subset of $\{X \leq x_2\}$. Hence, $F_x(x_1) \leq F_x(x_2)$. This implies that the CDF is a monotonic nondecreasing function. Also, we can break the event $\{X \leq x_2\}$ into the union of two mutually exclusive events, $\{X \leq x_2\} = \{X \leq x_1\} \cup \{x_1 < X \leq x_2\}$. Hence, $F_X(x_2) = F_X(x_1) + \Pr(x_1 < X \leq x_2)$, or equivalently, $\Pr(x_1 < X \leq x_2) = F_X(x_2) - F_X(x_1)$. Thus, the CDF can also be used to measure the probability that a random variable takes on a value in a certain interval. These properties of CDFs are summarized as follows:

$$(1)\, F_X(-\infty) = 0, F_X(\infty) = 1, \tag{3.4a}$$

$$(2)\, 0 \leq F_X(x) \leq 1, \tag{3.4b}$$

$$(3)\, \text{For } x_1 < x_2, F_X(x_1) \leq F_X(x_2), \tag{3.4c}$$

$$(4)\, \text{For } x_1 < x_2, \Pr(x_1 < X \leq x_2) = F_X(x_2) - F_X(x_1). \tag{3.4d}$$

■ Example 3.1:

Which of the following mathematical functions could be the CDF of some random variable?

(a) $F_X(x) = \dfrac{1}{2} + \dfrac{1}{\pi}\tan^{-1}(x)$,

(b) $F_X(x) = [1 - e^{-x}]u(x)$,   ($u(x)$ is the unit step function),

(c) $F_X(x) = e^{-x^2}$,

(d) $F_X(x) = x^2 u(x)$.

To determine this, we need to check that the function starts at zero when $x = -\infty$, ends at one when $x = \infty$, and is monotonic increasing in between. The first two functions satisfy these properties and thus are valid CDFs while the last two do not. The function in (c) is decreasing for positive values of x, while the function in (d) takes on values greater than 1 and $F_X(\infty) \neq 1$.

∎

To more carefully illustrate the behavior of the CDF, let us return to the computer random number generator which generates N possible values from the set $\{0, 1/N, 2/N, ..., (N-1)/N\}$ with equal probability. The CDF for this particular random variable can be described as follows. First, $F_X(x) = 0$ for all $x < 0$, since the random variable cannot take on negative values. Similarly, $F_X(x) = 1$ for all $x \geq (N-1)/N$ since the random variable cannot be greater than $(N-1)/N$. Next, consider a value of x in the range $0 \leq x < 1/N$. In this case, $\Pr(X \leq x) = \Pr(X = 0)$ since the only value in the specified range that this random variable can take on is $X = 0$. Hence $F_X(x) = \Pr(X = 0) = 1/N$ for $0 \leq x < 1/N$. Similarly, for $1/N \leq x < 2/N$, $F_X(x) = \Pr(X = 0) + \Pr(X = 1/N) = 2/N$.

Following this same reasoning, it is seen that, in general, for an integer k such that $0 < k < N$ and $(k-1)/N \leq x < k/N$, $F_X(x) = k/N$. A plot of $F_X(x)$ as a function of x would produce the general staircase-type function shown in Figure 3.1. In Figure 3.2a and b, the CDF is shown for specific values of $N = 10$ and $N = 50$, respectively. It should be clear from these plots that in the limit as N passes to infinity, the CDF of Figure 3.2c results. The functional form of this CDF is

$$F_X(x) = \begin{cases} 0, & x \leq 0, \\ x, & 0 < x \leq 1, \\ 1, & x > 1. \end{cases} \qquad (3.5)$$

Figure 3.1
General CDF of the random variable X.

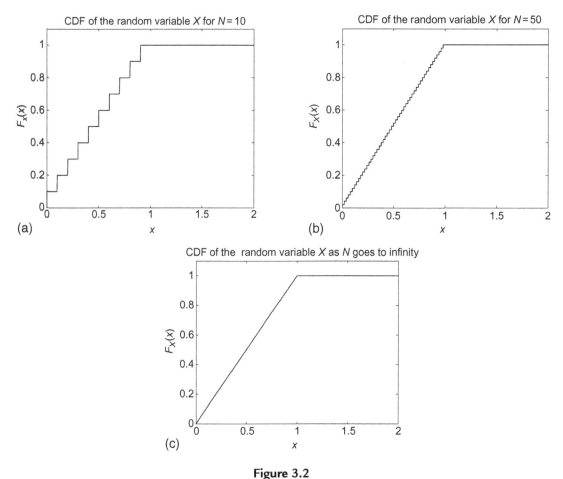

Figure 3.2

CDF of the random variable X for (a) $N = 10$, (b) $N = 50$, and (c) $N \to \infty$.

In this limiting case, the random variable X is a continuous random variable and takes on values in the range $[0, 1)$ with equal probability. Later in the chapter, this will be referred to as a uniform random variable. Note that when the random variable was discrete, the CDF was discontinuous and had jumps at the specific values that the random variable could take on whereas, for the continuous random variable, the CDF was a continuous function (although its derivative was not always continuous). This last observation turns out to be universal in that continuous random variables have a continuous CDF, while discrete random variables have a discontinuous CDF with a staircase type of function.

Occasionally, one also needs to work with a random variable whose CDF is continuous in some ranges and yet also has some discontinuities. Such a random variable is referred to as a *mixed random variable*.

■ Example 3.2:

Suppose we are interested in observing the occurrence of certain events and noting the time of first occurrence. The event might be the emission of a photon in our optical photodetector at the end of Chapter 2, the arrival of a message at a certain node in a computer communications network, or perhaps the arrival of a customer in a store. Let X be a random variable that represents the time that the event first occurs. We would like to find the CDF of such a random variable, $F_X(t) = \Pr(X \le t)$. Since the event could happen at any point in time and time is continuous, we expect X to be a continuous random variable. To formulate a reasonable CDF for this random variable, suppose we divide the time interval $(0, t]$ into many, tiny nonoverlapping time intervals of length Δt. Assume that the probability that our event occurs in a time interval of length Δt is proportional to Δt and take λ to be the constant of proportionality, That is

$$\Pr(\text{event occurs in } (k\Delta t, (k+1)\Delta t)) = \lambda Dt .$$

We also assume that the event occurring in one interval is independent of the event occurring in another nonoverlapping time interval. With these rather simple assumptions, we can develop the CDF of the random variable X as follows:

$$F_X(t) = \Pr(X \le t) = 1 - \Pr(X > t),$$

$$\Pr(X > t) = \Pr(X \notin (0, t]) = \Pr(\{X \notin (0, \Delta t]\} \cap \{X \notin (\Delta t, 2\Delta t]\} \cap \dots \cap \{X \notin ((k-1)\Delta t, k\Delta t]\}).$$

In this equation, it is assumed that the time interval $(0, t]$ has been divided into k intervals of length Δt. Since each of the events in the expression are independent, the probability of the intersection is just the product of the probabilities, so that

$$\Pr(X > t) = \Pr(X \notin (0, \Delta t]) \Pr(X \notin (\Delta t, 2\Delta t]) \dots \Pr(X \notin ((k-1)\Delta t, k\Delta t])$$

$$= (1 - \lambda \Delta t)^k = \left(1 - \frac{\lambda t}{k}\right)^k .$$

Finally, we pass to the limit as $\Delta t \to 0$ or, equivalently, $k \to \infty$ to produce

$$\Pr(X > t) = e^{-\lambda t} u(t) \Rightarrow F_X(t) = (1 - e^{-\lambda t}) u(t) .$$

■

■ Example 3.3:

Suppose a random variable has a CDF given by $F_X(x) = (1 - e^{-x})u(x)$. Find the following quantities:

(a) $\Pr(X > 5)$,

(b) $\Pr(X < 5)$,

(c) $\Pr(3 < X < 7)$,

(d) $\Pr(X > 5 | X < 7)$.

For part (a), we note that $\Pr(X > 5) = 1 - \Pr(X \le 5) = 1 - F_X(5) = e^{-5}$. In part (b), we note that $F_X(5)$ gives us $\Pr(X \le 5)$, which is not quite what we want. However, we note that

$$F_X(5) = \Pr(\{X < 5\} \cup \{X = 5\}) = \Pr(X < 5) + \Pr(X = 5).$$

Hence,

$$\Pr(X < 5) = F_X(5) - \Pr(X = 5).$$

In this case, since X is a continuous random variable, $\Pr(X = 5) = 0$ and so there is no need to make a distinction between $\Pr(X < 5)$ and $\Pr(X \le 5)$; however, for discrete random variables, we would need to be careful. Accordingly, $\Pr(X < 5) = F_X(5) = 1 - \exp(-5)$. For part (c), we note that in general $F_X(7) - F_X(3) = \Pr(3 < X \le 7)$. Again, for this continuous random variable, $\Pr(X = 7) = 0$, so we can also write $\Pr(3 < X < 7) = F_X(7) - F_X(3) = e^{-3} - e^{-7}$. Finally, for part (d) we invoke the definition of conditional probability to write the required quantity in terms of the CDF of X:

$$\Pr(X > 5 | X < 7) = \frac{\Pr(\{X > 5\} \cap \{X < 7\})}{\Pr(X < 7)} = \frac{\Pr(5 < X < 7)}{\Pr(X < 7)} = \frac{F_X(7) - F_X(5)}{F_X(7)} = \frac{e^{-5} - e^{-7}}{1 - e^{-7}}.$$

■

For discrete random variables, the CDF can be written in terms of the probability mass function defined in Chapter 2. Consider a general random variable, X, which can take on values from the discrete set $\{x_1, x_2, x_3, \ldots\}$. The CDF for this random variable is

$$F_X(x) = \sum_{i=1}^{k} P_X(x_i), \quad \text{for } x_k \le x < x_{k+1}. \tag{3.6}$$

The constraint in this equation can be incorporated using unit step functions, in which case the CDF of a discrete random variable can be written as

$$F_X(x) = \sum_{i=1}^{k} P_X(x_i)u(x - x_i). \tag{3.7}$$

In conclusion, if we know the PMF of a discrete random variable, we can easily construct its CDF.

3.2 The Probability Density Function

While the CDF introduced in the last section represents a mathematical tool to statistically describe a random variable, it is often quite cumbersome to work with CDFs. For example, we will see later in this chapter that the most important and commonly used random variable, the Gaussian random variable, has a CDF which cannot be expressed in closed form. Furthermore, it can often be difficult to infer various properties of a random variable from its CDF. To help circumvent these problems, an alternative and often more convenient description known as the probability density function is often used.

Definition 3.2: The PDF of the random variable X evaluated at the point x is

$$f_X(x) = \lim_{\varepsilon \to 0} \frac{\Pr(x \le X < x + \varepsilon)}{\varepsilon}. \tag{3.8}$$

As the name implies, the PDF is the probability that the random variable X lies in an infinitesimal interval about the point $X = x$, normalized by the length of the interval.

Note that the probability of a random variable falling in an interval can be written in terms of its CDF as specified in Equation (3.4d). For continuous random variables,

$$\Pr(x \le X < x + \varepsilon) = F_X(x + \varepsilon) - F_X(x) \tag{3.9}$$

so that

$$f_X(x) = \lim_{\varepsilon \to 0} \frac{F_X(x + \varepsilon) - F_X(x)}{\varepsilon} = \frac{dF_X(x)}{dx}. \tag{3.10}$$

Hence, it is seen that the PDF of a random variable is the derivative of its CDF. Conversely, the CDF of a random variable can be expressed as the integral of its PDF. This property is illustrated in Figure 3.3. From the definition of the PDF in Equation (3.8), it is apparent that

Figure 3.3
Relationship between the PDF and CDF of a random variable.

the PDF is a nonnegative function although it is not restricted to be less than unity as with the CDF. From the properties of the CDFs, we can also infer several important properties of PDFs. Some properties of PDFs are

$$(1) \quad f_X(x) \geq 0; \tag{3.11a}$$

$$(2) \quad f_X(x) = \frac{dF_X(x)}{dx}; \tag{3.11b}$$

$$(3) \quad F_X(x) = \int_{-\infty}^{x} f_X(y)dy; \tag{3.11c}$$

$$(4) \quad \int_{-\infty}^{\infty} f_X(x)dx = 1; \tag{3.11d}$$

$$(5) \quad \int_{a}^{b} f_X(x)dx = \Pr(a < X \leq b). \tag{3.11e}$$

■ Example 3.4:

Which of the following are valid PDFs?

(a) $f_X(x) = e^{-x}u(x);$

(b) $f_X(x) = e^{-|x|};$

(c) $f_X(x) = \begin{cases} \dfrac{3}{4}(x^2 - 1), & |x| < 2, \\ 0, & \text{otherwise}; \end{cases}$

(d) $f_X(x) = \begin{cases} 1, & 0 \leq x < 1, \\ 0, & \text{otherwise}; \end{cases}$

(e) $f_X(x) = 2xe^{-x^2}u(x).$

To verify the validity of a potential PDF, we need to verify only that the function is nonnegative and normalized so that the area underneath the function is equal to unity. The function in part (c) takes on negative values, while the function in part (b) is not properly normalized, and therefore these are not valid PDFs. The other three functions are valid PDFs.

■

■ Example 3.5:

A random variable has a CDF given by $F_X(x) = (1 - e^{-\lambda x})u(x)$. Its PDF is then given by

$$f_X(x) = \frac{dF_X(x)}{dx} = \lambda e^{-\lambda x}u(x).$$

Likewise, if a random variable has a PDF given by $f_X(x) = 2xe^{-x^2}u(x)$, then its CDF is given by

$$F_X(x) = \int_{-\infty}^{x} f_X(y)dy = \int_{-\infty}^{x} 2ye^{-y^2}u(y)dy = \int_{0}^{x} 2ye^{-y^2}dyu(x) = (1 - e^{-x^2})u(x).$$

■

■ Example 3.6:

The MATLAB function rand generates random numbers that are uniformly distributed in the interval (0, 1) using an algorithm which is discussed in Chapter 12. For the present, consider the algorithm to select a number from a table in a random like manner. To construct a histogram for the random numbers generated by rand we write a script that calls rand repeatedly. Since we can do this only a finite number of times, we quantize the range of the random numbers into increments of 0.1. We then calculate the number of times a random number falls in each quantized interval and divide by the total number of numbers generated for the example. If we plot this ratio of relative frequencies using a bar graph, the resulting plot is called a histogram. The MATLAB script for this example follows and the histogram is shown in Figure 3.4, where the total number of values generated is 10,000. Try changing the value of *N* or the number and width of the bins in this example to see how results vary.

```
N=10,000;                          % do N times
x=rand(1,N);                       % produce N random numbers
bins=[0.05:0.1:0.95];              % create 10 bins,
                                   % with centers at 0.05, 0.15, ...
[yvalues,xvalues]=hist(x,bins);    % define xvalues and yvalues
yvalues=yvalues/N;                 % Normalize to produce
                                   % relative frequencis
bar(xvalues,yvalues);              % plot bar graph
xlabel('x')
ylabel('Relative Frequencies')
```

■

3.3 The Gaussian Random Variable

In the study of random variables, the Gaussian random variable is the most commonly used and of most importance. As will be seen later in the text, many physical phenomenon can be modeled as Gaussian random variables, including the thermal noise encountered in electronic

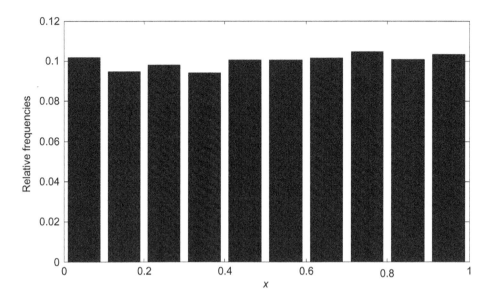

Figure 3.4
Histogram of relative frequencies for a uniform random variable generated by MATLAB's
rand function using 10,000 trials.

circuits. Although many students may not realize it, they are probably quite familiar with the
Gaussian random variable, for it is this random variable which leads to the so-called curve on
which many students are graded.

Definition 3.3: A Gaussian random variable is one whose PDF can be written in the
general form

$$f_X(x) = \frac{1}{\sqrt{2\pi\sigma^2}}\exp\left(-\frac{(x-m)^2}{2\sigma^2}\right). \tag{3.12}$$

The PDF of the Gaussian random variable has two parameters, m and σ, which have
the interpretation of the mean and standard deviation, respectively.[1] The parameter σ^2
is referred to as the variance.

An example of a Guassian PDF is shown in Figure 3.5. In general, the Gaussian PDF is
centered about the point $x = m$ and has a width that is proportional to σ.

---

[1] The terms *mean, standard deviation,* and *variance* will be defined and explained carefully in the next
chapter.

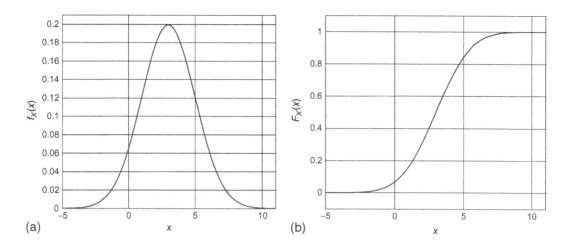

Figure 3.5
(a) PDF and (b) CDF of a Gaussian random variable with $m = 3$ and $\sigma = 2$.

It should be pointed out that in the mathematics and statistics literature, this random variable is referred to as a "normal" random variable. Furthermore, for the special case when $m = 0$ and $\sigma = 1$, it is called a "standard normal" random variable. However, in the engineering literature the term Gaussian is much more common, so this nomenclature will be used throughout this text.

Because Gaussian random variables are so commonly used in such a wide variety of applications, it is standard practice to introduce a shorthand notation to describe a Gaussian random variable, $X \sim N(m, \sigma^2)$. This is read "X is distributed normally (or Gaussian) with mean, m, and variance, σ^2."

The first topic to be addressed in the study of Gaussian random variables is to find its CDF. The CDF is required whenever we want to find the probability that a Gaussian random variable lies above or below some threshold or in some interval. Using the relationship in Equation (3.11c), the CDF of a Gaussian random variable is written as

$$F_X(x) = \int_{-\infty}^{x} \frac{1}{\sqrt{2\pi\sigma^2}} \exp\left(-\frac{(y-m)^2}{2\sigma^2}\right) dy . \tag{3.13}$$

It can be shown that it is impossible to express this integral in closed form. While this is unfortunate, it does not stop us from extensively using the Gaussian random variable. Two approaches can be taken to dealing with this problem. As with other important integrals that cannot be expressed in closed form (e.g., Bessel functions), the Gaussian CDF has been extensively tabulated and one can always look up values of the required CDF in a table, such as the one provided in Appendix E. However, it is often a better option to use one of several numerical routines which can approximate the desired integral to any desired accuracy.

The same sort of situation exists with many more commonly known mathematical functions. For example, what if a student was asked to find the tangent of 1.23 radians? While the student could look up the answer in a table of trig functions, that seems like a rather archaic approach. Any scientific calculator, high-level programming language, or math package will have internally generated functions to evaluate such standard mathematical functions. While not all scientific calculators and high-level programming languages have internally generated functions for evaluating Gaussian CDFs, most mathematical software packages do, and in any event, it is a fairly simple thing to write short program to evaluate the required function. Some numerical techniques for evaluating functions related to the Gaussian CDF are discussed specifically in Appendix F.

Whether the Gaussian CDF is to be evaluated by using a table or a program, the required CDF must be converted into one of a few commonly used standard forms. A few of these common forms are

- error function integral, $\text{erf}(x) = \dfrac{2}{\sqrt{\pi}}\displaystyle\int_0^x \exp(-t^2)\,\mathrm{d}t$,

- complementary error function integral, $\text{erfc}(x) = 1 - \text{erf}(x) = \dfrac{2}{\sqrt{\pi}}\displaystyle\int_x^\infty \exp(-t^2)\,\mathrm{d}t$,

- Φ-function, $\Phi(x) = \dfrac{1}{\sqrt{2\pi}}\displaystyle\int_{-\infty}^x \exp\left(-\dfrac{t^2}{2}\right)\mathrm{d}t$,

- Q-function, $Q(x) = \dfrac{1}{\sqrt{2\pi}}\displaystyle\int_x^\infty \exp\left(-\dfrac{t^2}{2}\right)\mathrm{d}t$.

The error function and its complement are most commonly used in the mathematics community; however, in this text, we will primarily use the Q-function. Nevertheless, students at least need to be familiar with the relationship between all of these functions because most math packages will have internally defined routines for the error function integral and perhaps its complement as well, but usually not the Φ-function or the Q-function. So, why not just use error functions? The reason is that if one compares the integral expression for the Gaussian CDF in Equation (3.13) with the integral functions defined in our list, it is a more straightforward thing to express the Gaussian CDF in terms of a Φ-function or a Q-function. Also, the Q-function seems to be enjoying the most common usage in the engineering literature in recent years.

Perhaps the advantage of not working with error function integrals is clearer if we note that the Φ-function is simply the CDF of a standard normal random variable. For general Gaussian random variables which are not in the normalized form, the CDF can be expressed in terms of a Φ-function using a simple transformation. Starting with the Gaussian CDF in Equation (3.13), make the transformation $t = (y - m)/\sigma$, resulting in

$$F_X(x) = \int_{-\infty}^{x} \frac{1}{\sqrt{2\pi\sigma^2}} \exp\left(-\frac{(y-m)^2}{2\sigma^2}\right) dy = \int_{-\infty}^{\frac{x-m}{\sigma}} \frac{1}{\sqrt{2\pi}} \exp\left(-\frac{t^2}{2}\right) dt = \Phi\left(\frac{x-m}{\sigma}\right) . \quad (3.14)$$

Hence, to evaluate the CDF of a Gaussian random variable, we just evaluate the Φ-function at the points $(x-m)/\sigma$.

The Q-function is more natural for evaluating probabilities of the form $\Pr(X > x)$. Following a line of reasoning identical to the previous paragraph, it is seen that if $X \sim N(m, \sigma^2)$, then

$$\Pr(X > x) = \int_{\frac{x-m}{\sigma}}^{\infty} \frac{1}{\sqrt{2\pi}} \exp\left(-\frac{t^2}{2}\right) dt = Q\left(\frac{x-m}{\sigma}\right). \quad (3.15)$$

Furthermore, since we have shown that $\Pr(X > x) = Q((x-m)/\sigma)$ and $\Pr(X \leq x) = \Phi((x-m)/\sigma)$, it is apparent that the relationship between the Φ-function and the Q-function is

$$Q(x) = 1 - \Phi(x). \quad (3.16)$$

This and other symmetry relationships can be visualized using the graphical definitions of the Φ-function (phi function) and the Q-function shown in Figure 3.6. Note that the CDF of a Gaussian random variable can be written in terms of a Q-function as

$$F_X(x) = 1 - Q\left(\frac{x-m}{\sigma}\right). \quad (3.17)$$

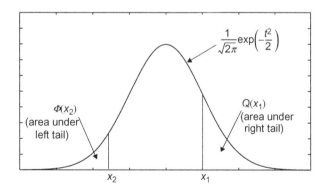

Figure 3.6

Standardized integrals related to the Gaussian CDF: the $\Phi(\cdot)$ and $Q(\cdot)$ functions.

Once the desired CDF has been expressed in terms of a Q-function, the numerical value can be looked up in a table or calculated with a numerical program. Other probabilities can be found in a similar manner as shown in the next example. It should be noted that some internally defined programs for Q and related functions may expect a positive argument. If it is required to evaluate the Q-function at a negative value, the relationship $Q(x) = 1 - Q(-x)$ can be used. That is, to evaluate $Q(-2)$, for example, $Q(2)$ can be evaluated and then use $Q(-2) = 1 - Q(2)$.

■ Example 3.7:

A random variable has a PDF given by

$$f_X(x) = \frac{1}{\sqrt{8\pi}} \exp\left(-\frac{(x+3)^2}{8}\right)$$

Find each of the following probabilities and express the answers in terms of Q-functions.

(a) $\Pr(X \le 0)$,

(b) $\Pr(X > 4)$,

(c) $\Pr(|X + 3| < 2)$,

(d) $\Pr(|X - 2| > 1)$.

For the given Gaussian PDF, $m = -3$ and $\sigma = 2$. For part (a),

$$\Pr(X \le 0) = \Phi((0-(-3))/2) = \Phi(1.5).$$

This can be rewritten in terms of a Q-function as

$$\Pr(X \le 0) = 1 - Q(1.5).$$

The probability in part (b) is easier to express directly in terms of a Q-function.

$$\Pr(X > 4) = Q((4-(-3))/2) = Q(3.5).$$

In part (c), the probability of the random variable X falling in an interval is required. This event can be rewritten as

$$\Pr(|X + 3| < 2) = \Pr(-5 < X < -1) = \Pr(X > -5) - \Pr(X > -1) = Q(-1) - Q(1) = 1 - 2Q(1).$$

We can proceed in a similar manner for part (d).

$$\Pr(|X - 2| > 1) = \Pr(X < 1) + \Pr(X > 3) = \Phi(2) + Q(3) = 1 - Q(2) + Q(3).$$

■

■ Example 3.8:

MATLAB has a built-in function, randn, which generates random variables according to a Gaussian or normal distribution. In particular, randn(k,n) creates an $k \times n$ matrix whose elements are randomly chosen according to a standard normal distribution. This example constructs a histogram of the numbers generated by the randn function similar to what was done in Example 3.6 using the rand function. Note that by multiplying the output of the randn function by σ and adding m, the Gaussian random variable produced by randn now has mean m and variance σ^2. We will elaborate on such transformations and others in the next chapter. Note that the MATLAB code that follows is similar to that of Example 3.6 with the exception that we are now using randn instead of rand. Also the Gaussian PDF has a domain which is infinite and, thus, in principle we would need an infinite number of bins in our histogram. Since this is impractical, we chose a sufficiently large number of bins such that those not included would represent relatively insignificant values. Note also that in this histogram we are plotting probability densities rather than relative frequencies so that a direct comparison can be made between the histogram and the true PDF. The histogram obtained using the following code is shown in Figure 3.7.

```
N=10,000;                                  % do N times
m=5; sigma=2;                              % set mean and variance
x=m+sigma*randn(1,N);                      % produce N random numbers
left=-4.5; width=1; right=14.5;            % set bin parameters
bins=[left:width:right];                   % create bins with centers at
                                           % left, left+width, ..., right

[yvalues,xvalues]=hist(x,bins);            % define xvalues and yvalues
yvalues=yvalues/(N*width);                 % Normalize to produce
                                           % probability densities

bar(xvalues,yvalues);                      % plot bar graph
z=[left-width/2:width/10:right+width/2];
```

Figure 3.7

Histogram of random numbers produces by randn along with a Gaussian PDF; $m = 5$, $\sigma = 2$.

(Continued)

```
pdf=exp(-(z-m).^2/(2*sigma^2));          % Compute true PDF.
pdf=pdf/sqrt(2*pi*sigma^2);
hold on                                  % place plot of true PDF on
plot(z,pdf)                              % top of histogram.
xlabel('x')
ylabel('f_X(x)')
```

3.4 Other Important Random Variables

This section provides a summary of some other important random variables that find use in various engineering applications. For each random variable, an indication is given as to the sorts of applications that find use for these random variables.

3.4.1 Uniform Random Variable

The uniform PDF is constant over an interval $[a, b)$. The PDF and its corresponding CDF are

$$f_X(x) = \begin{cases} \dfrac{1}{b-a}, & a \le x < b, \\[2mm] 0, & \text{elsewhere,} \end{cases} \tag{3.18a}$$

$$F_X(x) = \begin{cases} 0, & x < a, \\[2mm] \dfrac{x-a}{b-a}, & a \le x < b, \\[2mm] 1, & x \ge b. \end{cases} \tag{3.18b}$$

Since this is a continuous random variable, the interval over which the PDF is nonzero can be open or closed on either end. A plot of the PDF and CDF of a uniform random variable is shown in Figure 3.8. Most computer random number generators will generate

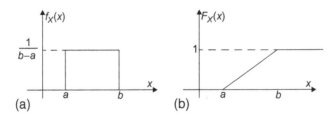

Figure 3.8
(a) Probability density function and (b) cumulative distribution function of a uniform random variable.

a random variable which closely approximates a uniform random variable over the interval $(0, 1)$. We will see in the next chapter that by performing a transformation on this uniform random variable, we can create many other random variables of interest. An example of a uniform random variable would be the phase of a radio frequency sinusoid in a communication system. Although the transmitter knows the phase of the sinusoid, the receiver may have no information about the phase. In this case, the phase at the receiver could be modeled as a random variable uniformly distributed over the interval $[0, 2\pi)$.

3.4.2 Exponential Random Variable

The exponential random variable has a PDF and CDF given by (for any $b > 0$)

$$f_X(x) = \frac{1}{b}\exp\left(-\frac{x}{b}\right)u(x), \tag{3.19a}$$

$$F_X(x) = \left[1 - \exp\left(-\frac{x}{b}\right)\right]u(x). \tag{3.19b}$$

A plot of the PDF and the CDF of an exponential random variable is shown in Figure 3.9. The parameter b is related to the width of the PDF and the PDF has a peak value of $1/b$ which occurs at $x = 0$. The PDF and CDF are nonzero over the semi-infinite interval $(0, \infty)$, which may be either open or closed on the left endpoint.

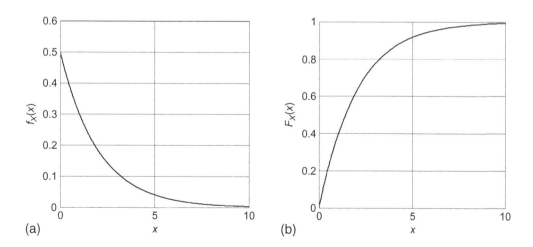

Figure 3.9
(a) Probability density function and (b) cumulative distribution function of an exponential random variable, $b = 2$.

Exponential random variables are commonly encountered in the study of queueing systems. The time between arrivals of customers at a bank, for example, is commonly modeled as an exponential random variable, as is the duration of voice conversations in a telephone network.

3.4.3 Laplace Random Variable

A Laplace random variable has a PDF which takes the form of a two-sided exponential. The functional forms of the PDF and CDF are given by (for any $b > 0$)

$$f_X(x) = \frac{1}{2b}\exp\left(-\frac{|x|}{b}\right),$$ (3.20a)

$$F_X(x) = \begin{cases} \frac{1}{2}\exp\left(\frac{x}{b}\right), & x < 0, \\ 1 - \frac{1}{2}\exp\left(-\frac{x}{b}\right), & x \geq 0. \end{cases}$$ (3.20b)

A plot of these functions is shown in Figure 3.10. The width of the PDF is determined by the parameter b, while the peak value of the PDF is $1/2b$. Note that this peak value is half of what it is in the case of the (one-sided) exponential shown in Figure 3.9. This makes sense since the Laplace random variable has two sides and the area under the curve must remain constant (and equal to unity). The Laplace random variable has been used to model the amplitude of a speech (voice) signal.

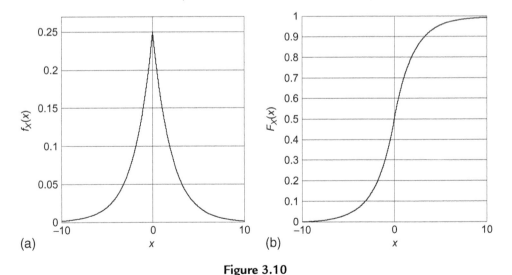

Figure 3.10
(a) Probability density function and (b) cumulative distribution function of a Laplace random variable, $b = 2$.

3.4.4 Gamma Random Variable

A random variable that follows a gamma distribution has a PDF and CDF given (for any $b > 0$ and any $c > 0$) by

$$f_X(x) = \frac{(x/b)^{c-1} \exp(-x/b)}{b\Gamma(c)} u(x),$$ (3.21a)

$$F_X(x) = \frac{\gamma(c, x/b)}{\Gamma(c)} u(x).$$ (3.21b)

In these two equations, the gamma function is a generalization of the factorial function defined by

$$\Gamma(\alpha) = \int_0^\infty e^{-t} t^{\alpha-1} dt,$$ (3.22)

and the incomplete gamma function is given by

$$\gamma(\alpha, \beta) = \int_0^\beta e^{-t} t^{\alpha-1} dt.$$ (3.23)

The gamma random variable is used in queueing theory and has several other random variables as special cases. If the parameter c is an integer, the resulting random variable is also known as an Erlang random variable, whereas, if $b = 2$ and c is a half integer, a chi-squared (χ^2) random variable results. Finally, if $c = 1$, the gamma random variable reduces to an exponential random variable.

3.4.5 Erlang Random Variable

As we have mentioned, the Erlang random variable is a special case of the gamma random variable. The PDF and CDF are given (for positive integer n and any $b > 0$) by

$$f_X(x) = \frac{(x/b)^{n-1} \exp(-x/b)}{b(n-1)!} u(x),$$ (3.24)

$$F_X(x) = \left[1 - \exp\left(-\frac{x}{b}\right) \sum_{m=0}^{n-1} \frac{(x/b)^m}{m!}\right] u(x).$$ (3.25)

The Erlang distribution plays a fundamental role in the study of wireline telecommunication networks. In fact, this random variable plays such an important role in the analysis of trunked telephone systems that the amount of traffic on a telephone line is measured in Erlangs.

3.4.6 Chi-Squared Random Variable

Another special case of the gamma random variable, the chi-squared (χ^2) random variable has a PDF and CDF given (for positive integer or half-integer values of c) by

$$f_X(x) = \frac{x^{c-1}\exp(-x/2)}{2^c \Gamma(c)} u(x), \tag{3.26}$$

$$F_X(x) = \frac{\gamma(c, x/2)}{\Gamma(c)} u(x). \tag{3.27}$$

Many engineering students are probably familiar with the χ^2 random variable from previous studies of statistics. It also commonly appears in various detection problems.

3.4.7 Rayleigh Random Variable

A Rayleigh random variable, like the exponential random variable, has a one-sided PDF. The functional form of the PDF and CDF is given (for any $\sigma > 0$) by

$$f_X(x) = \frac{x}{\sigma^2}\exp\left(-\frac{x^2}{2\sigma^2}\right)u(x), \tag{3.28a}$$

$$F_X(x) = \left(1 - \exp\left(-\frac{x^2}{2\sigma^2}\right)\right)u(x). \tag{3.28b}$$

Plots of these functions are shown in Figure 3.11. The Rayleigh distribution is described by a single parameter, σ^2, which is related to the width of the Rayleigh PDF. In this case, the parameter σ^2 is not to be interpreted as the variance of the Rayleigh random variable. It will be shown later that the Rayleigh distribution arises when studying the magnitude of a complex number whose real and imaginary parts both follow a zero-mean Gaussian distribution. The Rayleigh distribution arises often in the study of noncoherent communication systems and also in the study of wireless communication channels, where the phenomenon known as fading is often modeled using Rayleigh random variables.

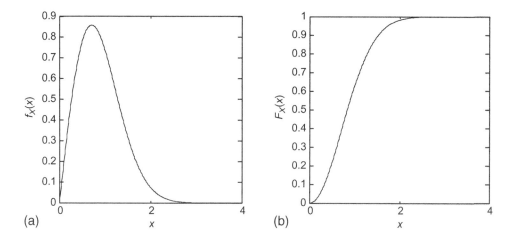

Figure 3.11
(a) Probability density function and (b) cumulative distribution function of a Rayleigh random variable, $\sigma^2 = 1/2$.

3.4.8 Rician Random Variable

A Rician random variable is closely related to the Rayleigh random variable (in fact the Rayleigh distribution is a special case of the Rician distribution). The functional form of the PDF for a Rician random variable is given (for any $a > 0$ and any $\sigma > 0$) by

$$f_X(x) = \frac{x}{\sigma^2} \exp\left(-\frac{x^2 + a^2}{2\sigma^2}\right) I_0\left(\frac{ax}{\sigma^2}\right) u(x). \tag{3.29}$$

In this expression, the function $I_0(x)$ is the modified Bessel function of the first kind of order zero which is defined by

$$I_0(x) = \frac{1}{2\pi}\int_0^{2\pi} e^{x\cos(\theta)} d\theta . \tag{3.30}$$

Like the Gaussian random variable, the CDF of a Rician random variable cannot be written in closed form. Similar to the Q-function that is used to describe the Gaussian CDF, there is another function known as *Marcum's Q-function* which describes the CDF of a Rician random variable. It is defined by

$$Q(\alpha, \beta) = \int_\beta^\infty z\exp\left(-\frac{(z^2 + \alpha^2)}{2}\right) I_0(\alpha z) dz . \tag{3.31}$$

The CDF of the Rician random variable is then given by:

$$F_X(x) = 1 - Q\left(\frac{a}{\sigma}, \frac{x}{\sigma}\right). \tag{3.32}$$

Tables of the Marcum Q-function can be found as well as efficient numerical routines for calculating it. A plot of the Rician PDF is shown in Figure 3.12. The Rician distribution is described by two parameters, a and σ^2, which are related to the center and width, respectively, of the PDF. As with the Rayleigh random variable, the parameter σ^2 is not to be interpreted as the variance of the Rician random variable. The Rician distribution arises in the study of noncoherent communication systems and also in the study of satellite communication channels, where fading is modeled using Rician random variables.

3.4.9 Cauchy Random Variable

The Cauchy random variable has a PDF and CDF given (for any a and any $b > 0$) by

$$f_X(x) = \frac{b/\pi}{b^2 + (x-a)^2},$$ (3.33)

$$F_X(x) = \frac{1}{2} + \frac{1}{\pi}\tan^{-1}\left(\frac{x-a}{b}\right).$$ (3.34)

The Cauchy random variable occurs when observing the tangent of a random variable which is uniformly distributed over $[0, 2\pi)$. The PDF is centered around $x = a$ and its width is determined by the parameter b. Unlike most of the other random variables where the PDFs decrease exponentially in the tails, the Cauchy PDF decays quadratically as $|x-a|$ increases. Hence, there is a greater amount of probability in the tails of the Cauchy PDF than in many of the other commonly used random variables. We say that this type of distribution is "heavy-tailed."

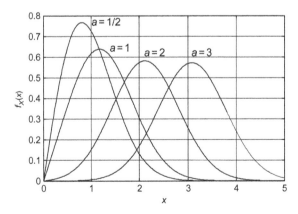

Figure 3.12

PDF of a Rician random variable, $\sigma^2 = 1/2$, $a = 1/2, 1, 2, 3$.

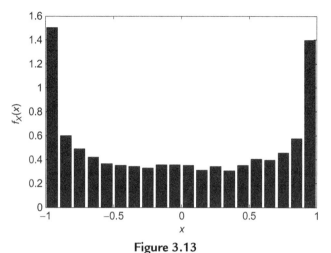

Figure 3.13

Histogram from Example 3.10, sine of a uniform phase.

■ Example 3.9:

One can construct many new types of random variables by making functions of random variables. In this example, we construct a random variable which is the sine of a uniform random phase. That is, we construct a random variable Θ which is uniformly distributed over $[0, 2\pi)$ and then form a new random variable according to $X = \sin(\Theta)$. In the next chapter, we will develop the tools to analytically figure out what the distribution of X should be, but for now we will simply observe its PDF by plotting a histogram. The MATLAB code below was used to accomplish this and the result is illustrated in Figure 3.13.

```
N=10000;
Theta=2*pi*rand(1,N);                  % Uniform phase
X=sin(Theta);                          % sinusoidal transformation
bins=[-0.95:0.1:0.95];                 % histogram bins
[yvalues,xvalues]=hist(X,bins);        % histogram values
pdf_estimate=yvalues/(N*0.1);          % normalize prob. densities
bar(xvalues,pdf_estimate)              % plot PDF histogram
xlabel('x'); ylabel('f_X(x)')          % label plot
```
■

3.5 Conditional Distribution and Density Functions

In Chapter 2, we defined the notion of conditional probability. In a similar manner, it is quite common to talk about the distribution or density of a random variable conditioned on some event, A. As with the initial study of random variables in the beginning of this chapter, it is convenient to start with the notion of a conditional CDF.

Definition 3.4: The *conditional CDF* of a random variable, X, conditioned on the event A having occurred is

$$F_{X|A}(x) = \Pr(X \le x | A) = \frac{\Pr(\{X \le x\}, A)}{\Pr(A)}. \tag{3.35}$$

Naturally, this definition requires the caveat that the probability of the event A must not be zero.

The properties of CDFs listed in Equations (3.4a)–(3.4d) also apply to conditional CDFs, resulting in the following properties of conditional CDFs:

(1) $F_{X|A}(-\infty) = 0, F_{X|A}(\infty) = 1,$ $\hspace{5cm}$ (3.36a)

(2) $0 \le F_{X|A}(x) \le 1,$ $\hspace{6cm}$ (3.36b)

(3)   For $x_1 < x_2$, $F_{X|A}(x_1) \le F_{X|A}(x_2),$ $\hspace{3.5cm}$ (3.36c)

(4)   For $x_1 < x_2$, $\Pr(x_1 < X \le x_2 | A) = F_{X|A}(x_2) - F_{X|A}(x_1).$ $\hspace{1cm}$ (3.36d)

It is left as an exercise for the reader (see Exercise 3.13) to prove that these properties of CDFs do indeed apply to conditional CDFs as well.

■ Example 3.10:

Suppose a random variable X is uniformly distributed over the interval $[0, 1)$ so that its CDF is given by

$$F_X(x) = \begin{cases} 0, & x < 0, \\ x, & 0 \le x \le 1, \\ 1, & x > 1. \end{cases}$$

Suppose further that we want to find the conditional CDF of X given that $X < 1/2$. Here, the event $A = \{X < 1/2\}$ is related to a numerical condition on the random variable itself. From the definition of a conditional CDF,

$$F_{X|\{X < 1/2\}}(x) = \frac{\Pr(X \le x, X < 1/2)}{\Pr(X < 1/2)}.$$

For $x < 0$, the event $X \le x$ has probability zero and hence $F_{X|\{X < 1/2\}}(x) = 0$ for $x < 0$. When $0 \le x \le 1/2$, the intersection of the events $X \le x$ and $X < 1/2$ is simply the event $X \le x$, so that

$$F_{X|\{X < 1/2\}}(x) = \frac{\Pr(X \le x)}{\Pr(X < 1/2)} = \frac{x}{1/2} = 2x \text{ for } 0 \le x \le 1/2.$$

Finally, for $x > 1/2$, the intersection of the events $X \leq x$ and $X < 1/2$ is simply the event $X < 1/2$ and the conditional CDF reduces to one. Putting this all together, the desired conditional CDF is

$$F_{X|\{X < 1/2\}}(x) = \begin{cases} 0, & x < 0, \\ 2x, & 0 \leq x \leq 1/2, \\ 1, & x > 1/2. \end{cases}$$

∎

In order to generalize the result of the previous example, suppose that for some arbitrary random variable X, the conditioning event is of the form $A = a < X \leq b$ for some constants $a < b$. Then

$$F_{X|\{a < X \leq b\}}(x) = \frac{\Pr(X \leq x, a < X \leq b)}{\Pr(a < X \leq b)}. \tag{3.37}$$

If $x \leq a$, then the events $\{X \leq x\}$ and $a < X \leq b$ are mutually exclusive and the conditional CDF is zero. For $x > b$ the event $a < X \leq b$ is a subset of $\{X \leq x\}$ and hence $\Pr(X \leq x, a < X \leq b) = \Pr(a < X \leq b)$ so that the conditional CDF is 1. When $a < x \leq b$, then $\{X \leq x\} \cap \{a < X \leq b\} = a < X \leq x$ and $\Pr(X \leq x, a < X \leq b) = \Pr(a < X \leq x)$. This can be written in terms of the CDF (unconditional) of X as $\Pr(a < X \leq x) = F_X(x) - F_X(a)$. Likewise, $\Pr(a < X \leq b) = F_X(b) - F_X(a)$. Putting these results together gives

$$F_{X|\{a \leq X < b\}}(x) = \begin{cases} 0, & x \leq a, \\ \dfrac{F_X(x) - F_X(a)}{F_X(b) - F_X(a)}, & a < x \leq b, \\ 1, & x > b. \end{cases} \tag{3.38}$$

This result could also be extended to conditioning events where X is conditioned on being in more extravagant regions.

As with regular random variables, it is often more convenient to work with a conditional PDF rather than a conditional CDF. The definition of the conditional PDF is a straightforward extension of the previous definition given for a PDF.

Definition 3.5: The *conditional PDF* of a random variable X conditioned on some event A is

$$f_{X|A}(x) = \lim_{\varepsilon \to 0} \frac{\Pr(x \leq X < x + \varepsilon | A)}{\varepsilon}. \tag{3.39}$$

As with the conditional CDF, it is not difficult to show that all of the properties of regular PDFs apply to conditional PDFs as well. In particular,

(1) $f_{X|A}(x) \geq 0$,   (3.40a)

(2) $f_{X|A}(x) = \dfrac{dF_{X|A}(x)}{dx}$,   (3.40b)

(3) $F_{X|A}(x) = \displaystyle\int_{-\infty}^{x} f_{X|A}(y)dy.$ (3.40c)

(4) $\displaystyle\int_{-\infty}^{\infty} f_{X|A}(x)dx = 1,$ (3.40d)

(5) $\displaystyle\int_{a}^{b} f_{X|A}(x)dx = \Pr(a < X \leq b|A).$ (3.40e)

Furthermore, the result in Equation (3.38) can be extended to the conditional PDF by applying Equation (3.40b). This results in the following general formula for the conditional PDF of a random variable, X, when the conditioning event is of the nature $A = \{a \leq X < b\}$:

$$f_{X|\{a \leq X < b\}}(x) = \begin{cases} \dfrac{f_X(x)}{\Pr(a \leq X < b)}, & a \leq x < b, \\ 0, & \text{otherwise.} \end{cases}$$ (3.41)

To summarize, the conditional PDF takes on the same functional form (but is scaled by the probability of the conditioning event) over the range of x where the condition is satisfied, and the conditional PDF is zero wherever the conditioning event is not true. This result is illustrated in Figure 3.14.

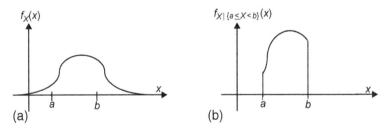

Figure 3.14
(a) A PDF and (b) the corresponding conditional PDF .

■ Example 3.11:

Let X be a random variable representing the length of time we spend waiting in the grocery store checkout line. Suppose the random variable X has an exponential PDF given by $f_X(x) = (1/c)\exp(-x/c)u(x)$, where $c = 3$ min. What is the PDF for the amount of time we spend waiting in line given that we have already been waiting for 2 min? Here the conditioning event is of the form $X > 2$. We can use the result in Equation (3.41) by taking $a = 2$ and $b = \infty$. The probability of the conditioning event is $\Pr(X > 2) = 1 - F_X(2) = \exp(-2/3)$. Therefore, the conditional PDF is

$$f_{X|\{X>2\}}(x) = \exp(2/3)f_X(x)u(x-2) = \frac{1}{3}\exp\left(-\frac{x-2}{3}\right)u(x-2).$$

It is curious to note that for this example, $f_{X|\{X>2\}}(x) = f_X(x-2)$. That is, given that we have been waiting in line for 2 min, the PDF of the total time we must wait in line is simply shifted by 2 min. This interesting behavior is unique to the exponential distribution and we might not have seen the same result if we had started with a different distribution. For example, try working the same problem starting with a Rayleigh distribution.

■

Up to this point, we have primarily looked at conditioning events that impose a numerical constraint. It is also common to consider conditioning events of a qualitative, or nonnumerical, nature. Consider, for example, a random variable X which represents a student's score on a certain standardized test (e.g., the SAT or GRE test). We might be interested in determining if there is any gender bias in the test. To do so, we could compare the distribution of the variable X given that the student is female, $F_{X|F}(x)$, with the distribution of the same variable given that the student is male, $F_{X|M}(x)$. If these distributions are substantially different, then we might conclude a gender bias and work to fix the bias in the exam. Naturally, we could work with conditional PDFs $f_{X|F}(x)$ and $f_{X|M}(x)$ as well. Here, the conditioning event is a characteristic of the experiment that may affect the outcome rather than a restriction on the outcome itself.

In general, consider a set of mutually exclusive and exhaustive conditioning events, $A_1, A_2, ..., A_N$. Suppose we had access to the conditional CDFs, $F_{X|A_n}(x)$, $n = 1, 2, ..., N$, and wanted to find the unconditional CDF, $F_X(x)$. According to the Theorem of Total Probability (Theorem 2.10),

$$F_X(x) = \Pr(X \le x) = \sum_{n=1}^{N} \Pr(X \le x | A_n)\Pr(A_n) = \sum_{n=1}^{N} F_{X|A_n}(x)\Pr(A_n). \qquad (3.42)$$

Hence, the CDF of X (unconditional) can be found by forming a weighted sum of the conditional CDFs with the weights determined by the probabilities that each of the conditioning events is true. By taking derivatives of both sides of the previous equation, a similar result is obtained for conditional PDFs, namely

$$f_X(x) = \sum_{n=1}^{N} f_{X|A_n}(x)\Pr(A_n). \tag{3.43}$$

We might also be interested in looking at things in the reverse direction. That is, suppose we observe that the random variable has taken on a value of $X = x$. Does the probability of the event A_n change? To answer this we need to compute $\Pr(A_n|(X = x))$. If X were a discrete random variable, we could do this by invoking the results of Theorem 2.5

$$\Pr(A_n| X = x) = \frac{\Pr(X = x|A_n)\Pr(A_n)}{\Pr(X = x)}. \tag{3.44}$$

In the case of continuous random variables, greater care must be taken since both $\Pr(X = x|A_n)$ and $\Pr(X = x)$ will be zero, resulting in an indeterminate expression. To avoid that problem, rewrite the event $\{X = x\}$ as $\{x \le X < x + \varepsilon\}$ and consider the result in the limit as $\varepsilon \to 0$:

$$\Pr(A_n|x \le X < x + \varepsilon) = \frac{\Pr(x \le X < x + \varepsilon|A_n)\Pr(A_n)}{\Pr(x \le X < x + \varepsilon)}. \tag{3.45}$$

Note that for infinitesimal ε, $\Pr(x \le X < x + \varepsilon) = f_X(x)\varepsilon$ and similarly $\Pr(x \le X < x + \varepsilon|A_n) = f_{X|A_n}(x)\varepsilon$. Hence,

$$\Pr(A_n|x \le X < x + \varepsilon) = \frac{f_{X|A_n}(x)\varepsilon\Pr(A_n)}{f_X(x)\varepsilon} = \frac{f_{X|A_n}(x)\Pr(A_n)}{f_X(x)}. \tag{3.46}$$

Finally, passing to the limit as $\varepsilon \to 0$ gives the desired result:

$$Pr(A_n| X = x) = \lim_{\varepsilon \to 0} \Pr(A_n|x \le X < x + \varepsilon) = \frac{f_{X|A_n}(x)\Pr(A_n)}{f_X(x)}. \tag{3.47}$$

We could also combine this result with Equation (3.43) to produce an extension to Bayes's theorem:

$$\Pr(A_n|X = x) = \frac{f_{X|A_n}(x)\Pr(A_n)}{\displaystyle\sum_{n=1}^{N} f_{X|A_n}(x)\Pr(A_n)}. \tag{3.48}$$

■ Example 3.12:

In a certain junior swimming competition, swimmers are placed into one of two categories based on their previous times so that all children can compete against others of their own abilities. The fastest swimmers are placed in the A category, while the slower swimmers are put in the B group. Let X be a random variable representing a child's time (in seconds) in the 50-m freestyle race. Suppose that it is determined that for those swimmers in group A, the PDF of a child's time is given by $f_{X|A}(x) = (4\pi)^{-1/2}\exp(-(x-40)^2/4)$ while for those in the B group the PDF is given by $f_{X|B}(x) = (4\pi)^{-1/2}\exp(-(x-45)^2/4)$. Furthermore, assume that 30% of the swimmers are in the A group and 70% are in the B group. If a child swims the race with a time of 42 s, what is the probability that the child was in the B group? Applying Equation (3.48) we get

$$\Pr(B|X = 42) = \frac{0.7 f_{X|B}(42)}{0.3 f_{X|A}(42) + 0.7 f_{X|B}(42)} = \frac{0.7\exp(-9/4)}{0.3\exp(-1) + 0.7\exp(-9/4)} = 0.4007 \, .$$

Naturally, the probability of the child being from group A must then be

$\Pr(A|X = 42) = 1 - \Pr(B|X = 42) = 0.5993.$

■

3.6 Engineering Application: Reliability and Failure Rates

The concepts of random variables presented in this chapter are used extensively in the study of system reliability. Consider an electronic component that is to be assembled with other components as part of a larger system. Given a probabilistic description of the lifetime of such a component, what can we say about the lifetime of the system itself. The concepts of reliability and failure rates are introduced in this section to provide tools to answer such questions.

> **Definition 3.6:** Let X be a random variable which represents the lifetime of a device. That is, if the device is turned on at time zero, X would represent the time at which the device fails. The *reliability function* of the device, $R_X(t)$, is simply the probability that the device is still functioning at time t:

$$R_X(t) = \Pr(X > t) \, . \tag{3.49}$$

Note that the reliability function is just the complement of the CDF of the random variable. That is, $R_X(t) = 1 - F_X(t)$. As it is often more convenient to work with PDFs rather than CDFs, we note that the derivative of the reliability function can be related to the PDF of the random variable X by $R_X'(t) = -f_X(t)$.

With many devices, the reliability changes as a function of how long the device has been functioning. Suppose we observe that a particular device is still functioning at some point in time t. The remaining lifetime of the device may behave (in a probabilistic sense) very differently from when it was first turned on. The concept of failure rate is used to quantify this effect.

Definition 3.7: Let X be a random variable which represents the lifetime of a device. The *failure rate function* is

$$r(t) = f_{X|\{X>t\}}(x)\Big|_{x=t}. \tag{3.50}$$

To give this quantity some physical meaning, we note that $\Pr(t < X < t + dt | X > t) = r(t)dt$. Thus, $r(t)dt$ is the probability that the device will fail in the next time instant of length dt, given the device has survived up to now (time t). Different types of "devices" have failure rates that behave in different manners. Our pet goldfish, Elvis, might have an increasing failure rate function (as do most biological creatures). That is, the chances of Elvis "going belly up" in the next week is greater when Elvis is 6 months old than when he is just 1 month old. We could also imagine devices that have a decreasing failure rate function (at least for part of their lifetime). For example, an integrated circuit might be classified into one of two types, those which are fabricated correctly and hence are expected to have a quite long lifetime and those with defects which will generally fail fairly quickly. When we select an IC, we may not know which type it is. Once the device lives beyond that initial period when the defective ICs tend to fail, the failure rate may go down (at least for awhile). Finally, there may be some devices whose failure rates remain constant with time.

The failure rate of a device can be related to its reliability function. From Equation (3.41) it is noted that

$$f_{X|\{X>t\}}(x) = \frac{f_X(x)u(x-t)}{1 - F_x(t)}. \tag{3.51}$$

The denominator in the above expression is the reliability function, $R_X(t)$, while the PDF in the numerator is simply $-R_X'(x)$. Evaluating at $x = t$ produces the failure rate function

$$r(t) = \frac{-R_X'(t)}{R_X(t)}. \tag{3.52}$$

Conversely, given a failure rate function, $r(t)$, one can solve for the reliability function by solving the first-order differential equation:

$$\frac{d}{dt}R_x(t) = -r(t)R_X(t). \tag{3.53}$$

The general solution to this differential equation (subject to the initial condition $R_X(0) = 1$) is

$$R_X(t) = \exp\left[-\int_0^t r(u)\mathrm{d}u\right]u(t). \tag{3.54}$$

It is interesting to note that a failure rate function completely specifies the PDF of a device's lifetime:

$$f_X(t) = -R_X'(t) = r(t)\exp\left(-\int_0^t r(u)\mathrm{d}u\right)u(t). \tag{3.55}$$

For example, suppose a device had a constant failure rate function, $r(t) = \lambda$. The PDF of the device's lifetime would then follow an exponential distribution, $f_X(t) = \lambda\exp(-\lambda t)u(t)$. The corresponding reliability function would also be exponential, $R_X(t) = \exp(-\lambda t)u(t)$. We say that the exponential random variable has the memoryless property. That is, it does not matter how long the device has been functioning, the failure rate remains the same.

■ Example 3.13:

Suppose the lifetime of a certain device follows a Rayleigh distribution given by $f_X(t) = 2bt\exp(-bt^2)u(t)$. What are the reliability function and the failure rate function? The reliability function is given by

$$R_X(t) = \Pr(X > t) = \left[\int_t^\alpha 2bu\exp(-bu^2)\mathrm{d}u\right]u(t) = \exp(-bt^2)u(t).$$

A straightforward application of Equation (3.52) produces the failure rate function, $r(t) = 2btu(t)$. In this case, the failure rate is linearly increasing in time.

■

Next, suppose we have a system which consists of N components, each of which has a lifetime described by the random variable X_n, $n = 1, 2, ..., N$. Furthermore, assume that for the system to function, all N components must each be functioning. In other words, if any of the individual components fail, the whole system fails. This is usually referred to as a series connection of components. If we can characterize the reliability and failure rate functions of each individual component, can we calculate the same functions for the entire system? The answer is yes, under some mild assumptions. Define X to be the random variable representing the lifetime of the system. Then

$$X = \min(X_1, X_2, ..., X_N). \tag{3.56}$$

Furthermore,

$$R_X(t) = \Pr(X > t)$$

$$= \Pr(\{X_1 > t\} \cap \{X_2 > t\} \cap ... \cap \{X_N > t\}). \tag{3.57}$$

We assume that all of the components fail independently. That is, the event $\{X_i > t\}$ is taken to be independent of $\{X_j > t\}$ for all $i \neq j$. Under this assumption,

$$R_X(t) = \Pr(X_1 > t)\Pr(X_2 > t)\ldots\Pr(X_N > t)$$

$$= R_{X_1}(t)R_{X_2}(t)\ldots R_{X_N}(t). \tag{3.58}$$

Furthermore, application of Equation (3.52) provides an expression for the failure rate function:

$$r(t) = \frac{-R_X'(t)}{R_X(t)} = -\frac{\frac{d}{dt}[R_{X_1}(t)R_{X_2}(t)\ldots R_{X_N}(t)]}{R_{X_1}(t)R_{X_2}(t)\ldots R_{X_N}(t)} \tag{3.59}$$

$$= -\sum_{n=1}^{N} \frac{R_{X_n}'(t)}{R_{X_n}(t)} = \sum_{n=1}^{N} r_n(t), \tag{3.60}$$

where $r_n(t)$ is the failure rate function of the nth component. We have shown that for a series connection of components, the reliability function of the system is the product of the reliability functions of each component and the failure rate function of the system is the sum of the failure rate functions of the individual components.

We may also consider a system which consists of a parallel interconnection of components. That is, the system will be functional as long as any of the components are functional. We can follow a similar derivation to compute the reliability and failure rate functions for the parallel interconnection system. First, the reliability function is written as

$$R_X(t) = \Pr(\{X_1 > t\} \cup \{X_2 > t\} \cup \ldots \cup \{X_N > t\}). \tag{3.61}$$

In this case, it is easier to work with the complement of the reliability function (the CDF of the lifetime). Since the reliability function represents the probability that the system is still functioning at time t, the complement of the reliability function represents the probability that the system is not working at time t. With the parallel interconnections, the system will fail only if all the individual components fail. Hence,

$$1 - R_X(t) = \Pr(X \leq t) = \Pr(\{X_1 \leq t\} \cap \{X_2 \leq t\} \cap \ldots \cap \{X_N \leq t\})$$

$$= \Pr(X_1 \leq t)\Pr(X_2 \leq t)\ldots\Pr(X_N \leq t)$$

$$= (1 - R_{X_1}(t))(1 - R_{X_2}(t))\ldots(1 - R_{X_N}(t)). \tag{3.62}$$

As a result, the reliability function of the parallel interconnection system is given by

$$R_X(t) = 1 - \prod_{n=1}^{N} (1 - R_{X_n}(t)) \ . \tag{3.63}$$

Unfortunately, the general formula for the failure rate function is not as simple as in the serial interconnection case. Application of Equation (3.52) to our preceding equation gives (after some straightforward manipulations)

$$r(t) = -\frac{1 - R_X(t)}{R_x(t)} \sum_{n=1}^{N} \frac{R_{X_n}'(t)}{1 - R_{X_n}(t)} , \tag{3.64}$$

or, equivalently,

$$r(t) = \left[\frac{1}{R_X(t)} - 1\right] \sum_{n=1}^{N} \frac{r_n(t)}{\frac{1}{R_{X_n}(t)} - 1} \ . \tag{3.65}$$

■ Example 3.14:

Suppose a system consists of N components each with a constant failure rate, $r_n(t) = \lambda_n$, $n = 1, 2, ..., N$. Find the reliability and failure rate functions for a series interconnection. Then find the same functions for a parallel interconnection. It was shown previously that a constant failure rate function corresponds to an exponential reliability function. That is, $R_{X_n}(t) = \exp(-\lambda_n t)u(t)$. For the serial interconnection we then have

$$R_X(t) = \prod_{n=1}^{N} R_{X_n}(t) = \prod_{n=1}^{N} \exp(-\lambda_n t)u(t) = \exp\left(-\left[\sum_{n=1}^{N} \lambda_n\right]t\right)u(t),$$

$$r(t) = \sum_{n=1}^{N} r_n(t) = \sum_{n=1}^{N} \lambda_n \ .$$

For the parallel interconnection,

$$R_X(t) = \left\{1 - \prod_{n=1}^{N} [1 - \exp(-\lambda_n t)]\right\}u(t) ,$$

$$r(t) = \frac{\prod_{n=1}^{N} [1 - \exp(-\lambda_n t)]}{1 - \prod_{n=1}^{N} [1 - \exp(-\lambda_n t)]} \sum_{n=1}^{N} \frac{\lambda_n}{\exp(\lambda_n t) - 1} \ .$$

■

Section 3.1: The Cumulative Distribution Function

3.1 Which of the following mathematical functions could be the CDF of some random variable?

(a) $F_X(x) = \begin{cases} 1-|x|, & |x| \le 1, \\ 0, & |x| < 1. \end{cases}$

(c) $F_X(x) = \begin{cases} \frac{1}{2}e^x, & x < 0, \\ 1 - \frac{1}{2}e^{-x}, & x \ge 0. \end{cases}$

(b) $F_X(x) = \begin{cases} 0, & x \le 0, \\ x^2, & 0 < x \le 1, \\ 1, & x > 1. \end{cases}$

(d) $F_X(x) = \log(x)u(x)$.

3.2 Suppose a random variable has a CDF given by

$$F_X(x) = \frac{x^2}{1+x^2}u(x).$$

Find the following quantities:
(a) $\Pr(X < 2)$,
(b) $\Pr(X > 4)$,
(c) $\Pr(1 < X < 3)$,
(d) $\Pr(X > 2 | X < 4)$.

3.3 Repeat Exercise 3.2 for the case where the random variable has the CDF

$$F_X(x) = \left(\frac{1}{2} + \frac{1}{\pi}\tan^{-1}(x)\right)u(x).$$

3.4 Suppose we flip a balanced coin five times and let the random variable X represent the number of times heads occurs.

(a) Sketch the CDF of X, $F_X(x)$.

(b) Write $F_X(x)$ analytically in terms of unit step functions.

3.5 A random variable is equally likely to take on any integer value in the set $\{0, 1, 2, 3\}$.

(a) Sketch the CDF of X, $F_X(x)$.

(b) Write $F_X(x)$ analytically in terms of unit step functions.

3.6 A certain discrete random variable has a CDF given by

$$F_X(x) = \sum_{k=0}^{\infty} \left(\frac{1}{2}\right)^{k+1} u(x-k).$$

(a) Find the probability mass function, $P_X(k)$, of this random variable.

(b) For a positive integer n, find $\Pr(X \geq n)$.

(c) For two positive integers, n_1 and n_2, such that $n_1 < n_2$, find $\Pr(n_1 \leq X < n_2)$.

3.7 In this problem, we generalize the results of Exercise 3.6. Suppose a discrete random variable takes on nonnegative integer values and has a CDF of the general form

$$F_X(x) = \sum_{k=0}^{\infty} a_k u(x-k).$$

(a) What conditions must the sequence a_k satisfy for this to be a valid CDF.

(b) For a positive integer n, find $\Pr(X \leq n)$ in terms of the a_k.

3.8 A random variable has a CDF given by $F_X(x) = (1 - e^{-x})u(x)$.

(a) Find $\Pr(X > 3)$.

(b) Find $\Pr(X < 5 | X > 3)$.

(c) Find $\Pr(X > 6 | X > 3)$.

(d) Find $\Pr(|X - 5| < 4 \,||X - 6| > 2)$.

3.9   A random variable has a CDF given by

$$
F_X(x) = \begin{cases} 0, & x < 0, \\ \dfrac{1+x}{2}, & 0 \le x \le 1, \\ 1, & x > 1. \end{cases}
$$

(a)   Find $\Pr(X=0)$ and $\Pr(X=1)$.

(b)   Find $\Pr(X < 0)$ and $\Pr\left(X > \dfrac{1}{2}\right)$.

(c)   Find $\Pr\left(X > \dfrac{1}{2} \middle| X > 0\right)$.

Section 3.2: The Probability Density Function

3.10 Suppose a random variable is equally likely to fall anywhere in the interval $[a, b]$. Then the PDF is of the form

$$
f_X(x) = \begin{cases} \dfrac{1}{b-a}, & a \le x \le b, \\ 0, & \text{otherwise.} \end{cases}
$$

Find and sketch the corresponding CDF.

3.11 Find and plot the CDFs corresponding to each of the following PDFs:

(a) $f_X(x) = \begin{cases} 1, & 0 \le x < 1, \\ 0, & \text{otherwise.} \end{cases}$

(b) $f_X(x) = \begin{cases} x, & 0 \le x < 1, \\ 2 - x, & 1 \le x < 2, \\ 0, & \text{otherwise.} \end{cases}$

3.12 A random variable has the following exponential PDF:

$$f_X(x) = \begin{cases} a^{-bx}, & x \le 0, \\ 0, & \text{otherwise}, \end{cases}$$

where a and b are constants.

(a) Determine the required relationship between a and b.

(b) Determine the corresponding CDF.

3.13 A certain random variable has a probability density function of the form $f_X(x) = ce^{-2x}u(x)$. Find the following:

(a) the constant c,

(b) $Pr(X > 2)$,

(c) $Pr(X < 3)$,

(d) $Pr(X < 3 | X > 2)$.

3.14 Repeat Exercise 3.13 using the PDF $f_X(x) = \dfrac{c}{x^2 + 4}$.

3.15 Repeat Exercise 3.13 using the PDF $f_X(x) = \dfrac{c}{\sqrt{25 - x^2}}$, $-5 < x < 5$.

3.16 The voltage of communication signal S is measured. However, the measurement procedure is corrupted by noise resulting in a random measurement with the PDF shown in the accompanying diagram. Find the probability that for any particular measurement, the error will exceed $\pm 0.75\%$ of the correct value if this correct value is 10 V.

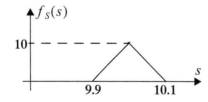

3.17 Which of the following mathematical functions could be the PDF of some random variable?

(a) $f_X(x) = \begin{cases} 1 - |x|, & |x| \leq 1, \\ 0, & |x| > 1. \end{cases}$

(c) $f_X(x) = x^3 \exp(-x^4)u(x).$

(b) $f_X(x) = \begin{cases} 1 - x^2, & |x| \leq 1, \\ 0, & |x| > 1. \end{cases}$

(d) $f_X(x) = \begin{cases} 4x^3, & |x| \leq 1, \\ 0, & |x| > 1. \end{cases}$

3.18 Find the value of the constant c that makes each of the following functions a properly normalized PDF.

(a) $f_X(x) = \begin{cases} c\left(1 - \left(\dfrac{x}{2}\right)^2\right), & |x| \leq 2, \\ 0, & |x| > 2. \end{cases}$

(b) $f_X(x) = \begin{cases} c\cos\left(\dfrac{\pi x}{2}\right), & |x| \leq 1, \\ 0, & |x| > 1. \end{cases}$

(c) $f_X(x) = cx^n \exp(-x^2)u(x)$ (where n is an odd integer).

(d) $f_X(x) = \dfrac{c}{(1 + x^2)^2}.$

Section 3.3: The Gaussian Random Variable

3.19 Prove the integral identity, $I = \displaystyle\int_{-\infty}^{\infty} \exp\left(-\dfrac{x^2}{2}\right)dx = \sqrt{2\pi}$. *Hint:* It may be easier to show that $I^2 = 2\pi$.

3.20 Using the normalization integral for a Gaussian random variable, find an analytical expression for the following integral:

$$ I = \int_{-\infty}^{\infty} \exp(-(ax^2 + bx + c))dx, $$

where $a > 0$, b, and c are constants.

3.21 A Gaussian random variable has a probability density function of the form
$$f_X(x) = c\exp(-(2x^2 + 3x + 1)) \ .$$
(a)   Find the value of the constant c.

(b)   Find the values of the parameters m and σ for this Gaussian random variable.

3.22 A Gaussian random variable has a PDF of the form

$$f_X(x) = \frac{1}{\sqrt{50\pi}}\exp\left(-\frac{(x-10)^2}{50}\right) \ .$$

Write each of the following probabilities in terms of Q-functions (with positive arguments) and also give numerical evaluations:

(a) $\Pr(X > 17)$,   (b) $\Pr(X > 4)$,

(c) $\Pr(X < 15)$,   (d) $\Pr(X < -2)$,

(e) $\Pr(|X - 10| > 7)$,   (f) $\Pr(|X - 10| < 3)$,

(g) $\Pr(|X - 7| > 5)$,   (h) $\Pr(|X - 4| < 7)$.

3.23 A Gaussian random variable has a PDF of the form

$$f_X(x) = \sqrt{\frac{2}{\pi}}\exp(-2(x+1)^2) \ .$$

Write each of the following probabilities in terms of Q-functions (with positive arguments) and also give numerical evaluations:

(a) $\Pr(X > 0)$,   (b) $\Pr(X > 2)$

(c) $\Pr(X < -3)$,   (d) $\Pr(X < -4)$,

(e) $\Pr(|X + 1| > 3)$,   (f) $\Pr(|X + 1| < 2)$

(g) $\Pr(|X + 2| > 1)$,   (h) $\Pr(|X - 1| < 2)$

3.24 Suppose we measure the noise in a resistor (with no applied voltage) and find that the noise voltage exceeds $10 \ \mu V$ 5% of the time. We have reason to believe the noise is well modeled as a Gaussian random variable, and furthermore, we expect the noise to be positive and negative equally often so we take the parameter, m, in the Gaussian PDF to be $m = 0$. Find the value of the parameter, σ^2. What units should be associated with σ^2 in this case.

3.25 Now suppose we modify Exercise 3.24 so that in addition to noise in the resistor there is also a weak (constant) signal present. Thus, when we measure the voltage across the resistor, we do not necessarily expect positive and negative measurements equally often, and therefore, we now allow the parameter m to be something other than zero. Suppose now we observe that the measure voltage exceeds 10 μV 40% of the time and is below -10 μV only 2% of the time. If we continue to model the voltage across the resistor as a Gaussian random variable, what are the appropriate values of the parameters m and σ^2. Give proper units for each.

3.26 The IQ of a randomly chosen individual is modeled using a Gaussian random variable. Given that 50% of the population have an IQ above 100 (and 50% have an IQ below 100) and that 50% of the population have an IQ in the range 90-100, what percentage of the population have an IQ of at least 140 and thereby are considered "genius?"

Section 3.4: Other Important Random Variables

3.27 The phase of a sinusoid, Θ, is uniformly distributed over $[0, 2\pi)$ so that its PDF is of the form

$$f_\Theta(\theta) = \begin{cases} \dfrac{1}{2\pi}, & 0 \le \theta < 2\pi, \\ 0, & \text{otherwise.} \end{cases}$$

(a) Find $\Pr\left(\theta > \dfrac{3\pi}{4}\right)$.

(b) Find $\Pr\left(\Theta < \pi | \Theta > \dfrac{3\pi}{4}\right)$.

(c) Find $\Pr\left(\cos(\Theta) < \dfrac{1}{2}\right)$.

3.28 Let X be an exponential random variable with PDF, $f_X(x) = e^{-x}u(x)$.

(a) Find $\Pr(3X < 5)$.

(b) Generalize your answer to part (a) to find $\Pr(3X < y)$ for some arbitrary constant y.

(c) Note that if we define a new random variable according to $Y = 3X$, then your answer to part (b) is the CDF of Y, $F_Y(y)$. Given your answer to part (b), find $f_Y(y)$.

3.29 Let W be a Laplace random variable with a PDF given by $f_W(w) = ce^{-2|w|}$.

    (a)   Find the value of the constant c.

    (b)   Find $\Pr(-1 < W < 2)$.

    (c)   Find $\Pr(W > 0 | -1 < W < 2)$.

3.30 Let Z be a random variable whose PDF is given by $f_Z(z) = cz^2 e^{-z} u(z)$.

    (a)   Find the value of the constant c.

    (b)   Find the form of the CDF of Z, $F_Z(z)$.

    (c)   Find $\Pr(\{Z < 2\} \cup (Z > 4))$.

3.31 Let R be a Rayleigh random variable whose PDF is $f_R(r) = cr\exp(-r^2)u(r)$.

    (a)   Find the value of the constant c.

    (b)   Find $\Pr(R > r)$ for an arbitrary constant r.

    (c)   Find $\Pr(R > 1 | R < 2)$.

3.32 A random variable, Y, has a PDF given by

$$f_Y(y) = \frac{c}{y^2 + ay + b}.$$

    (a)   List all constraints on the constants a, b, and c that must be satisfied for this to be a valid PDF.

    (b)   In terms of the constants a, b, and c, find $\Pr(Y > 0)$.

3.33 Prove the following properties of the Gamma function.

    (a) $\Gamma(n) = (n-1)!$ for $n = 1, 2, 3, \ldots$

    (b) $\Gamma(x + 1) = x\Gamma(x)$,

    (c) $\Gamma(1/2) = \sqrt{\pi}$.

Section 3.5: Conditional Distribution and Density Functions

3.34 Prove the following properties of conditional CDFs.

    (a) $F_{X|A}(-\infty) = 0$, $F_{X|A}(\infty) = 1$,

    (b) $0 \le F_{X|A}(x) \le 1$,

(c) For $x_1 < x_2$, $F_{X|A}(x_1) \le F_{X|A}(x_2)$,

(d) For $x_1 < x_2$, $\Pr(x_1 < X \le x_2 | A) = F_{X|A}(x_2) - F_{X|A}(x_1)$.

3.35 Let X be a Gaussian random variable such that $X \sim N(0, \sigma^2)$. Find and plot the following conditional PDFs.

(a) $f_{X|X>0}(x)$,

(b) $f_{X||X|<3}(x)$,

(c) $f_{X||X|>3}(x)$.

3.36 A digital communication system sends two messages, $M = 0$ or $M = 1$, with equal probability. A receiver observes a voltage which can be modeled as a Gaussian random variable, X, whose PDFs conditioned on the transmitted message are given by

$$f_{X|M=0}(x) = \frac{1}{\sqrt{2\pi\sigma^2}} \exp\left(-\frac{x^2}{2\sigma^2}\right) \text{ and } f_{X|M=1}(x) = \frac{1}{\sqrt{2\pi\sigma^2}} \exp\left(-\frac{(x-1)^2}{2\sigma^2}\right).$$

(a) Find and plot $\Pr(M = 0 | X = x)$ as a function of x for $\sigma^2 = 1$. Repeat for $\sigma^2 = 5$.

(b) Repeat part (a) assuming that the a priori probabilities are $\Pr(M = 0) = 1/4$ and $\Pr(M = 1) = 3/4$.

3.37 In Exercise 3.36, suppose our receiver must observe the random variable X and then make a decision as to what message was sent. Furthermore, suppose the receiver makes a three-level decision as follows:

Decide 0 was sent if $\Pr(M = 0 | X = x) \ge 0.9$,

Decide 1 was sent if $\Pr(M = 1 | X = x) \ge 0.9$,

Erase the symbol (decide not to decide) if both $\Pr(M = 0 | X = x) < 0.9$ and $\Pr(M = 1 | X = x) < 0.9$.

Assuming the two messages are equally probable, $\Pr(M = 0) = \Pr(M = 1) = 1/2$, and that $\sigma^2 = 1$, find

(a) the range of x over which each of the three decisions should be made,

(b) the probability that the receiver erases a symbol,

(c) the probability that the receiver makes an error (i.e., decides a "0" was sent when a "1" was actually sent, or vice versa).

3.38 In this problem, we extend the results of Exercise 3.36 to the case when there are more than two possible messages sent. Suppose now that the communication system sends one of four possible messages $M = 0$, $M = 1$, $M = 2$, and $M = 3$, each with equal probability. The corresponding conditional PDFs of the voltage measured at the receiver are of the form

$$f_{X|M=m}(x) = \frac{1}{\sqrt{2\pi\sigma^2}} \exp\left(-\frac{(x-m)^2}{2\sigma^2}\right), \quad m = 0, 1, 2, 3.$$

(a) Find and plot $\Pr(M=m|X=x)$ as a function of x for $\sigma^2 = 1$.

(b) Determine the range of x for which $\Pr(M=0|X=x) > \Pr(M=m|X=x)$ for all $m \neq 0$. This will be the range of x for which the receiver will decide in favor of the message $M = 0$.

(c) Determine the range of x for which $\Pr(M=1|X=x) > \Pr(M=m|X=x)$ for all $m \neq 1$. This will be the range of x for which the receiver will decide in favor of the message $M = 1$.

(d) Based on your results of parts (b) and (c) and the symmetry of the problem, can you infer the ranges of x for which the receiver will decide in favor of the other two messages, $M = 2$ and $M = 3$?

3.39 Suppose V is a uniform random variable,

$$f_V(v) = \begin{cases} \frac{1}{2}, & 0 \leq v < 2, \\ 0, & \text{otherwise.} \end{cases}$$

(a) Find the conditional PDF, $f_{V|\{V>1\}}(v)$.

(b) Find the conditional PDF, $f_{V|\{1/2<V<3/2\}}(v)$.

(c) Find the conditional CDF, $F_{V|\{1/2<V<3/2\}}(v)$.

3.40 Repeat Exercise 3.39 if V is a Rayleigh random variable with PDF,
$f_V(v) = v\exp(-v^2/2)u(v)$.

Section 3.6: Reliability and Failure Rates

3.41 Recalling Example 3.14, suppose that a serial connection system has 10 components and the failure rate function is the same constant for all components and is 1 per 100 days.
  (a) Determine the probability that the lifetime of the system exceeds 10 days.
  (b) What is the probability that the lifetime of one component exceeds 10 days?
  (c) What is the reliability function of each component and the system as a whole?

Miscellaneous Exercises

3.42 Mr. Hood is a good archer. He can regularly hit a target having a 3-ft. diameter and often hits the bull's-eye, which is 0.5 ft. in diameter, from 50 ft. away. Suppose the miss is measured as the radial distance from the center of the target and, furthermore, that the radial miss distance is a Rayleigh random variable with the constant in the Rayleigh PDF being $\sigma^2 = 4$ (sq-ft).
  (a) Determine the probability of Mr. Hood's hitting the target.
  (b) Determine the probability of Mr. Hood's hitting the bull's-eye.
  (c) Determine the probability of Mr. Hood's hitting the bull's-eye given that he hits the target.

3.43 In this problem, we revisit the light bulb problem of Exercise 2.74. Recall that there were two types of light bulbs, long-life (L) and short-life (S) and we were given an unmarked bulb and needed to identify which type of bulb it was by observing how long it functioned before it burned out. Suppose we modify the problem so that the lifetime of the bulbs are modeled with a continuous random variable. In particular, suppose the two conditional PDFs are now given by

$$f_{X|S}(x) = \frac{1}{100}\exp\left(-\frac{x}{100}\right)u(x) \text{ and } f_{X|L}(x) = \frac{1}{1000}\exp\left(-\frac{x}{1000}\right)u(x),$$

where X is the random variable that measures the lifetime of the bulb in hours. The a priori probabilities of the bulb type were $\Pr(S) = 0.75$ and $\Pr(L) = 0.25$.
  (a) If a bulb is tested and it is observed that the bulb burns out after 200 h, which type of bulb was most likely tested?
  (b) What is the probability that your decision in part (a) was incorrect?

3.44 Consider the light bulb problem in Exercise 3.43. Suppose we do not necessarily want to wait for the light bulb to burn out before we make a decision as to which type of bulb is being tested. Therefore, a modified experiment is proposed. The light bulb to be tested will be turned on at 5 pm on Friday and will be allowed to burn all weekend. We will come back and check on it Monday morning at 8 am and at that point it will either still be lit or it will have burned out. Note that since there are a total of 63 h between the time we start and end the experiment and we will not be watching the bulb at any point in time in between, there are only two possible observations in this experiment, (the bulb burnt out $\Leftrightarrow \{X < 63\}$) or (the bulb is still lit $\Leftrightarrow \{X > 63\}$).

(a)   Given it is observed that the bulb burnt out over the weekend, what is the probability that the bulb was an S-type bulb?

(b)   Given it is observed that the bulb is still lit at the end of the weekend, what is the probability that the bulb was an S type bulb?

3.45   Suppose we are given samples of the CDF of a random variable. That is, we are given $F_n = F_X(x_n)$ at several points, $x_n \in \{x_1, x_2, x_3, \ldots, x_k\}$. After examining a plot of the samples of the CDF, we determine that it appears to follow the functional form of a Rayleigh CDF,

$$F_X(x; \sigma^2) = \left(1 - \exp\left(-\frac{x^2}{2\sigma^2}\right)\right)u(x).$$

The object of this problem is to determine what value of the parameter, σ^2, in the Rayleigh CDF best fits the given data.

(a)   Define the error between the nth sample point and the model to be

$$e_n = F_n - F_X(x_n; \sigma^2) = F_n - \left(1 - \exp\left(-\frac{x_n^2}{2\sigma^2}\right)\right).$$

Find an equation that the parameter σ^2 must satisfy if it is to minimize the sum of the squared errors,

$$\text{SSE} = \sum_{n=1}^{k} e_n^2.$$

Note, you probably will not be able to solve this equation, you just need to set up the equation.

(b)   Next, we will show that the optimization works out to be analytically simpler if we do it in the log domain and if we work with the complement of the CDF. That is, suppose we redefine the error between the nth sample point and the model to be

$$e_n = \log(1 - F_n) - \log(1 - F_X(x_n; \sigma^2)) = \log(1 - F_n) + \exp\left(-\frac{x_n^2}{2\sigma^2}\right).$$

Find an equation that the parameter σ^2 must satisfy if it is to minimize the sum of the squared errors. In this case, you should be able to solve the equation and find an expression for the optimum value of σ^2.

MATLAB Exercises

3.46 Write a MATLAB program to calculate the probability $\Pr(x_1 \leq X \leq x_2)$ if X is a Gaussian random variable for an arbitrary x_1 and x_2. Note you will have to specify the mean and variance of the Gaussian random variable.

3.47 Write a MATLAB program to calculate the probability $\Pr(|X - a| < b)$ if X is a Gaussian random variable for an arbitrary a and $b > 0$. Note you will have to specify the mean and variance of the Gaussian random variable.

3.48 Use the MATLAB `rand` function to create a random variable X uniformly distributed over $(0, 1)$. Then create a new random variable according to $Y = -\ln(X)$. Repeat this procedure many times to create a large number of realizations of Y. Using these samples, estimate and plot the probability density function of Y. Find an analytical model that seems to fit your estimated PDF.

3.49 Use the MATLAB `randn` function to create a Gaussian distributed random variable X. Repeat this procedure and form a new random variable Y. Finally, form a random variable Z according to $Z = \sqrt{X^2 + Y^2}$. Repeat this procedure many times to create a large number of realizations of Z. Using these samples, estimate and plot the probability density function of Z. Find an analytical model that seems to fit your estimated PDF.

3.50 Use the MATLAB `randn` function to generate a large number of samples generated according to a Gaussian distribution. Let A be the event $A = \{$the sample is greater than 1.5$\}$. Of those samples that are members of the event A, what proportion (relative frequency) is greater than 2. By computing this proportion you will have estimated the conditional probability $\Pr(X > 2 | X > 1.5)$. Calculate the exact conditional probability analytically and compare it with the numerical results obtained through your MATLAB program.

3.51 Write a MATLAB program to evaluate the inverse of a Q-function. That is, if the program is input a number, x, subject to $0 \leq x \leq 1$, it will produce an output $y = Q^{-1}(x)$ which is the solution to $Q(y) = x$.

3.52 Use the MATLAB function `marcumq` to write a program to plot the CDF of a Rician random variable. Your program should take as inputs the two parameters a and σ of the Rician random variable and output a plot of the CDF. Be sure to correctly label the axes on your plot.

Operations on a Single Random Variable

In our study of random variables, we use the probability density function (PDF), the cumulative distribution function (CDF), or the probability mass function (PMF) to provide a complete statistical description of the random variable. From these functions, we could, in theory, determine just about anything we might want to know about the random variable. In many cases, it is of interest to distill this information down to a few parameters which describe some of the important features of the random variable. For example, we saw in Chapter 3 that the Gaussian random variable is described by two parameters, which were referred to as the mean and variance. In this chapter, we look at these parameters as well as several others that describe various characteristics of random variables. We see that these parameters can be viewed as the results of performing various operations on a random variable.

4.1 Expected Value of a Random Variable

To begin, we introduce the idea of an average or expected value of a random variable. This is perhaps the single most important characteristic of a random variable and also is a concept that is very familiar to most students. After taking a test, one of the most common questions a student will ask after they see their grade is "What was the average?" On the other hand, how often does a student ask "What was the PDF of the exam scores?" While the answer to the second question would provide the student with more information about how the class performed, the student may not want all that information. Just knowing the average may be sufficient to tell the student how he/she performed relative to the rest of the class.

> **Definition 4.1:** The *expected value* of a random variable X which has a PDF, $f_X(x)$, is
>
> $$E[X] = \int_{-\infty}^{\infty} x f_X(x) \, dx \, . \tag{4.1}$$
>
> The terms *average*, *mean*, *expectation*, and *first moment* are all alternative names for the concept of expected value and will be used interchangeably throughout the text. Furthermore, an overbar is often used to denote expected value so that the symbol $\bar{X}$ is to be interpreted as meaning the same thing as $E[X]$. Another commonly used notation is to write $\mu_X = E[X]$.

Probability and Random Processes. DOI: 10.116/B978-0-12-386981-4.00004-8

For discrete random variables, the PDF can be written in terms of the PMF,

$$f_X(x) = \sum_k P_X(x_k)\delta(x - x_k). \tag{4.2}$$

In that case, using the properties of delta functions, the definition of expected values for discrete random variables reduces to

$$E[X] = \sum_k x_k P_X(x_k). \tag{4.3}$$

Hence, the expected value of a discrete random variable is simply a weighted average of the values that the random variable can take on, weighted by the probability mass of each value. Naturally, the expected value of a random variable only exists if the integral in Equation (4.1) or the series in Equation (4.3) converges. One can dream up many random variables for which the integral or series does not converge and hence their expected values don't exist (or less formally, their expected value is infinite). To gain some physical insight into this concept of expected value, we may think of $f_X(x)$ as a mass distribution of an object along the x axis, then Equation (4.1) calculates the centroid or center of gravity of the mass.

■ Example 4.1:

Consider a random variable that has an exponential PDF given by $f_X(x) = \frac{1}{b}\exp\left(-\frac{x}{b}\right)u(x)$. Its expected value is calculated as follows:

$$E[X] = \int_0^\infty \frac{x}{b}\exp\left(-\frac{x}{b}\right)dx = b\int_0^\infty y\exp(-y)dy = b.$$

The last equality in the series is obtained by using integration by parts once. It is seen from this example that the parameter b which appears in this exponential distribution is in fact the mean (or expected value) of the random variable.

■

■ Example 4.2:

Next, consider a Poisson random variable whose PMF is given by $P_X(k) = \alpha^k e^{-\alpha}/k!$, $k = 0, 1, 2, \ldots$. Its expected value is found in a similar manner.

$$E[X] = \sum_{k=0}^\infty k\frac{\alpha^k e^{-\alpha}}{k!} = e^{-\alpha}\sum_{k=1}^\infty \frac{\alpha^k}{(k-1)!} = \alpha e^{-\alpha}\sum_{k=1}^\infty \frac{\alpha^{k-1}}{(k-1)!} = \alpha e^{-\alpha}\sum_{m=0}^\infty \frac{\alpha^m}{m!} = \alpha e^{-\alpha}e^\alpha = \alpha.$$

Once again, we see that the parameter α in the Poisson distribution is equal to the mean.

■

■ Example 4.3:

In the last two examples, we saw random variables whose PDF or PMF was described by a single parameter which in both cases turned out to be the mean. We work one more example here to show that this does not always have to be the case. Consider a Rayleigh random variable with PDF

$$f_X(x) = \frac{x}{\sigma^2}\exp\left(-\frac{x^2}{2\sigma^2}\right)u(x).$$

The mean is calculated as follows:

$$E[X] = \int_0^\infty \frac{x^2}{\sigma^2}\exp\left(-\frac{x^2}{2\sigma^2}\right)dx = \sqrt{2}\sigma\int_0^\infty y^{1/2}\exp(-y)dy = \sqrt{2}\sigma\Gamma(3/2) = \sqrt{\frac{\pi}{2}}\sigma.$$

The last equality is obtained using the fact that $\Gamma(3/2) = \sqrt{\pi}/2$. Alternatively (for those students not familiar with the properties of Γ functions), one could obtain this result using integration by parts once on the original integral (setting $u = x$ and $dv = (x/\sigma^2)\exp(-x^2/(2\sigma^2))$). In this case, neither the parameter σ nor σ^2 is equal to the expected value of the random variable. However, since the mean is proportional to σ, we could, if we wanted to, rewrite the Rayleigh PDF in terms of its expected value, μ_X, as follows:

$$f_X(x) = \frac{\pi x}{2\mu_X^2}\exp\left(-\frac{\pi x^2}{4\mu_X^2}\right)u(x) \ .$$

■

4.2 Expected Values of Functions of Random Variables

The concept of expectation can be applied to the functions of random variables as well as to the random variable itself. This will allow us to define many other parameters that describe various aspects of a random variable.

Definition 4.2: Given a random variable X with PDF $f_X(x)$, the expected value of a function, $g(X)$, of that random variable is given by

$$E[g(X)] = \int g(x)f_X(x)dx. \tag{4.4}$$

For a discrete random variable, this definition reduces to

$$E[g(X)] = \sum_k g(x_k)P_X(x_k). \tag{4.5}$$

To start with, we demonstrate one extremely useful property of expectations in the following theorem.

Theorem 4.1: For any constants a and b,

$$E[aX + b] = aE[X] + b. \tag{4.6}$$

Furthermore, for any function $g(x)$ which can be written as a sum of several other functions (i.e., $g(x) = g_1(x) + g_2(x) + \ldots + g_N(x)$),

$$E\left[\sum_{k=1}^{N} g_k(X)\right] = \sum_{k=1}^{N} E[g_k(X)] \tag{4.7}$$

In other words, expectation is a linear operation and the expectation operator can be exchanged (in order) with any other linear operation.

Proof: The proof follows directly from the linearity of the integration operator.

$$E[aX + b] = \int_{-\infty}^{\infty} (ax + b)f_X(x)\mathrm{d}x$$

$$= a\int_{-\infty}^{\infty} xf_X(x)\mathrm{d}x + b = aE[X] + b. \tag{4.8}$$

The second part of the theorem is proved in an identical manner:

$$E\left[\sum_{k=1}^{N} g_k(X)\right] = \int_{-\infty}^{\infty}\left[\sum_{k=1}^{N} g_k(x)\right]f_X(x)\mathrm{d}x = \sum_{k=1}^{N}\int_{-\infty}^{\infty} g_k(x)f_X(x)\mathrm{d}x$$

$$= \sum_{k=1}^{N} E[g_k(X)] . \quad\square \tag{4.9}$$

Different functional forms of $g(X)$ lead to various different parameters which describe the random variable and are known by special names. A few of the more common ones are listed in Table 4.1. In the following sections, selected parameters will be studied in more detail.

Table 4.1: Expected values of various functions of random variables

Name	Function of X	Expected Value, Notation
Mean, average, expected value, expectation, first moment	$g(x) = x$	$\mu_X = \bar{X} = E[X]$
nth moment	$g(x) = x^n$	$\overline{X^n} = E[X^n]$
nth central moment	$g(x) = (x - \mu_X)^n$	$\overline{(x - \mu_X)^n} = E[(X - \mu_X)^n]$
Variance	$g(x) = (x - \mu_X)^2$	$\sigma_X^2 = E[(X - \mu_X)^2]$
Coefficient of skewness	$g(x) = \left(\dfrac{x - \mu_X}{\sigma_X}\right)^3$	$c_s = E\left[\left(\dfrac{X - \mu_X}{\sigma_X}\right)^3\right]$
Coefficient of kurtosis	$g(x) = \left(\dfrac{x - \mu_X}{\sigma_X}\right)^4$	$c_k = E\left[\left(\dfrac{X - \mu_X}{\sigma_X}\right)^4\right]$
Characteristic function	$g(x) = e^{j\omega x}$	$\Phi_X(\omega) = E[e^{j\omega X}]$
Moment-generating function	$g(x) = e^{sx}$	$M_X(s) = E[e^{sX}]$
Probability-generating function	$g(x) = z^x$	$H_X(z) = E[z^X]$

4.3 Moments

Definition 4.3: The nth *moment* of a random variable X is defined as

$$E[X^n] = \int x^n f_X(x)\,dx. \tag{4.10}$$

For a discrete random variable, this definition reduces to

$$E[X^n] = \sum_k x_k^n P_X(x_k). \tag{4.11}$$

The zeroth moment is simply the area under the PDF and must be one for any random variable. The most commonly used moments are the first and second moments. The first moment is what we previously referred to as the mean, while the second moment is the mean-squared value. For some random variables, the second moment might be a more meaningful characterization than the first. For example, suppose X is a sample of a noise waveform. We might expect that the distribution of the noise is symmetric about zero (i.e., just as likely to be positive as negative) and hence the first moment will be zero. So if we are told that X has a zero mean, all this says is merely that the noise does not have a bias. However, the second moment of the random noise sample is in some sense a measure of the strength of the noise. In fact, in Chapter 10, we will associate the second moment of a noise process with the power in the process. Thus, specifying the second moment can give us some useful physical insight into the noise process.

■ Example 4.4:

Consider a discrete random variable that has a binomial distribution. Its probability mass function is given by

$$P_X(k) = \binom{n}{k} p^k (1-p)^{n-k}, \ k = 0, 1, 2, ..., n.$$

The first moment is calculated as follows:

$$E[X] = \sum_{k=0}^{n} k \binom{n}{k} p^k (1-p)^{n-k} = \sum_{k=1}^{n} \frac{kn!}{k!(n-k)!} p^k (1-p)^{n-k} = \sum_{k=1}^{n} \frac{n!}{(k-1)!(n-k)!} p^k (1-p)^{n-k}$$

$$= np \sum_{k=1}^{n} \frac{(n-1)!}{(k-1)!(n-k)!} p^{k-1} (1-p)^{n-k} = np \sum_{k=1}^{n} \binom{n-1}{k-1} p^{k-1} (1-p)^{n-k} = np \sum_{m=0}^{n-1} \binom{n-1}{m} p^m (1-p)^{n-1-m}.$$

In this last expression, the summand is a valid PMF (i.e., that of a binomial random variable with parameters p and $n-1$) and therfore must sum to unity. Thus, $E[X] = np$. To calculate the second moment, we employ a helpful little trick. Note that we can write $k^2 = k(k-1) + k$. Then

$$E[X^2] = \sum_{k=0}^{n} k^2 \binom{n}{k} p^k (1-p)^{n-k} = \sum_{k=0}^{n} k(k-1) \binom{n}{k} p^k (1-p)^{n-k} + \sum_{k=0}^{n} k \binom{n}{k} p^k (1-p)^{n-k}.$$

The second sum is simply the first moment, which has already been calculated. The first sum is evaluated in a manner similar to that used to calculate the mean.

$$\sum_{k=0}^{n} k(k-1) \binom{n}{k} p^k (1-p)^{n-k} = \sum_{k=2}^{n} \frac{n!}{(k-2)!(n-k)!} p^k (1-p)^{n-k}$$

$$= n(n-1)p^2 \sum_{k=2}^{n} \frac{(n-2)!}{(k-2)!(n-k)!} p^{k-2} (1-p)^{n-k} = n(n-1)p^2 \sum_{k=2}^{n} \binom{n-2}{k-2} p^{k-2} (1-p)^{n-k}$$

$$= n(n-1)p^2 \sum_{m=0}^{n-2} \binom{n-2}{m} p^m (1-p)^{n-2-m} = n(n-1)p^2$$

Putting these two results together gives

$$E[X^2] = n(n-1)p^2 + np = n^2 p^2 + np(1-p) .$$

■

■ Example 4.5:

Consider a random variable with a uniform PDF given as

$$f_X(x) = \begin{cases} 1/a, & 0 \le x \le a, \\ 0, & \text{otherwise.} \end{cases}$$

The mean is given by

$$E[X] = \int_0^a \frac{x}{a} dx = \frac{x^2}{2a}\Big|_0^a = \frac{a}{2},$$

while the second moment is

$$E[X^2] = \int_0^a \frac{x^2}{a} dx = \frac{x^3}{3a}\Big|_0^a = \frac{a^2}{3}.$$

In fact, it is not hard to see that in general, the nth moment of this uniform random variable is given by

$$E[X^n] = \int_0^a \frac{x^n}{a} dx = \frac{x^{n+1}}{(n+1)a}\Big|_0^a = \frac{a^n}{n+1}.$$

■

4.4 Central Moments

Consider a random variable Y which could be expressed as the sum, $Y = a + X$ of a deterministic (i.e., not random) part a and a random part X. Furthermore, suppose that the random part tends to be very small compared to the fixed part. That is, the random variable Y tends to take small fluctuations about a constant value, a. Such might be the case in a situation where there is a fixed signal corrupted by noise. In this case, we might write $Y^n = (a + X)^n \approx a^n$. As such, the nth moment of Y would be dominated by the fixed part. That is, it is difficult to characterize the randomness in Y by looking at the moments. To overcome this, we can use the concept of central moments.

Definition 4.4: The nth *central moment* of a random variable X is defined as

$$E[(X - \mu_X)^n] = \int (x - \mu_X)^n f_X(x) dx. \tag{4.12}$$

In the above equation, μ_X is the mean (first moment) of the random variable. For discrete random variables, this definition reduces to

$$E[(X - \mu_X)^n] = \sum_k (x_k - \mu_X)^k P_X(x_k). \tag{4.13}$$

With central moments, the mean is subtracted from the variable before the moment is taken in order to remove the bias in the higher moments due to the mean. Note that, like regular moments, the zeroth central moment is $E[(X - \mu_X)^0] = E[1] = 1$. Furthermore, the first central moment is $E[X - \mu_X] = E[X] - \mu_X = \mu_X - \mu_X = 0$. Hence, the lowest central moment of any real interest is the second central moment. This central moment is given a special name, the *variance*, and we quite often use the notation σ_X^2 to represent the variance of the random variable X. Note that

$$\sigma_X^2 = E[(X - \mu_X)^2] = E[X^2 - 2\mu_X X + \mu_X^2] = E[X^2] - 2\mu_X E[X] + \mu_X^2$$

$$= E[X^2] - \mu_X^2. \tag{4.14}$$

In many cases, the best way to calculate the variance of a random variable is to calculate the first two moments and then form the second moment minus the first moment squared.

■ Example 4.6:

For the binomial random variable in Example 4.4, recall that the mean was $E[X] = np$ and the second moment was $E[X^2] = n^2 p^2 + np(1 - p)$. Therefore, the variance is given by $\sigma_X^2 = np(1 - p)$. Similarly, for the uniform random variable in Example 4.5, $E[X] = a/2$, $E[X^2] = a^2/3$, and therfore $\sigma_X^2 = a^2/3 - a^2/4 = a^2/12$. Note that if the moments have not previously been calculated, it may be just as easy to compute the variance directly. In the case of the uniform random variable, once the mean has been calculated, the variance can be found as

$$\sigma_X^2 = \int_0^a (x - a/2)^2 \frac{1}{a} dx = \int_{-\frac{a}{2}}^{\frac{a}{2}} \frac{x^2}{a} dx = \frac{x^3}{3a} \Big|_{-a/2}^{a/2} = \frac{a^2}{12}.$$

■

Another common quantity related to the second central moment of a random variable is the *standard deviation,* which is defined as the square root of the variance,

$$\sigma_X = \sqrt{E[(X - \mu_X)^2]}. \tag{4.15}$$

Both the variance and the standard deviation serve as a measure of the width of the PDF of a random variable. Some of the higher-order central moments also have special names, although they are much less frequently used. The third central moment is known as the *skewness* and is a measure of the symmetry of the PDF about the mean. The fourth central moment is called the *kurtosis* and is a measure of the peakedness of a random variable near the mean. Note that not all random variables have finite moments and/or central moments. We give an example of this later for the Cauchy random variable. Some quantities related to these higher-order central moments are given in the following definition.

Definition 4.5: The *coefficient of skewness* is

$$c_\mathrm{s} = \frac{E[(X - \mu_X)^3]}{\sigma_X^3}. \tag{4.16}$$

This is a dimensionless quantity that is positive if the random variable has a PDF skewed to the right and negative if skewed to the left. The *coefficient of kurtosis* is also dimensionless and is given as

$$c_\mathrm{k} = \frac{E[(X - \mu_X)^4]}{\sigma_X^4}. \tag{4.17}$$

The more the PDF is concentrated near its mean, the larger the coefficient of kurtosis. In other words, a random variable with a large coefficient of kurtosis will have a large peak near the mean.

■ Example 4.7:

An exponential random variable has a PDF given by

$$f_X(x) = b\exp(-bx)u(x).$$

The mean value of this random variable is $\mu_X = 1/b$. The nth central moment is given by

$$E[(X - \mu_X)^n] = \int_0^\infty (x - 1/b)^n b\exp(-bx)\mathrm{d}x$$

$$= \frac{b}{e}\int_{-1/b}^\infty y^n \exp(-by)\mathrm{d}y$$

$$= \frac{1}{b^n}\sum_{m=0}^n \frac{n!}{m!}(-1)^m.$$

In the preceding expression, it is understood that $0! = 1$. As expected, it is easily verified from the above expression that the 0th central moment is 1 and the first central moment is 0. Beyond these, the second central moment is $\sigma_X^2 = 1/b^2$, the third central moment is $E[(X\ 1/b)^3] = -2/b^3$, and the fourth central moment is $E[(X - 1/b)^4] = 9/b^4$. The coefficients of skewness and kurtosis are given by

$$c_\mathrm{s} = \frac{E[(X - \mu_X)^3]}{\sigma_X^3} = \frac{-2/b^3}{1/b^3} = -2,$$
$$c_\mathrm{k} = \frac{E[(X - \mu_X)^4]}{\sigma_X^4} = \frac{9/b^4}{1/b^4} = 9.$$

The fact that the coefficient of skewness is negative shows that the exponential PDF is skewed to the left of its mean.

■

■ Example 4.8:

Next consider a Laplace (two-sided exponential) random variable with a PDF given by

$$f_X(x) = \frac{b}{2}\exp(-b|x|).$$

Since this PDF is symmetric about zero, its mean is zero and therefore in this case the central moments are simply the moments,

$$E[(X-\mu_X)^n] = E[X^n] = \int_{-\infty}^{\infty} \frac{bx^n}{2}\exp(-b|x|)dx.$$

Since the two-sided exponential is an even function and x^n is an odd function for any odd n, the integrand is then an odd function for any odd n. The integral of any odd function over an interval symmetric about zero is equal to zero, and hence all odd moments of the Laplace random variable are zero. The even moments can be calculated individually:

$$\sigma_X^2 = E[X^2] = \int_{-\infty}^{\infty} \frac{bx^2}{2}\exp(-b|x|)dx = \int_0^{\infty} bx^2\exp(-bx)dx = \frac{2}{b^2},$$

$$E[(X-\mu_X)^4] = E[X^4] = \int_{-\infty}^{\infty} \frac{bx^4}{2}\exp(-b|x|)dx = \int_0^{\infty} bx^4\exp(-bx)dx = \frac{24}{b^4}.$$

The coefficient of skewness is zero (since the third central moment is zero) and the coefficient of kurtosis is

$$c_k = \frac{E[(X-\mu_X)^4]}{\sigma_X^4} = \frac{24/b^4}{4/b^4} = 6.$$

Note that the Laplace distribution has a sharp peak near its mean as evidenced by a large coefficient of kurtois. The fact that the coefficient of skewness is zero is consistent with the fact that the distribution is symmetric about its mean.

■

■ Example 4.9:

It is often the case that the PDF of random variables of practical interest may be too complicated to allow us to compute various moments and other important parameters of the distribution in an analytic fashion. In those cases, we can use a computer to calculate the needed quantities numerically. Take for example a Rician random variable whose PDF is given by

$$f_X(x) = \frac{x}{\sigma^2}\exp\left(-\frac{x^2+a^2}{2\sigma^2}\right)I_0\left(\frac{ax}{\sigma^2}\right)u(x).$$

Suppose we wanted to know the mean of this random variable. This requires us to evaluate

$$\mu_X = \int_0^{\infty} \frac{x^2}{\sigma^2}\exp\left(-\frac{x^2+a^2}{2\sigma^2}\right)I_0\left(\frac{ax}{\sigma^2}\right)dx.$$

Note that the parameter σ^2 which shows up in the Rician PDF is *not* the variance of the Rician random variable. While analytical evaluation of this integral looks formidable, given numerical values for a and σ^2, this integral can be evaluated (at least approximately) using standard numerical integration techniques. In order to use the numerical integration routines built into MATLAB, we must first write a function which evaluates the integrand. For evaluating the mean of the Rician PDF, this can be accomplished as follows (see Appendix D, Equation (D.52) for an analytic expression for the mean):

```
function pdf=Rician_pdf(x,a,sigma)
% Evaluate the integrand needed for calculating the mean of a
% Rician random variable with parameters a and sigma.
pdf=(x./sigma).^2.*exp(-(x.^2+a^2)/(2*sigma^2));
pdf=pdf.*besseli(0,a*x/sigma^2);
```

Once this function is defined, the MATLAB function `quad8` can be called upon to perform the numerical integration. For example, if $a = 2$ and $\sigma = 3$, the mean could be calculated as follows:

```
a=2; sigma=3;                          % set parameters
limit1=0; limit2=20;                   % set limits of integration.
mean=quad8('Rician_pdf',limit1,limit2,[],[],a,sigma);
```

Executing this code produced an answer of $\mu_X = 4.1665$. Note that in order to calculate the mean, the upper limit of the integral should be infinite. However using `limit2=Inf` in the above code would have led MATLAB to produce a result of `NaN` ("not a number"). Instead, we must use an upper limit sufficiently large that for the purposes of evaluating the integral it is essentially infinite. This can be done by observing the integrand and seeing at what point the integrand dies off. The reader is encouraged to execute the code in this example using different values of the upper integration limit to see how large the upper limit must be to produce accurate results.

■

4.5 Conditional Expected Values

Another important concept involving expectation is that of conditional expected value. As specified in Definition 4.6, the conditional expected value of a random variable is a weighted average of the values the random variable can take on, weighted by the conditional PDF of the random variable.

Definition 4.6: The expected value of a random variable X, conditioned on some event A is

$$E[X|A] = \int x f_{X|A}(x)\mathrm{d}x. \tag{4.18}$$

For a discrete random variable, this definition reduces to

$$E[X|A] = \sum_k x_k P_{X|A}(x_k).$$

(4.19)

Similarly, the conditional expectation of a function, $g(\cdot)$, of a random variable, conditioned on the event A is

$$E[g(X)|A] = \int g(x) f_{X|A}(x) dx \text{ or } E[g(X)|A] = \sum_k g(x_k) P_{X|A}(x_k),$$

(4.20)

depending on whether the random variable is continuous or discrete.

Conditional expected values are computed in the same manner as regular expected values with the PDF or PMF replaced by a conditional PDF or conditional PMF.

■ Example 4.10:

Consider a Gaussian random variable of the form

$$f_X(x) = \frac{1}{\sqrt{2\pi}} \exp\left(-\frac{x^2}{2}\right).$$

Suppose the event A is the event that the random variable X is positive, $A = \{X > 0\}$. Then

$$f_{X|A}(x) = \frac{f_X(x)}{\Pr(X > 0)} u(x) = \sqrt{\frac{2}{\pi}} \exp\left(-\frac{x^2}{2}\right) u(x).$$

The conditional expected value of X given that $X > 0$ is then

$$E[X|X > 0] = \int x f_{X|X > 0}(x) dx = \sqrt{\frac{2}{\pi}} \int_0^\infty x \exp\left(-\frac{x^2}{2}\right) dx = -\sqrt{\frac{2}{\pi}} \exp\left(-\frac{x^2}{2}\right)\Bigg|_0^\infty = \sqrt{\frac{2}{\pi}}.$$

■

4.6 Transformations of Random Variables

Consider a random variable X with a PDF and CDF given by $f_X(x)$ and $F_X(x)$, respectively. Define a new random variable Y such that $Y = g(X)$ for some function $g(\cdot)$. What is the PDF, $f_Y(y)$, (or CDF) of the new random variable? This problem is often encountered in the study of systems where the PDF for the input random variable X is known and the PDF for the output random variable Y needs to be determined. In such a case, we say that the input random variable has undergone a transformation.

4.6.1 Monotonically Increasing Functions

To begin our exploration of transformations of random variables, let us assume that the function is continuous, one-to-one, and monotonically increasing. A typical function of this

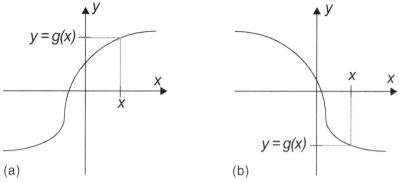

Figure 4.1

(a) A monotonic increasing function and (b) a monotonic decreasing function.

form is illustrated in Figure 4.1a. This assumption will be lifted later when more general functions are considered, but for now this simpler case applies. Under these assumptions, the inverse function, $X = g^{-1}(Y)$, exists and is well behaved. In order to obtain the PDF of Y, we first calculate the CDF. Recall that $F_Y(y) = \Pr(Y \leq y)$. Since there is a one-to-one relationship between values of Y and their corresponding values of X, this CDF can be written in terms of X according to

$$F_Y(y) = \Pr(g(X) \leq y) = \Pr(X \leq g^{-1}(y)) = F_X(g^{-1}(y)). \tag{4.21}$$

Note, this can also be written as

$$F_X(x) = F_Y(g(x)). \tag{4.22}$$

Differentiating Equation (4.21) with respect to y produces

$$f_Y(y) = f_X(g^{-1}(y))\frac{\mathrm{d}g^{-1}(y)}{\mathrm{d}y} = \left. f_X(x)\frac{\mathrm{d}x}{\mathrm{d}y} \right|_{x = g^{-1}(y)}, \tag{4.23}$$

while differentiating Equation (4.22) with respect to x gives

$$f_X(x) = f_Y(g(x))\frac{\mathrm{d}y}{\mathrm{d}x} \Rightarrow f_Y(y) = \left. \frac{f_X(x)}{\dfrac{\mathrm{d}y}{\mathrm{d}x}} \right|_{x = g^{-1}(y)}. \tag{4.24}$$

Either Equation (4.23) or (4.24) can be used (whichever is more convenient) to compute the PDF of the new random variable.

■ Example 4.11:

Suppose X is a Gaussian random variable with mean, μ, and variance, σ^2. A new random variable is formed according to $Y = aX + b$, where $a > 0$ (so that the transformation is monotonically increasing). Since $dy/dx = a$, then applying Equation (4.24) produces

$$f_Y(y) = \frac{1}{a} f_X\left(\frac{y-b}{a}\right).$$

Furthermore, plugging in the Gaussian PDF of X results in

$$f_Y(y) = \frac{1}{a\sqrt{2\pi\sigma^2}} \exp\left(-\frac{\left(\frac{y-b}{a} - \mu\right)^2}{2\sigma^2}\right) = \frac{1}{\sqrt{2\pi(a\sigma)^2}} \exp\left(-\frac{(y-(b+a\mu))^2}{2(a\sigma)^2}\right).$$

Note that the PDF of Y still has a Gaussian form. In this example, the transformation did not change the form of the PDF; it merely changed the mean and variance.

■

■ Example 4.12:

Let X be an exponential random variable with $f_X(x) = 2e^{-2x}u(x)$ and let the transformation be $Y = X^3$. Then $dy/dx = 3x^2$ and

$$f_Y(y) = \left.\frac{f_X(x)}{3x^2}\right|_{x = \sqrt[3]{y}} = \frac{2}{3}y^{-2/3}\exp(-2y^{1/3})u(y).$$

■

■ Example 4.13:

Suppose a phase angle Θ is uniformly distributed over $(-\pi/2, \pi/2)$ and the transformation is $Y = \sin(\Theta)$. Note that in general $y = \sin(\theta)$ is not a monotonic transformation, but under the restriction $-\pi/2 < \theta < \pi/2$, this transformation is indeed monotonically increasing. Also note that with this transformation the resulting random variable, Y, must take on values in the range $(-1, 1)$. Therefore, whatever PDF is obtained for Y, it must be understood that the PDF is zero outside $(-1, 1)$. Applying Equation (4.24) gives

$$f_Y(y) = \left.\frac{f_\Theta(\theta)}{\cos(\theta)}\right|_{\theta = \sin^{-1}(y)} = \frac{1}{\pi\cos(\sin^{-1}(y))} = \frac{1}{\pi\sqrt{1-y^2}}, \quad -1 < y < 1.$$

This is known as an *arcsine distribution*.

■

4.6.2 Monotonically Decreasing Functions

If the transformation is monotonically decreasing rather than increasing, a simple modification to the previous derivations can lead to a similar result. First, note that for monotonic decreasing functions, the event $\{Y \leq y\}$ is equivalent to the event $X \geq g^{-1}(y)$, giving us

$$F_Y(y) = \Pr(Y \leq y) = \Pr(X \geq g^{-1}(y)) = 1 - F_X(g^{-1}(y)). \qquad (4.25)$$

Differentiating with respect to y gives

$$f_Y(y) = -f_X(x)\frac{dx}{dy}\bigg|_{x = g^{-1}(y)}. \qquad (4.26)$$

Similarly, writing $F_Y(g(x)) = 1 - F_X(x)$ and differentiating with respect to x results in

$$f_Y(y) = -\frac{f_X(x)}{\dfrac{dy}{dx}}\bigg|_{x = g^{-1}(y)}. \qquad (4.27)$$

Equations (4.23), (4.24), (4.26), and (4.27) can be consolidated into the following compact form:

$$f_Y(y) = f_X(x)\left|\frac{dx}{dy}\right|\bigg|_{x = g^{-1}(y)} = \frac{f_X(x)}{\left|\dfrac{dy}{dx}\right|}\bigg|_{x = g^{-1}(y)}, \qquad (4.28)$$

where now the sign differences have been accounted for by the absolute value operation. This equation is valid for any monotonic function, either monotonic increasing or monotonic decreasing.

4.6.3 Nonmonotonic Functions

Finally, we consider a general function which is not necessarily monotonic. Figure 4.2 illustrates one such example. In this case, we cannot associate the event $\{Y \leq y\}$ with events

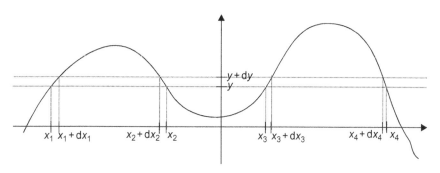

Figure 4.2
A nonmonotonic function; the inverse function may have multiple roots.

of the form $\{X \leq g^{-1}(y)\}$ or $\{X \geq g^{-1}(y)\}$ because the transformation is not monotonic. To avoid this problem, we calculate the PDF of Y directly, rather than first finding the CDF. Consider an event of the form $\{y \leq Y < y + dy\}$ for an infinitesimal dy. The probability of this event is $\Pr(y \leq Y < y + dy) = f_Y(y)dy$. In order to relate the PDF of Y to the PDF of X, we relate the event $\{y \leq Y < y + dy\}$ to events involving the random variable X. Because the transformation is not monotonic, there may be several values of x which map to the same value of y. These are the roots of the equation $x = g^{-1}(y)$. Let us refer to these roots as $x_1, x_2, \ldots, x_N$. Furthermore, let X^+ be the subset of these roots at which the function $g(x)$ has a positive slope and similarly let X^- be the remaining roots for which the slope of the function is negative. Then

$$\{y \leq Y < y + dy\} = \left[\bigcup_{i:x_i \in X^+} \{x_i \leq X < x_i + dx_i\}\right] \cup \left[\bigcup_{i:x_i \in X^-} \{x_i + dx_i < X \leq x_i\}\right]. \quad (4.29)$$

Since each of the events on the right-hand side is mutually exclusive, the probability of the union is simply the sum of the probabilities so that

$$f_Y(y)dy = \sum_{x_i \in X^+} \Pr(x_i \leq X < x_i + dx_i) + \sum_{x_i \in X^-} \Pr(x_i + dx_i < X \leq x_i)$$

$$= \sum_{x_i \in X^+} f_X(x_i)dx_i + \sum_{x_i \in X^-} f_X(x_i)(-dx_i) \quad (4.30)$$

Again, invoking absolute value signs to circumvent the need to have two separate sums and dividing by dy, the following result is obtained:

$$f_Y(y) = \sum_{x_i} f_X(x)\left|\frac{dx}{dy}\right|\Bigg|_{x_i = g^{-1}(y)}. \quad (4.31)$$

When it is more convenient, the equivalent expression

$$f_Y(y) = \sum_{x_i} \frac{f_X(x)}{\left|\frac{dy}{dx}\right|}\Bigg|_{x_i = g^{-1}(y)}, \quad (4.32)$$

can also be used. The following theorem summarizes the general formula for transformations of random variables.

Theorem 4.2: Given a random variable X with known PDF, $f_X(x)$, and a transformation $Y = g(X)$. The PDF of Y is

$$f_Y(y) = \sum_{x_i} f_X(x) \left| \frac{dx}{dy} \right| \Bigg|_{x_i = g^{-1}(y)} = \sum_{x_i} \frac{f_X(x)}{\left| \frac{dy}{dx} \right|} \Bigg|_{x_i = g^{-1}(y)}, \qquad (4.33)$$

where the x_i are the roots of the equation $y = g(x)$. The proof precedes the theorem.

▪ Example 4.14:

Suppose X is a Gaussian random variable with zero mean and variance σ^2 together with a quadratic transformation, $Y = X^2$. For any positive value of y, $y = x^2$ has two real roots, namely $x = \pm\sqrt{y}$ (for negative values of y there are no real roots). Application of Equation (4.33) gives

$$f_Y(y) = \left[\frac{f_X(+\sqrt{y})}{2|+\sqrt{y}|} + \frac{f_X(-\sqrt{y})}{2|-\sqrt{y}|} \right] u(y) = \frac{f_X(+\sqrt{y}) + f_X(-\sqrt{y})}{2\sqrt{y}} u(y).$$

For a zero-mean Gaussian PDF, $f_X(x)$ is an even function so that $f_X(+\sqrt{y}) = f_X(-\sqrt{y})$. Therefore,

$$f_Y(y) = \frac{1}{\sqrt{y}} f_X(\sqrt{y}) u(y) = \sqrt{\frac{1}{2\pi y \sigma^2}} \exp\left(-\frac{y}{2\sigma^2}\right) u(y).$$

Hence, Y is a Gamma random variable.

▪

▪ Example 4.15:

Suppose the same Gaussian random variable from the previous example is passed through a half-wave rectifier which is described by the input-output relationship

$$y = g(x) = \begin{cases} x, & x \geq 0, \\ 0, & x \leq 0. \end{cases}$$

For $x > 0$, $dy/dx = 1$ so that $f_Y(y) = f_X(y)$. However, when $x < 0$, $dy/dx = 0$, which will create a problem if we try to insert this directly into Equation (4.33). To treat this case, we note that the event $X < 0$ is equivalent to the event $Y = 0$; therefore, $\Pr(Y = 0) = \Pr(X < 0)$. Since the input Gaussian PDF is symmetric about zero, $\Pr(X < 0) = 1/2$. Basically, the random variable Y is a mixed random variable. It has a continuous part over the region $y > 0$ and a discrete part at $y = 0$. Using a delta function, we can write the PDF of Y as

$$f_Y(y) = \frac{1}{\sqrt{2\pi\sigma^2}} \exp\left(-\frac{y^2}{2\sigma^2}\right) u(y) + \frac{1}{2}\delta(y).$$

▪

Example 4.15 illustrates how to deal with transformations that are flat over some interval of nonzero length. In general, suppose the transformation $y = g(x)$ is such that $g(x) = y_0$ for any x in the interval $x_1 \le x \le x_2$. Then the PDF of Y will include a discrete component (a delta function) of height $\Pr(Y=y_0) = \Pr(x_1 \le x \le x_2)$ at the point $y = y_0$. One often encounters transformations that have several different flat regions. One such "staircase" function is shown in Figure 4.3. Hence, a random variable X that may be continuous will be converted into a discrete random variable. The classical example of this is analog-to-digital conversion of signals. Suppose the transformation is of a general staircase form,

$$
y = \begin{cases}
y_0, & x < x_1, \\
y_i, & x_i \le x < x_{i+1}, \quad i = 1, 2, ..., N-1, \\
y_N, & x \ge x_N.
\end{cases}
\tag{4.34}
$$

Then Y will be a discrete random variable whose PMF is

$$
P(Y=y_i) = \begin{cases}
\Pr(X < x_1), & i = 0, \\
\Pr(x_i \le x < x_{i+1}), & i = 1, 2, ..., N-1, \\
\Pr(X \ge x_N), & i = N.
\end{cases}
\tag{4.35}
$$

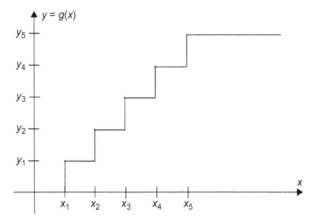

Figure 4.3
A staircase (quantizer) transformation: a continuous random variable will be converted into a discrete random variable.

■ Example 4.16:

Suppose X is an exponential random variable with a PDF $f_X(x) = \exp(-x)u(x)$ and we form a new random variable Y by rounding X down to the nearest integer. That is,

$$Y = g(X) = \text{floor}(X) = k \, , \, k \leq X < k + 1 \, .$$

Then, the PMF of Y is

$$P(Y = k) = \Pr(k \leq X < k + 1) = \int_k^{k+1} e^{-x}dx = e^{-k} - e^{-(k+1)} = e^{-k}(1 - 1/e) \, , \, k = 0, 1, 2, \dots \, .$$

Hence, quantization of an exponential random variable produces a geometric random variable.

■

■ Example 4.17:

Let X be a random variable uniformly distributed over $(-a/2, a/2)$. Accordingly, its PDF is of the form

$$f_X(x) = \frac{1}{a}\left(u\left(x + \frac{a}{2}\right) - u\left(x - \frac{a}{2}\right)\right) \, .$$

A new random variable is to be formed according to the square law transformation $Y = X^2$. Applying the theory developed in this section you should be able to demonstrate that the new random variable has a PDF given by

$$f_Y(y) = \frac{1}{a\sqrt{y}}(u(y) - u(y - a^2/4)) \, .$$

Using MATLAB, we create a large number of samples of the uniform random variable, pass these samples through the square law transformation, and then construct a histogram of the resulting probability densities. The MATLAB code to do so follows and the results of running this code are shown in Figure 4.4.

```
clear
N=10000;
a=5; ymax=a^2/4;                          % set parameters.
x=a*(rand(1,N)-0.5);                      % generate uniform RVs
y=x.^2;                                   % square law transformation
bw=0.25;                                  % bin width
bins=[bw/2:bw:ymax];                      % histogram bins.
[yvals,xvals]=hist(y,bins);               % compute histogram values
pdf_est=yvals/(N*bw);                     % convert to prob. densities
bar(xvals,pdf_est)                        % plot histogram

% Compare true PDF with histogram.
y=[0.01:0.01:ymax];
pdf=1./(a*sqrt(y));
hold on
plot(y,pdf)
xlabel('y'); ylabel('f_Y(y)')
hold off
```

(Continued)

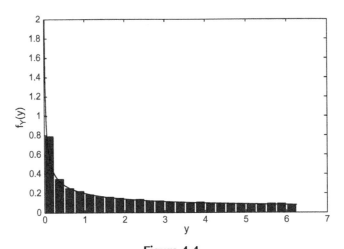

Figure 4.4
Comparison of estimated and true PDF for Example 4.17. (For color version of this figure, the reader is refered to the web version of this chapter).

4.7. Characteristic Functions

In this section, we introduce the concept of a characteristic function. The characteristic function of a random variable is closely related to the Fourier transform of the PDF of that random variable. Thus, the characteristic function provides a sort of "frequency domain" representation of a random variable, although in this context there is no connection between our frequency variable ω and any physical frequency. In studies of deterministic signals, it was found that the use of Fourier transforms greatly simplified many problems, especially those involving convolutions. We will see in future chapters the need for performing convolution operations on PDFs of random variables, and hence frequency domain tools will become quite useful. Furthermore, we will find that characteristic functions have many other uses. For example, the characteristic function is quite useful for finding moments of a random variable. In addition to the characteristic function, two other related functions, namely, the moment-generating function (analogous to the Laplace transform) and the probability-generating function (analogous to the z-transform), will also be studied in the following sections.

Definition 4.7: The *characteristic function* of a random variable, X, is given by

$$\Phi_X(\omega) = E[e^{j\omega X}] = \int_{-\infty}^{\infty} e^{j\omega x} f_X(x)\,\mathrm{d}x. \tag{4.36}$$

Note the similarity between this integral and the Fourier transform. In most of the electrical engineering literature, the Fourier transform of the function $f_X(x)$ would be $\Phi(-\omega)$. Given this relationship between the PDF and the characteristic function, it should be clear that one can get the PDF of a random variable from its characteristic function through an inverse Fourier transform operation:

$$f_X(x) = \frac{1}{2\pi}\int_{-\infty}^{\infty} e^{-j\omega x}\Phi_X(\omega)d\omega. \tag{4.37}$$

The characteristic functions associated with various random variables can be easily found using tables of commonly used Fourier transforms, but one must be careful since the Fourier integral used in Equation (4.36) may be different from the definition used to generate common tables of Fourier transforms. In addition, various properties of Fourier transforms can also be used to help calculate characteristic functions as shown in the following example.

■ Example 4.18:

An exponential random variable has a PDF given by $f_X(x) = \exp(-x)u(x)$. Its characteristic function is found to be

$$\Phi_X(\omega) = \int_{-\infty}^{\infty} e^{j\omega x}f_X(x)dx = \int_{0}^{\infty} e^{j\omega x}e^{-x}dx = -\frac{e^{-(1-j\omega)x}}{1-j\omega}\bigg|_{0}^{\infty} = \frac{1}{1-j\omega}.$$

This result assumes that ω is a real quantity. Now suppose another random variable Y has a PDF given by $f_Y(y) = a\exp(-ay)u(y)$. Note that $f_Y(y) = af_X(ay)$, thus using the scaling property of Fourier transforms, the characteristic function associated with the random variable Y is given by

$$\Phi_Y(\omega) = a\frac{1}{|a|}\Phi_X\left(\frac{\omega}{a}\right) = \frac{1}{1-j\omega/a} = \frac{a}{a-j\omega},$$

assuming a is a positive constant (which it must be for Y to have a valid PDF). Finally, suppose that Z has a PDF given by $f_Z(z) = a\exp(-a(z-b))u(z-b)$. Since $f_Z(z) = f_Y(z-b)$, the shifting property of Fourier transforms can be used to help find the characteristic function associated with the random variable Z:

$$\Phi_Z(\omega) = \Phi_Y(\omega)e^{-j\omega b} = \frac{ae^{-j\omega b}}{a-j\omega}.$$

■

The next example demonstrates that the characteristic function can also be computed for discrete random variables. In Section 4.8, the probability-generating function will be introduced which is preferred by some when dealing with discrete random variables.

■ Example 4.19:

A binomial random variable has a PDF which can be expressed as

$$f_X(x) = \sum_{k=0}^{n} \binom{n}{k} p^k (1-p)^{n-k} \delta(x-k).$$

Its characteristic function is computed as follows:

$$\Phi_X(\omega) = \int_{-\infty}^{\infty} e^{j\omega x} \left(\sum_{k=0}^{n} \binom{n}{k} p^k (1-p)^{n-k} \delta(x-k) \right) dx = \sum_{k=0}^{n} \binom{n}{k} p^k (1-p)^{n-k} \int_{-\infty}^{\infty} \delta(x-k) e^{j\omega x} dx$$

$$= \sum_{k=0}^{n} \binom{n}{k} p^k (1-p)^{n-k} e^{j\omega k} = \sum_{k=0}^{n} \binom{n}{k} (pe^{j\omega})^k (1-p)^{n-k} = (1-p+pe^{j\omega})^n.$$

■

Since the Gaussian random variable plays such an important role in so many studies, we derive its characteristic function in Example 4.20. We recommend that the student commit the result of this example to memory. The techniques used to arrive at this result are also important and should be carefully understood.

■ Example 4.20:

For a standard normal random variable, the characteristic function can be found as follows:

$$\Phi_X(\omega) = \int_{-\infty}^{\infty} \frac{1}{\sqrt{2\pi}} e^{-\frac{x^2}{2}} e^{j\omega x} dx = \int_{-\infty}^{\infty} \frac{1}{\sqrt{2\pi}} \exp\left(-\frac{(x^2 - 2j\omega x)}{2}\right) dx.$$

To evaluate this integral, we complete the square in the exponent.

$$\Phi_X(\omega) = \exp\left(-\frac{\omega^2}{2}\right) \int_{-\infty}^{\infty} \frac{1}{\sqrt{2\pi}} \exp\left(-\frac{(x^2 - 2j\omega x - \omega^2)}{2}\right) dx = \exp\left(-\frac{\omega^2}{2}\right) \int_{-\infty}^{\infty} \frac{1}{\sqrt{2\pi}} \exp\left(-\frac{(x-j\omega)^2}{2}\right) dx.$$

The integrand in the above expression looks like the properly normalized PDF of a Gaussian random variable, and since the integral is over all values of x, the integral must be unity. However, close examination of the integrand reveals that the "mean" of this Gaussian integrand is complex. It is left to the student to rigorously verify that this integral still evaluates to unity even though the integrand is not truly a Gaussian PDF (since it is a complex function and hence not a PDF at all). The resulting characteristic function is then

$$\Phi_X(\omega) = \exp\left(-\frac{\omega^2}{2}\right).$$

For a Gaussian random variable whose mean is not zero or whose standard deviation is not unity (or both), the shifting and scaling properties of Fourier transforms can be used to show that

$$f_X(x) = \frac{1}{\sqrt{2\pi\sigma^2}} e^{-\frac{(x-\mu)^2}{2\sigma^2}} \leftrightarrow \Phi_X(\omega) = \exp\left(j\mu\omega - \frac{\omega^2\sigma^2}{2}\right).$$

■

Theorem 4.3: For any random variable whose characteristic function is differentiable at $\omega = 0$,

$$E[X] = -j \frac{\mathrm{d}}{\mathrm{d}\omega} \Phi_X(\omega) \bigg|_{\omega = 0}. \tag{4.38}$$

Proof: The proof follows directly from the fact that the expectation and differentiation operations are both linear and consequently the order of these operations can be exchanged.

$$\frac{\mathrm{d}}{\mathrm{d}\omega} \Phi_X(\omega) = \frac{\mathrm{d}}{\mathrm{d}\omega} (E[e^{j\omega X}]) = E\left[\frac{\mathrm{d}}{\mathrm{d}\omega}(e^{j\omega X})\right] = E[jX e^{j\omega X}] = jE[X e^{j\omega X}].$$

Multiplying both sides by $-j$ and evaluating at $\omega = 0$ produces the desired result. □

Theorem 4.3 demonstrates a very powerful use of the characteristic function. Once the characteristic function of a random variable has been found, it is generally a very straightforward thing to produce the mean of the random variable. Furthermore, by taking the kth derivative of the characteristic function and evaluating at $\omega = 0$, an expression proportional to the kth moment of the random variable is produced. In particular,

$$E[X^k] = (-j)^k \frac{\mathrm{d}^k}{\mathrm{d}\omega^k} \Phi_X(\omega) \bigg|_{\omega = 0}. \tag{4.39}$$

Hence, the characteristic function represents a convenient tool to easily determine the moments of a random variable.

■ Example 4.21:

Consider the exponential random variable of Example 4.18 where $f_Y(y) = a\exp(-ay)u(y)$. The characteristic function was found to be

$$\Phi_Y(\omega) = \frac{a}{a - j\omega}.$$

The derivative of the characteristic function is

$$\frac{\mathrm{d}}{\mathrm{d}\omega} \Phi_Y(\omega) = \frac{ja}{(a - j\omega)^2},$$

and thus the first moment of Y is

$$E[Y] = -j\frac{\mathrm{d}}{\mathrm{d}\omega} \Phi_Y(\omega) \bigg|_{\omega = 0} = \frac{a}{(a - j\omega)^2} \bigg|_{\omega = 0} = \frac{1}{a}.$$

(Continued)

For this example, it is not difficult to show that the kth derivative of the characteristic function is

$$\frac{d^k}{d\omega^k}\Phi_Y(\omega) = \frac{j^k k! a}{(a-j\omega)^{k+1}},$$

and from this, the kth moment of the random variable is found to be

$$E[Y^k] = (-j)^k\frac{d^k}{d\omega^k}\Phi_Y(\omega)\Big|_{\omega=0} = \frac{k!a}{(a-j\omega)^{k+1}}\Big|_{\omega=0} = \frac{k!}{a^k}.$$

For random variables that have a more complicated characteristic function, evaluating the kth derivative in general may not be an easy task. However, Equation (4.39) only calls for the kth derivative evaluated at a single point ($\omega = 0$), which can be extracted from the Taylor series expansion of the characteristic function. To see this, note that from Taylor's theorem, the characteristic function can be expanded in a power series as

$$\Phi_X(\omega) = \sum_{k=0}^{\infty}\frac{1}{k!}\left(\frac{d^k}{d\omega^k}\Phi_X(\omega)\Big|_{\omega=0}\right)\omega^k. \tag{4.40}$$

If one can obtain a power series expansion of the characteristic function, then the required derivatives are proportional to the coefficients of the power series. Specifically, suppose an expansion of the form

$$\Phi_X(\omega) = \sum_{k=0}^{\infty}\phi_k\omega^k, \tag{4.41}$$

is obtained. Then the derivatives of the characteristic function are given by

$$\frac{d^k}{d\omega^k}\Phi_X(\omega)\Big|_{\omega=0} = k!\,\phi_k. \tag{4.42}$$

The moments of the random variable are then given by

$$E[X^k] = (-j)^k k!\,\phi_k. \tag{4.43}$$

This procedure is illustrated using a Gaussian random variable in the next example.

■ Example 4.22:

Consider a Gaussian random variable with a mean of $\mu = 0$ and variance σ^2. Using the result of Example 4.20, the characteristic function is $\Phi_X(\omega) = \exp(-\omega^2\sigma^2/2)$. Using the well-known Taylor series expansion of the exponential function, the characteristic function is expressed as

$$\Phi_X(\omega) = \sum_{n=0}^{\infty} \frac{(-\omega^2\sigma^2/2)^n}{n!} = \sum_{n=0}^{\infty} \frac{(-1)^n\sigma^{2n}}{2^n n!}\omega^{2n}.$$

The coefficients of the general power series as expressed in Equation (4.41) are given by

$$\phi_k = \begin{cases} \dfrac{j^k(\sigma/\sqrt{2})^k}{(k/2)!}, & k \text{ even}, \\ 0, & k \text{ odd}. \end{cases}$$

Hence, the moments of the zero-mean Gaussian random variable are

$$E[X^k] = \begin{cases} \dfrac{k!}{(k/2)!}\left(\dfrac{\sigma}{\sqrt{2}}\right)^k, & k \text{ even}, \\ 0, & k \text{ odd}. \end{cases}$$

As expected, $E[X^0] = 1$, $E[X] = 0$ (since it was specified that $\mu = 0$), and $E[X^2] = \sigma^2$ (since in the case of zero-mean variables, the second moment and variance are one and the same). Now, we also see that $E[X^3] = 0$ (as are all odd moments), $E[X^4] = 3\sigma^4$, $E[X^6] = 15\sigma^6$, and so on. We can also conclude from this that for Gaussian random variables, the coefficient of skewness is $c_s = 0$ while the coefficient of kurtosis is $c_k = 3$. ■

In many cases of interest, the characteristic function has an exponential form. The Gaussian random variable is a typical example. In such cases, it is convenient to deal with the natural logarithm of the characteristic function.

Definition 4. 8: In general, we can write a series expansion of $\ln[\Phi_X(\omega)]$ as

$$\ln[\Phi_X(\omega)] = \sum_{n=1}^{\infty} \lambda_n \frac{(j\omega)^n}{n!}, \tag{4.44}$$

where the coefficients, λ_n, are called the *cumulants* and are given as

$$\lambda_n = \frac{d^n}{d(j\omega)^n}\{\ln[\Phi_X(\omega)]\}\Big|_{\omega = 0}, \quad n = 1, 2, 3, \dots \tag{4.45}$$

The cumulants are related to the moments of the random variable. By taking the derivatives specified in Equation (4.45) we obtain

$$\lambda_1 = \mu_X, \tag{4.46}$$

$$\lambda_2 = E[X^2] - \mu_X^2 = \sigma_X^2, \tag{4.47}$$

$$\lambda_3 = E[X^3] - 3\mu_X E[X^2] + 2\mu_X^3 = E[(X - \mu_X)^3]. \tag{4.48}$$

Thus, λ_1 is the mean, λ_2 is the second central moment (or the variance), and λ_3 is the third central moment. However, higher-order cumulants are not as simply related to the central moments.

4.8. Probability-Generating Functions

In the world of signal analysis, we often use Fourier transforms to describe continuous time signals, but when we deal with discrete time signals, it is common to use a z-transform instead. In the same way, the characteristic function is a useful tool for working with continuous random variables, but when discrete random variables are concerned, it is often more convenient to use a device similar to the z-transform which is known as the probability-generating function.

> **Definition 4.9:** For a discrete random variable with a PMF, $P_X(k)$, defined on the nonnegative integers,[1] $k = 0, 1, 2, \ldots$, the *probability-generating function*, $H_X(z)$, is defined as
>
> $$H_X(z) = \sum_{k=0}^{\infty} P_X(k) z^k. \tag{4.49}$$

Note the similarity between the probability-generating function and the unilateral z-transform of the PMF.

Since the PMF is seen as the coefficients of the Taylor series expansion of $H_X(z)$, it should be apparent that the PMF can be obtained from the probability-generating function through

$$P_X(k) = \frac{1}{k!} \frac{d^k}{dz^k} H_X(z) \Big|_{z=0}. \tag{4.50}$$

---

[1] Note that this definition assumes that the discrete random variable, X, is defined on nonnegative integer values, k. One could also define a probability-generating function based on a bilateral z-transform which would allow for random variables which can take on negative integer values as well. However, since this is less common, we do not consider it further here.

The derivatives of the probability-generating function evaluated at zero return the PMF and not the moments as with the characteristic function. However, the moments of the random variable can be obtained from the derivatives of the probability-generating function at $z = 1$.

Theorem 4.4: The mean of a discrete random variable can be found from its probability-generating function according to

$$E[X] = \frac{d}{dz}H_X(z)\bigg|_{z=1}. \qquad (4.51)$$

Furthermore, the higher-order derivatives of the probability-generating function evaluated at $z = 1$ lead to quantities which are known as the *factorial moments*,

$$h_k = \frac{d^k}{dz^k}H_X(z)\bigg|_{z=1} = E[X(X-1)(X-2)...(X-k+1)]. \qquad (4.52)$$

Proof: The result follows directly from differentiating Equation (4.49). The details are left to the reader. □

It is a little unfortunate that these derivatives do not produce the moments directly, but the moments can be calculated from the factorial moments. For example,

$$h_2 = E[X(X-1)] = E[X^2] - E[X] = E[X^2] - h_1,$$

$$\Rightarrow E[X^2] = h_2 + h_1. \qquad (4.53)$$

Hence, the second moment is simply the sum of the first two factorial moments. Furthermore, if we were interested in the variance as well, we would obtain

$$\sigma_X^2 = E[X^2] - \mu_X^2 = h_2 + h_1 - h_1^2. \qquad (4.54)$$

■ Example 4.23:

Consider the binomial random variable of Example 4.4 whose PMF is

$$P_X(k) = \binom{n}{k}p^k(1-p)^{n-k}, \; k = 0, 1, 2, ..., n.$$

The corresponding probability-generating function is

$$H_X(z) = \sum_{k=0}^{n}\binom{n}{k}p^k(1-p)^{n-k}z^k = \sum_{k=0}^{n}\binom{n}{k}(pz)^k(1-p)^{n-k} = (1-p+pz)^n.$$

(Continued)

Evaluating the first few derivatives at $z = 1$ produces

$$h_1 = \frac{d}{dz}H_X(z)\Big|_{z=1} = np(1-p+pz)^{n-1}\Big|_{z=1} = np \, ,$$

$$h_2 = \frac{d^2}{dz^2}H_X(z)\Big|_{z=1} = n(n-1)p^2(1-p+pz)^{n-2}\Big|_{z=1} = n(n-1)p^2 \, .$$

From these factorial moments, we calculate the mean, second moment, and variance of a binomial random variable as

$$\mu_X = h_1 = np \, , \; E[X^2] = h_1 + h_2 = (np)^2 + np(1-p) \, , \; \sigma_X^2 = h_2 + h_1 - h_1^2 = np(1-p) \, .$$ ∎

In order to gain an appreciation for the power of these "frequency domain" tools, compare the amount of work used to calculate the mean and variance of the binomial random variable using the probability-generating function in Example 4.23 with the direct method used in Example 4.4.

As was the case with the characteristic function, we can compute higher-order factorial moments without having to take many derivatives, by expanding the probability-generating function into a Taylor series. In this case, the Taylor series must be about the point $z = 1$.

$$H_X(z) = \sum_{k=0}^{\infty} \frac{1}{k!}\left(\frac{d^k}{dz^k}H_X(z)\Big|_{z=1}\right)(z-1)^k = \sum_{k=0}^{\infty}\frac{1}{k!}h_k(z-1)^k. \tag{4.55}$$

Once this series is obtained, one can easily identify all of the factorial moments. This is illustrated using a geometric random variable in Example 4.24.

■ Example 4.24:

A geometric random variable has a PMF given by $P_X(k) = (1-p)p^k$, $k = 0, 1, 2, \ldots$ The probability-generating function is found to be

$$H_X(z) = \sum_{k=0}^{\infty}(1-p)(pz)^k = \frac{1-p}{1-pz}.$$

In order to facilitate forming a Taylor series expansion of this function about the point $z \approx 1$, it is written explicitly as a function of $z - 1$. From there, the power series expansion is fairly simple.

$$H_X(z) = \frac{1-p}{1-p-p(z-1)} = \frac{1}{1-\frac{p}{1-p}(z-1)} = \sum_{k=0}^{\infty}\left(\frac{p}{1-p}\right)^k(z-1)^k \, .$$

Comparing the coefficients of this series with the coefficients given in Equation (4.55) leads to immediate identification of the factorial moments,

$$h_k = \frac{k!p^k}{(1-p)^k} \, .$$ ∎

4.9 Moment-Generating Functions

In many problems, the random quantities we are studying are often inherently nonnegative. Examples include the magnitude of a random signal, the time between arrivals of successive customers in a queueing system, or the number of points scored by your favorite football team. The resulting PDFs of these quantities are naturally one-sided. For such one-sided waveforms, it is common to use Laplace transforms as a frequency domain tool. The moment-generating function is the equivalent tool for studying random variables.

> **Definition 4.10:** The *moment-generating function*, $M_X(u)$, of a nonnegative[2] random variable, X, is
>
> $$M_X(u) = E[e^{uX}] = \int_0^\infty f_X(x)e^{ux}dx. \qquad (4.56)$$
>
> Note the similarity between the moment-generating function and the Laplace transform of the PDF.

The PDF can in principle be retrieved from the moment-generating function through an operation similar to an inverse Laplace transform,

$$f_X(x) = \frac{1}{2\pi j}\int_{c-j\infty}^{c+j\infty} M_X(u)e^{-ux}du. \qquad (4.57)$$

Because the sign in the exponential term in the integral in Equation (4.56) is the opposite of the traditional Laplace transform, the contour of integration (the so-called Bromwich contour) in the integral specified in Equation (4.57) must now be placed to the left of all poles of the moment-generating function. As with the characteristic function, the moments of the random variable can be found from the derivatives of the moment-generating function (hence, its name) according to

$$E[X^k] = \frac{d^k}{du^k}M_X(u)\bigg|_{u=0}. \qquad (4.58)$$

---

[2] One may also define a moment-generating function for random variables which are not necessarily nonnegative. In that case, a two-sided Laplace transform would be appropriate. This would be identical to the characteristic function with the association $u = j\omega$.

It is also noted that if the moment-generating function is expanded in a power series of the form

$$M_X(u) = \sum_{k=0}^{\infty} m_k u^k , \qquad (4.59)$$

then the moments of the random variable are given by $E[X^k] = k! m_k$.

■ Example 4.25:

Consider an Erlang random variable with a PDF of the form

$$f_X(x) = \frac{x^{n-1}\exp(-x)u(x)}{(n-1)!}$$

The moment-generating function is calculated according to

$$M_X(u) = \int_0^{\infty} f_X(x)e^{ux}dx = \int_0^{\infty} \frac{x^{n-1}\exp(-(1-u)x)}{(n-1)!}dx .$$

To evaluate this function, we note that the integral looks like the Laplace transform of the function $x^{n-1}/(n-1)!$ evaluated at $s = 1-u$. Using standard tables of Laplace transforms (or using integration by parts several times) we get

$$M_X(u) = \frac{1}{(1-u)^n} .$$

The first two moments are then found as follows:

$$E[X] = \frac{d}{du}(1-u)^{-n}\Big|_{u=0} = n(1-u)^{-(n+1)}\Big|_{u=0} = n ,$$

$$E[X^2] = \frac{d^2}{du^2}(1-u)^{-n}\Big|_{u=0} = n(n+1)(1-u)^{-(n+2)}\Big|_{u=0} = n(n+1) .$$

From this, we could also infer that the variance is $\sigma_X^2 = n(n+1) - n^2 = n$. If we wanted a general expression for the kth moment, it is not hard to see that

$$E[X^k] = \frac{d^k}{du^k}(1-u)^{-n}\Big|_{u=0} = n(n+1)...(n+k-1) = \frac{(n+k-1)!}{(n-1)!} .$$

■

4.10 Evaluating Tail Probabilities

A common problem encountered in a variety of applications is the need to compute the probability that a random variable exceeds a threshold, $\Pr(X > x_o)$. Alternatively we might want to know, $\Pr(|X - \mu_X| > x_o)$. These quantities are referred to as tail probabilities. That is, we are asking, what is the probability that the random variable takes on a value that is in the tail of the distribution? While this can be found directly from the CDF of the random variable, quite often, the CDF may be difficult or even impossible to find. In those cases, one can always resort to numerical integration of the PDF. However, this involves a numerical integration over a semi-infinite region, which in some cases may be problematic. Then, too, in some situations, we might not even have the PDF of the random variable, but rather the random variable may be described in some other fashion. For example, we may only know the mean, or the mean and variance, or the random variable may be described by one of the frequency domain functions discussed in the previous sections. Obviously, if we are only given partial information about the random variable, we would not expect to be able to perfectly evaluate the tail probabilities, but we can obtain bounds on these probabilities. In this section, we present several techniques for obtaining various bounds on tail probabilities based on different information about the random variable. We then conclude the section by showing how to exactly evaluate the tail probabilities directly from one of the frequency domain descriptions of the random variables.

Theorem 4.5 (Markov's inequality): Suppose that X is a nonnegative random variable (i.e., one whose PDF is nonzero only over the range $[0, \infty)$). Then,

$$\Pr(X \geq x_0) \leq \frac{E[X]}{x_0}. \tag{4.60}$$

Proof: For nonnegative random variables, the expected value is

$$E[X] = \int_0^\infty x f_X(x)\,dx = \int_0^{x_0} x f_X(x)\,dx + \int_{x_0}^\infty x f_X(x)\,dx \geq \int_{x_0}^\infty x f_X(x)\,dx \geq x_0 \int_{x_0}^\infty f_X(x)\,dx . \tag{4.61}$$

Dividing both sides by x_0 gives the desired result. □

Markov's inequality provides a bound on the tail probability. The bound requires only knowledge of the mean of the random variable. Because the bound uses such limited information, it has the potential of being very loose. In fact, if $x_0 < E[X]$, then the Markov inequality states that $\Pr(X \geq x_0)$ is bounded by a number that is greater than 1. While this is true, in this case the Markov inequality gives us no useful information. Even in less extreme cases, the result can still be very loose as shown by the next example.

■ **Example 4.26:**

Suppose the average life span of a person was 78 years. The probability of a human living to be 110 years would then be bounded by

$$\Pr(X \geq 110) \leq \frac{78}{110} = 0.7091 .$$

Of course, we know that in fact very few people live to be 110 years old, and hence this bound is almost useless to us.

■

If we know more about the random variable than just its mean, we can obtain a more precise estimate of its tail probability. In Example 4.26, we know that the bound given by the Markov's inequality is ridiculously loose because we know something about the variability of the human life span. The next result allows us to use the variance as well as the mean of a random variable to form a different bound on the tail probability.

Theorem 4.6 (Chebyshev's inequality): Suppose that X is a random variable with mean, μ_X, and variance, σ_X^2. The probability that the random variable takes on a value that is removed from the mean by more than x_o is given by

$$\Pr(|X - \mu_X| \geq x_o) \leq \frac{\sigma_X^2}{x_o^2}. \tag{4.62}$$

Proof: Chebyshev's inequality is a direct result of Markov's inequality. Note that the event $\{|X - \mu_X| \geq x_o\}$ is equivalent to the event $\{(X - \mu_X)^2 \geq x_o^2\}$. Applying Markov's inequality to the later event results in

$$\Pr((X - \mu_X)^2 \geq x_o^2) \leq \frac{E[(X - \mu_X)^2]}{x_o^2} = \frac{\sigma_X^2}{x_o^2}. \ \square \tag{4.63}$$

Chebyshev's inequality gives a bound on the two-sided tail probability, whereas the Markov inequality applies to the one-sided tail probability. Also, the Chebyshev inequality can be applied to any random variable, not just those that are nonnegative.

■ Example 4.27:

Continuing the previous example, suppose that in addition to a mean of 78 years, the human life span had a standard deviation of 15 years. In this case
$$\Pr(X \geq 110) \leq \Pr(X \geq 110) + \Pr(X \leq 46) = \Pr(|X - 78| \geq 32) \ .$$
Now the Chebyshev inequality can be applied to give
$$\Pr(|X - 78| \geq 32) \leq \left(\frac{15}{32}\right)^2 = 0.2197 \ .$$
While this result may still be quite loose, by using the extra piece of information provided by the variance, a better bound is obtained.

■

Theorem 4.7 (Chernoff bound): Suppose X is a random variable whose moment generating function is $M_X(s)$. Then

$$\Pr(X \geq x_o) \leq \min_{s \geq 0} \ e^{-sx_o} M_X(s). \tag{4.64}$$

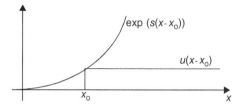

Figure 4.5
The unit step function and an exponential upper bound.

Proof: First, note that

$$\Pr(X \geq x_0) = \int_{x_0}^{\infty} f_X(x)dx = \int_{-\infty}^{\infty} f_X(x)u(x - x_0)dx . \qquad (4.65)$$

Next, upper bound the unit step function in the above integrand by an exponential function of the form $u(x - x_0) \leq \exp(s(x - x_0))$. This bound is illustrated in Figure 4.5. Note that the bound is valid for any real $s \geq 0$. The tail probability is then upper bounded by

$$\Pr(X \geq x_0) \leq e^{-sx_0} \int_{-\infty}^{\infty} f_X(x)e^{sx}dx = e^{-sx_0}M_X(s) . \qquad (4.66)$$

Since this bound is valid for any $s \geq 0$, it can be tightened by finding the value of s that minimizes the right-hand side. In this expression, a two-sided Laplace transform must be used to obtain the moment-generating function if the random variable is not nonnegative (see footnote 2 associated with Definition 4.10).   □

■ Example 4.28:

Consider a standard normal random variable whose moment-generating function is given by $M_X(u) = \exp(u^2/2)$ (see the result of Example 4.20, where the characteristic function is found and replace ω with $-ju$). The tail probability, $\Pr(X \geq x_0)$, in this case is simply the Q-function, $Q(x_0)$. According to (4.66), this tail probability can be bounded by

$$Q(x_0) \leq \exp\left(-ux_0 + \frac{u^2}{2}\right)$$

for any $u \geq 0$. Minimizing with respect to u, we get

$$\frac{d}{du}\exp\left(-ux_0 + \frac{u^2}{2}\right) = (-x_0 + u)\exp\left(-ux_0 + \frac{u^2}{2}\right) = 0 \Rightarrow u = x_0 .$$

(Continued)

Hence, the Chernoff bound on the tail probability for a standard normal random variable is

$$Q(x_o) \le \exp\left(-\frac{x_o^2}{2}\right).$$

The result of this example provides a convenient upper bound on the Q-function.

∎

Evaluating the Chernoff bound requires knowledge of the moment-generating function of the random variable. This information is sufficient to calculate the tail probability exactly since, in theory, one can obtain the PDF from the moment-generating function, and from there the exact tail probability can be obtained. However, in cases where the moment-generating function is of a complicated analytical form determining the PDF may be exceedingly difficult. Indeed, in some cases, it may not be possible to express the tail probability in closed form (like with the Gaussian random variable). In these cases, the Chernoff bound will often provide an analytically tractable expression which can give a crude bound on the tail probability. If a precise expression for the tail probability is required, the result of Theorem 4.8 will show how this can be obtained directly from the moment-generating function (without having to explicitly find the PDF).

Theorem 4.8: For a random variable, X, with a moment-generating function, $M_X(u)$, an exact expression for the tail probability, $\Pr(X \ge x_o)$, is given by

$$\Pr(X \ge x_o) = \frac{1}{2\pi j} \int_{c-j\infty}^{c+j\infty} \frac{M_X(u)}{u} e^{-ux_o} du, \qquad (4.67)$$

where the contour of integration is to the right of the origin, but to the left of all singularities of the moment-generating function in the right half plane.

Proof: The right tail probability is given in general by

$$\Pr(X \ge x_o) = \int_{x_o}^{\infty} f_X(x) dx. \qquad (4.68)$$

Then, replace the PDF in this integral, with an inverse transform of the moment-generating function as specified in Equation (4.57).

$$\Pr(X \ge x_o) = \int_{x_o}^{\infty} \frac{1}{2\pi j} \int_{c-j\infty}^{c+j\infty} M_X(u) e^{-ux} du dx = \frac{1}{2\pi j} \int_{c-j\infty}^{c+j\infty} M_X(u) \int_{x_o}^{\infty} e^{-ux} dx du. \quad (4.69)$$

Evaluating the inner integral results in

$$\Pr(X \geq x_0) = \frac{1}{2\pi j} \int_{c-j\infty}^{c+j\infty} \frac{M_X(u)}{u} e^{-ux_0} du, \text{ for } Re[u] > 0. \quad \square \qquad (4.70)$$

The integral specified in Equation (4.67) can be evaluated numerically or, when convenient to do so, it can also be evaluated by computing the appropriate residues.[3] For $x_0 > 0$ the contour of integration can be closed to the right. The resulting closed contour will encompass all the singularities of the moment-generating function in the right half plane. According to Cauchy's residue theorem, the value of the integral will then be $-2\pi j$ times the sum of the residues of all the singularities encompassed by the contour. Hence,

$$\Pr(X \geq x_0) = \sum_{\substack{\text{right half plane}}} \text{Residues}\left\{ \frac{M_X(u)}{u} e^{-ux_0} \right\}. \qquad (4.71)$$

If a precise evaluation of the tail probability is not necessary, several approximations to the integral in Equation (4.67) are available. Perhaps the simplest and most useful is known as the *saddle point approximation*. To develop the saddle point approximation, define $\psi(u) = \ln(M_X(u))$ and

$$\lambda(u) = \ln\left(\frac{M_X(u)}{u} e^{-ux_0}\right) = \psi(u) - ux_0 - \ln(u). \qquad (4.72)$$

Furthermore, consider a Taylor series expansion of the function $\lambda(u)$ about some point $u = u_0$,

$$\lambda(u) = \lambda(u_0) + \lambda'(u_0)(u - u_0) + \frac{1}{2}\lambda''(u_0)(u - u_0)^2 + \dots. \qquad (4.73)$$

In particular, if u_0 is chosen so that $\lambda'(u_0) = 0$, then near the point $u = u_0$, the integrand in Equation (4.67) behaves approximately like

$$\frac{M_X(u)}{u} e^{-ux_0} = e^{\lambda(u)} \approx \exp(\lambda(u_0)) \exp\left(\frac{1}{2}\lambda''(u_0)(u - u_0)^2\right). \qquad (4.74)$$

In general, the point $u = u_0$ will be a minima of the integrand as viewed along the real axis. This follows from the fact that the integrand is a concave function of u. A useful property of complex (analytic) functions tells us that if the function has a minima at some point u_0 as u

---

[3] The remainder of this section assumes the student is familiar with the concepts of contour integration and residue calculus. For those students not familiar with these topics, the remainder of this section can be skipped without any loss in continuity.

passes through it in one direction, the function will also have a maxima as u passes through the same point in the orthogonal direction. Such a point is called a "saddle point" since the shape of the function resembles a saddle near that point. If the contour of integration is selected to pass through the saddle point, the integrand will reach a local maximum at the saddle point. As just seen in Equation (4.74), the integrand also has a Gaussian behavior at and around the saddle point. Hence, using the approximation of (4.74) and running the contour of integration through the saddle point so that $u = u_0 + j\omega$ along the integration contour, the tail probability is approximated by

$$\Pr(X \geq x_0) \approx \frac{\exp(\lambda(u_0))}{2\pi j} \int_{u_0 - j\infty}^{u_0 + j\infty} \exp\left(\frac{1}{2}\lambda''(u_0)(u - u_0)^2\right) du$$

$$= \frac{\exp(\lambda(u_0))}{2\pi} \int_{-\infty}^{\infty} \exp\left(-\frac{1}{2}\lambda''(u_0)\omega^2\right) d\omega$$

$$= \frac{\exp(\lambda(u_0))}{\sqrt{2\pi\lambda''(u_0)}}$$

$$= \frac{M_X(u_0)\exp(-u_0 x_0)}{u_0\sqrt{2\pi\lambda''(u_0)}}. \tag{4.75}$$

The third step is accomplished using the normalization integral for Gaussian PDFs.

The saddle point approximation is usually quite accurate provided that $x_0 \gg E[X]$. That is, the farther we go out into the tail of the distribution, the better the approximation. If it is required to calculate $\Pr(X \geq x_0)$ for $x_0 < E[X]$, it is usually better to calculate the left tail probability in which case the saddle point approximation is

$$\Pr(X \leq x_0) \approx -\frac{M_X(u_0)\exp(-u_0 x_0)}{u_0\sqrt{2\pi\lambda''(u_0)}}, \tag{4.76}$$

where in this case, the saddle point, u_0, must be negative.

■ Example 4.29:

In this example, we form the saddle point approximation to the Q-function which is the right tail probability for a standard normal random variable. The corresponding moment-generating function is $M_X(u) = \exp(u^2/2)$. To find the saddle point, we note that

$$\lambda(u) = \frac{u^2}{2} - \ln(u) - ux_0.$$

We will need the first two derivatives of this function:

$$\lambda'(u) = u - \frac{1}{u} - x_o, \quad \lambda''(u) = 1 + \frac{1}{u^2}.$$

The saddle point is the solution to $\lambda'(u_o) = 0$. This results in a quadratic equation whose roots are

$$u_o = \frac{x_o \pm \sqrt{x_o^2 + 4}}{2}.$$

When calculating the right tail probability, the saddle point must be to the right of the imaginary axis, hence the positive root must be used:

$$u_o = \frac{x_o + \sqrt{x_o^2 + 4}}{2}.$$

The saddle point approximation then becomes

$$Q(x_o) \approx \frac{M_x(u_o)\exp(-u_o x_o)}{u_o\sqrt{2\pi\lambda''(u_o)}} = \frac{\exp\left(\frac{u_o^2}{2} - u_o x_o\right)}{\sqrt{2\pi(1 + u_o^2)}}.$$

The exact value of the Q-function and the saddle point approximation are compared in Figure 4.6. As long as x_o is not close to zero, this approximation is quite accurate.

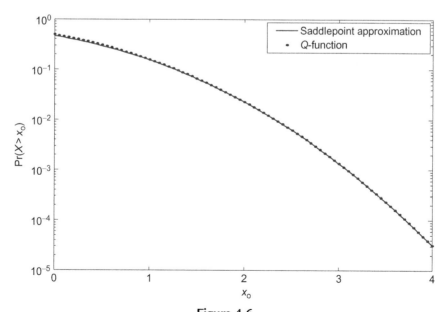

Figure 4.6
The Q-function and its saddle point approximation.

4.11 Engineering Application—Scalar Quantization

In many applications, it is convenient to convert a signal which is analog in nature to a digital one. This is typically done in three steps. First, the signal is sampled, which converts the signal from continuous time to discrete time. Then samples of the signal are quantized. This second action converts the signal from one with a continuous amplitude to one whose amplitude can only take on discrete values. Once the signal is converted to discrete time/discrete amplitude, the signal can then easily be represented by a sequence of bits. In this third step, each discrete amplitude level is represented by a binary codeword. While the first step (sampling) and the third step (encoding) are invertible, the second step (quantization) is not. That is, we can perfectly recover the analog signal from its discrete time samples (provided the samples are taken at a rate above the Nyquist rate), and the discrete amplitude levels can easily be recovered from the codewords which represent them (provided a lossless source code is used). However, the act of quantization causes distortion of the signal which cannot be undone. Given the discrete amplitude of a sample, it is not possible to determine the exact value of the original (continuous amplitude) sample. For this reason, careful attention is paid to the quantization process in order to minimize the amount of distortion.

In order to determine efficient ways to quantize signals, we must first quantify this concept of signal distortion. Suppose a signal is sampled and we focus attention on one of those samples. Let the random variable X represent the value of that sample, which in general will draw from a continuous sample space. Now suppose that sample is quantized (using some quantization function $q(x)$) to form a new (discrete) random variable $Y = q(X)$. The difference between the original sample value and its quantized value, $X - q(X)$, is the error caused by the quantizer, or the quantizer noise. It is common to measure signal distortion as the mean-squared quantizer error,

$$d = E[(X - q(X))^2] = \int_{-\infty}^{\infty} (x - q(x))^2 f_X(x) dx. \tag{4.77}$$

We will see in Chapter 10 that the mean-squared value of a signal has the physical interpretation of the signal's power. Thus, the quantity d can be interpreted as the quantization noise power. Often the fidelity of a quantized signal is measured in terms of the ratio of the original signal power, $E[X^2]$, to the quantization noise power. This is referred to as the signal-to-quantization-noise power ratio (SQNR)

$$\text{SQNR} = \frac{E[X^2]}{E[(X - q(X))^2]}. \tag{4.78}$$

The goal of the quantizer design is to choose a quantization function which minimizes the distortion, d. Normally, the quantizer maps the sample space of X into one of $M = 2^n$ levels. Then each quantization level can be represented by a unique n-bit codeword. We refer to this as an n-bit quantizer. As indicated in Equation (4.77), the expected value is with respect to the PDF of X. Hence, the function $q(x)$ that minimizes the distortion will depend on the distribution of X.

To start with consider a random variable X which is uniformly distributed over the interval $(-a/2, a/2)$. Since the sample X is equally likely to fall anywhere in the region, it would make sense for the quantizer to divide that region into M equally spaced subintervals of width $\Delta = a/M$. For each subinterval, the quantization level (i.e., the value of $q(x)$ for that subinterval) should be chosen as the midpoint of the subinterval. This is referred to as a uniform quantizer. A 3-bit uniform quantizer is illustrated in Figure 4.7. For example, if $X \in (0, a/8)$, then $q(X) = a/16$. To measure the distortion for this signal together with the uniform quantizer, condition on the event that the signal falls within one of the quantization intervals, and then use the theorem of total probability:

$$d = E[(X - q(X))^2] = \sum_{k=1}^{\infty} E[(X - q(X))^2 | X \in X_k] \Pr(X \in X_k), \tag{4.79}$$

where X_k refers to the kth quantization interval. Consider, for example, $X_5 = (0, a/8)$, so that

$$E[(X - q(X))^2 | X \in X_5] = E[(X - a/16)^2 | X \in (0, a/8)]. \tag{4.80}$$

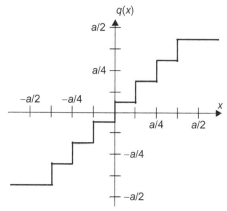

Figure 4.7
A 3-bit uniform quantizer on the interval $(-a/2, a/2)$.

To calculate this conditional expected value requires the conditional PDF of X. From Equation (3.41) this is

$$f_X(x|X \in (0, a/8)) = \frac{f_X(x)}{\Pr(X \in (0, a/8))} = \frac{1/a}{1/8} = \frac{8}{a}, \; x \in (0, a/8). \quad (4.81)$$

Not surprisingly, conditioned on $X \in (0, a/8)$, X is uniformly distributed over $(0, a/8)$. The conditional expected value is then

$$E[(X - a/16)^2 | X \in (0, a/8)] = \frac{8}{a} \int_0^{a/8} (x - a/16)^2 dx = \frac{a^2}{768}. \quad (4.82)$$

Due to the symmetry of the uniform distribution, this conditional distortion is the same regardless of what quantization interval the signal falls in. Hence, Equation (4.82) is also the unconditional distortion. Note that the power of the original signal is

$$E[X^2] = \frac{1}{a} \int_{-a/2}^{a/2} x^2 dx = \frac{a^2}{12}. \quad (4.83)$$

The resulting SQNR is then

$$\text{SQNR} = \frac{E[X^2]}{d} = \frac{a^2/12}{a^2/768} = 64 = 18.06 \text{ dB}. \quad (4.84)$$

The preceding result for the three-bit uniform quantizer can be generalized to any uniform quantizer. In general, for an n-bit uniform quantizer, there will be $M = 2^n$ quantization intervals of width $\Delta = a/M$. Consider the quantization interval $(0, \Delta)$ and suppose the quantization level for that interval is chosen to be the midpoint, $\Delta/2$. Then the distortion for that interval (and hence the distortion for the quantizer) is

$$E[(X - q(X))^2 | X \in (0, \Delta)] = \frac{1}{\Delta} \int_0^{\Delta} (x - \Delta/2)^2 dx = \Delta^2/12. \quad (4.85)$$

The SQNR for an n-bit uniform quantizer with a uniformly distributed input is

$$\text{SQNR} = \frac{E[X^2]}{d} = \frac{a^2/12}{\Delta^2/12} = M^2 = 2^{2n} \text{ or SQNR (dB)} = 2n\log_{10}(2) = 6.02n \text{ dB}. (4.86)$$

This is the so-called 6 dB rule whereby the SQNR is increased by approximately 6 dB for each extra bit added to the quantizer. For example, in wireline digital telephony, 8-bit quantization is used which would result in an SQNR of approximately 48 dB.

The previous results assumed that the input to the quantizer followed a uniform probability distribution. This is rarely the case. Speech signals, for example, are commonly modeled using a Laplace (two-sided exponential) distribution. For such signals, small sample values are much more frequent than larger values. In such a situation, it would make sense to use finer resolution (i.e., narrower quantization intervals) for the most frequent smaller amplitudes in order to keep the distortion minimized in those regions, at the cost of more distortion in the less frequent larger amplitude regions.

Given that an n-bit quantizer is to be used, the design of an optimum quantizer involves two separate problems. First, the ranges for each quantization interval must be specified; then the quantization level for each interval must be chosen. The following theorem specifies how each of these two tasks should be accomplished.

Theorem 4.9: A random variable X with PDF $f_X(x)$ is to be quantized with an M-level quantizer that produces a discrete random variable Y according to

$$y = q(x) = y_i, \text{ for } x_{i-1} < x < x_i, i = 1, 2, ..., M, \tag{4.87}$$

where it is assumed that the lower limit of the first quantization interval is $x_0 = -\infty$ and the upper limit of the last quantization interval is $x_M = \infty$. The (mean-squared) distortion is minimized by choosing the quantization intervals (i.e., the x_i) and the quantization levels (i.e., the y_i) according to

(i) $y_i = E[X|x_{i-1} < X < x_i], i = 1, 2, ..., M,$ (the conditional mean criterion), (4.88)

(ii) $x_i = \dfrac{y_i + y_{i+1}}{2}, i = 1, 2, ..., M-1,$ (the midpoint criterion). (4.89)

These two criteria provide a system of $2M-1$ equations with which to solve for the $2M-1$ quantities $(x_1, x_2, ..., x_{M-1}, y_1, y_2, ..., y_M)$ which specify the optimum M-level quantizer.

Proof: The distortion is given by

$$d = \sum_{i=1}^{M} \int_{x_{i-1}}^{x_i} (x - y_i)^2 f_X(x) dx. \tag{4.90}$$

To minimize d with respect to y_j,

$$\frac{\partial d}{\partial y_j} = -2 \int_{x_{i-1}}^{x_i} (x - y_i) f_X(x) dx = 0. \tag{4.91}$$

Solving for y_j in this equation establishes the conditional mean criterion. Similarly differentiating with respect to x_j gives

$$\frac{\partial d}{\partial x_j} = (x_j - y_j)^2 f_X(x_j) - (x_j - y_{j+1})^2 f_X(x_j) = 0 .$$ (4.92)

Solving for x_j produces the midpoint criterion. □

■ Example 4.30:

Using the criteria set forth in Theorem 4.9, an ideal nonuniform 2-bit quantizer will be designed for a signal whose samples have a Laplace distribution, $f_X(x) = (1/2)\exp(-|x|)$. A 2-bit quantizer will have four quantization levels, $\{y_1, y_2, y_3, y_4\}$, and four corresponding quantization intervals which can be specified by three boundary points, $\{x_1, x_2, x_3\}$. The generic form of the 2-bit quantizer is illustrated in Figure 4.8. Due to the symmetry of the Laplace distribution, it seems reasonable to expect that the quantizer should have a negative symmetry about the y-axis. That is, $x_1 = -x_3$, $x_2 = 0$, $y_1 = -y_4$, and $y_2 = -y_3$. Hence, it is sufficient to determine just three unknowns, such as $\{x_3, y_3, y_4\}$. The rest can be inferred from the symmetry. Application of the conditional mean criterion and the midpoint criterion leads to the following set of three equations:

$$y_3 = E[X|0 < X < x_3] = \frac{\int_0^{x_3} \frac{x}{2}\exp(-x)dx}{\int_0^{x_3} \frac{1}{2}\exp(-x)dx} = 1 - \frac{x_3}{e^{x_3} - 1},$$

$$y_4 = E[X|X > x_3] = \frac{\int_{x_3}^{\infty} \frac{x}{2}\exp(-x)dx}{\int_{x_3}^{\infty} \frac{1}{2}\exp(-x)dx} = x_3 + 1,$$

$$x_3 = \frac{y_3 + y_4}{2} .$$

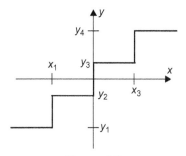

Figure 4.8
A 2-bit quantizer.

Plugging the expressions for y_3 and y_4 into the last equation results in a single equation to solve for the variable x_3. Unfortunately, the equation is transcendental and must be solved numerically. Doing so results in the solution $\{x_3, y_3, y_4\} = \{1.594, 0.594, 2.594\}$. The (mean-squared) distortion of this 2-bit quantizer is given by

$$d = 2\int_0^{y_3} (x - y_3)^2 f_X(x)\,dx + 2\int_{x_3}^{\infty} (x - y_4)^2 f_X(x)\,dx = 0.3524.$$

Note that the power in the original (unquantized) signal is $E[X^2] = 2$ so that the SQNR of this quantizer is

$$\text{SQNR} = \frac{E[X^2]}{d} = \frac{2}{0.3524} = 5.675 = 7.54 \text{ dB}.$$

∎

■ Example 4.31:

In this example, we generalize the results of the last example for an arbitrary number of quantization levels. When the number of quantization levels get large, the number of equations to solve becomes too difficult to do by hand, so we use MATLAB to help us with this task. Again, we assume that the random variable X follows a Laplace distribution given by $f_X(x) = (1/2)\exp(-|x|)$. Because of the symmetry of this distribution, we again take advantage of the fact that the optimum quantizer will be symmetric. We design a quantizer with M levels for positive X (and hence M levels for negative X as well, for a total of $2M$ levels). The M quantization levels are at $y_1, y_2, ..., y_M$ and the quantization bin edges are at $x_0 = 0, x_1, x_2, ..., x_{N-1}, x_N = \infty$. We compute the optimum quantizer in an iterative fashion. We start by arbitrarily setting the quantization bins in a uniform fashion. We choose to start with $x_i = 2i/M$. We then iterate between computing new quantization levels according to Equation (4.88) and new quantization bin edges according to Equation (4.89). After going back and forth between these two equations several times, the results converge toward a final optimum quantizer. For the Laplace distribution, the conditional mean criterion results in

$$y_i = \frac{(x_{i-1} + 1)\exp(-x_{i-1}) - (x_i + 1)\exp(-x_i)}{\exp(-x_{i-1}) - \exp(-x_i)}.$$

At each iteration stage (after computing new quantization levels), we also compute the the SQNR. By observing the SQNR at each stage, we can verify that this iterative process is in fact improving the quantizer design at each iteration. For this example, the SQNR is computed according to (see Exercise 4.82).

$$\text{SQNR} = \frac{1}{1 - \frac{1}{2}\sum_{i=1}^{M} p_i y_i^2}, \quad p_i = \exp(-x_{i-1}) - \exp(-x_i).$$

(Continued)

The MATLAB code we used to implement this process is included below. Figure 4.9 shows the results of running this code for the case of $M = 8$ (16 level, 4-bit quantizer).

```
M=8;
x=[0 2*[1:M-1]/M];                        % Initialize quantization bin edges
iterations=50;
for k=1:iterations
    % Update quantization levels
    x1=x(1:M-1);
    x2=x(2:M);
    y=(x1+1).*exp(-x1)-(x2+1).*exp(-x2);
    y=y./(exp(-x1)-exp(-x2));
    y=[y x(length(x))+1];
    % Calculate SQNR
    p=exp(-x1)-exp(-x2);
    p=[p exp(-x(length(x)))];
    SQNR(k)=1/(1-(y.^2)*p'/2);
    % Update quantization bin edges
    y1=y(1:M-1);
    y2=y(2:M);
    x=[0 (y1+y2)/2];
end
plot(10*log10(SQNR),'o')
```

Figure 4.9
SQNR measurements for the iterative quantizer design in Example 4.31. (For color version of this figure, the reader is refered to the web version of this chapter).

4.12 Engineering Application—Entropy and Source Coding

The concept of information is something we hear about frequently. After all, we supposedly live in the "information age" with the internet is often referred to as the "information superhighway." But, what is information? In this section, we give a quantitative definition of information and show how this concept is used in the world of digital communications.

To motivate the forthcoming definition, imagine a situation where we had a genie who could tell us about certain future events. Suppose that this genie told us that on July 15th of next year, the high temperature in the state of Texas would be above $90°F$. Anyone familiar with the weather trends in Texas would know that our genie has actually given us very little information. Since we know that the temperature in July in Texas is above $90°F$ with probability approaching one, the statement made by the genie does not tell us anything new. Next, suppose that the genie tells us that on July 15th of next year, the high temperature in the state of Texas would be below $80°F$. Since this event is improbable, in this case the genie would be giving us a great deal of information.

To define a numerical quantity which we will call information, we note from the previous discussion that

- Information should be a function of various events. Observing (or being told) that some event A occurs (or will occur) provides a certain amount of information, $I(A)$.
- The amount of information associated with an event should be inversely related to the probability of the event. Observing highly probable events provides very little information, while observing very unlikely events gives a large amount of information.

At this point, there are many definitions which could satisfy the two previous bullet items. We include one more observation that will limit the possibilities:

- If it is observed that two events, A and B, have occurred and if those two events are independent, then we expect that the information $I(A \cap B) = I(A) + I(B)$.

Since we observed that information should be a function of the probability of an event, the last bullet item requires us to define information as a function of probability that satisfies

$$I(p_A p_B) = I(p_A) + I(p_B). \tag{4.93}$$

Since a logarithmic function satisfies this property, we obtain the following definition.

Definition 4.11: If some event A occurs with probability p_A, then observing the event A provides an amount of *information* given by

$$I(A) = -\log(p_A). \tag{4.94}$$

The units associated with this measure of information depend on the base of the logarithm used. If base 2 logs are used, then the unit of information is a "bit"; if natural logs are used, the unit of information is the "nat."

Note that with this definition, an event which is sure to happen ($p_A = 1$) provides $I(A) = 0$ bits of information. This makes sense since if we know the event must happen, observing that it does happen provides us with no information.

Next, suppose we conduct some experiment which has a finite number of outcomes. The random variable X will be used to map those outcomes into the set of integers, $0, 1, 2, ..., n-1$. How much information do we obtain when we observe the outcome of the experiment? Since information is a function of the probability of each outcome, the amount of information is random and depends on which outcome occurs. We can, however, talk about the average information associated with the observation of the experiment.

> **Definition 4.12:** Suppose a discrete random variable X takes on the values $0, 1, 2, ..., n-1$ with probabilities $p_0, p_1, ..., p_{n-1}$. The *average information* or *(Shannon) entropy* associated with observing a realization of X is
>
> $$H(X) = \sum_{k=0}^{n-1} \Pr(X=k)I(X=k) = \sum_{k=0}^{n-1} p_k \log\left(\frac{1}{p_k}\right). \qquad (4.95)$$

Entropy provides a numerical measure of how much randomness or uncertainty there is in a random variable. In the context of a digital communication system, the random variable might represent the output of a data source. For example, suppose a binary source outputs the letters $X = 0$ and $X = 1$ with probabilities p and $1-p$, respectively. The entropy associated with each letter the source outputs is

$$H(X) = p\log\left(\frac{1}{p}\right) + (1-p)\log\left(\frac{1}{1-p}\right) = \mathcal{H}(p). \qquad (4.96)$$

The function $\mathcal{H}(p)$ is known as the binary entropy function and is plotted in Figure 4.10. Note that this function has a maximum value of 1 bit when $p = 1/2$. Consequently, to maximize the information content of a binary source, the source symbols should be equally likely.

Next, suppose a digital source described by a discrete random variable X periodically outputs symbols. Then the information rate of the source is given by $H(X)$ bits/source symbol. Furthermore, suppose we wish to represent the source symbols with binary codewords such that the resulting binary representation of the source outputs uses r bits/symbol. A fundamental result of source coding, which we will not attempt to prove here, is that if we desire the source code to be lossless (that is, we can always recover the original source

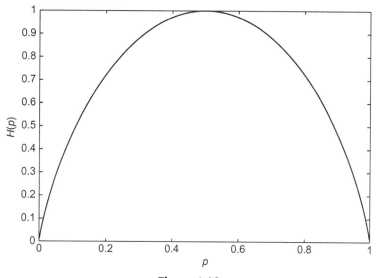

Figure 4.10
The binary entropy function.

symbols from their binary representation), then the source code rate must satisfy $r \geq H(X)$. In other words, the entropy of a source provides a lower bound on the average number of bits that are needed to represent each source output.

■ Example 4.32:

Consider a source that outputs symbols from a four-letter alphabet. That is, suppose $X \in \{a, b, c, d\}$. Let $p_a = 1/2$, $p_b = 1/4$, and $p_c = p_d = 1/8$ be the probability of each of the four source symbols. Given this source distribution, the source has an entropy of

$$H(X) = \frac{1}{2}\log(2) + \frac{1}{4}\log(4) + 2*\frac{1}{8}\log(8) = 1.75 \text{bits/source symbol}.$$

Table 4.2 shows several different possible binary representations of this source. This first code is the simplest and most obvious representation. Since there are four letters, we can always assign a unique 2-bit codeword to represent each source letter. This results in a code rate of $r_1 = 2$ bits/symbol which is indeed greater than the entropy of the source. The second code uses variable length codewords. The average codeword length is

$$r_2 = \frac{1}{2}*1 + \frac{1}{4}*2 + \frac{1}{4}*3 = 1.75 \text{ bits/symbol}.$$

Therefore, this code produces the most efficient representation of any lossless source coded since the code rate is equal to the source entropy. Note that Code 3 from Table 4.2 produces a code rate of

(Continued)

$$r_3 = \frac{3}{4}*1 + \frac{1}{4}*2 = 1.25 \text{ bits/symbol},$$

which is lower than the entropy, but this code is not lossless. This can easily be seen by noting that the source sequences "*d*" and "*a,b*" both lead to the same encoded sequence "01."

Table 4.2: Three possible codes for a four-letter source

	Source Letters			
	a	*b*	*c*	*d*
Code 1	00	01	10	11
Code 2	0	10	110	111
Code 3	0	1	10	01

■

Section 4.1: Expected Values of a Random Variable

4.1   Find the mean of the random variables described by each of the following probability density functions:

(a) $f_W(w) = \dfrac{1}{b-a}$, $a < w < b$;   (c) $f_Y(y) = y(1-y)$, $0 < y < 1$;

(b) $f_X(x) = 3x^2$, $0 < x < 1$;   (d) $f_Z(z) = \dfrac{2}{\pi}\dfrac{1}{1+z^2}$, $z > 0$.

Note: Each of the PDFs described above is zero outside the intervals indicated.

4.2   Find the mean of the random variables described by each of the following probability mass functions:

(a) $P_K(k) = \dfrac{1}{k_0+1}$, $k = 0, 1, 2, ..., k_0$;

(b) $P_M(m) = \dfrac{2m}{m_0(m_0+1)}$, $m = 0, 1, 2, ..., m_0$;

(c) $P_N(n) = \left(\dfrac{1}{2}\right)^n$, $n = 1, 2, 3, ...$;

(d) $P_Q(q) = \dbinom{q-1}{q_0-1}\left(\dfrac{1}{2}\right)^q$, $q = q_0, q_0+1, q_0+2, ...$.

Note: Each of the PMFs described above is zero outside the intervals indicated.

4.3   Find the mean of the random variables described by each of the following cumulative distribution functions:

(a) $F_W(w) = \begin{cases} 0, & w < 0, \\ w/10, & 0 < w < 10 \\ 1, & w > 10; \end{cases}$ (c) $F_Y(y) = \begin{cases} 0, & y < 0, \\ \frac{1}{4}y^2, & 0 \le y \le 2, \\ 1, & y > 2; \end{cases}$

(b) $F_X(x) = [1 - \exp(-2x)]u(x)$;   (d) $F_Z(z) = [1 - \exp(-z^2)]u(z)$.

4.4 In each of the following cases, find the value of the parameter a which causes the indicated random variable to have a mean value of 10.

(a) PDF: $f_W(w) = \begin{cases} \dfrac{2}{a}, & \dfrac{a}{2} < w < a, \\ 0, & \text{otherwise} \end{cases}$

(b) PMF: $P_M(m) = \left(\dfrac{1}{2}\right)^{m-a+1}, \quad m = a, a+1, a+2, \ldots$

(c) CDF: $F_X(x) = \begin{cases} 0, & x < 0, \\ ax^2, & 0 \le x \le 10, \\ \dfrac{1-a}{10}x + 2a - 1, & 10 \le x \le 20, \\ 1, & x > 20. \end{cases}$

4.5 Suppose a random variable X has a PDF which is nonzero only on the interval $[0, \infty)$. That is, the random variable cannot take on negative values. Prove that

$$E[X] = \int_0^\infty [1 - F_X(x)]dx.$$

Section 4.2: Expected Values of Functions of a Random Variable

4.6 Two players compete against each other in a game of chance where Player A wins with probability 1/3 and Player B wins with probability 2/3. Every time Player A loses he must pay Player B \$1, while every time Player B loses he must pay Player A \$3. Each time the two play the game, the results are independent of any other game. If the two players repeat the game 10 times, what is the expected value of Player A's winnings?

4.7 The current flowing through a 75 Ω resistor is modelled as a Gaussian random variable with parameters, $m = 0\text{A}$ and $\sigma = 15\text{ mA}$. Find the average value of the power consumed in the resistor.

4.8 The received voltage in a 75 Ω antenna of a wireless communication system is modeled as a Rayleigh random variable, $f_V(v) = \dfrac{v}{\sigma^2}\exp\left(-\dfrac{v^2}{2\sigma^2}\right)u(v)$. What does the value of the parameter σ need to be for the received power to be $10\,\mu\text{W}$?

4.9 Suppose X is a Gaussian random variable with a mean of μ and a variance of σ^2 (i.e., $X \sim N(\mu, \sigma^2)$). Find an expression for $E[|X|]$.

4.10 Prove *Jensen's inequality,* which states that for any convex function $g(x)$ and any random variable X,
$$E[g(X)] \geq g(E[X]) .$$

Section 4.3: Moments

4.11 Find an expression for the m th moment of an Erlang random variable whose PDF is given by $f_X(x) = \dfrac{b^n}{(n-1)!} x^{n-1} e^{-bx} u(x)$ for some positive integer n and positive constant b.

4.12 Find an expression for the even moments of a Rayleigh random variable. That is, find $E[Y^{2m}]$ for any positive integer m if the random variable, Y, has a PDF given by
$$f_Y(y) = \frac{y}{\sigma^2} \exp\left(-\frac{y^2}{2\sigma^2}\right) u(y) .$$

4.13 Find the first three moments of a geometric random variable whose PMF is $P_N(n) = (1-p)p^n$, $n = 0, 1, 2, \dots$.

4.14 Find the first three moments of a Poisson random variable whose PMF is
$$P_M(m) = \frac{\alpha^m e^{-\alpha}}{m!}, \quad m = 0, 1, 2, \dots .$$

4.15 For the Rayleigh random variable described in Exercise 4.12, find a relationship between the n th moment, $E[Y^n]$, and the n th moment of a standard normal random variable.

4.16 Suppose X is a random variable whose n th moment is g_n, $n = 1, 2, 3 \dots$. In terms of the g_n, find an expression for the m th moment of the random variable $Y = aX + b$ for constants a and b.

4.17 Suppose X is a random variable whose n th moment is g_n, $n = 1, 2, 3 \dots$. In terms of the g_n, find an expression for $E[e^X]$.

Section 4.4: Central Moments

4.18 Calculate the mean value, second moment, and variance of each of the following random variables:

(a)   Binomial, $P_X(k) = \binom{n}{k} p^k (1-p)^{n-k}$, $k = 0, 1, 2, ..., n$;

(b)   Poisson, $P_X(k) = \frac{\alpha^k}{k!} e^{-\alpha}$, $k = 0, 1, 2, ...$;

(c)   Laplace, $f_X(x) = \frac{1}{2b} \exp\left(-\frac{|x|}{b}\right)$;

(d)   Gamma, $f_X(x) = \dfrac{(x/b)^{c-1} \exp\left(-\frac{x}{b}\right)}{b\Gamma(c)} u(x)$;

4.19 For a Gaussian random variable, derive expressions for the coefficient of skewness and the coefficient of kurtosis in terms of the mean and variance, μ and σ^2.

4.20 Prove that all odd central moments of a Gaussian random variable are equal to zero. Furthermore, develop an expression for all even central moments of a Gaussian random variable.

4.21 Show that the variance of a Cauchy random variable is undefined (infinite).

4.22 Let c_n be the nth central moment of a random variable and μ_n be its nth moment. Find a relationship between c_n and μ_k, $k = 0, 1, 2, ..., n$.

4.23 Let X be a random variable with $E[X] = 1$ and $\text{var}(X) = 4$. Find the following:
(a) $E[2X-4]$;
(b) $E[X^2]$;
(c) $E[(2X-4)^2]$.

4.24 A random variable X has a uniform distribution over the interval $(-a/2, a/2)$ for some positive constant a.
(a)   Find the coefficient of skewness for X;
(b)   Find the coefficient of kurtosis for X;
(c)   Compare the results of (a) and (b) with the same quantities for a standard normal random variable.

4.25 Suppose Θ is a random variable uniformly distributed over the interval $[0, 2\pi)$.
(a) Find the PDF of $Y = \sin(\Theta)$.
(b) Find the PDF of $Z = \cos(\Theta)$.
(c) Find the PDF of $W = \tan(\Theta)$.

4.26 A random variable has a CDF given by

$$F_X(x) = \begin{cases} 0, & x < 0, \\ x^2, & 0 \le x \le 1, \\ 1, & x > 1 \end{cases}$$

(a) Find the mean of X;
(b) Find the variance of X;
(c) Find the coefficient of skewness of X;
(d) Find the coefficient of kurtosis of X.

4.27 Find the variance and coefficient of skewness for a geometric random variable whose PMF is $P_N(n) = (1-p)p^n$, $n = 0, 1, 2, \ldots$. *Hint:* You may want to use the results of Exercise 4.13.

4.28 Find the variance and coefficient of skewness for a Poisson random variable whose PMF is $P_M(m) = \dfrac{\alpha^m e^{-\alpha}}{m!}$, $m = 0, 1, 2, \ldots$. *Hint:* You may want to use the results of Exercise 4.14.

Section 4.5: Conditional Expected Values

4.29 Show that the concept of total probability can be extended to expected values. That is, if $\{A_i\}$, $i = 1, 2, 3, \ldots, n$ is a set of mutually exclusive and exhaustive events, then

$$E[X] = \sum_{k=1}^{n} E[X|A_k]Pr(A_k).$$

4.30 An exponential random variable has a PDF given by $f_X(x) = \exp(-x)u(x)$.
(a) Find the mean and variance of X.
(b) Find the conditional mean and the conditional variance given that $X > 1$.

4.31 A uniform random variable has a PDF given by $f_X(x) = u(x) - u(x-1)$.
   (a) Find the mean and variance of X.
   (b) Find the conditional mean and the conditional variance given that $\frac{1}{2} < X < \frac{3}{4}$.

4.32 Consider a Gaussian random variable, X, with mean μ and variance σ^2.

   (a) Find $E[X|X > \mu + \sigma]$.
   (b) Find $E[X||X - \mu| < \sigma]$.

4.33 A professor is studying the performance of various student groups in his class. Let the random variable X represent a students final score in the class and define the following conditioning events:
   - F = {student is a freshman}
   - So = {student is a sophomore}
   - J = {student is a junior}
   - Se = {student is a senior}
   - M = {student is a male}
   - F = {student is a female}
   (a) Suppose the average score among males is $E[X|M] = 73.2$ and the average score among females is $E[X|F] = 75.8$. If the overall class average score is $E[X] = 74.6$, what percentage of the class is female?
   (b) Suppose the conditional average scores by class are $E[X|F] = 65.8$, $E[X|So] = 71.2$, $E[X|J] = 75.4$, $E[X|Se] = 79.1$. If the overall class average score is $E[X] = 72.4$, what can we say about the percentage of freshmen, sophomore, juniors and seniors in the class?
   (c) Given the class statistics in part (b). If it is known that there are 10 freshmen, 12 sophomores, and 9 juniors in the class, how many seniors are in the class?

Section 4.6: Transformations of Random Variables

4.34 Suppose X is uniformly distributed over $(-a, a)$, where a is some positive constant. Find the PDF of $Y = X^2$.

4.35 Suppose X is a random variable with an exponential PDF of the form $f_X(x) = 2e^{-2x}u(x)$. A new random variable is created according to the transformation $Y = 1 - X$.
   (a) Find the domain for X and Y.
   (b) Find $f_Y(y)$.

4.36 Let X be a standard normal random variable (i.e., $X \sim N(0, 1)$). Find the PDF of $Y = |X|$.

4.37 Repeat Exercise 4.36 if the transformation is

$$Y = \begin{cases} X, & X > 0, \\ 0, & X \le 0. \end{cases}$$

4.38 Suppose a random variable, X, has a Gaussian PDF with zero mean and variance σ_X^2. The random variable is transformed by the device whose input–output relationship is shown in the accompanying figure. Find and sketch the PDF of the transformed random variable, Y.

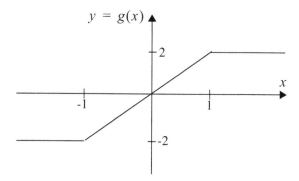

4.39 Let X be a Gaussian random variable with zero mean and arbitrary variance, σ^2. Given the transformation $Y = X^3$, find $f_Y(y)$.

4.40 A real number between 0 and 100 is randomly selected according to a uniform distribution and rounded off to the nearest integer. For example, 36.5001 is rounded off to 37; $\sqrt{3}$ is rounded off to 2; and 69.49 is rounded off to 69. Define a random variable to be $X =$ (number selected) $-$ (nearest integer).
   (a) What is the domain of this random variable?
   (b) Determine the PDF for X.
   (c) Determine the mean square value of X.

4.41 A Gaussian random variable with zero mean and variance σ_X^2 is applied to a device that has only two possible outputs, 0 or 1. The output 0 occurs when the input is negative, and the output 1 occurs when the input is nonnegative.
   (a) What is the probability mass function of the output?
   (b) Rework the problem when $\mu_X = 1/2$ and $\sigma_X^2 = 1$.

4.42 Let X be a Cauchy random variable whose PDF is given by

$$f_X(x) = \frac{b/\pi}{b^2 + x^2}$$

Find the PDF of $Y = 1/X$.

4.43 Let X be a Chi-square random variable with a PDF given by

$$f_X(x) = \frac{x^{c-1}\exp(-x/2)u(x)}{2^c\Gamma(c)},$$

where $c = n/2$ for any positive integer n. Find the PDF of $Y = \sqrt{X}$.

4.44 Suppose a random variable has some PDF given by $f_X(x)$. Find a function $g(x)$ such that $Y = g(X)$ is a uniform random variable over the interval $(0, 1)$. Next, suppose that X is a uniform random variable. Find a function $g(x)$ such that $Y = g(X)$ has some specified PDF, $f_Y(y)$.

4.45 Suppose X is uniformly distributed over $(0, 1)$. Using the results of Exercise 4.44, find transformations $Y = g(X)$ to produce random variables with the following distributions:
   (a)   Exponential,
   (b)   Rayleigh,
   (c)   Cauchy,
   (d)   Geometric,
   (e)   Poisson.

4.46 Suppose X is a binomial random variable with parameters n and p. That is, the PMF of X is given by

$$P_X(k) = \binom{n}{k}p^k(1-p)^{n-k}, \quad k = 0, 1, 2, \ldots, n.$$

Find the PMF of a new random variable generated through the transformation, $Y = n - X$.

4.47 Suppose X is a Gaussian random variable with mean μ_X and variance σ_X^2. Suppose we form a new random variable according to $Y = aX + b$ for constants a and b.
   (a)   Prove that Y is also Gaussian for any $a \neq 0$.
   (b)   What values for the constants a and b will lead to the new random variable Y having zero mean and unit variance?
   (c)   What values for the constants a and b will lead to the new random variable Y having a mean of μ_Y and a variance of σ_Y^2?

4.48 The amplitude, A, of a received signal in a wireless communication system is often modelled as a Rayleigh random variable whose PDF is of the form, $f_A(a) = \dfrac{a}{\sigma^2} \exp\left(-\dfrac{a^2}{2\sigma^2}\right) u(a)$.

The (normalized) power of the same signal is given by $P = A^2$. Find the PDF of P.

4.49 Suppose X is an exponential random variable with PDF, $f_X(x) = \exp(-x)u(x)$. Find a transformation, $Y = g(X)$ so that the new random variable Y has a Cauchy PDF given by $f_Y(y) = \dfrac{1/\pi}{1+x^2}$. *Hint:* Use the results of Exercise 4.44.

Section 4.7: Characteristic Functions

4.50 A random variable X has a characteristic function, $\phi_X(\omega)$. Write the characteristic function of $Y = aX + b$ in terms of $\phi_X(\omega)$ and the constants a and b.

4.51 Prove that the characteristic function of any random variable must satisfy the following properties.

(a) $\phi_X^*(\omega) = \phi_X(-\omega)$.

(b) $\phi_X(0) = 1$.

(c) For real ω, $|\phi_X(\omega)| \leq 1$.

(d) If the PDF is symmetric about the origin (i.e, an even function), then $\phi_X(\omega)$ is real.

(e) $\phi_X(\omega)$ cannot be purely imaginary.

4.52 Suppose X is an integer-valued random variable. Show that in this case, $\phi_X(2\pi n) = 1$ for any integer, n. Likewise, prove the reverse is also true. That is, show that if $\phi_X(2\pi n) = 1$ for any integer, n, the random variable X must be integer-valued.

4.53 For a Laplace random variable whose PDF is given by $f_X(x) = \dfrac{1}{2b} \exp\left(-\dfrac{|x|}{b}\right)$, find the following:

(a) the characteristic function, $\phi_X(\omega)$,

(b) the Taylor series expansion of $\phi_X(\omega)$,

(c) a general expression for the kth moment of X.

4.54 An Erlang Random variable has a PDF of the form $f_X(x) = \dfrac{b^n}{(n-1)!} x^{n-1} e^{-bx} u(x)$.

(a) Find the characteristic function, $\phi_X(\omega)$.

(b) Find the Taylor series expansion of $\phi_X(\omega)$.

(c) Find a general expression for the kth moment of X.

4.55 A Cauchy random variable has a PDF $f_X(x) = \dfrac{b/\pi}{b^2 + x^2}$.

   (a)   Find the characteristic function, $\phi_X(\omega)$.

   (b)   Show that the derivatives $\dfrac{d^k}{d\omega^k}(\phi_X(\omega))$ do not exist at $\omega = 0$. What does this mean?

4.56 A Poisson random variable has a PMF of the form $P_X(k) = \dfrac{\alpha^k}{k!}\exp(-\alpha)$, $k = 0, 1, 2, \ldots$.

   (a)   Find the characteristic function, $\phi_X(\omega)$.

   (b)   Find the first three nonzero terms in the Taylor series expansion of $\ln[\phi_X(\omega)]$.

   (c)   Use the results of part (b) to find the mean, variance, and skewness of the Poisson random variable.

4.57 A certain random variable has a characteristic function given by

$$\phi_X(\omega) = \frac{\sin^3(\pi\omega)}{(\pi\omega)^3}.$$

Find the PDF of this random variable.

4.58 Which of the following functions could be the characteristic function of a random variable? See Appendix E, Section 5 for definitions of these functions.

   (a) $\phi_a(\omega) = \text{rect}(\omega)$.

   (b) $\phi_b(\omega) = \text{tri}(\omega)$.

   (c) $\phi_c(\omega) = \text{sinc}(\omega)$.

   (d) $\phi_d(\omega) = \text{sinc}^2(\omega)$.

Section 4.8: Probability-Generating Functions

4.59 Derive a formula expressing the variance of a random variable in terms of its factorial moments.

4.60 Derive a relationship between the kth factorial moment for a nonnegative, integer-valued random variable and the coefficients of the Taylor series expansion of its probability-generating function, $H_X(z)$, about the point $z = 1$.

4.61 For a Poisson random variable whose PMF is given by $P_X(k) = \dfrac{\alpha^k}{k!} e^{-\alpha}$, $k = 0, 1, 2, \ldots$, find the following:

    (a)   the probability-generating function, $H_X(z)$,

    (b)   the Taylor series expansion of $H_X(z)$ about the point $z = 1$,

    (c)   a general expression for the kth factorial moment.

4.62 A certain random variable has a probability-generating function given by

$$H_X(z) = \frac{1}{n}\frac{1 - z^n}{1 - z}.$$

Find the PMF for this random variable.

4.63 Show that for any probability-generating function, $H(z)$, $H(1) = 1$.

4.64 Suppose $H_X(z)$ is the probability-generating function of some random variable X with PMF $P_X(k)$. In terms of $P_X(k)$, find the PMF of the random variable Y if its probability-generating function is given as in each of the following cases.

    (a) $H_Y(z) = H_X\left(\dfrac{z}{2}\right)$.

    (b) $H_Y(z) = H_X^2(z)$.

    (c) $H_Y(z) = z^m H_X(z)$.

    (d) $H_Y(z) = H_X(z^{-1})$.

Section 4.9: Moment-Generating Functions

4.65 Derive an expression for the moment-generating function of a Rayleigh random variable whose PDF is

$$f_X(x) = x \exp\left(-\frac{x^2}{2}\right) u(x).$$

4.66 Suppose X is a Rician random variable with a PDF given by

$$f_X(x) = x \exp\left(-\frac{x^2 + a^2}{2}\right) I_0(ax) u(x).$$

Derive an expression for $E[e^{uX^2}]$. Note that this is not quite the moment-generating function, but it can be used in a similar way.

4.67 A Gaussian mixture is a random variable whose PDF is a linear combination of two Gaussian PDFs,

$$f_X(x) = \frac{p}{\sqrt{2\pi\sigma_1^2}}\exp\left(-\frac{(x-\mu_1)^2}{2\sigma_1^2}\right) + \frac{1-p}{\sqrt{2\pi\sigma_2^2}}\exp\left(-\frac{(x-\mu_2)^2}{2\sigma_2^2}\right).$$

(a) Find the moment-generating function, $M_X(u)$, of the Gaussian mixture.

(b) Use the moment-generating function to find the mean of the Gaussian mixture.

4.68 A random variable has a moment-generating function given by

$$M_X(u) = \frac{2}{(1-u)(2-u)}.$$

(a) Find the PDF of the random variable.

(b) Use the moment-generating function to find an expression for the kth moment of the random variable.

4.69 Consider a moment-generating function of the general form

$$M_X(u) = \frac{c}{au^2 + bu + c},$$

for constants a, b, and c. Find constraints that the constants a, b, and c must satisfy so that $M_X(u)$ is the MGF of a valid random variable.

4.70 Prove the following properties of moment-generating functions.

(a) $M_X(0) = 1$.

(b) For a nonnegative random variable X, and for real $u < 0$, $M_X(u) \leq 1$.

4.71 Find the mean of the random variable whose moment-generating function is

$$M_X(u) = \frac{6}{(1-u)^2(2-u)(3-u)}.$$

Section 4.10: Evaluating Tail Probabilities

4.72 Prove that for a random variable X with mean μ_X,

$$\Pr(|X - \mu_X| > \varepsilon) < \frac{E[|X - \mu_X|^n]}{\varepsilon^n},$$

where n is any positive integer.

4.73 Suppose we are interested in finding the left tail probability for a random variable, X. That is, we want to find $\Pr(X \leq x_o)$. Rederive an expression for the Chernoff bound for the left tail probability.

4.74 Suppose X is a Poisson random variable with PMF, $P_X(k) = \dfrac{\alpha^k}{k!} \exp(-\alpha)$, $k = 0, 1, 2, \ldots$. Find the Chernoff bound for the tail probability, $\Pr(X \geq n_o)$.

4.75 Suppose X is a Gamma random variable with PDF,

$$f_X(x) = \frac{(x/b)^{c-1} \exp(-x/b)}{b \Gamma(c)} u(x).$$ Find the Chernoff bound for the tail probability,

$\Pr(X > x_o)$.

4.76 Let X be an Erlang random variable with PDF, $f_X(x) = \dfrac{x^{n-1} e^{-x} u(x)}{(n-1)!}$. Derive a saddle point approximation for the left tail probability, $\Pr(X < x_o)$. Compare your result with the exact value for $0 \leq x_o < E[X]$.

4.77 In Exercise 4.66, an expression was derived for $E[e^{uX^2}]$ for a Rician random variable. Use this function to obtain a saddle point approximation for the tail probability of a Rician random variable, $\Pr(X \geq x_o)$.

Hint: For one-sided random variables, $\Pr(X \geq x_o) = \Pr(X^2 \geq x_o^2)$.

4.78 The average number of cars per hour arriving at a toll booth is 57 while the standard deviation is 15.
  (a) Use Markov's inequality to find an upper bound on the probability of having more than 200 cars arrive in an hour.
  (b) Use Chebyshev's inequality to find an upper bound on the probability of having more than 200 cars arrive in an hour.

4.79 In a certain communication system, bits are encoded into blocks of length 128 bits. Error correction techniques are used such that the block will be decoded correctly as long as no more than 7 of the 128 bits are received in error. We observe that 2% of our blocks are decoded incorrectly. From Markov's inequality, what can we say about the average number of bit errors that occur in each 128 bit block?

4.80 A nonnegative random variable X has moments which are known to be $E[X] = 1$, $E[X^2] = 2$, $E[X^3] = 5$, $E[X^4] = 9$, $E[X^5] = 14$, $E[X^6] = 33$.
  (a) Show that for any nonnegative random variable, $\Pr(X \geq x_0) \leq \dfrac{E[X^n]}{x_o^n}$.

(b)   Using the result of part (a) and the values of the moments given, find the tightest bound on $\Pr(X \geq 2)$.

(c)   Using the result of part (a) and the values of the moments given, find the tightest bound on $\Pr(X \geq 3)$.

4.81   Since the Q-function represents the tail probability of a Gaussian random variable, we can use the various bounds on tail probabilities to produce bounds on the Q-function.

(a)   Use Markov's inequality to produce an upper bound on the Q-function. *Hint:* a Gaussian random variable has a two-sided PDF, and Markov's inequality requires the random variable to be one-sided. You will need to work with absolute values to resolve this issue.

(b)   Use Chebyshev's inequality to produce an upper bound on the Q-function.

(c)   Plot your results from parts (a) and (b) along with the bound obtained from the Chernoff bound from Example 4.28. In order to determine how tight (or loose) these various bounds are, also include on your plot the exact value of the Q-function.

Section 4.11: Scalar Quantization

4.82   Consider a quantizer that is designed to minimize the mean square quantization error. That is, the quantization levels, y_i, are chosen according to the conditional mean criterion and the bin edges, x_i, are chosen according to the midpoint criterion. Show that the distortion is given by

$$d = E[(X - q(X))^2] = E[X^2] - \sum_{i=1}^{M} p_i y_i^2 ,$$

and therefore the SQNR can be written as

$$\text{SQNR} = \frac{E[X^2]}{d} = \left(1 - \frac{\displaystyle\sum_{i=1}^{M} p_i y_i^2}{E[X^2]} \right)^{-1} .$$

4.83   Following the lead of Example 4.30, design an optimum 2-bit quantizer for a signal whose samples follow a triangular PDF,

$$f_X(x) = \begin{cases} 1 - |x|, & |x| \leq 1, \\ 0, & |x| \geq 1. \end{cases}$$

(a) Find the four quantization levels, $\{y_1, y_2, y_3, y_4\}$, and the three boundary points, $\{x_1, x_2, x_3\}$.

(b) Find the mean-squared distortion, d.

(c) Find the signal-to-quantization-noise ratio in dB.

Section 4.12: Entropy and Source Coding

4.84 Suppose a source sends symbols from a three letter alphabet with $X \in \{a, b, c\}$ and $p_a = 1/2$, $p_b = 1/4$, $p_c = 1/4$ are the source symbol probabilities.

(a) Determine the entropy of this source.

(b) Give a source code that has an average codeword length that matches the entropy.

4.85 Suppose a random variable, X, has $N = 2^n$ equally likely outcomes. What is the entropy of X in bits?

4.86 Suppose a fair coin is flipped n times and the random variable Y counts the number of times heads occurs. What is the entropy of Y in bits? Compare your answer to that of Exercise 4.85 and explain any difference.

4.87 Consider an N-letter source with probabilities, p_i, $i = 1, 2, 3, \ldots, N$. The entropy of the source is given by $H(\boldsymbol{p}) = -\sum_{i=1}^{N} p_i \log(p_i)$. Prove that the discrete distribution that maximizes the entropy is a uniform distribution. *Hint:* You need to perform a constrained optimization since the source probabilities must form a valid probability mass function, and thus $p_1 + p_2 + \ldots + p_N = 1$.

4.88 Consider a geometric random variable, Z, whose PMF is $P_Z(k) = (1-p)p^k$, $k = 0, 1, 2, \ldots$. Find the entropy of this random variable as a function of p.

Miscellaneous Exercises

4.89 Imagine that you are trapped in a circular room with three doors symmetrically placed around the perimeter. You are told by a mysterious voice that one door leads to the outside after a 2-h trip through a maze. However, the other two doors lead to mazes that terminate back in the room after a 2-h trip at which time you are unable to tell through which door you exited or entered. What is the average time for escape to the outside? Can you guess the answer ahead of time? If not, can you provide a physical explanation for the answer you calculate?

4.90 A communication system sends data in the form of packets of fixed length. Noise in the communication channel may cause a packet to be received incorrectly. If this happens, then the packet is retransmitted. Let the probability that a packet is received incorrectly be q. Determine the average number of transmissions that are necessary before a packet is received correctly.

4.91 In Exercise 4.90 let the transmission time be T_t seconds for a packet. If the packet was received incorrectly, then a message is sent back to the transmitter that states that the message was received incorrectly. Let the time for sending such a message be T_i. Assume that if the packet is received correctly that we do not send an acknowledgement. What is the average time for a successful transmission?

4.92 Use the characteristic function (or the moment-generating function or the probability-generating function) to show that a Poisson PMF is the limit of a binomial PMF with n approaching infinity and p approaching zero in such a way that $np = \mu = \text{constant}$.

MATLAB Exercises

4.93 Let X be a random variable that is uniformly distributed over the interval $(0, 100)$. Form a new random variable Y by rounding X to the nearest integer. In MATLAB code, this could be represented by `Y=round(X)`. Finally, form the random roundoff error according to $Z = X - Y$.
   (a) Using analytical methods, find the PDF of Z as well as the mean-squared value, $E[Z^2]$.
   (b) Using MATLAB, create a histogram for the probability densities for the random variable Z. Compare with the PDF found analytically in part (a).

4.94 Suppose you have a random variable X with PDF, $f_X(x) = 2x(u(x) - u(x-1))$ and that this random variable is transformed as $Y = 2 - X$. Calculate $f_Y(y)$. Repeat this problem using MATLAB. Compare the estimate of the PDF from MATLAB with the analytically determined PDF. Note that for this problem there is no function in MATLAB that provides a sequence of data samples that has the PDF specified in this problem for X. Thus you must find an appropriate way to transform a uniform random variable to produce the desired X. The results of Exercise 4.44 will be helpful here.

4.95 Use MATLAB to generate a large number of samples from a Gaussian distribution with mean $\mu = 20$ and variance $\sigma = 4$. *Hint:* the MATLAB command `sigma*randn(1,N)+mu` will create N such numbers with mean `mu` and standard deviation `sigma`. Let $x_1, x_2, ..., x_N$ represent the samples you generated. Compute each of the following "mean" values:

(a) Sample mean, $\hat{\mu}_{sm} = \dfrac{1}{N} \sum_{k=1}^{N} x_k$;

(b) Geometric mean, $\hat{\mu}_{gm} = \left(\prod_{k=1}^{N} x_k \right)^{1/N}$;

(c) Harmonic mean, $\hat{\mu}_{hm} = \left(\dfrac{1}{N} \sum_{k=1}^{N} \dfrac{1}{x_k} \right)^{-1}$;

(d) Quadratic mean (root mean square), $\hat{\mu}_{qm} = \sqrt{\dfrac{1}{N} \sum_{k=1}^{N} x_k^2}$.

Which of these "estimates" give a decent estimate of the true mean?

4.96 Write a MATLAB program to simulate the problem described in Exercise 4.89. Estimate the average time until escape. Do your MATLAB results agree with your analytically determined results in Exercise 4.89?

4.97 Copy a segment of text into MATLAB as a string (you choose the source of the text). Then write a MATLAB program to count the relative frequency of each character (ignore all characters that do not correspond to one of the 26 letters and do not make a distinction between upper and lower case). Using the results of your program, calculate the entropy of a source that outputs the 26 English characters with the probabilities you calculated.

4.98 Suppose a random variable has a PDF given by

$$f_X(x) = \begin{cases} 1+x, & -1 < x < 0, \\ 1-x, & 0 \leq x < 1, \\ 0, & |x| \geq 1. \end{cases}$$

Following the procedure laid out in Example 4.31, write a MATLAB program to design an optimum 4-bit (16 level) quantizer for this random variable. Compute the SQNR in decibels of the quantizer you designed. How does this SQNR compare with that obtained in Example 4.31. Can you explain any differences?

Pairs of Random Variables

The previous two chapters dealt with the theory of single random variables. However, many problems of practical interest require the modeling of random phenomenon using two or maybe even more random variables. This chapter extends the theory of Chapters 3 and 4 to consider pairs of random variables. Chapter 6 then generalizes these results to include an arbitrary number of random variables. A common example that involves two random variables is the study of a system with a random input. Due to the randomness of the input, the output will naturally be random as well. Quite often it is necessary to characterize the relationship between the input and the output. A pair of random variables can be used to characterize this relationship: one for the input and another for the output.

Another class of examples involving random variables are those involving spatial coordinates in two dimensions. A pair of random variables can be used to probabilistically describe the position of an object which is subject to various random forces. There are endless examples of situations where we are interested in two random quantities that may or may not be related to one another, for example, the height and weight of a student, or the grade point average and GRE scores of a student, or the temperature and relative humidity at a certain place and time.

To start with, consider an experiment E whose outcomes lie in a sample space, S. A two-dimensional random variable is a mapping of the points in the sample space to ordered pairs $\{x, y\}$. Usually, when dealing with a pair of random variables, the sample space naturally partitions itself so that it can be viewed as a combination of two simpler sample spaces. For example, suppose the experiment was to observe the height and weight of a typical student. The range of student heights could fall within some set which we call sample space S_1, while the range of student weights could fall within the space S_2. The overall sample space of the experiment could then be viewed as $S = S_1 \times S_2$. For any outcome $s \in S$ of this experiment, the pair of random variables (X, Y) is merely a mapping of the outcome s to a pair of numerical values $(x(s), y(s))$. In the case of our height/weight experiment, it would be natural to choose $x(s)$ to be the height of the student (in inches perhaps), while $y(s)$ is the weight of the student (in pounds). Note that it is probably not sufficient to consider two separate experiments, one where the student's height is measured and assigned to the random variable X and another where a student's weight is measured and assigned to the random variable Y.

While the density functions $f_X(x)$ and $f_Y(y)$ do partially characterize the experiment, they do not completely describe the situation. It would be natural to expect that the height and

Probability and Random Processes. DOI: 10.1016/B978-0-12-386981-4.00005-X

weight are somehow related to each other. While it may not be very rare to have a student 74 in. tall nor unusual to have a student who weighs 120 pounds, it is probably rare indeed to have a student who is both 74 in. tall and weighs 120 pounds. A careful reading of the wording in the previous sentence makes it clear that in order to characterize the relationship between a pair of random variables, it is necessary to look at the joint probabilities of events relating to both random variables. We accomplish this through the joint cumulative distribution function (CDF) and the joint probability density function (PDF) in the next two sections.

5.1 Joint Cumulative Distribution Functions

When introducing the idea of random variables in Chapter 3, we started with the notion of a CDF. In the same way, to probabilistically describe a pair of random variables, $\{X, Y\}$, we start with the notion of a joint CDF.

> **Definition 5.1:** The *joint CDF* of a pair of random variables, $\{X, Y\}$, is $F_{X,Y}(x, y) = \Pr(X \le x, Y \le y)$. That is, the joint CDF is the joint probability of the two events $\{X \le x\}$ and $\{Y \le y\}$.

As with the CDF of a single random variable, not any function can be a joint CDF. The joint CDF of a pair of random variables will satisfy properties similar to those satisfied by the CDFs of single random variables. First of all, since the joint CDF is a probability, it must take on a value between 0 and 1. Also, since the random variables X and Y are real valued, it is impossible for either to take on a value less than $-\infty$ and both must be less than ∞. Hence, $F_{X,Y}(x, y)$ evaluated at either $x = -\infty$ or $y = -\infty$ (or both) must be zero and $F_{X,Y}(\infty, \infty)$ must be one. Next, for $x_1 \le x_2$ and $y_1 \le y_2$, $\{X \le x_1\} \cap \{Y \le y_1\}$ is a subset of $\{X \le x_2\} \cap \{Y \le y_2\}$ so that $F_{X,Y}(x_1, y_1) \le F_{X,Y}(x_2, y_2)$. That is, the CDF is a monotonic, nondecreasing function of both x and y. Note that since the event $\{\{X \le \infty\}$ must happen, then $\{X \le \infty\} \cap \{Y \le y\} = \{Y \le y\}$ so that $F_{X,Y}(\infty, y) = F_Y(y)$. Likewise, $F_{X,Y}(x, \infty) = F_X(x)$. In the context of joint CDFs, $F_X(x)$ and $F_Y(y)$ are referred to as the *marginal* CDFs of X and Y, respectively.

Finally, consider using a joint CDF to evaluate the probability that the pair of random variables (X, Y) falls into a rectangular region bounded by the points (x_1, y_1), (x_2, y_1), (x_1, y_2), and (x_2, y_2). This calculation is illustrated in Figure 5.1. The desired rectangular region is the lightly shaded area. Evaluating $F_{X,Y}(x_2, y_2)$ gives the probability that the random variable falls anywhere below or to the left of the point (x_2, y_2); this includes all of the area in the desired rectangle, but it also includes everything below and to the left of the desired rectangle. The probability of the random variable falling to the left of the rectangle can be subtracted off using $F_{X,Y}(x_1, y_2)$. Similarly, the region below the rectangle can be subtracted off using $F_{X,Y}(x_2, y_1)$; these are the two medium-shaded regions in Figure 5.1. In subtracting off these two quantities, we have subtracted twice the probability of the pair falling both below and to the left of the desired rectangle (the dark-shaded region). Hence we must

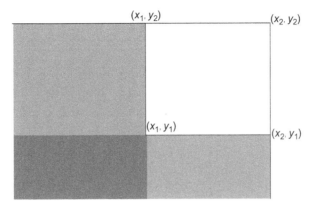

Figure 5.1
Illustrating the evaluation of the probability of a pair of random variables
falling in a rectangular region.

add back this probability using $F_{X,Y}(x_1, y_1)$. All of these properties of joint CDFs are summarized as follows:

(1) $F_{X,Y}(-\infty, -\infty) = F_{X,Y}(-\infty, y) = F_{X,Y}(x, -\infty) = 0$;   (5.1a)

(2) $F_{X,Y}(\infty, \infty) = 1$;   (5.1b)

(3) $0 \le F_{X,Y}(x, y) \le 1$;   (5.1c)

(4) $F_{X,Y}(x, \infty) = F_X(x)$, $F_{X,Y}(\infty, y) = F_Y(y)$;   (5.1d)

(5) $\Pr(x_1 < X_1 \le x_2, y_1 < Y_1 \le y_2)$ (5.1e)
$= F_{X,Y}(x_2, y_2) - F_{X,Y}(x_1, y_2) - F_{X,Y}(x_2, y_1) + F_{X,Y}(x_1, y_1) \ge 0$.

With the exception of property (4), all of these properties are analogous to the ones listed in Equation (3.3) for CDFs of single random variables.

Property (5) tells us how to calculate the probability of the pair of random variables falling in a rectangular region. Often, we are interested in also calculating the probability of the pair of random variables falling in a region which is not rectangular (e.g., a circle or triangle). This can be done by forming the required region using many infinitesimal rectangles and then repeatedly applying property (5). In practice, however, this task is somewhat overwhelming, and hence we do not go into the details here.

■ Example 5.1:

One of the simplest examples (conceptually) of a pair of random variables is one which is uniformly distributed over the unit square (i.e., $0<x<1$, $0<y<1$). The CDF of such a random variable is

$$F_{X,Y}(x,y) = \begin{cases} 0, & x<0 \text{ or } y<0, \\ x, & 0\leq x\leq 1, y>1, \\ y, & x>1, 0\leq y\leq 1, \\ xy, & 0\leq x\leq 1, 0\leq y\leq 1, \\ 1, & x>1, y>1. \end{cases}$$

Even this very simple example leads to a rather cumbersome function. Nevertheless, it is straightforward to verify that this function does indeed satisfy all the properties of a joint CDF. From this joint CDF, the marginal CDF of X can be found to be

$$F_X(x) = F_{X,Y}(x,\infty) = \begin{cases} 0, & x<0, \\ x, & 0\leq x\leq 1, \\ 1, & x>1. \end{cases}$$

Hence, the marginal CDF of X is also a uniform distribution. The same statement holds for Y as well.

■

5.2 Joint Probability Density Functions

As seen in Example 5.1, even the simplest joint random variables can lead to CDFs which are quite unwieldy. As a result, working with joint CDFs can be difficult. In order to avoid extensive use of joint CDFs, attention is now turned to the two dimensional equivalent of the PDF.

Definition 5.2: The *joint probability density function* of a pair of random variables (X, Y) evaluated at the point (x, y) is

$$f_{X,Y}(x,y) = \lim_{\varepsilon_x \to 0, \, \varepsilon_y \to 0} \frac{\Pr(x \leq X < x + \varepsilon_x, y \leq Y < y + \varepsilon_y)}{\varepsilon_x \varepsilon_y}. \tag{5.2}$$

Similar to the one-dimensional case, the joint PDF is the probability that the pair of random variables (X, Y) lies in an infinitesimal region defined by the point (x, y) normalized by the area of the region.

For a single random variable, the PDF was the derivative of the CDF. By applying Equation (5.1e) to the definition of the joint PDF, a similar relationship is obtained.

Theorem 5.1: The joint PDF $f_{X,Y}(x,y)$ can be obtained from the joint CDF $F_{X,Y}(x,y)$ by taking a partial derivative with respect to each variable. That is,

$$f_{X,Y}(x,y) = \frac{\partial^2}{\partial x \partial y} F_{X,Y}(x,y).$$
(5.3)

Proof: Using Equation (5.1e),

$$\begin{aligned}
&\Pr(x \leq X < x + \varepsilon_x, y \leq Y < y + \varepsilon_y) \\
&= F_{X,Y}(x + \varepsilon_x, y + \varepsilon_y) - F_{X,Y}(x, y + \varepsilon_y) - F_{X,Y}(x + \varepsilon_x, y) + F_{X,Y}(x,y) \\
&= [F_{X,Y}(x + \varepsilon_x, y + \varepsilon_y) - F_{X,Y}(x, y + \varepsilon_y)] - [F_{X,Y}(x + \varepsilon_x, y) - F_{X,Y}(x,y)].
\end{aligned}$$
(5.4)

Dividing by ε_x and taking the limit as $\varepsilon_x \to 0$ results in

$$\begin{aligned}
&\lim_{\varepsilon_x \to 0} \frac{\Pr(x \leq X < x + \varepsilon_x, y \leq Y < y + \varepsilon_y)}{\varepsilon_x} \\
&= \lim_{\varepsilon_x \to 0} \frac{F_{X,Y}(x + \varepsilon_x, y + \varepsilon_y) - F_{X,Y}(x, y + \varepsilon_y)}{\varepsilon_x} - \lim_{\varepsilon_x \to 0} \frac{F_{X,Y}(x + \varepsilon_x, y) - F_{X,Y}(x,y)}{\varepsilon_x} \\
&= \frac{\partial}{\partial x} F_{X,Y}(x, y + \varepsilon_y) - \frac{\partial}{\partial x} F_{X,Y}(x,y).
\end{aligned}$$
(5.5)

Then dividing by ε_y and taking the limit as $\varepsilon_y \to 0$ gives the desired result:

$$\begin{aligned}
f_{X,Y}(x,y) &= \lim_{\varepsilon_x \to 0, \varepsilon_y \to 0} \frac{\Pr(x \leq X < x + \varepsilon_x, y \leq Y < y + \varepsilon_y)}{\varepsilon_x \varepsilon_y} \\
&= \lim_{\varepsilon_y \to 0} \frac{\frac{\partial}{\partial x} F_{X,Y}(x, y + \varepsilon_y) - \frac{\partial}{\partial x} F_{X,Y}(x,y)}{\varepsilon_y} = \frac{\partial^2}{\partial x \partial y} F_{X,Y}(x,y). \quad \square
\end{aligned}$$
(5.6)

This theorem shows that we can obtain a joint PDF from a joint CDF by differentiating with respect to each variable. The converse of this statement would be that we could obtain a joint CDF from a joint PDF by integrating with respect to each variable. Specifically,

$$F_{X,Y}(x,y) = \int_{-\infty}^{y} \int_{-\infty}^{x} f_{X,Y}(u,v) \, du \, dv.$$
(5.7)

■ Example 5.2:

From the joint CDF given in Example 5.1, it is found (by differentiating the joint CDF with respect to both x and y) that the joint PDF for a pair of random variables uniformly distributed over the unit square is

$$f_{X,Y}(x,y) = \begin{cases} 1, & 0 < x < 1, 0 < y < 1, \\ 0, & \text{otherwise.} \end{cases}$$

Note how much simpler the joint PDF is to specify than is the joint CDF.   ■

From the definition of the joint PDF in Equation (5.2) as well as the relationships specified in Equations (5.3) and (5.7), several properties of joint PDFs can be inferred. These properties are summarized as follows:

$$(1) \quad f_{X,Y}(x,y) \geq 0 ; \tag{5.8a}$$

$$(2) \quad \int_{-\infty}^{\infty} \int_{-\infty}^{\infty} f_{X,Y}(x,y) \mathrm{d}x \mathrm{d}y = 1 ; \tag{5.8b}$$

$$(3) \quad F_{X,Y}(x,y) = \int_{-\infty}^{y} \int_{-\infty}^{x} f_{X,Y}(u,v) \mathrm{d}u \mathrm{d}v ; \tag{5.8c}$$

$$(4) \quad f_{X,Y}(x,y) = \frac{\partial^2}{\partial x \partial y} F_{X,Y}(x,y) ; \tag{5.8d}$$

$$(5) \quad f_X(x) = \int_{-\infty}^{\infty} f_{X,Y}(x,y) \mathrm{d}y, \quad f_Y(y) = \int_{-\infty}^{\infty} f_{X,Y}(x,y) \mathrm{d}x ; \tag{5.8e}$$

$$(6) \quad \Pr(x_1 < X_1 \leq x_2, y_1 < Y_1 \leq y_2) = \int_{y_1}^{y_2} \int_{x_1}^{x_2} f_{X,Y}(x,y) \mathrm{d}x \mathrm{d}y . \tag{5.8f}$$

Property (1) follows directly from the definition of the joint PDF in Equation (5.2) since both the numerator and denominator there are nonnegative. Property (2) results from the relationship in Equation (5.7) together with the fact that $F_{X,Y}(\infty, \infty) = 1$. This is the normalization integral for joint PDFs. These first two properties form a set of sufficient conditions for a function of two variables to be a valid joint PDF. Properties (3) and (4) have already been developed. Property (5) is obtained by first noting that the marginal CDF of X is $F_X(x) = F_{X,Y}(x, \infty)$. Using Equation (5.7) then results in $F_X(x) = \int_{-\infty}^{\infty} \int_{-\infty}^{x} f_{X,Y}(u,y) \mathrm{d}u \mathrm{d}y$. Differentiating this expression with respect to x produces the expression in property (5) for the marginal PDF of x. A similar derivation produces the marginal PDF of y. Hence, the marginal PDFs are obtained by integrating out the unwanted variable in the joint PDF. The last property is obtained by combining Equations (5.1e) and (5.7).

■ Example 5.3:

Suppose two random variables are jointly uniformly distributed over the unit circle. That is, the joint PDF $f_{X,Y}(x,y)$ is constant anywhere such that $x^2+y^2<1$:

$$f_{X,Y}(x,y) = \begin{cases} c, & x^2+y^2<1, \\ 0, & \text{otherwise.} \end{cases}$$

The constant c can be determined using the normalization integral for joint PDFs:

$$\iint\limits_{x^2+y^2<1} c\, dx\, dy = 1 \Rightarrow c = \frac{1}{\pi}.$$

The marginal PDF of X is found by integrating y out of the joint PDF:

$$f_X(x) = \int_{-\infty}^{\infty} f_{X,Y}(x,y)dy = \int_{-\sqrt{1-x^2}}^{\sqrt{1-x^2}} \frac{1}{\pi}dy = \frac{2}{\pi}\sqrt{1-x^2}, \text{ for } -1\leq x\leq 1 .$$

By symmetry, the marginal PDF of Y would have the same functional form:

$$f_Y(y) = \frac{2}{\pi}\sqrt{1-y^2}, \text{ for } -1\leq y\leq 1 .$$

Although X and Y were jointly uniformly distributed, the marginal distributions are not uniform. Stated another way, suppose we are given just the marginal PDFs of X and Y as just specified. This information alone is not enough to determine the joint PDF. One may be able to form many joint PDFs that produce the same marginal PDFs. For example, suppose we form

$$f_{X,Y}(x,y) = \begin{cases} \dfrac{4}{\pi^2}\sqrt{(1-x^2)(1-y^2)}, & -1\leq x\leq 1, -1\leq y\leq 1, \\ 0, & \text{otherwise.} \end{cases}$$

It is easy to verify that this is a valid joint PDF and leads to the same marginal PDFs. Yet, this is clearly a completely different joint PDF than the uniform distribution with which we started. This reemphasizes the need to specify the joint distributions of random variables and not just their marginal distributions.

■

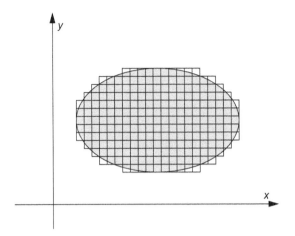

Figure 5.2
Approximation of an arbitrary region by a series of infinitesimal rectangles.

Property (6) of joint PDFs given in Equation (5.8f) specifies how to compute the probability that a pair of random variables takes on a value in a rectangular region. Often, we are interested in computing the probability that the pair of random variables falls in a region which is not rectangularly shaped. In general, suppose we wish to compute $\Pr((X, Y) \in A)$, where A is the region illustrated in Figure 5.2. This general region can be approximated as a union of many nonoverlapping rectangular regions as shown in the figure. In fact, as we make the rectangles ever smaller, the approximation improves to the point where the representation becomes exact in the limit as the rectangles get infinitely small. That is, any region can be represented as an infinite number of infinitesimal rectangular regions so that $A = \bigcup R_i$, where R_i represents the ith rectangular region. The probability that the random pair falls in A is then computed as

$$\Pr((X, Y) \in A) = \sum_i \Pr((X, Y) \in R_i) = \sum_i \int\!\!\int_{R_i} f_{X, Y}(x, y) \mathrm{d}x \mathrm{d}y. \tag{5.9}$$

The sum of the integrals over the rectangular regions can be replaced by an integral over the original region A:

$$\Pr((X, Y) \in A) = \int\!\!\int_A f_{X, Y}(x, y) \mathrm{d}x \mathrm{d}y. \tag{5.10}$$

This important result shows that the probability of a pair of random variables falling in some two-dimensional region A is found by integrating the joint PDF of the two random variables over the region A.

■ Example 5.4:

Suppose a pair of random variables has the joint PDF given by

$$f_{X,Y}(x,y) = \frac{1}{2\pi}\exp\left(-\frac{x^2+y^2}{2}\right).$$

The probability that the point (X, Y) falls inside the unit circle is given by

$$\Pr(X^2+Y^2 < 1) = \iint\limits_{x^2+y^2 < 1} \frac{1}{2\pi}\exp\left(-\frac{x^2+y^2}{2}\right)dxdy.$$

Converting this integral to polar coordinates results in

$$\Pr(X^2+Y^2 < 1) = \int_0^{2\pi}\int_0^1 \frac{r}{2\pi}\exp\left(-\frac{r^2}{2}\right)dr d\theta = \int_0^1 r\exp\left(-\frac{r^2}{2}\right)dr = -\exp\left(-\frac{r^2}{2}\right)\Big|_0^1 = 1 - \exp\left(-\frac{1}{2}\right).$$

■

■ Example 5.5:

Now suppose that a pair of random variables has the joint PDF given by

$$f_{X,Y}(x,y) = c\exp\left(-x-\frac{y}{2}\right)u(x)u(y).$$

First, the constant c is found using the normalization integral

$$\int_0^\infty\int_0^\infty c\exp\left(-x-\frac{y}{2}\right)dxdy = 1 \Rightarrow c = \frac{1}{2}.$$

Next, suppose we wish to determine the probability of the event $\{X > Y\}$. This can be viewed as finding the probability of the pair (X, Y) falling in the region A that is now defined as $A = \{(x,y):x>y\}$. This probability is calculated as

$$\Pr(X > Y) = \iint\limits_{x>y} f_{X,Y}(x,y)dxdy = \int_0^\infty\int_y^\infty \frac{1}{2}\exp\left(-x-\frac{y}{2}\right)dxdy = \int_0^\infty \frac{1}{2}\exp\left(-\frac{3y}{2}\right)dy = \frac{1}{3}.$$

■

■ Example 5.6:

In many cases, evaluating the probability of a pair a random variables falling in some region may be quite difficult to calculate analytically. For example, suppose we modify Example 5.4 so that the joint PDF is now of the form

$$f_{X,Y}(x,y) = \frac{1}{2\pi}\exp\left(-\frac{((x-2)^2+(y-3)^2)}{2}\right).$$

Again, we would like to evaluate the probability that the pair (X, Y) falls in the unit circle. To do this analytically we must evaluate

(Continued)

$$\iint\limits_{x^2+y^2<1} \frac{1}{2\pi}\exp\left(-\frac{((x-2)^2+(y-3)^2)}{2}\right)dx\,dy \ .$$

Converting to polar coordinates the integral becomes

$$\int_0^1 \int_0^{2\pi} \frac{r}{2\pi}\exp\left(-\frac{((r\cos(\theta)-2)^2+(r\sin(\theta)-3)^2)}{2}\right)d\theta\,dr \ .$$

Either way the double integral looks formidable. We can enlist MATLAB to help in one of two ways. First, we could randomly generate many samples of the pair of random variables according to the specified distribution and count the relative frequency of the number that falls within the unit circle. Alternatively, we can get MATLAB to calculate one of the preceding double integrals numerically. We will take the latter approach here and evaluate the double integral in polar coordinates. First, we must define a MATLAB function to evaluate the integrand:

```
function out=dblintegrand(q,r)
out=r.*exp(-((r*cos(q)-2).^2+(r*sin(q)-3).^2)/2);
```

MATLAB will then evaluate the integral by executing the command

```
dblquad('dblintegrand',0,2*pi,0,1)/(2*pi).
```

By executing these MATLAB commands, we find the value of the integral to be 0.002072. ∎

5.3 Joint Probability Mass Functions

When the random variables are discrete rather than continuous, it is often more convenient to work with probability mass functions (PMFs) rather than PDFs or CDFs. It is straightforward to extend the concept of the PMF to a pair of random variables.

> **Definition 5.3:** The *joint PMF* for a pair of discrete random variables X and Y is given by $P_{X,Y}(x,y) = \Pr(\{X=x\}\cap\{Y=y\})$.

In particular, suppose the random variable X takes on values from the set $\{x_1, x_2, ..., x_M\}$ and the random variable Y takes on values from the set $\{y_1, y_2, ..., y_N\}$. Here, either M or N could be potentially infinite, or both could be finite. Several properties of the joint PMF analogous to those developed for joint PDFs should be apparent.

$$(1)\quad 0 \le P_{X,Y}(x_m, y_n) \le 1 ; \tag{5.11a}$$

$$(2)\quad \sum_{m=1}^{M}\sum_{n=1}^{N} P_{X,Y}(x_m, y_n) = 1 ; \tag{5.11b}$$

(3) $\displaystyle\sum_{n=1}^{N} P_{X,Y}(x_m, y_n) = P_X(x_m), \quad \sum_{m=1}^{M} P_{X,Y}(x_m, y_n) = P_Y(y_n);$ (5.11c)

(4) $\mathrm{Pr}((X, Y) \in A) = \displaystyle\sum\sum_{(x,y) \in A} P_{X,Y}(x,y).$ (5.11d)

Furthermore, the joint PDF or the joint CDF of a pair of discrete random variables can be related to the joint PMF through the use of delta functions or step functions by

$$f_{X,Y}(x,y) = \sum_{m=1}^{M} \sum_{n=1}^{N} P_{X,Y}(x_m, y_n)\delta(x - x_m)\delta(y - y_m),$$ (5.12)

$$F_{X,Y}(x,y) = \sum_{m=1}^{M} \sum_{n=1}^{N} P_{X,Y}(x_m, y_n)u(x - x_m)u(y - y_m).$$ (5.13)

Usually, it is most convenient to work with PMFs when the random variables are discrete. However, if the random variables are mixed (i.e., one is discrete and one is continuous), then it becomes necessary to work with PDFs or CDFs since the PMF will not be meaningful for the continuous random variable.

■ Example 5.7:

Two discrete random variables N and M have a joint PMF given by

$$P_{N,M}(n,m) = \frac{(n+m)!}{n!m!}\frac{a^n b^m}{(a+b+1)^{n+m+1}}, \qquad m = 0, 1, 2, 3, \ldots, \qquad n = 0, 1, 2, 3, \ldots.$$

The marginal PMF of N can be found by summing over m in the joint PMF:

$$P_N(n) = \sum_{m=0}^{\infty} P_{N,M}(n,m) = \sum_{m=0}^{\infty} \frac{(n+m)!}{n!m!}\frac{a^n b^m}{(a+b+1)^{n+m+1}}.$$

To evaluate this series, the following identity is used:

$$\sum_{m=0}^{\infty} \frac{(n+m)!}{n!m!}x^m = \left(\frac{1}{1-x}\right)^{n+1}.$$

The marginal PMF then reduces to

(Continued)

$$P_N(n) = \frac{a^n}{(a+b+1)^{n+1}} \sum_{m=0}^{\infty} \frac{(n+m)!}{n!m!} \frac{b^m}{(a+b+1)^m} = \frac{a^n}{(a+b+1)^{n+1}} \left(\frac{1}{1 - \frac{b}{a+b+1}}\right)^{n+1} = \frac{a^n}{(1+a)^{n+1}}.$$

Likewise, by symmetry, the marginal PMF of M is

$$P_M(m) = \frac{b^m}{(1+b)^{m+1}}.$$

Hence, the random variables M and N both follow a geometric distribution. ∎

5.4 Conditional Distribution, Density, and Mass Functions

The notion of conditional distribution functions and conditional density functions was first introduced in Chapter 3. In this section, those ideas are extended to the case where the conditioning event is related to another random variable. For example, we might want to know the distribution of a random variable representing the score a student achieves on a test given the value of another random variable representing the number of hours the student studied for the test. Or, perhaps we want to know the probability density function of the outside temperature given that the humidity is known to be below 50%.

To start with, consider a pair of discrete random variables X and Y with a PMF, $P_{X,Y}(x,y)$. Suppose we would like to know the PMF of the random variable X given that the value of Y has been observed. Then, according to the definition of conditional probability

$$\Pr(X=x|Y=y) = \frac{\Pr(X=x, Y=y)}{\Pr(Y=y)} = \frac{P_{X,Y}(x,y)}{P_Y(y)}. \qquad (5.14)$$

We refer to this as the conditional PMF of X given Y. By way of notation we write $P_{X|Y}(x|y) = P_{X,Y}(x,y)/P_Y(y)$.

∎ Example 5.8:

Using the joint PMF given in Example 5.7 along with the marginal PMF found in that example, it is found that

$$P_{N|M}(n|m) = \frac{P_{M,N}(m,n)}{P_M(m)} = \frac{(n+m)!}{n!m!} \frac{a^n b^m}{(a+b+1)^{n+m+1}} \frac{(1+b)^{m+1}}{b^m} = \frac{(n+m)!}{n!m!} \frac{a^n(1+b)^{m+1}}{(a+b+1)^{n+m+1}}.$$

Note that the conditional PMF of N given M is quite different than the marginal PMF of N. That is, knowing M changes the distribution of N. ∎

The simple result developed in Equation (5.14) can be extended to the case of continuous random variables and PDFs. The following theorem shows that the PMFs in (5.14) can simply be replaced by PDFs.

Theorem 5.2: The conditional PDF of a random variable X given that $Y = y$ is

$$f_{X|Y}(x|y) = \frac{f_{X,Y}(x,y)}{f_Y(y)}.$$ (5.15)

Proof: Consider the conditioning event $A = y \leq Y < y + \mathrm{d}y$. Then

$$f_{X|A}(x)\mathrm{d}x = \Pr(x \leq X < x + \mathrm{d}x | y \leq Y < y + \mathrm{d}y) = \frac{\Pr(x \leq X < x + \mathrm{d}x, y \leq Y < y + \mathrm{d}y)}{\Pr(y \leq Y < y + \mathrm{d}y)}$$

$$= \frac{f_{X,Y}(x,y)\mathrm{d}x\mathrm{d}y}{f_Y(y)\mathrm{d}y} = \frac{f_{X,Y}(x,y)\mathrm{d}x}{f_Y(y)}.$$

Passing to the limit as $\mathrm{d}y \to 0$, the event A becomes the event $\{Y = y\}$, producing the desired result.

Integrating both sides of this equation with respect to x produces the appropriate result for CDFs:

$$F_{X|Y}(x|y) = \frac{\int_{-\infty}^{x} f_{X,Y}(x',y)\mathrm{d}x'}{f_Y(y)}. \qquad \square$$ (5.16)

Usually, the conditional PDF is much easier to work with, so the conditional CDF will not be discussed further.

■ Example 5.9:

A certain pair of random variables has a joint PDF given by

$$f_{X,Y}(x,y) = \frac{2abc}{(ax + by + c)^3} u(x)u(y)$$

for some positive constants a, b, and c. The marginal PDFs are easily found to be

$$f_X(x) = \int_0^{\infty} f_{X,Y}(x,y)\mathrm{d}y = \frac{ac}{(ax+c)^2}u(x) \text{ and } f_Y(y) = \int_0^{\infty} f_{X,Y}(x,y)\mathrm{d}x = \frac{bc}{(by+c)^2}u(y).$$

The conditional PDF of X given Y then works out to be

$$f_{X|Y}(x|y) = \frac{f_{X,Y}(x,y)}{f_Y(y)} = \frac{2a(by+c)^2}{(ax+by+c)^3}u(x).$$

The conditional PDF of Y given X could also be determined in a similar way:

$$f_{Y|X}(y|x) = \frac{f_{X,Y}(x,y)}{f_X(x)} = \frac{2b(ax+c)^2}{(ax+by+c)^3}u(y).$$

■

■ Example 5.10:

This example involves two Gaussian random variables. Suppose X and Y have a joint PDF given by

$$f_{X,Y}(x,y) = \frac{1}{\pi\sqrt{3}}\exp\left(-\frac{2}{3}(x^2 - xy + y^2)\right).$$

The marginal PDF is found as follows:

$$f_X(x) = \int_{-\infty}^{\infty} f_{X,Y}(x,y)dy = \frac{1}{\pi\sqrt{3}}\exp\left(-\frac{2}{3}x^2\right)\int_{-\infty}^{\infty}\exp\left(-\frac{2}{3}(y^2 - xy)\right)dy.$$

In order to evaluate the integral, complete the square in the exponent:

$$f_X(x) = \frac{1}{\pi\sqrt{3}}\exp\left(-\frac{2}{3}x^2\right)\exp\left(\frac{x^2}{6}\right)\int_{-\infty}^{\infty}\exp\left(-\frac{2}{3}\left(y^2 - xy + \frac{x^2}{4}\right)\right)dy = \frac{1}{\pi\sqrt{3}}\exp\left(-\frac{x^2}{2}\right)\int_{-\infty}^{\infty}\exp\left(-\frac{2}{3}\left(y - \frac{x}{2}\right)^2\right)dy.$$

Now the integrand is a Gaussian-looking function. If the appropriate constant is added to the integrand, the integrand will be a valid PDF and hence must integrate out to one. In this case, the constant we need to add to the integrand to make the integral unity is $\sqrt{2/(3\pi)}$. Stated another way, the integral as just written must evaluate to $\sqrt{3\pi/2}$. Hence, the marginal PDF of X is

$$f_X(x) = \frac{1}{\sqrt{2\pi}}\exp\left(-\frac{x^2}{2}\right),$$

and we see that X is a zero-mean, unit-variance, Gaussian (i.e., standard normal) random variable. By symmetry, the marginal PDF of Y must also be of the same form. The conditional PDF of X given Y is

$$f_{X|Y}(x|y) = \frac{f_{X,Y}(x,y)}{f_Y(y)} = \frac{\frac{1}{\pi\sqrt{3}}\exp\left(-\frac{2}{3}(x^2 - xy + y^2)\right)}{\frac{1}{\sqrt{2\pi}}\exp\left(-\frac{y^2}{2}\right)} = \sqrt{\frac{2}{3\pi}}\exp\left(-\frac{2}{3}\left(x^2 - xy + \frac{y^2}{4}\right)\right) = \sqrt{\frac{2}{3\pi}}\exp\left(-\frac{2}{3}\left(x - \frac{y}{2}\right)^2\right)$$

So, the conditional PDF of X given Y is also Gaussian. But, given that it is known that $Y = y$, the mean of X is now $y/2$ (instead of zero), and the variance of X is $3/4$ (instead of one). In this example, knowledge of Y has shifted the mean and reduced the variance of X. ■

In addition to conditioning on a random variable taking on a point value such as $Y = y$, the conditioning can also occur on an interval of the form $y_1 \le Y \le y_2$. To simplify notation, let the conditioning event A be $A = \{y_1 \le Y \le y_2\}$. The relevant conditional PMF, PDF, and CDF are then given, respectively, by

$$P_{X|A}(x) = \frac{\sum_{y=y_1}^{y_2} P_{X,Y}(x,y)}{\sum_{y=y_1}^{y_2} P_Y(y)} \; ; \qquad (5.17)$$

$$f_{X|A}(x) = \frac{\int_{y_1}^{y_2} f_{X,Y}(x,y)\mathrm{d}y}{\int_{y_1}^{y_2} f_Y(y)\mathrm{d}y} \; ; \qquad (5.18)$$

$$F_{X|A}(x) = \frac{F_{X,Y}(x,y_2) - F_{X,Y}(x,y_1)}{F_Y(y_2) - F_Y(y_1)} \; . \qquad (5.19)$$

It is left as an exercise for the reader to derive these expressions.

■ Example 5.11:

Using the joint PDF of Example 5.10, suppose we want to determine the conditional PDF of X given that $Y > y_0$. The numerator in Equation (5.18) is calculated according to

$$\int_{y_0}^{\infty} f_{X,Y}(x,y)\mathrm{d}y = \int_{y_0}^{\infty} \frac{1}{\pi\sqrt{3}} \exp\left(-\frac{2}{3}(x^2 - xy + y^2)\right)\mathrm{d}y = \frac{1}{\sqrt{2\pi}} \exp\left(-\frac{x^2}{2}\right) \int_{y_0}^{\infty} \sqrt{\frac{2}{3\pi}} \exp\left(-\frac{2}{3}\left(y - \frac{x}{2}\right)^2\right)\mathrm{d}y$$

$$= \frac{1}{\sqrt{2\pi}} \exp\left(-\frac{x^2}{2}\right) Q\left(\frac{2y_0 - x}{\sqrt{3}}\right)$$

Since the marginal PDF of Y is a zero-mean, unit-variance Gaussian PDF, the denominator of Equation (5.18) becomes

$$\int_{y_0}^{\infty} f_Y(y)\mathrm{d}y = \int_{y_0}^{\infty} \frac{1}{\sqrt{2\pi}} \exp\left(-\frac{y^2}{2}\right)\mathrm{d}y = Q(y_0) \; .$$

Therefore, the PDF of X conditioned on $Y > y_0$ is

$$f_{X|Y>y_0}(x) = \frac{1}{\sqrt{2\pi}} \exp\left(-\frac{x^2}{2}\right) \frac{Q\left(\frac{2y_0 - x}{\sqrt{3}}\right)}{Q(y_0)} \; .$$

Note that when the conditioning event was a point condition on Y, the conditional PDF of X was Gaussian; yet, when the conditioning event is an interval condition on Y, the resulting conditional PDF of X is not Gaussian at all.

■

5.5 Expected Values Involving Pairs of Random Variables

The notion of expected value is easily generalized to pairs of random variables. To begin, we define the expected value of an arbitrary function of two random variables.

> **Definition 5.4:** Let $g(x, y)$ be an arbitrary two-dimensional function. The expected value of $g(X, Y)$, where X and Y are random variables, is
>
> $$E[g(X, Y)] = \int\int g(x, y) f_{X, Y}(x, y) \, dx \, dy. \tag{5.20}$$

For discrete random variables, the equivalent expression in terms of the joint PMF is

$$E[g(X, Y)] = \sum_m \sum_n g(x_m, y_n) P_{X, Y}(x_m, y_n). \tag{5.21}$$

If the function $g(x, y)$ is actually a function of only a single variable, say x, then this definition reduces to the definition of expected values for functions of a single random variable as given in Definition 4.2.

$$E[g(X)] = \int_{-\infty}^{\infty} \int_{-\infty}^{\infty} g(x) f_{X, Y}(x, y) \, dx \, dy = \int_{-\infty}^{\infty} g(x) \left[\int_{-\infty}^{\infty} f_{X, Y}(x, y) \, dy \right] dx. \tag{5.22}$$

$$= \int_{-\infty}^{\infty} g(x) f_X(x) \, dx$$

To start with, consider an arbitrary linear function of the two variables $g(x, y) = ax + by$, where a and b are constants. Then

$$E[aX + bY] = \int_{-\infty}^{\infty} \int_{-\infty}^{\infty} [ax + by] f_{X, Y}(x, y) \, dx \, dy$$

$$= a \int_{-\infty}^{\infty} \int_{-\infty}^{\infty} x f_{X, Y}(x, y) \, dx \, dy + b \int_{-\infty}^{\infty} \int_{-\infty}^{\infty} y f_{X, Y}(x, y) \, dx \, dy$$

$$= aE[X] + bE[Y]. \tag{5.23}$$

This result merely states that expectation is a linear operation.

In addition to the functions considered in Chapter 4 which led to statistics such as means, variances, and the like, functions involving both variables x and y will be considered here.

These new functions will lead to statistics that will partially characterize the relationships between the two random variables.

Definition 5.5: The *correlation* between two random variables is defined as

$$R_{X,Y} = E[XY] = \iint xy f_{X,Y}(x,y) \, dx \, dy.$$ (5.24)

Furthermore, two random variables which have a correlation of zero are said to be *orthogonal*.

One instance in which the correlation appears is in calculating the second moment of a sum of two random variables. That is, consider finding the expected value of $g(X,Y) = (X+Y)^2$.

$$E[(X+Y)^2] = E[X^2 + 2XY + Y^2] = E[X^2] + E[Y^2] + 2E[XY].$$ (5.25)

Hence the second moment of the sum is the sum of the second moments plus twice the correlation.

Definition 5.6: The *covariance* between two random variables is

$$\text{Cov}(X,Y) = E[(X-\mu_X)(Y-\mu_Y)] = \iint (x-\mu_X)(y-\mu_Y) f_{X,Y}(x,y) \, dx \, dy.$$ (5.26)

If two random variables have a covariance of zero, they are said to be *uncorrelated*.

The correlation and covariance are strongly related to one another as shown by the following theorem.

Theorem 5.3: $\text{Cov}(X,Y) = R_{X,Y} - \mu_X \mu_Y$. (5.27)

Proof: $\text{Cov}(X,Y) = E[(X-\mu_X)(Y-\mu_Y)] = E[XY - \mu_X Y - \mu_Y X + \mu_X \mu_Y]$

$$= E[XY] - \mu_X E[Y] - \mu_Y E[X] + \mu_X \mu_Y = E[XY] - \mu_X \mu_Y. \quad \square$$

As a result, if either X or Y (or both) has a mean of zero, correlation and covariance are equivalent. The covariance function occurs when calculating the variance of a sum of two random variables.

$$\text{Var}(X+Y) = \text{Var}(X) + \text{Var}(Y) + 2\text{Cov}(X,Y).$$ (5.28)

This result can be obtained from Equation (5.25) by replacing X with $X-\mu_X$ and Y with $Y-\mu_Y$.

Another statistical parameter related to a pair of random variables is the correlation coefficient, which is nothing more than a normalized version of the covariance.

Definition 5.7: The *correlation coefficient* of two random variables X and Y, ρ_{XY}, is defined as

$$\rho_{XY} = \frac{\text{Cov}(X, Y)}{\sqrt{\text{Var}(X)\text{Var}(Y)}} = \frac{E[(X - \mu_X)(Y - \mu_Y)]}{\sigma_X \sigma_Y}. \tag{5.29}$$

The next theorem quantifies the nature of the normalization. In particular, it shows that a correlation coefficient can never be more than 1 in absolute value.

Theorem 5.4: The correlation coefficient is less than 1 in magnitude.

Proof: Consider taking the second moment of $X + aY$, where a is a real constant:

$$E[(X + aY)^2] = E[X^2] + 2aE[XY] + a^2 E[Y^2] \geq 0.$$

Since this is true for any a, we can tighten the bound by choosing the value of a that minimizes the left-hand side. This value of a turns out to be

$$a = \frac{-E[XY]}{E[Y^2]}.$$

Plugging in this value gives

$$E[X^2] + \frac{(E[XY])^2}{E[Y^2]} - \frac{2(E[XY])^2}{E[Y^2]} \geq 0$$

$$\Rightarrow (E[XY])^2 \leq E[X^2]E[Y^2]$$

If we replace X with $X - \mu_X$ and Y with $Y - \mu_Y$ the result is

$$(\text{Cov}(X, Y))^2 \leq \text{Var}(X)\text{Var}(Y).$$

Rearranging terms then gives the desired result:

$$|\rho_{XY}| = \left| \frac{\text{Cov}(X, Y)}{\sqrt{\text{Var}(X)\text{Var}(Y)}} \right| \leq 1. \quad \square \tag{5.30}$$

Note that we can also infer from the proof that equality holds if Y is a constant times X. That is, a correlation coefficient of 1 (or -1) implies that X and Y are completely correlated (knowing Y determines X). Furthermore, uncorrelated random variables will have a correlation coefficient of zero. Therefore, as its name implies, the correlation coefficient is a quantitative measure of the correlation between two random variables. It should be emphasized at this point that zero correlation is not to be confused with independence. These two concepts are not the same (more on this later).

The significance of the correlation, covariance, and correlation coefficient will be discussed further in the next two sections. For now, we present an example showing how to compute these parameters.

■ Example 5.12:

Consider once again the joint PDF of Example 5.10. The correlation for these random variables is

$$E[XY] = \int_{-\infty}^{\infty} \int_{-\infty}^{\infty} \frac{xy}{\pi\sqrt{3}} \exp\left(-\frac{2}{3}(x^2 - xy + y^2)\right) dy dx .$$

In order to evaluate this integral, the joint PDF is rewritten $f_{X,Y}(x,y) = f_{Y|X}(y|x)f_X(x)$ and then those terms involving only x are pulled outside the inner integral over y.

$$E[XY] = \int_{-\infty}^{\infty} \frac{x}{\sqrt{2\pi}} \exp\left(-\frac{x^2}{2}\right) \left[\int_{-\infty}^{\infty} y \sqrt{\frac{2}{3\pi}} \exp\left(-\frac{2}{3}\left(y - \frac{x}{2}\right)^2\right) dy \right] dx .$$

The inner integral (in square brackets) is the expected value of a Gaussian random variable with a mean of $x/2$ and variance of 3/4 which thus evaluates to $x/2$. Hence,

$$E[XY] = \frac{1}{2} \int_{-\infty}^{\infty} \frac{x^2}{\sqrt{2\pi}} \exp\left(-\frac{x^2}{2}\right) dx .$$

The remaining integral is the second moment of a Gaussian random variable with zero-mean and unit variance which integrates to 1. The correlation of these two random variables is therefore $E[XY] = 1/2$. Since both X and Y have zero means, $\text{Cov}(X, Y)$ is also equal to 1/2. Finally, the correlation coefficient is also $\rho_{XY} = 1/2$ due to the fact that both X and Y have unit variance.

■

The concepts of correlation and covariance can be generalized to higher-order moments as given in the following definition.

Definition 5.8: The (m, n)th joint moment of two random variables X and Y is

$$E[X^m Y^n] = \iint x^m y^n f_{X,Y}(x,y) dx dy . \tag{5.31}$$

The (m, n)th joint central moment is similarly defined as

$$E[(X-\mu_X)^m(Y-\mu_Y)^n] = \int\int(x-\mu_X)^m(y-\mu_Y)^n f_{X,Y}(x,y)dxdy. \qquad (5.32)$$

These higher-order joint moments are not frequently used and therefore are not considered further here.

As with single random variables, a conditional expected value can also be defined for which the expectation is carried out with respect to the appropriate conditional density function.

Definition 5.9: The conditional expected value of a function $g(X)$ of a random variable X given that $Y = y$ is

$$E[g(X)|Y] = \int_{-\infty}^{\infty} g(x)f_{X|Y}(x|y)dx. \qquad (5.33)$$

Conditional expected values can be particularly useful in calculating expected values of functions of two random variables that can be factored into the product of two one-dimensional functions. That is, consider a function of the form $g(x,y) = g_1(x)g_2(y)$. Then

$$E[g_1(X)g_2(Y)] = \int_{-\infty}^{\infty}\int_{-\infty}^{\infty} g_1(x)g_2(y)f_{X,Y}(x,y)dxdy. \qquad (5.34)$$

From Equation (5.15) the joint PDF is rewritten as $f_{X,Y}(x,y) = f_{Y|X}(y|x)f_X(x)$, resulting in

$$E[g_1(X)g_2(Y)] = \int_{-\infty}^{\infty} g_1(x)f_X(x)\left[\int_{-\infty}^{\infty} g_2(y)f_{Y|X}(y|x)dy\right]dx$$

$$= \int_{-\infty}^{\infty} g_1(x)f_X(x)E_Y[g_2(Y)|X]dx = E_X[g_1(X)E_Y[g_2(Y)|X]]. \qquad (5.35)$$

Here, the subscripts on the expectation operator have been included for clarity to emphasize that the outer expectation is with respect to the random variable X, while the inner expectation is with respect to the random variable Y (conditioned on X). This result allows us to break a two-dimensional expectation into two one-dimensional expectations. This technique was used in Example 5.12, where the correlation between two variables was essentially written as

$$R_{X,Y} = E_X[XE_Y[Y|X]]. \qquad (5.36)$$

In that example, the conditional PDF of Y given X was Gaussian, thus finding the conditional mean was accomplished by inspection. The outer expectation then required finding the second moment of a Gaussian random variable, which is also straightforward.

5.6 Independent Random Variables

The concept of independent events was introduced in Chapter 2. In this section, we extend this concept to the realm of random variables. To make that extension, consider the events $A = \{X \leq x\}$ and $B = \{Y \leq y\}$ related to the random variables X and Y. The two events A and B are statistically independent if $\Pr(A, B) = \Pr(A)\Pr(B)$. Restated in terms of the random variables, this condition becomes

$$\Pr(X \leq x, Y \leq y) = \Pr(X \leq x)\Pr(Y \leq y) \quad \Rightarrow \quad F_{X, Y}(x, y) = F_X(x)F_Y(y). \tag{5.37}$$

Hence, two random variables are statistically independent if their joint CDF factors into a product of the marginal CDFs. Differentiating both sides of this equation with respect to both x and y reveals that the same statement applies to the PDF as well. That is, for statistically independent random variables, the joint PDF factors into a product of the marginal PDFs:

$$f_{X, Y}(x, y) = f_X(x)f_Y(y). \tag{5.38}$$

It is not difficult to show that the same statement applies to PMFs as well. The preceding condition can also be restated in terms of conditional PDFs. Dividing both sides of Equation (5.38) by $f_X(x)$ results in

$$f_{Y|X}(y|x) = f_Y(y). \tag{5.39}$$

A similar result involving the conditional PDF of X given Y could have been obtained by dividing both sides by the PDF of Y. In other words, if X and Y are independent, knowing the value of the random variable X should not change the distribution of Y and vice versa.

■ Example 5.13:

Returning once again to the joint PDF of Example 5.10, we saw in that example that the marginal PDF of X is

$$f_X(x) = \frac{1}{\sqrt{2\pi}}\exp\left(-\frac{x^2}{2}\right),$$

while the conditional PDF of X given Y is

$$f_{X|Y}(x|y) = \sqrt{\frac{2}{3\pi}}\exp\left(-\frac{2}{3}\left(x - \frac{y}{2}\right)^2\right).$$

Clearly, these two random variables are not independent.

■

■ Example 5.14:

Suppose the random variables X and Y are uniformly distributed on the square defined by $0 \leq x, y \leq 1$. That is

$$f_{X,Y}(x,y) = \begin{cases} 1, & 0 \leq x, y \leq 1, \\ 0, & \text{otherwise.} \end{cases}$$

The marginal PDFs of X and Y work out to be

$$f_X(x) = \begin{cases} 1, & 0 \leq x \leq 1, \\ 0, & \text{otherwise,} \end{cases} \qquad f_Y(y) = \begin{cases} 1, & 0 \leq y \leq 1, \\ 0, & \text{otherwise.} \end{cases}$$

These random variables are statistically independent since $f_{X,Y}(x,y) = f_X(x)f_Y(y)$.

■

Theorem 5.5: Let X and Y be two independent random variables and consider forming two new random variables $U = g_1(X)$ and $V = g_2(Y)$. These new random variables U and V are also independent.

Proof: To show that U and V are independent, consider the events $A = \{U \leq u\}$ and $B = \{V \leq v\}$. Next define the region R_u to be the set of all points x such that $g_1(x) \leq u$. Similarly, define R_v to be the set of all points y such that $g_2(y) \leq v$. Then

$$\Pr(U \leq u, V \leq v) = \Pr(X \in R_u, Y \in R_v) = \int_{R_v} \int_{R_u} f_{X,Y}(x,y) dx dy.$$

Since X and Y are independent, their joint PDF can be factored into a product of marginal PDFs resulting in

$$\Pr(U \leq u, V \leq v) = \int_{R_u} f_X(x) dx \int_{R_v} f_Y(y) dy = \Pr(X \in R_u) \Pr(Y \in R_v).$$
$$= \Pr(U \leq u) \Pr(V \leq v)$$

Since we have shown that $F_{U,V}(u,v) = F_U(u)F_V(v)$, the random variables U and V must be independent. □

Another important result deals with the correlation, covariance, and correlation coefficients of independent random variables.

Theorem 5.6: If X and Y are independent random variables, then $E[XY] = \mu_X \mu_Y$, $\text{Cov}(X, Y) = 0$, and $\rho_{X,Y} = 0$.

Proof: $E[XY] = \int \int xy f_{X,Y}(x,y) dx dy = \int x f_X(x) dx \int y f_Y(y) dy = \mu_X \mu_Y$. □

The conditions involving covariance and correlation coefficient follow directly from this result.

Therefore, independent random variables are necessarily uncorrelated, but the converse is not always true. Uncorrelated random variables do not have to be independent as demonstrated by the next example.

■ Example 5.15:

Consider a pair of random variables X and Y that are uniformly distributed over the unit circle so that

$$f_{X,Y}(x,y) = \begin{cases} \dfrac{1}{\pi}, & x^2 + y^2 \leq 1, \\ 0, & \text{otherwise.} \end{cases}$$

The marginal PDF of X can be found as follows:

$$f_X(x) = \int_{-\infty}^{\infty} f_{X,Y}(x,y)\mathrm{d}y = \int_{-\sqrt{1-x^2}}^{\sqrt{1-x^2}} \frac{1}{\pi}\mathrm{d}y = \frac{2}{\pi}\sqrt{1-x^2}, \quad -1 \leq x \leq 1 .$$

By symmetry, the marginal PDF of Y must take on the same functional form. Hence, the product of the marginal PDFs is

$$f_X(x)f_Y(y) = \frac{4}{\pi^2}\sqrt{(1-x^2)(1-y^2)}, \quad -1 \leq x,y \leq 1 .$$

Clearly, this is not equal to the joint PDF, and therefore, the two random variables are dependent. This conclusion could have been determined in a simpler manner. Note that if we are told that $X = 1$, then necessarily $Y = 0$, whereas if we know that $X = 0$, then Y can range anywhere from -1 to 1. Therefore, conditioning on different values of X leads to different distributions for Y.

Next, the correlation between X and Y is calculated.

$$E[XY] = \int\int_{x^2+y^2 \leq 1} \frac{xy}{\pi}\mathrm{d}x\mathrm{d}y = \frac{1}{\pi}\int_{-1}^{1} x\left[\int_{-\sqrt{1-x^2}}^{\sqrt{1-x^2}} y\mathrm{d}y\right]\mathrm{d}x .$$

Since the inner integrand is an odd function (of y) and the limits of integration are symmetric about zero, the integral is zero. Hence, $E[XY] = 0$. Note from the marginal PDFs just found that both X and Y are zero-mean. So, it is seen for this example that while the two random variables are uncorrelated, they are not independent.

■

■ Example 5.16:

Suppose we wish to use MATLAB to generate samples of a pair of random variables (X, Y) that are uniformly distributed over the unit circle. That is, the joint PDF is

$$f_{X, Y}(x, y) = \begin{cases} \dfrac{1}{\pi}, & x^2 + y^2 < 1, \\ 0, & \text{otherwise.} \end{cases}$$

If we generated two random variables independently according to the MATLAB code: X=rand(1); Y=rand(1); this would produce a pair of random variables uniformly distributed over the square $0 < x < 1, 0 < y < 1$. One way to achieve the desired result is to generate random variables uniformly over some region which includes the unit circle and then only keep those pairs of samples which fall inside the unit circle. In this case, it is straightforward to generate random variables which are uniformly distributed over the square, $-1 < x < 1, -1 < y < 1$, which circumscribes the unit circle. Then we keep only those samples drawn from within this square that also fall within the unit circle. The code that follows illustrates this technique. We also show how to generate a three-dimensional plot of an estimate of the joint PDF from the random data generated. To get a decent estimate of the joint PDF, we need to generate a rather large number of samples (we found that 100,000 worked pretty well). This requires that we create and perform several operations on some very large vectors. Doing so tends to make the program run slowly. In order to speed up the operation of the program, we choose to create shorter vectors of random variables (1000 in this case) and then repeat the procedure several times (100 in this case). Although this makes the code a little longer and probably a little harder to follow, by avoiding the creation of very long vectors, it substantially speeds up the program. The results of this program are shown in Figure 5.3.

```
clear
N=1000;                         % number of samples per iteration
bw=0.1;                         % bin widths for histogram
xbins=[-1.4:bw:1.4];
ybins=[-1.4:bw:1.4];            % histogram bins
iterations=100;                 % number of iterations
M=length(xbins);
Nsamples=zeros(M);              % initialize matrix for storing data
count=0;                        % initialize counter.
for ii=1:iterations
x=2*rand(1,N)-1; y=2*rand(1,N)-1;   % generate variables over square
% keep only those within the unit circle.
X=[]; Y=[];
```

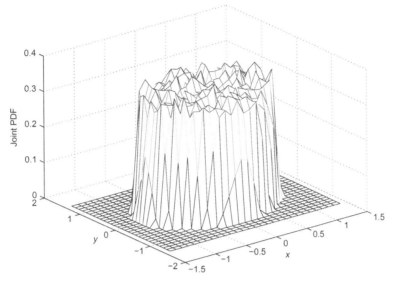

Figure 5.3
Estimate of the joint PDF of a pair of random variables uniformly distributed over the unit circle
from the data generated in Example 5.16. (For color version of this figure, the reader is referred
to the web version of this chapter.)

```
for k=1:N
    if x(k)^2+y(k)^2<1
        X=[X x(k)];
        Y=[Y y(k)];
    end                                % end if statement
end                                    % end k loop
count=count+length(X);                 % count random samples generated

% Compute number of samples that fall within each bin.
for m=1:length(xbins)
    for n=1:length(ybins)
        temp1=(abs(X-xbins(m))<bw/2);
        temp2=(abs(Y-ybins(n))<bw/2);
        Nsamples(m,n)=Nsamples(m,n)+sum(temp1.*temp2);
    end                                % end n loop
end                                    % end m loop
end                                    % end iterations

PDFest=Nsamples/(count*bw^2);          % convert to prob. densities
mesh(xbins,ybins,PDFest)               % plot estimate of joint PDF
xlabel('x'); ylabel('y');              % label plot axes
zlabel('Joint PDF');
```

∎

5.7 Jointly Gaussian Random Variables

As with single random variables, the most common and important example of a two-dimensional probability distribution is that of a joint Gaussian distribution. We begin by defining what is meant by a joint Gaussian distribution.

Definition 5.10: A pair of random variables X and Y is said to be *jointly Gaussian* if their joint PDF is of the general form

$$f_{X,Y}(x,y) = \frac{1}{2\pi\sigma_X\sigma_Y\sqrt{1-\rho_{XY}^2}}\exp\left(-\frac{\left[\left(\frac{x-\mu_X}{\sigma_X}\right)^2 - 2\rho_{XY}\left(\frac{x-\mu_X}{\sigma_X}\right)\left(\frac{y-\mu_Y}{\sigma_Y}\right) + \left(\frac{y-\mu_Y}{\sigma_Y}\right)^2\right]}{2(1-\rho_{XY}^2)}\right),$$

(5.40)

where μ_X and μ_Y are the means of X and Y, respectively; σ_X and σ_Y are the standard deviations of X and Y, respectively; and ρ_{XY} is the correlation coefficient of X and Y.

It is left as an exercise for the reader (see Exercise 5.35) to verify that this joint PDF results in marginal PDFs that are Gaussian. That is,

$$f_X(x) = \int_{-\infty}^{\infty} f_{X,Y}(x,y)\mathrm{d}y = \frac{1}{\sqrt{2\pi\sigma_X^2}}\exp\left(-\frac{(x-\mu_X)^2}{2\sigma_X^2}\right),$$

$$f_Y(y) = \int_{-\infty}^{\infty} f_{X,Y}(x,y)\mathrm{d}x = \frac{1}{\sqrt{2\pi\sigma_Y^2}}\exp\left(-\frac{(y-\mu_Y)^2}{2\sigma_Y^2}\right).$$

(5.41)

It is also left as an exercise for the reader (see Exercise 5.36) to demonstrate that if X and Y are jointly Gaussian, then the conditional PDF of X given $Y = y$ is also Gaussian, with a mean of $\mu_X + \rho_{XY}(\sigma_X/\sigma_Y)(y-\mu_Y)$ and a variance of $\sigma_X^2(1-\rho_{XY}^2)$. An example of this was shown in Example 5.10, and the general case can be proven following the same steps shown in that example.

Figure 5.4 shows the joint Gaussian PDF for three different values of the correlation coefficient. In Figure 5.4a, the correlation coefficient is $\rho_{XY} = 0$ and thus the two random variables are uncorrelated (and as we will see shortly, independent). Figure 5.4b shows the joint PDF when the correlation coefficient is large and positive, $\rho_{XY} = 0.9$. Note how the surface has become taller and thinner and largely lies above the line $y = x$. In Figure 5.4c, the correlation is now large and negative, $\rho_{XY} = -0.9$. Note that this is the same picture as in Figure 5.4b, except that it has been rotated by $90°$. Now the surface lies largely above the line $y = -x$. In all three figures, the means of both X and Y are zero and the variances of both X and Y are 1. Changing the means would simply translate the surface but would not change the shape. Changing the variances would expand or contract the surface along either the X- or Y-axis depending on which variance was changed.

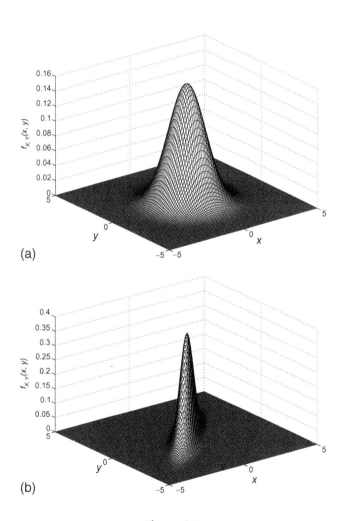

Figure 5.4

The joint Gaussian PDF: (a) $\mu_X = \mu_Y = 0$, $\sigma_X = \sigma_Y = 1$, $\rho_{XY} = 0$; (b) $\mu_X = \mu_Y = 0$, $\sigma_X = \sigma_Y = 1$, $\rho_{XY} = 0.9$;

(Continued)

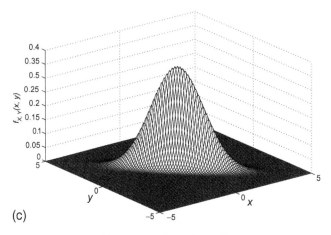

(c)

Figure 5.4 (Continued)
(c) $\mu_X = \mu_Y = 0$, $\sigma_X = \sigma_Y = 1$, $\rho_{XY} = -0.9$.

■ Example 5.17:

The joint Gaussian PDF is given by

$$f_{X,Y}(x,y) = \frac{1}{2\pi\sigma_X\sigma_Y\sqrt{1-\rho_{XY}^2}}$$
$$\exp\left(-\frac{1}{2(1-\rho_{XY}^2)}\left[\left(\frac{x-\mu_X}{\sigma_X}\right)^2 - 2\rho_{XY}\left(\frac{x-\mu_X}{\sigma_X}\right)\left(\frac{y-\mu_Y}{\sigma_Y}\right) + \left(\frac{y-\mu_Y}{\sigma_Y}\right)^2\right]\right).$$

Suppose we equate the portion of this equation that is within the square brackets to a constant. That is,

$$\left(\frac{x-\mu_X}{\sigma_X}\right)^2 - 2\rho_{XY}\left(\frac{x-\mu_X}{\sigma_X}\right)\left(\frac{y-\mu_Y}{\sigma_Y}\right) + \left(\frac{y-\mu_Y}{\sigma_Y}\right)^2 = c^2.$$

This is the equation for an ellipse. Plotting these ellipses for different values of c results in what is known as a contour plot. Figure 5.5 shows such plots for the two-dimensional joint Gaussian PDF. The following code can be used to generate such plots. The reader is encouraged to try creating similar plots for different values of the parameters in the Gaussian distribution.

```
clear
[X,Y]=meshgrid(-8:0.1:8);          % generate x and y array to
                                   % be used for contour plot
mux=0; muy=0;                      % set means
stdx=3; stdy=3;                    % set standard deviations
```

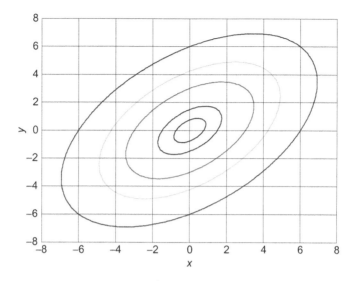

Figure 5.5
Contour plots for Example 5.17. (For color version of this figure, the reader is referred to the web version of this chapter.)

```
varx=stdx^2; vary=stdy^2;          % compute variances
rho=0.5;                           % set correlation coefficient
                                   % Compute exponent of 2-D Gaussian PDF.
X1=(X-mux)/stdx;
Y1=(Y-muy)/stdy;
Z=X1.^2-2*rho*X1.*Y1+Y1.^2;
c=[1/16 1/4 1 2 4];                % set contour levels
contour(X,Y,Z,c)                   % produce contour plot grid
                                   % turn on grid lines
xlabel('x'); ylabel('y')           % label axes
```

■

Theorem 5.7: Uncorrelated Gaussian random variables are independent.

Proof: Uncorrelated Gaussian random variables have a correlation coefficient of zero. Plugging $\rho_{XY} = 0$ into the general joint Gaussian PDF results in

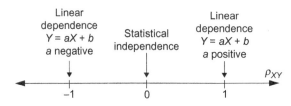

Figure 5.6
Interpretation of the correlation coefficient for jointly Gaussian random variables.

$$f_{X,Y}(x,y) = \frac{1}{2\pi\sigma_X\sigma_Y}\exp\left(-\frac{\left[\left(\frac{x-\mu_X}{\sigma_X}\right)^2 + \left(\frac{y-\mu_Y}{\sigma_Y}\right)^2\right]}{2}\right).$$

This clearly factors into the product of the marginal Gaussian PDFs.

$$f_{X,Y}(x,y) = \frac{1}{\sqrt{2\pi\sigma_x^2}}\exp\left(-\frac{(x-\mu_x)^2}{2\sigma_x^2}\right)\frac{1}{\sqrt{2\pi\sigma_y^2}}\exp\left(-\frac{(y-\mu_y)^2}{2\sigma_y^2}\right) = f_X(x)f_Y(y). \quad \square$$

While Example 5.15 demonstrated that this property does not hold for all random variables, it is true for Gaussian random variables. This allows us to give a stronger interpretation to the correlation coefficient when dealing with Gaussian random variables. Previously, it was stated that the correlation coefficient is a quantitative measure of the amount of correlation between two variables. While this is true, it is a rather vague statement. After all, what does "correlation" mean? In general, we cannot equate correlation and statistical dependence. Now, however, we see that in the case of Gaussian random variables, we can make the connection between correlation and statistical dependence. Hence, for jointly Gaussian random variables, the correlation coefficient can indeed be viewed as a quantitative measure of statistical dependence. This relationship is illustrated in Figure 5.6.

5.8 Joint Characteristic and Related Functions

When computing the joint moments of random variables, it is often convenient to use characteristic functions, moment-generating functions, or probability-generating functions. Since a pair of random variables is involved, the "frequency domain" function must now be two dimensional. We start with a description of the joint characteristic function which is similar to a two-dimensional Fourier transform of the joint PDF.

Definition 5.11: Given a pair of random variables X and Y with a joint PDF, $f_{X,Y}(x,y)$, the *joint characteristic function* is

$$\Phi_{X,Y}(\omega_1, \omega_2) = E[e^{j(\omega_1 X + \omega_2 Y)}] = \int_{-\infty}^{\infty}\int_{-\infty}^{\infty} e^{j(\omega_1 x + \omega_2 y)} f_{X,Y}(x,y) \mathrm{d}x\mathrm{d}y \qquad (5.42)$$

The various joint moments can be evaluated from the joint characteristic function using techniques similar to those used for single random variables. It is left as an exercise for the reader to establish the following relationship:

$$E[X^m Y^n] = (-j)^{m+n}\frac{\partial^m}{\partial \omega_1^m}\frac{\partial^n}{\partial \omega_2^n}\Phi_{X,Y}(\omega_1, \omega_2)\bigg|_{\omega_1 = \omega_2 = 0} \qquad (5.43)$$

■ Example 5.18:

Consider a pair of zero-mean, unit-variance, jointly Gaussian random variables whose joint PDF is

$$f_{X,Y}(x,y) = \frac{1}{2\pi\sqrt{1-\rho^2}}\exp\left(-\frac{(x^2 - 2\rho xy + y^2)}{2(1-\rho^2)}\right).$$

One way to calculate the joint characteristic function is to break the problem into two one-dimensional problems.

$$\Phi_{X,Y}(\omega_1, \omega_2) = E[e^{j(\omega_1 X + \omega_2 Y)}] = E_Y[e^{j\omega_2 Y}E_X[e^{j\omega_1 X}|Y]].$$

Conditioned on Y, X is a Gaussian random variable with a mean of ρY and a variance of $1-\rho^2$. The general form of the characteristic function (one-dimensional) of a Gaussian random variable with mean μ_X and variance σ_X^2 is (see Example 4.20)

$$\Phi_X(\omega) = \exp\left(j\mu_X\omega - \frac{\omega^2\sigma_X^2}{2}\right).$$

Therefore, the inner expectation above evaluates to

$$E_X[e^{j\omega_1 X}|Y] = \exp\left(j\rho Y\omega_1 - \frac{\omega_1^2(1-\rho^2)}{2}\right).$$

The joint characteristic function is then

$$\Phi_{X,Y}(\omega_1, \omega_2) = E_Y\left[\exp\left(j\omega_2 Y + j\rho Y\omega_1 - \frac{\omega_1^2(1-\rho^2)}{2}\right)\right] = \exp\left(-\frac{\omega_1^2(1-\rho^2)}{2}\right)E_Y[e^{j(\rho\omega_1 + \omega_2)Y}].$$

The remaining expectation is the characteristic function of a zero-mean, unit-variance Gaussian random variable evaluated at $\omega = \rho\omega_1 + \omega_2$. The resulting joint characteristic function is then found to be

$$\Phi_{X,Y}(\omega_1, \omega_2) = \exp\left(-\frac{\omega_1^2(1-\rho^2)}{2}\right)\exp\left(-\frac{(\rho\omega_1 + \omega_2)^2}{2}\right) = \exp\left(-\frac{\omega_1^2 + 2\rho\omega_1\omega_2 + \omega_2^2}{2}\right).$$

(Continued)

From this expression, various joint moments can be found. For example, the correlation is

$$E[XY] = -\frac{\partial}{\partial \omega_1 \partial \omega_2} \exp\left(-\frac{\omega_1^2 + 2\rho\omega_1\omega_2 + \omega_2^2}{2}\right)\Bigg|_{\omega_1 = \omega_2 = 0} = \frac{\partial}{\partial \omega_2}\rho\omega_2\exp\left(-\frac{\omega_2^2}{2}\right)\Bigg|_{\omega_2 = 0} = \rho.$$

Since the two random variables were zero mean, $\mathrm{Cov}(X, Y) = \rho$. Furthermore, since the two random variables were unit variance, ρ is also the correlation coefficient. We have proved therefore that the parameter ρ that shows up in the joint Gaussian PDF is indeed the correlation coefficient.

We could easily compute higher-order moments as well. For example, suppose we needed to compute $E[X^2 Y^2]$. It can be computed in a similar manner to the preceding:

$$E[X^2 Y^2] = \frac{\partial^2}{\partial \omega_1^2 \partial \omega_2^2} \exp\left(-\frac{\omega_1^2 + 2\rho\omega_1\omega_2 + \omega_2^2}{2}\right)\Bigg|_{\omega_1 = \omega_2 = 0} = -\frac{\partial^2}{\partial \omega_2^2}[1 - (\rho\omega_2)^2]\exp\left(-\frac{\omega_2^2}{2}\right)\Bigg|_{\omega_2 = 0} = 1 + 2\rho^2.$$

∎

Definition 5.12: For a pair of discrete random variables defined on a two-dimensional lattice of nonnegative integers, one can define a *joint probability-generating function* as

$$H_{X, Y}(z_1, z_2) = E[z_1^X z_2^Y] = \sum_{m=0}^{\infty} \sum_{n=0}^{\infty} P_{X, Y}(m, n)z_1^m z_2^n. \tag{5.44}$$

The reader should be able to show that the joint partial derivatives of the joint probability-generating function evaluated at zero are related to the terms in the joint PMF, whereas those same derivatives evaluated at 1 lead to joint factorial moments. Specifically:

$$P_{X, Y}(k, l) = \frac{1}{k!l!}\frac{\partial^k}{\partial z_1^k}\frac{\partial^l}{\partial z_2^l}H_{X, Y}(z_1, z_2)\Bigg|_{z_1 = z_2 = 0}, \tag{5.45}$$

$$E[X(X-1)\ldots(X-k+1)Y(Y-1)\ldots(Y-l+1)] = \frac{\partial^k}{\partial z_1^k}\frac{\partial^l}{\partial z_2^l}H_{X, Y}(z_1, z_2)\Bigg|_{z_1 = z_2 = 1}. \tag{5.46}$$

■ **Example 5.19:**

Consider the joint PMF given in Example 5.7:

$$P_{N, M}(n, m) = \frac{(n+m)!}{n!m!}\frac{a^n b^m}{(a+b+1)^{n+m+1}}.$$

It is not too difficult to work out the joint probability-generating function for this pair of discrete random variables.

$$H_{N,M}(z_1,z_2) = \sum_{n=0}^{\infty}\sum_{m=0}^{\infty} \frac{(n+m)!}{n!m!}\frac{(az_1)^n(bz_2)^m}{(a+b+1)^{n+m+1}} = \sum_{n=0}^{\infty}\frac{(az_1)^n}{(a+b+1)^{n+1}}\sum_{m=0}^{\infty}\frac{(n+m)!}{n!m!}\left(\frac{bz_2}{a+b+1}\right)^m$$

$$= \sum_{n=0}^{\infty}\frac{(az_1)^n}{(a+b+1)^{n+1}}\left(\frac{1}{1-\dfrac{bz_2}{a+b+1}}\right)^{n+1} = \sum_{n=0}^{\infty}\frac{(az_1)^n}{(a+b(1-z_2)+1)^{n+1}}$$

$$= \frac{1}{a+b(1-z_2)+1}\sum_{n=0}^{\infty}\left(\frac{az_1}{a+b(1-z_2)+1}\right)^n = \frac{1}{1+a(1-z_1)+b(1-z_2)}.$$

It should be noted that the closed form expression used for the various series preceding limits the range in the (z_1,z_2) plane for which these expressions are valid; thus, care must be taken when evaluating this function and its derivatives at various points. However, for this example, the expression is valid in and around the points of interest (i.e., $(z_1,z_2)=(0,0)$ and $(z_1,z_2)=(1,1)$).

Now that the joint probability-generating function has been found, joint moments are fairly easy to compute. For example,

$$E[NM] = \frac{\partial}{\partial z_1}\frac{\partial}{\partial z_2}\frac{1}{1+a(1-z_1)+b(1-z_2)}\bigg|_{z_1=z_2=1} = \frac{\partial}{\partial z_1}\frac{b}{[1+a(1-z_1)]^2}\bigg|_{z_1=1} = 2ab,$$

$$E[N(N-1)M] = \frac{\partial^2}{\partial z_1^2}\frac{\partial}{\partial z_2}\frac{1}{1+a(1-z_1)+b(1-z_2)}\bigg|_{z_1=z_2=1} = \frac{\partial^2}{\partial z_1^2}\frac{b}{[1+a(1-z_1)]^2}\bigg|_{z_1=1} = 6a^2b.$$

Putting these two results together, it is found that

$$E[N^2M] = E[N(N-1)M] + E[NM] = 6a^2b + 2ab.$$

By symmetry, we can also conclude that $E[NM(M-1)] = 6ab^2$ and $E[NM^2] = 6ab^2+2ab$. As one last example, we note that

$$E[N(N-1)M(M-1)] = \frac{\partial^2}{\partial z_1^2}\frac{\partial^2}{\partial z_2^2}\frac{1}{1+a(1-z_1)+b(1-z_2)}\bigg|_{z_1=z_2=1} = \frac{\partial^2}{\partial z_1^2}\frac{2b^2}{[1+a(1-z_1)]^3}\bigg|_{z_1=1} = 24a^2b^2.$$

From this and the previous results, we can find $E[N^2M^2]$ as follows:

$$E[N^2M^2] = E[N(N-1)M(M-1)] + E[NM^2] + E[N^2M] - E[NM] = 24a^2b^2 + 6ab^2 + 6a^2b - 2ab.$$

∎

The moment-generating function can also be generalized in a manner virtually identical to what was done for the characteristic function. We leave the details of this extension to the reader.

5.9 Transformations of Pairs of Random Variables

In this section, we consider forming a new random variable as a function of a pair of random variables. When a pair of random variables is involved, there are two classes of such transformations. The first class of problems deals with the case when a single new variable is created as a function of two random variables. The second class of problems involves creating two new random variables as two functions of two random variables. These two distinct, but related, problems are treated in this section.

Consider first a single function of two random variables, $Z = g(X, Y)$. If the joint PDF of X and Y is known, can the PDF of the new random variable Z be found? Of course, the answer is yes, and there are a variety of techniques to solve these types of problems depending on the nature of the function $g(\cdot)$. The first technique to be developed is an extension of the approach we used in Chapter 4 for functions of a single random variable.

The CDF of Z can be expressed in terms of the variables X and Y as

$$F_Z(z) = \Pr(Z \le z) = \Pr(g(X, Y) \le z) = \iint\limits_{g(x,y) \le z} f_{X,Y}(x,y)\mathrm{d}x\mathrm{d}y. \tag{5.47}$$

The inequality $g(x, y) \le z$ defines a region in the (x, y) plane. By integrating the joint PDF of X and Y over that region, the CDF of Z is found. The PDF can then be found by differentiating with respect to z. In principle, one can use this technique with any transformation; however, the integral to be computed may or may not be analytically tractable, depending on the specific joint PDF and the transformation.

To illustrate, consider a simple, yet very important example where the transformation is just the sum of the random variables, $Z = X + Y$. Then,

$$F_Z(z) = \iint\limits_{x+y \le z} f_{X,Y}(x,y)\mathrm{d}x\mathrm{d}y = \int_{-\infty}^{\infty} \int_{-\infty}^{z-y} f_{X,Y}(x,y)\mathrm{d}x\mathrm{d}y. \tag{5.48}$$

Differentiating to form the PDF results in

$$f_Z(z) = \frac{\mathrm{d}}{\mathrm{d}z} \int_{-\infty}^{\infty} \int_{-\infty}^{z-y} f_{X,Y}(x,y)\mathrm{d}x\mathrm{d}y = \int_{-\infty}^{\infty} f_{X,Y}(z-y,y)\mathrm{d}y. \tag{5.49}$$

The last step in the previous equation is completed using Liebnitz's rule.[1] An important special case results when X and Y are independent. In that case, the joint PDF factors into the product of the marginals producing

---

[1] Liebnitz's rule states that: $\frac{\partial}{\partial x}\int_{a(x)}^{b(x)} f(x,y)\mathrm{d}y = \frac{\partial b}{\partial x}f(x,b(x)) - \frac{\partial a}{\partial x}f(x,a(x)) + \int_{a(x)}^{b(x)} \frac{\partial}{\partial x}f(x,y)\mathrm{d}y.$

$$f_Z(z) = \int_{-\infty}^{\infty} f_X(z-y)f_Y(y)\mathrm{d}y\,. \tag{5.50}$$

Note that this integral is a convolution. Thus, the following important result has been proven:

> **Theorem 5.8:** If X and Y are statistically independent random variables, then the PDF of $Z = X + Y$ is given by the convolution of the PDFs of X and Y, $f_Z(z) = f_X(z)*f_Y(z)$.

■ Example 5.20:

Suppose X and Y are independent and both have exponential distributions,

$$f_X(x) = a\exp(-ax)u(x)\ ,\ \ f_Y(y) = b\exp(-by)u(y)\,.$$

The PDF of $Z = X + Y$ is then found by performing the necessary convolution:

$$f_Z(z) = \int_{-\infty}^{\infty} f_X(z-y)f_Y(y)\mathrm{d}y = ab\int_{-\infty}^{\infty} \exp(-a(z-y))\exp(-by)u(z-y)u(y)\mathrm{d}y$$

$$= abe^{-az}\int_0^z \exp((a-b)y)\mathrm{d}y\,u(z) = \frac{ab}{a-b}\left[e^{-az}e^{(a-b)y}\Big|_{y=0}^{y=z}\right]u(z) = \frac{ab}{a-b}[e^{-bz}-e^{-az}]u(z)\,.$$

The above result is valid assuming that $a \neq b$. If $a = b$, then the convolution works out to be

$$f_Z(z) = a^2 z e^{-az}u(z)\,.$$

■

Students familiar with the study of signals and systems should recall that the convolution integral appears in the context of passing signals through linear time invariant systems. In that context, most students develop a healthy respect for the convolution and will realize that quite often the convolution can be a cumbersome operation. To avoid difficult convolutions, these problems can often by solved using a frequency domain approach in which a Fourier or Laplace transform is invoked to replace the convolution with a much simpler multiplication. In the context of probability, the characteristic function or the moment generating function can fulfill the same role. Instead of finding the PDF of $Z = X + Y$ directly via convolution, suppose we first find the characteristic function of Z:

$$\Phi_Z(\omega) = E[e^{j\omega Z}] = E[e^{j\omega(X+Y)}] = E[e^{j\omega X}e^{j\omega Y}]\,. \tag{5.51}$$

If X and Y are independent, then the expected value of the product of a function of X times a function of Y factors into the product of expected values:

$$\Phi_Z(\omega) = E[e^{j\omega X}]E[e^{j\omega Y}] = \Phi_X(\omega)\Phi_Y(\omega)\,. \tag{5.52}$$

Once the characteristic function of Z is found, the PDF can be found using an inverse Fourier Transform.

Again, the characteristic function can be used to simplify the amount of computation involved in calculating PDFs of sums of independent random variables. Furthermore, we have also developed a new approach to find the PDFs of a general function of two random variables. Returning to a general transformation of the form $Z = g(X, Y)$, one can first find the characteristic function of Z according to

$$\Phi_Z(\omega) = E[e^{j\omega g(X, Y)}] = \int_{-\infty}^{\infty} \int_{-\infty}^{\infty} e^{j\omega g(x, y)} f_{X, Y}(x, y) \, dx \, dy. \tag{5.53}$$

An inverse transform of this characteristic function will then produce the desired PDF. In some cases, this method will provide a simpler approach to the problem, while in other cases the direct method may be easier.

■ Example 5.21:

Suppose X and Y are independent, zero-mean, unit-variance Gaussian random variables. The PDF of $Z = X^2 + Y^2$ can be found using either of the methods described thus far. Using characteristic functions,

$$\Phi_Z(\omega) = E[e^{j\omega(X^2 + Y^2)}] = E[e^{j\omega X^2}]E[e^{j\omega Y^2}].$$

The expected values are evaluated as follows:

$$E[e^{j\omega X^2}] = \int_{-\infty}^{\infty} \frac{1}{\sqrt{2\pi}} e^{j\omega x^2} e^{-x^2/2} dx = \frac{1}{\sqrt{1 - 2j\omega}} \int_{-\infty}^{\infty} \sqrt{\frac{1 - 2j\omega}{2\pi}} e^{-(1 - 2j\omega)x^2/2} dx = \frac{1}{\sqrt{1 - 2j\omega}}.$$

The last step is accomplished using the normalization integral for Gaussian functions. The other expected value is identical to the first since X and Y have identical distributions. Hence,

$$\Phi_Z(\omega) = \left(\frac{1}{\sqrt{1 - 2j\omega}}\right)^2 = \frac{1}{1 - 2j\omega}.$$

The PDF is found from the inverse Fourier transform to be

$$f_Z(z) = \frac{1}{2} \exp\left(-\frac{z}{2}\right) u(z).$$

The other approach is to find the CDF as follows:

$$F_Z(z) = \Pr(X^2 + Y^2 \le z) = \iint_{x^2 + y^2 \le z} \frac{1}{2\pi} \exp\left(-\frac{x^2 + y^2}{2}\right) dx \, dy.$$

Converting to polar coordinates,

$$F_Z(z) = \int_0^{\sqrt{z}} \int_0^{2\pi} \frac{r}{2\pi} \exp\left(-\frac{r^2}{2}\right) d\theta dr u(z) = \int_0^{\sqrt{z}} r \exp\left(-\frac{r^2}{2}\right) dr u(z) = \left[1 - \exp\left(-\frac{z}{2}\right)\right] u(z).$$

Finally, differentiating with respect to z results in

$$f_Z(z) = \frac{d}{dz}\left[1 - \exp\left(-\frac{z}{2}\right)\right] u(z) = \frac{1}{2} \exp\left(-\frac{z}{2}\right) u(z).$$

■

Another approach to solving these types of problems uses conditional distributions. Consider a general transformation, $Z = g(X, Y)$. Next, suppose we condition on one of the two variables, say $X = x$. Conditioned on $X = x$, $Z = g(x, Y)$ is now a single variable transformation. Hence, the conditional PDF of Z given X can be found using the general techniques presented in Chapter 4. Once $f_{Z|X}(z|x)$ is known, the desired (unconditional) PDF of Z can be found according to

$$f_Z(z) = \int f_{Z, X}(z, x)\, dx = \int f_{Z|X}(z|x) f_X(x) dx. \tag{5.54}$$

■ Example 5.22:

Suppose X and Y are independent zero-mean, unit-variance Gaussian random variables and we want to find the PDF of $Z = Y/X$. Conditioned on $X = x$, the transformation $Z = Y/x$ is a simple linear transformation and

$$f_{Z|X}(z|x) = |x| f_Y(xz) = \frac{|x|}{\sqrt{2\pi}} \exp\left(-\frac{x^2 z^2}{2}\right).$$

Multiplying the conditional PDF by the marginal PDF of X and integrating out x gives the desired marginal PDF of Z.

$$f_Z(z) = \int_{-\infty}^{\infty} f_{Z|X}(z|x) f_X(x) dx = \int_{-\infty}^{\infty} \frac{|x|}{\sqrt{2\pi}} \exp\left(-\frac{x^2 z^2}{2}\right) \frac{1}{\sqrt{2\pi}} \exp\left(-\frac{x^2}{2}\right) dx$$

$$= \frac{1}{2\pi} \int_{-\infty}^{\infty} |x| \exp\left(-\frac{(1 + x^2)z^2}{2}\right) dx = \frac{1}{\pi} \int_0^{\infty} x \exp\left(-\frac{(1 + z^2)x^2}{2}\right) dx = \frac{1}{\pi} \frac{1}{1 + z^2}.$$

Evaluating the integral in the last step can be accomplished by making the substitution $u = (1 + z^2)x^2/2$. Thus, the quotient of two independent Gaussian random variables follows a Cauchy distribution.

■

Up to this point, three methods have been developed for finding the PDF of $Z = g(X, Y)$ given the joint PDF of X and Y. They can be summarized as follows:

- Method 1—CDF approach

  Define a set $R(z) = \{(x, y): g(x, y) \le z\}$. The CDF of Z is the integral of the joint PDF of X and Y over the region $R(z)$. The PDF is then found by differentiating the expression for the CDF:

$$f_Z(z) = \frac{d}{dz} \int \int_{R(z)} f_{X, Y}(x, y) \, dx \, dy . \tag{5.55}$$

- Method 2—Characteristic function approach

  First, find the characteristic function of Z according to:

$$\Phi_Z(\omega) = E[e^{j\omega g(X, Y)}] . \tag{5.56}$$

  Then compute the inverse transform to get the PDF of Z.

- Method 3—Conditional PDF approach

  Fix either $X = x$ or $Y = y$ (whichever is more convenient). The conditional PDF of Z can then be found using the techniques developed for single random variables in Chapter 4. Once the conditional PDF of Z is found, the unconditional PDF is given by

$$f_Z(z) = \int f_{Z|Y}(z|y) f_Y(y) \, dy \quad \text{or} \quad f_Z(z) = \int f_{Z|X}(z|x) f_X(x) \, dx . \tag{5.57}$$

Next, our attention moves to solving a slightly more general class of problems. Given two random variables X and Y, suppose we now create two new random variables W and Z according to some 2×2 transformation of the general form

$$Z = g_1(X, Y), \\ W = g_2(X, Y). \tag{5.58}$$

The most common example of this type of problem involves changing coordinate systems. Suppose, for example, the variables X and Y represent the random position of some object in Cartesian coordinates. In some problems, it may be easier to view the object in a polar coordinate system, in which case, two new variables R and Θ could be created to describe the location of the object in polar coordinates. Given the joint PDF of X and Y, how can we find the joint PDF of R and Θ?

The procedure for finding the joint PDF of Z and W for a general transformation of the form given in Equation (5.58) is an extension of the technique used for a 1×1 transformation. First, recall the definition of the joint PDF given in Equation (5.2) which says that for an infinitesimal region $A_{x,y} = (x, x + \varepsilon_x) \times (y, y + \varepsilon_y)$, the joint PDF, $f_{X,Y}(x, y)$, has the interpretation

$$\Pr((X, Y) \in A_{x,y}) = f_{X,Y}(x, y)\varepsilon_x \varepsilon_y = f_{X,Y}(x, y)(\text{Area of } A_{x,y}). \tag{5.59}$$

Assume for now that the transformation is invertible. In that case, the transformation maps the region $A_{x,y}$ into a corresponding region $A_{z,w}$ in the (z, w)-plane. Furthermore,

$$\Pr((X, Y) \in A_{x,y}) = \Pr((Z, W) \in A_{z,w}) = f_{Z,W}(z, w)(\text{Area of } A_{z,w}). \tag{5.60}$$

Putting the two previous equations together results in

$$f_{Z,W}(z, w) = f_{X,Y}(x, y)\frac{\text{Area of } A_{x,y}}{\text{Area of } A_{z,w}}. \tag{5.61}$$

A fundamental result of multi-variable calculus states that if a transformation of the form in Equation (5.58) maps an infinitesimal region $A_{x,y}$, to a region $A_{z,w}$, then the ratio of the areas of these regions is given by the absolute value of the Jacobian of the transformation,

$$\frac{\text{Area of } A_{x,y}}{\text{Area of } A_{z,w}} = \left| J\begin{pmatrix} x & y \\ z & w \end{pmatrix} \right| = \left| \det \begin{bmatrix} \dfrac{\partial x}{\partial z} & \dfrac{\partial y}{\partial z} \\ \dfrac{\partial x}{\partial w} & \dfrac{\partial y}{\partial w} \end{bmatrix} \right|. \tag{5.62}$$

The PDF of Z and W is then given by

$$f_{Z,W}(z, w) = f_{X,Y}(x, y)\left| J\begin{pmatrix} x & y \\ z & w \end{pmatrix} \right|. \tag{5.63}$$

If it is more convenient to take derivatives of z and w with respect to x and y rather than vice-versa, we can alternatively use

$$\frac{\text{Area of } A_{z,w}}{\text{Area of } A_{x,y}} = \left| J\begin{pmatrix} z & w \\ x & y \end{pmatrix} \right| = \left| \det \begin{bmatrix} \dfrac{\partial z}{\partial x} & \dfrac{\partial z}{\partial y} \\ \dfrac{\partial w}{\partial x} & \dfrac{\partial w}{\partial y} \end{bmatrix} \right|, \tag{5.64}$$

$$f_{Z,W}(z, w) = \frac{f_{X,Y}(x, y)}{\left| J\begin{pmatrix} z & w \\ x & y \end{pmatrix} \right|}. \tag{5.65}$$

Whether Equation (5.63) or (5.65) is used, any expressions involving x or y must be replaced with the corresponding functions of z and w. Let the inverse transformation of Equation (5.58) be written as

$$X = h_1(Z, W),$$
$$Y = h_2(Z, W) \ . \tag{5.66}$$

Then these results can be summarized as

$$f_{Z, W}(z, w) = \frac{f_{X, Y}(x, y)}{\left| J\begin{pmatrix} z & w \\ x & y \end{pmatrix} \right|}\Bigg|_{\substack{x = h_1(z, w) \\ y = h_2(z, w)}} = f_{X, Y}(x, y) \left| J\begin{pmatrix} x & y \\ z & w \end{pmatrix} \right|\Bigg|_{\substack{x = h_1(z, w) \\ y = h_2(z, w)}} . \tag{5.67}$$

If the original transformation is not invertible, then the inverse transformation may have multiple roots. In this case, as with transformations involving single random variables, the expression in Equation (5.67) must be evaluated at each root of the inverse transformation and the results summed together. This general procedure for transforming pairs of random variables is demonstrated next through a few examples.

■ Example 5.23:

A classical example of this type of problem involves the transformation of two independent Gaussian random variables from cartesian to polar coordinates. Suppose

$$f_{X, Y}(x, y) = \frac{1}{2\pi\sigma^2}\exp\left(-\frac{x^2 + y^2}{2\sigma^2}\right).$$

We seek the PDF of the polar magnitude and phase given by

$$R = \sqrt{X^2 + Y^2},$$
$$\Theta = \tan^{-1}(Y/X) \ .$$

The inverse transformation is

$$X = R\cos(\Theta)$$
$$Y = R\sin(\Theta) \ .$$

In this case, the inverse transformation takes on a simpler functional form and so we elect to use this form to compute the Jacobian.

$$J\begin{pmatrix} x & y \\ r & \theta \end{pmatrix} = \det\begin{bmatrix} \frac{\partial x}{\partial r} & \frac{\partial x}{\partial \theta} \\ \frac{\partial y}{\partial r} & \frac{\partial y}{\partial \theta} \end{bmatrix} = \det\begin{bmatrix} \cos(\theta) & -r\sin(\theta) \\ \sin(\theta) & r\cos(\theta) \end{bmatrix} = r\cos^2(\theta) + r\sin^2(\theta) = r \ .$$

The joint PDF of R and Θ is then

$$f_{R,\,\Theta}(r,\theta) = f_{X,\,Y}(x,y)\left\|J\begin{pmatrix} x\ y \\ r\ \theta \end{pmatrix}\right\|\Bigg|_{\substack{x=h_1(r,\theta) \\ y=h_2(r,\theta)}} = \frac{r}{2\pi\sigma^2}\exp\left(-\frac{x^2+y^2}{2\sigma^2}\right)\Bigg|_{\substack{x=r\cos(\theta) \\ y=r\sin(\theta)}} = \frac{r}{2\pi\sigma^2}\exp\left(-\frac{r^2}{2\sigma^2}\right), \quad r\ge 0$$
$$0\le\theta<2\pi.$$

Note that in these calculations, we do not have to worry about taking the absolute value of the Jacobian since for this problem the Jacobian ($=r$) is always nonnegative. If we were interested, we could also find the marginal distributions of R and Θ to be

$$f_R(r) = \frac{r}{\sigma^2}\exp\left(-\frac{r^2}{2\sigma^2}\right)u(r) \text{ and } f_\Theta(\theta) = \frac{1}{2\pi}, \ 0\le\theta<2\pi.$$

The magnitude follows a Rayleigh distribution while the phase is uniformly distributed over $(0,2\pi)$.   ■

■ Example 5.24:

Suppose X and Y are independent and both uniformly distributed over $(0,1)$, so that

$$f_{X,\,Y}(x,y) = \begin{cases} 1, & 0\le x,y<1, \\ 0, & \text{otherwise.} \end{cases}$$

Consider forming the two new random variables

$$Z = \sqrt{-2\ln(X)}\cos(2\pi Y),$$
$$W = \sqrt{-2\ln(X)}\sin(2\pi Y).$$

The inverse transformation in this case is found to be

$$X = \exp\left(\frac{Z^2+W^2}{2}\right),$$
$$Y = \frac{1}{2\pi}\tan^{-1}\left(\frac{W}{Z}\right).$$

In this example, we compute the Jacobian by taking derivatives of z and w with respect to x and y to produce

$$J\begin{pmatrix} z\ w \\ x\ y \end{pmatrix} = \det\begin{bmatrix} \dfrac{\partial z}{\partial x} & \dfrac{\partial z}{\partial y} \\ \dfrac{\partial w}{\partial x} & \dfrac{\partial w}{\partial y} \end{bmatrix} = \det\begin{bmatrix} -\dfrac{1}{x}\dfrac{\cos(2\pi y)}{\sqrt{-2\ln(x)}} & -2\pi\sqrt{-2\ln(X)}\sin(2\pi Y) \\ -\dfrac{1}{x}\dfrac{\sin(2\pi y)}{\sqrt{-2\ln(x)}} & 2\pi\sqrt{-2\ln(X)}\cos(2\pi Y) \end{bmatrix}$$
$$= \left(-\frac{2\pi}{x}[\cos^2(2\pi y)+\sin^2(2\pi y)]\right) = -\frac{2\pi}{x}.$$

Note that since x is always nonnegative, the absolute value of the Jacobian will just be $2\pi/x$. The joint PDF of Z and W is then found to be

(Continued)

$$f_{Z,W}(z,w) = \frac{f_{X,Y}(x,y)}{\left| J\left(\begin{array}{c} z\ w \\ x\ y \end{array} \right) \right|} \Bigg|_{\substack{x = h_1(z,w) \\ y = h_2(z,w)}} = \frac{x}{2\pi} \Bigg|_{\substack{x = \exp\left(-\frac{z^2 + w^2}{2}\right) \\ y = \frac{1}{2\pi}\tan^{-1}\left(\frac{w}{z}\right)}} = \frac{1}{2\pi}\exp\left(-\frac{z^2 + w^2}{2}\right).$$

This transformation is known as the Box-Muller transformation. It transforms a pair of independent uniform random variables into a pair of independent Gaussian random variables. This transformation has application in the world of computer simulations. Techniques for generating uniform random variables are well known. This transformation then allows us to generate Gaussian random variables as well. More material on this subject is given in Chapter 12.

∎

■ Example 5.25:

Suppose X and Y are independent Gaussian random variables, both with zero-mean and unit variance. Two new random variables Z and W are formed through a linear transformation of the form

$$Z = aX + bY$$
$$W = cX + dY .$$

The inverse transformation is given by

$$X = \frac{d}{ad - bc}Z - \frac{b}{ad - bc}W,$$

$$Y = -\frac{c}{ad - bc}Z + \frac{a}{ad - bc}W.$$

With this general linear transformation, the various partial derivatives are trivial to compute and the resulting Jacobian is

$$J\left(\begin{array}{c} z\ w \\ x\ y \end{array} \right) = \det\begin{bmatrix} a\ b \\ c\ d \end{bmatrix} = ad - bc .$$

Plugging these results into the general formula results in

$$f_{Z,W}(z,w) = \frac{f_{X,Y}(x,y)}{\left| J\left(\begin{array}{c} z\ w \\ x\ y \end{array} \right) \right|} \Bigg|_{\substack{x = h_1(z,w) \\ y = h_2(z,w)}} = \frac{\frac{1}{2\pi}\exp\left(-\frac{x^2 + y^2}{2}\right)}{|ad - bc|} \Bigg|_{\substack{x = \frac{d}{ad-bc}z - \frac{b}{ad-bc}w \\ y = -\frac{c}{ad-bc}z + \frac{a}{ad-bc}w}}$$

$$= \frac{1}{2\pi\sqrt{(ad-bc)^2}}\exp\left(-\frac{(c^2 + d^2)z^2 - 2(bd + ac)zw + ((a^2 + b^2)w)^2}{2(ad - bc)^2}\right)$$

With a little algebraic manipulation, it can be shown that this joint PDF fits the general form of a joint Gaussian PDF. In particular,

$$f_{Z,W}(z,w) = \frac{1}{2\pi\sigma_Z\sigma_W\sqrt{1-\rho_{ZW}^2}}\exp\left(-\frac{(z/\sigma_Z)^2 - 2\rho_{ZW}(z/\sigma_Z)(w/\sigma_w) + (w/\sigma_w)^2}{2(1-\rho_{ZW}^2)}\right),$$

where $\sigma_Z^2 = a^2 + b^2$, $\sigma_W^2 = c^2 + d^2$, and $\rho_{WZ}^2 = (ac+bd)^2(a^2+b^2)^{-1}(c^2+d^2)^{-1}$.

A few remarks about the significance of the result of Example 5.25 are appropriate. First, we have performed an arbitrary linear transformation on a pair of independent Gaussian random variables and produced a new pair of Gaussian random variables (which are no longer independent). In the next chapter, it will be shown that a linear transformation of any number of jointly Gaussian random variables always produces jointly Gaussian random variables. Second, if we look at this problem in reverse, two correlated Gaussian random variables Z and W can be transformed into a pair of uncorrelated Gaussian random variables X and Y using an appropriate linear transformation. More information will be given on this topic in the next chapter as well.

5.10 Complex Random Variables

In engineering practice, it is common to work with quantities which are complex. Usually, a complex quantity is just a convenient shorthand notation for working with two real quantities. For example, a sinusoidal signal with amplitude, A, frequency, ω, and phase, θ, can be written as

$$s(t) = A\cos(\omega t + \theta) = Re[Ae^{j\theta}e^{j\omega t}], \tag{5.68}$$

where $j = \sqrt{-1}$. The complex number $Z = Ae^{j\theta}$ is known as a phasor representation of the sinusoidal signal. It is a complex number with real part of $X = Re[Z] = A\cos(\theta)$ and imaginary part of $Y = Im[Z] = A\sin(\theta)$. The phasor Z can be constructed from two real quantities (either A and θ or X and Y).

Suppose a complex quantity we are studying is composed of two real quantities which happen to be random. For example, the sinusoidal signal above might have a random amplitude and/or a random phase. In either case, the complex number Z will also be random. Unfortunately, our formulation of random variables does not allow for complex quantities. When we began to describe a random variable via its CDF in the beginning of Chapter 3, the CDF was defined as $F_Z(z) = \Pr(Z \leq z)$. This definition makes no sense if Z is a complex number: what does it mean for a complex number to be less that another number? Nevertheless, the engineering literature is filled with complex random variables and their distributions.

The concept of a complex random variable can often be the source of great confusion to many students, but it does not have to be as long as we realize that a complex random variable is nothing more than a shorthand representation of two real random variables. To motivate the concept of a complex random variable, we use the most common example of a pair of independent, equal variance, jointly Gaussian random variables, X and Y. The joint PDF is of the form

$$f_{X,Y}(x,y) = \frac{1}{2\pi\sigma^2}\exp\left(-\frac{(x-\mu_X)^2+(y-\mu_Y)^2}{2\sigma^2}\right). \tag{5.69}$$

This joint PDF (of two real random variables) naturally lends itself to be written in terms of some complex variables. Define $Z = X+jY$, $z = x+jy$ and $\mu_Z = \mu_X+j\mu_Y$. Then,

$$f_{X,Y}(x,y) = f_Z(z) = \frac{1}{2\pi\sigma^2}\exp\left(-\frac{|z-\mu_Z|^2}{2\sigma^2}\right). \tag{5.70}$$

We reemphasize at this point that this is not to be interpreted as the PDF of a complex random variable (since such an interpretation would make no sense); rather, this is just a compact representation of the joint PDF of two real random variables. This density is known as the *circular Gaussian density function* (since the contours of $f_Z(z)$ = constant form circles in the complex z-plane).

Note that the PDF in Equation (5.70) has two parameters, μ_Z and σ. The parameter μ_Z is interpreted as the mean of the complex quantity, $Z = X+jY$,

$$\mu_Z = E[Z] = E[X+jY] = \int(x+jy)f_{X,Y}(x,y)\mathrm{d}x\mathrm{d}y = \mu_X+j\mu_Y. \tag{5.71}$$

But what about σ^2? We would like to be able to interpret it as the variance of $Z = X+jY$. To do so, we need to redefine what we mean by variance of a complex quantity. If we used the definition we are used to (for real quantities) we would find

$$E[(Z-\mu_Z)^2] = E\{[(X-\mu_X)+j(Y-\mu_Y)]^2\} = \mathrm{Var}(X)-\mathrm{Var}(Y)+2j\mathrm{Cov}(X,Y). \tag{5.72}$$

In the case of our independent Gaussian random variables, since $\mathrm{Cov}(X,Y) = 0$ and $\mathrm{Var}(X) = \mathrm{Var}(Y)$, this would lead to $E[(Z-\mu_Z)^2] = 0$. To overcome this inconsistency, we redefine the variance for a complex quantity as follows.

Definition 5.13: For a complex random quantity, $Z = X+jY$, the variance is defined as

$$\mathrm{Var}(Z) = \frac{1}{2}E[|Z-\mu_Z|^2] = \frac{1}{2}\mathrm{Var}(X)+\frac{1}{2}\mathrm{Var}(Y). \tag{5.73}$$

We emphasize at this point that this definition is somewhat arbitrary and was chosen so that the parameter σ^2 which shows up in Equation (5.70) can be interpreted as the variance of Z. Many textbooks do not include the factor of 1/2 in the definition, while many others (besides this one) do include the 1/2. Hence, there seems to be no way to avoid a little bit of confusion here. The student just needs to be aware that there are two inconsistent definitions prevalent in the literature.

Definition 5.14: For two complex random variables $Z_1 = X_1 + jY_1$ and $Z_2 = X_2 + jY_2$, the correlation and covariance are defined as

$$R_{1,2} = \frac{1}{2}E[Z_1 Z_2^*] = \frac{1}{2}\{E[X_1 X_2] + E[Y_1 Y_2] - jE[X_1 Y_2] + jE[X_2 Y_1]\}, \qquad (5.74)$$

$$C_{1,2} = \frac{1}{2}E[(Z_1 - \mu_{Z_1})(Z_2 - \mu_{Z_2})^*]. \qquad (5.75)$$

As with real random variables, complex quantities are said to be orthogonal if their correlation is zero, whereas they are uncorrelated if their covariance is zero.

5.11 Engineering Application: Mutual Information, Channel Capacity, and Channel Coding

In Section 4.12, we introduced the idea of the entropy of a random variable which is a quantitative measure of how much randomness there is in a specific random variable. If the random variable represents the output of a source, the entropy tells us how much mathematical information there is in each source symbol. We can also construct similar quantities to describe the relationships between random variables. Consider two random variables X and Y that are statistically dependent upon one another. Each random variable has a certain entropy associated with it, $H(X)$ and $H(Y)$, respectively. Suppose it is observed that $Y = y$. Since X and Y are related, knowing Y will tell us something about X and hence the amount of randomness in X will be changed. This could be quantified using the concept of conditional entropy.

Definition 5.15: The *conditional entropy* of a discrete random variable X given knowledge of a particular realization of a related random variable $Y = y$ is

$$H(X|Y=y) = \sum_x \Pr(X=x|Y=y)\log\left(\frac{1}{\Pr(X=x|Y=y)}\right). \qquad (5.76)$$

Averaging over all possible conditioning events produces

$$H(X|Y) = \sum_x \sum_y \Pr(Y=y)\Pr(X=x|Y=y)\log\left(\frac{1}{\Pr(X=x|Y=y)}\right)$$

$$= \sum_x \sum_y \Pr(X=x, Y=y)\log\left(\frac{1}{\Pr(X=x|Y=y)}\right). \qquad (5.77)$$

The conditional entropy tells how much uncertainty remains in the random variable X after we observe the random variable Y. The amount of information provided about X by observing Y can be determined by forming the difference between the entropy in X before and after observing Y.

Definition 5.16: The *mutual information* between two discrete random variables X and Y is

$$I(X;Y) = H(X) - H(X|Y) = \sum_x \sum_y \Pr(X=x, Y=y) \log\left(\frac{\Pr(X=x|Y=y)}{\Pr(X=x)}\right). \qquad (5.78)$$

We leave it as an exercise for the reader to prove the following properties of mutual information:

- *Nonnegative*: $I(X;Y) \geq 0$.
- *Independence*: $I(X;Y) = 0$ if and only if X and Y are independent.
- *Symmetry*: $I(X;Y) = I(Y;X)$.

Now we apply the concept of mutual information to a digital communication system. Suppose we have some digital communication system which takes digital symbols from some source (or from the output of a source encoder) and transmits them via some modulation format over some communications medium. At the receiver, a signal is received and processed and ultimately a decision is made as to what symbol(s) was most likely sent. We will not concern ourselves with the details of how the system operates, but rather we will model the entire process in a probabilistic sense. Let X represent the symbol to be sent, which is randomly drawn from some n-letter alphabet according to some distribution $\boldsymbol{p} = (p_0, p_1, \ldots, p_{n-1})$. Furthermore, let Y represent the decision made by the receiver, with Y taken to be a random variable on an m-letter alphabet. It is not unusual to have $m \neq n$, but in order to keep this discussion as simple as possible, we will only consider the case where $m = n$ so that the input and output of our communication system are taken from the same alphabet. Also, we assume the system to be memoryless so that decisions made on one symbol are not affected by previous decisions nor do they affect future decisions. In that case, we can describe the operation of the digital communication system using a transition diagram as illustrated in Figure 5.7 for a three-letter alphabet. Mathematically, the operation of this communication system can be described by a matrix $\boldsymbol{Q}$ whose elements are $q_{i,j} = \Pr(Y=i|X=j)$.

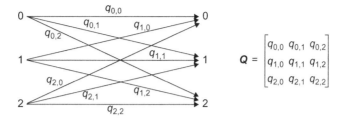

Figure 5.7
A transition diagram for a ternary (three-letter) communication channel.

We can now ask ourselves how much information does the communication system carry? Or, in other words, if we observe the output of the system, how much information does this give us about what was really sent? The mutual information answers this question. In terms of the channel (as described by Q) and the input (as described by p), the mutual information is

$$I(X, Y) = \sum_i \sum_j p_j q_{i,j} \log\left(\frac{q_{i,j}}{\sum_k q_{i,k} p_k}\right). \tag{5.79}$$

Note that the amount of information carried by the system is a function not only of the channel but also of the source. As an extreme example, suppose the input distribution were $p = (1, 0, \ldots, 0)$. In that case it is easy to show that $I(X, Y) = 0$; that is, the communication system carries no information. This is not because the communication system is incapable of carrying information, but because what we are feeding into the system contains no information. To describe the information carrying capability of a communication channel, we need a quantity which is a function of the channel only and not of the input to the channel.

Definition 5.17: Given a discrete communications channel described by a transition probability matrix Q, the channel capacity is given by

$$C = \max_{p} I(X; Y) = \max_{p} \sum_i \sum_j p_j q_{i,j} \log\left(\frac{q_{i,j}}{\sum_k q_{i,k} p_k}\right). \tag{5.80}$$

The maximization of the mutual information is with respect to any valid probability distribution p.

■ Example 5.26:

As a simple example, consider the so-called binary symmetric channel (BSC) described by the transition probability matrix

$$Q = \begin{bmatrix} 1-q & q \\ q & 1-q \end{bmatrix}.$$

The BSC is described by a single parameter q, which has the interpretation of the probability of bit error of the binary communications system. That is, q is the probability of the receiver deciding a 0 was sent when a 1 was actually sent and it is also the probability of the receiver deciding a 1 was sent when a 0 was actually sent. Since the input to this channel is binary, its distribution can also be described by a single parameter. That is, $p = (p, 1-p)$. Likewise, the output of the channel is also binary and thus can be described in terms of a single parameter, $r = (r, 1-r)$ where $r = \Pr(Y=0) = p(1-q) + q(1-p)$. The mutual information for the BSC is

$$I(X;Y) = p(1-q)\log\left(\frac{1-q}{r}\right) + pq\log\left(\frac{q}{r}\right) + (1-p)(1-q)\log\left(\frac{1-q}{1-r}\right) + (1-p)q\log\left(\frac{q}{1-r}\right).$$

Some straightforward algebraic manipulations reveal that the above expression can be simplified to $I(X;Y) = \mathcal{H}(r) - \mathcal{H}(q)$, where $\mathcal{H}(\cdot)$ is the binary entropy function. Maximization with respect to p is now straightforward. The mutual information is maximized when the output distribution is $r = (0.5, 0.5)$ and the resulting capacity is

$$C = 1 - \mathcal{H}(q).$$

Due to the symmetry of the channel, the output distribution will be symmetric when the input distribution is also symmetric, $p = (0.5, 0.5)$. This function is illustrated in Figure 5.8.

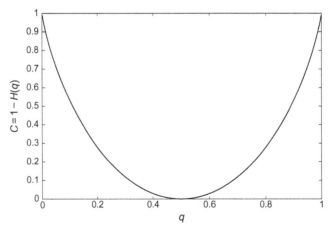

Figure 5.8
Capacity of a binary symmetric channel.

■

Figure 5.9
A functional block diagram of a digital communication system.

The channel capacity provides a fundamental limitation on the amount of information that can reliably be sent over a channel. For example, suppose we wanted to transmit information across the BSC of Example 5.26. Furthermore, suppose the error probability of the channel was $q = 0.1$. Then the capacity is $C = 1 - \mathcal{H}(0.1) = 0.53$ bits. That is, every physical bit that is transmitted across the channel must contain less than 0.53 bits of mathematical information. This is achieved through the use of redundancy via channel coding. Consider the block diagram of the digital communication system in Figure 5.9. The binary source produces independent bits which are equally likely to be "0" or "1." This source has an entropy of 1 bit/source symbol. Since the channel has a capacity of 0.53 bits, the information content of the source must be reduced before these symbols are sent across the channel. This is achieved by the channel coder which takes blocks of k information bits and maps them to n bit code words where $n > k$. Each code word contains k bits of information and so each coded bit contains k/n bits of mathematical information. By choosing the code rate, k/n, to be less than the channel capacity, C, we can assure that the information content of the symbols being input to the channel is no greater than the information carrying capability of the channel.

Viewed from a little more concrete perspective, the channel used to transmit physical bits has an error rate of 10%. The purpose of the channel code is to add redundancy to the data stream to provide the ability to correct the occasional errors caused by the channel. A fundamental result of information theory known as the *channel coding theorem* states that as k and n go to infinity in such a way that $k/n < C$, it is possible to construct a channel code (along with the appropriate decoder) which will provide error-free communication. That is, the original information bits will be provided to the destination with arbitrarily small probability of error. The channel coding theorem does not tell us how to construct such a code, but significant progress has been made in recent years towards finding practical techniques to achieve what information theory promises is possible.

Section 5.1: Joint CDFs

5.1 Recall the joint CDF given in Example 5.1,

$$F_{X,Y}(x,y) = \begin{cases} 0, & x<0 \text{ or } y<0, \\ x, & 0 \le x \le 1, y>1, \\ y, & x>1, 0 \le y \le 1, \\ xy, & 0 \le x \le 1, 0 \le y \le 1, \\ 1, & x>1, y>1. \end{cases}$$

(a) Find $\Pr(X < 3/4)$.
(b) Find $\Pr(X > 1/2)$.
(c) Find $\Pr(Y > 1/4)$.
(d) Find $\Pr(1/4 < X < 1/2, 1/2 < Y < 1)$.

5.2 A colleague of your proposes that a certain pair of random variables be modeled with a joint CDF of the form

$$F_{X,Y}(x,y) = [1 - ae^{-x} - be^{-y} + ce^{-(x+y)}]u(x)u(y).$$

(a) Find any restrictions on the constants a, b, and c needed for this to be a valid joint CDF.
(b) Find the marginal CDFs, $F_X(x)$ and $F_Y(y)$ under the restrictions found in part (a).

5.3 Consider again the joint CDF given in Exercise 5.2.
(a) For constants a and b, such that $0 < a < 1$, $0 < b < 1$ and $a < b$, find $\Pr(a < X < b)$.
(b) For constants c and d, such that $0 < c < 1$, $0 < d < 1$ and $c < d$, find $\Pr(c < Y < d)$.
(c) Find $\Pr(a < X < b | c < Y < d)$. Are the events $\{a < X < b\}$ and $\{c < Y < d\}$ statistically independent?

5.4 Suppose a random variable X has a CDF given by $F_X(x)$ and similarly, a random variable Y has a CDF, $F_Y(y)$. Prove that the function $F(x, y) = F_X(x)F_Y(y)$ satisfies all the properties required of joint CDFs and hence will always be a valid joint CDF.

5.5 For the joint CDF that is the product of two marginal CDFs, $F_{X, Y}(x, y) = F_X(x)F_Y(y)$, as described in Exercise 5.4, show that the events $\{a < X < b\}$ and $\{c < Y < d\}$ are always independent for any constants $a < b$ and $c < d$.

Section 5.2: Joint PDFs

5.6 For positive constants a and b, a pair of random variables has a joint PDF specified by

$$f_{X, Y}(x, y) = abe^{-(ax + by)}u(x)u(y).$$

(a) Find the joint CDF, $F_{X, Y}(x, y)$.
(b) Find the marginal PDFs, $f_X(x)$ and $f_Y(y)$.
(c) Find $\Pr(X > Y)$.
(d) Find $\Pr(X > Y^2)$.

5.7 For positive constants a, b, c, and positive integer n, a pair of random variables has a joint PDF specified by

$$f_{X, Y}(x, y) = \frac{d}{(ax + by + c)^n}u(x)u(y).$$

(a) Find the constant d in terms of a, b, c, and n.
(b) Find the marginal PDFs, $f_X(x)$ and $f_Y(y)$.
(c) Find $\Pr(X > Y)$.

5.8 A pair of random variables has a joint PDF specified by

$$f_{X, Y}(x, y) = d\exp(-(ax^2 + bxy + cy^2)).$$

(a) Find the constant d in terms of a, b, and c. Also, find any restrictions needed for a, b, and c themselves for this to be a valid PDF.
(b) Find the marginal PDFs, $f_X(x)$ and $f_Y(y)$.
(c) Find $\Pr(X > Y)$.

5.9 A pair of random variables has a joint PDF specified by

$$f_{X, Y}(x, y) = \begin{cases} c\sqrt{1 - x^2 - y^2}, & x^2 + y^2 \le 1, \\ 0, & \text{otherwise.} \end{cases}$$

(a) Find the constant c.

(b) Find $\Pr(X^2 + Y^2 > 1/4)$.

(c) Find $\Pr(X > Y)$.

5.10 A pair of random variables has a joint PDF specified by

$$f_{X,Y}(x, y) = \frac{1}{8\pi} \exp\left(-\frac{(x-1)^2 + (y+1)^2}{8}\right).$$

(a) Find $\Pr(X > 2, Y < 0)$.

(b) Find $\Pr(0 < X < 2, |Y+1| > 2)$.

(c) Find $\Pr(Y > X)$ *Hint:* Set up the appropriate double integral and then use the change of variables: $u = x - y$, $v = x + y$.

5.11 A pair of random variables, (X, Y), is equally likely to fall anywhere in the ellipse described by $9X^2 + 4Y^2 < 36$.

(a) Write the form of the joint PDF, $f_{X,Y}(x, y)$.

(b) Find the marginal PDFs, $f_X(x)$ and $f_Y(y)$.

(c) Find $\Pr(X > 1)$ and $\Pr(Y < 1)$

(d) Find $\Pr(Y < 1 | X > 1)$. Are the events $\{X > 1\}$ and $\{Y < 1\}$ independent?

5.12 A pair of random variables, (X, Y), is equally likely to fall anywhere within the region defined by $|X| + |Y| \le 1$.

(a) Write the form of the joint PDF, $f_{X,Y}(x, y)$.

(b) Find the marginal PDFs, $f_X(x)$ and $f_Y(y)$.

(c) Find $\Pr(X > 1/2)$ and $\Pr(Y < 1/2)$

(d) Find $\Pr(Y < 1/2 | X > 1/2)$. Are the events $\{X > 1/2\}$ and $\{Y < 1/2\}$ independent?

Section 5.3: Joint PMFs

5.13 For some integer L and constant c, two discrete random variables have a joint PMF given by

$$P_{M,N}(m, n) = \begin{cases} c, & m \ge 0, n \ge 0, m + n < L, \\ 0, & \text{otherwise.} \end{cases}$$

(a) Find the value of the constant c in terms of L.

(b) Find the marginal PMFs, $P_M(m)$ and $P_N(n)$.

(c) Find $\Pr(M + N < L/2)$.

5.14 Two discrete random variables have a joint PMF as described in the following table.

$P_{M,N}(m, n)$	$m = 1$	$m = 2$	$m = 3$
$n = 1$	1/5	7/45	1/9
$n = 2$	8/45	4/45	2/45
$n = 3$	2/15	1/15	1/45

(a)   Find the marginal PDFs, $P_M(m)$ and $P_N(n)$.
(b)   Find $\Pr(N = 1 | M = 2)$.
(c)   Find $\Pr(M = N)$.
(d)   Find $\Pr(M > N)$.

5.15 For a constant k, two discrete random variables have a joint PMF given by

$$P_{M,N}(m, n) = \begin{cases} c, & m, n = 0, 1, 2, \ldots, k - 1, \\ 0, & \text{otherwise.} \end{cases}$$

(a)   Find the value of the constant c in terms of k.
(b)   Find the marginal PMFs, $P_M(m)$ and $P_N(n)$.
(c)   Find $\Pr(M + N < k/2)$.

5.16 Let M be a random variable that follows a Poisson distribution, so that for some constant α, its PMF is

$$P_M(m) = \frac{\alpha^m}{m!}e^{-\alpha}, \quad m = 0, 1, 2, \ldots .$$

Let N be another random variable that, given $M = m$, is equally likely to take on any value in the set $\{0, 1, 2, \ldots, m\}$.
(a)   Find the joint PMF of M and N.
(b)   Find the marginal PMF of N, $P_N(n)$. Plot your result for $\alpha = 1$.

Section 5.4: Conditional Distribution, Density and Mass Functions

5.17 For the discrete random variables whose joint PMF is described by the table in Exercise 5.14, find the following conditional PMFs:
  (a) $P_M(m|N = 2)$;
  (b) $P_M(m|N \geq 2)$;
  (c) $P_N(n|M \neq 2)$.

5.18 Consider again the random variables in Exercise 5.11 that are uniformly distributed over an ellipse.
  (a) Find the conditional PDFs, $f_{X|Y}(x|y)$ and $f_{Y|X}(y|x)$.
  (b) Find $f_{X|Y>1}(x)$.
  (c) Find $f_{Y|\{|X|<1\}}(y)$.

5.19 Recall the random variables of Exercise 5.12 that are uniformly distributed over the region $|X| + |Y| \leq 1$.
  (a) Find the conditional PDFs, $f_{X|Y}(x|y)$ and $f_{Y|X}(y|x)$.
  (b) Find the conditional CDFs, $F_{X|Y}(x|y)$ and $F_{Y|X}(y|x)$.
  (c) Find $f_{X|\{Y>1/2\}}(x)$ and $F_{X|\{Y>1/2\}}(x)$.

5.20 Suppose a pair of random variables (X, Y) is uniformly distributed over a rectangular region, A: $x_1 < X < x_2$, $y_1 < Y < y_2$. Find the conditional PDF of (X, Y) given the conditioning event $(X, Y) \in B$, where the region B is an arbitrary region completely contained within the rectangle A as shown in the accompanying figure.

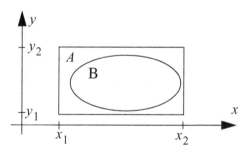

Section 5.5: Expected Values Involving Pairs of Random Variables

5.21 A pair of random variables has a joint PDF specified by

$$f_{X,Y}(x, y) = \frac{1}{2\pi\sqrt{3}} \exp\left(-\frac{x^2 + 2xy + 4y^2}{6}\right).$$

  (a) Find the marginal PDFs, $f_X(x)$ and $f_Y(y)$.
  (b) Based on the results of part (a), find $E[X]$, $E[Y]$, $\text{Var}(X)$, and $\text{Var}(Y)$.
  (c) Find the conditional PDF, $f_{X|Y}(x|y)$.
  (d) Based on the results of part (c), find $E[XY]$, $\text{Cov}(X, Y)$, and $\rho_{X,Y}$.

5.22 A pair of random variables is uniformly distributed over the ellipse defined by $x^2 + 4y^2 \le 1$.
(a) Find the marginal PDFs, $f_X(x)$ and $f_Y(y)$.
(b) Based on the results of part (a), find $E[X]$, $E[Y]$, $\text{Var}(X)$, and $\text{Var}(Y)$.
(c) Find the conditional PDFs, $f_{X|Y}(x|y)$ and $f_{Y|X}(y|x)$.
(d) Based on the results of part (c), find $E[XY]$, $\text{Cov}(X, Y)$, and $\rho_{X,Y}$.

5.23 Prove that if two random variables are linearly related (i.e., $Y = aX + b$ for constants $a \ne 0$ and b), then
$$\rho_{X,Y} = \text{sgn}(a) = \begin{cases} 1, & \text{if } a > 0, \\ -1, & \text{if } a < 0. \end{cases}$$
Also, prove that if two random variables have $|\rho_{X,Y}| = 1$, then they are linearly related.

5.24 Prove the triangle inequality which states that
$$\sqrt{E[(X+Y)^2]} \le \sqrt{E[X^2]} + \sqrt{E[Y^2]}.$$

5.25 Two random variables X and Y have, $\mu_X = 2$, $\mu_Y = -1$, $\sigma_X = 1$, $\sigma_Y = 4$, and $\rho_{X,Y} = 1/4$. Let $U = X + 2Y$ and $V = 2X - Y$. Find the following quantities:
(a) $E[U]$ and $E[V]$;
(b) $E[U^2]$, $E[V^2]$, $\text{Var}(U)$, and $\text{Var}(V)$;
(c) $E[UV]$, $\text{Cov}(U, V)$, and $\rho_{U,V}$.

5.26 Suppose two random variables are related by $Y = aX^2$ and assume that $f_X(x)$ is symmetric about the origin. Show that $\rho_{X,Y} = 0$.

5.27 Let X and Y be random variables with means μ_X and μ_Y, variances σ_X^2 and σ_Y^2, and correlation coefficient $\rho_{X,Y}$.
(a) Find the value of the constant a which minimizes $E[(Y - aX)^2]$.
(b) Find the value of $E[(Y - aX)^2]$ when a is given as determined in part (a).

5.28 For the discrete random variables whose joint PMF is described by the table in Exercise 5.14, compute the following quantities:
(a) $E[XY]$;
(b) $\text{Cov}(X, Y)$;
(c) $\rho_{X,Y}$;
(d) $E[Y|X]$.

5.29 Let Θ be a phase angle which is uniformly distributed over $(0, 2\pi)$. Suppose we form two new random variables according to $X = \cos(a\Theta)$ and $Y = \sin(a\Theta)$ for some constant a.

    (a)   For what values of the constant a are the two random variables X and Y orthogonal?

    (b)   For what values of the constant a are the two random variables X and Y uncorrelated?

5.30 Suppose two random variables X and Y are both zero mean and unit variance. Furthermore, assume they have a correlation coefficient of ρ. Two new random variables are formed according to:

$$W = aX + bY,$$

$$Z = cX + dY.$$

Determine under what conditions on the constants a, b, c, and d the random variables W and Z are uncorrelated.

Section 5.6: Independent Random Variables

5.31 Find an example (other than the one given in Example 5.15) of two random variables that are uncorrelated but not independent.

5.32 Determine whether or not each of the following pairs of random variables are independent:

    (a)   The random variables described in Exercise 5.6;

    (b)   The random variables described in Exercise 5.7;

    (c)   The random variables described in Exercise 5.14;

    (d)   The random variables described in Exercise 5.13.

5.33 Consider two discrete random variables X and Y which take on values from the set $\{1, 2, 3, ..., k\}$. Suppose we construct an $n \times n$ matrix $\boldsymbol{P}$ whose elements comprise the joint PMF of the two random variables. That is, if $p_{i,j}$ is the element in the ith row and jth column of $\boldsymbol{P}$, then $p_{i,j} = P_{X,Y}(i,j) = \Pr(X{=}i, Y{=}j)$.

    (a)   Show that if X and Y are independent random variables, then the matrix $\boldsymbol{P}$ can be written as an outer product of two vectors. What are the components of the outer product?

    (b)   Show that the converse is also true. That is, show that if $\boldsymbol{P}$ can be factored as an outer product, the two random variables are independent.

5.34 Two fair dice are rolled. Let one of the dice be red and the other green so that we can tell them apart. Let X be the sum of the two values shown on the dice and Y be the difference (red minus green) of the two values shown on the dice. Determine whether these two random variables are independent or not. Does you answer make sense?

Section 5.7: Joint Gaussian Random Variables

5.35 Starting from the general form of the joint Gaussian PDF in Equation (5.40), show that the resulting marginal PDFs are both Gaussian.

5.36 Starting from the general form of the joint Gaussian PDF in Equation (5.40) and using the results of Exercise 5.35, show that conditioned on $Y = y$, X is Gaussian with a mean of $\mu_X + \rho_{XY}(\sigma_X/\sigma_Y)(y - \mu_Y)$ and a variance of $\sigma_X^2(1 - \rho_{XY}^2)$.

5.37 Two random variables are jointly Gaussian with means of $\mu_X = 2$, $\mu_Y = -3$, variances of $\sigma_X^2 = 1$, $\sigma_Y^2 = 4$, and a covariance of $\text{Cov}(X, Y) = -1$.
(a)   Write the form of the joint PDF of these jointly Gaussian random variables.
(b)   Find the marginal PDFs, $f_X(x)$ and $f_Y(y)$.
(c)   Find $\Pr(X < 0)$ and $\Pr(Y > 0)$ and write both in terms of Q-functions.

5.38 Two random variables have a joint Gaussian PDF given by

$$f_{X, Y}(x, y) = \frac{9}{2\pi\sqrt{2}} \exp\left(-\frac{36x^2 - 36xy + 81y^2}{16}\right).$$

(a)   Identify σ_X^2, σ_Y^2, and $\rho_{X, Y}$.
(b)   Find the marginal PDFs, $f_X(x)$ and $f_Y(y)$.
(c)   Find the conditional PDFs, $f_{X|Y}(x|y)$ and $f_{Y|X}(y|x)$.

5.39 Two random variables have a joint Gaussian PDF given by

$$f_{X, Y}(x, y) = \frac{1}{6\pi\sqrt{5}} \exp\left(-\frac{3x^3 + 3y^2 - 4xy + 14x - 16y + 7}{30}\right).$$

Find $E[X]$, $E[Y]$, $\text{Var}(X)$, $\text{Var}(Y)$, $\rho_{X, Y}$, $\text{Cov}(X, Y)$, and $E[XY]$.

Section 5.8: Joint Characteristic and Related Functions

5.40 Let X and Y be zero-mean jointly Gaussian random variables with a correlation coefficient of ρ and unequal variances of σ_X^2 and σ_Y^2.
    (a) Find the joint characteristic function, $\Phi_{X,Y}(\omega_1, \omega_2)$.
    (b) Using the joint characteristic function, find the correlation, $E[XY]$.
    (c) Find $E[X^2 Y^2]$.

5.41 Find the general form of the joint characteristic function of two jointly Gaussian random variables.

5.42 A pair of random variables has a joint characteristic function given by

$$\Phi_{X,Y}(\omega_1, \omega_2) = \frac{1}{(1 - 2j\omega_1)(1 - 2j\omega_2)}.$$

    (a) Find $E[X]$ and $E[Y]$.
    (b) Find $E[XY]$ and $\text{Cov}(X, Y)$.
    (c) Find $E[X^2 Y^2]$ and $E[XY^3]$.

5.43 A pair of random variables has a joint characteristic function given by

$$\Phi_{X,Y}(\omega_1, \omega_2) = \exp\left(-\frac{1}{2}(4\omega_1^2 - \omega_1\omega_2 + 9\omega_2^2)\right).$$

    (a) Find $E[X]$ and $E[Y]$.
    (b) Find $E[XY]$ and $\text{Cov}(X, Y)$.
    (c) Find $E[X^2 Y^2]$ and $E[XY^3]$.

5.44
    (a) Find the joint PGF for the pair of discrete random variables given in Exercise 5.13.
    (b) From the result of part (a), find $E[M]$ and $E[N]$.
    (c) From the result of part (a), find $E[MN]$.

5.45 A pair of discrete random variables has a PGF given by

$$H_{M,N}(z_1, z_2) = \frac{4}{(4 - z_1 - z_2)^2}.$$

    (a) Find the means, $E[M]$ and $E[N]$.
    (b) Find the correlation, $E[MN]$.
    (c) Find the joint PMF, $P_{M,N}(m, n)$.

5.46 The joint moment-generating function (MGF) for two random variables, X and Y, is defined as

$$M_{X,Y}(u_1, u_2) = E[\exp(u_1 X + u_2 Y)] = \int_{-\infty}^{\infty} \int_{-\infty}^{\infty} f_{X,Y}(x,y) \exp(u_1 X + u_2 Y) \mathrm{d}x \mathrm{d}y.$$

Develop an equation to find the mixed moment $E[X^n Y^m]$ from the joint MGF.

5.47
  (a) Given the joint characteristic function of a pair of random variables, $\Phi_{X,Y}(\omega_1, \omega_2)$. How do we get a marginal characteristic function of one of the random variables, say $\Phi_X(\omega)$, from the joint characteristic function?
  (b) Repeat part (a) for the joint PGF, $H_{X,Y}(z_1, z_2)$.

Section 5.9: Transformations of Pairs of Random Variables

5.48 A quarterback throws a football at a target marked out on the ground 40 yards from his position. Assume that the PDF for the football's hitting the target is Gaussian within the plane of the target. Let the coordinates of the plane of the target be denoted by the x and y axes. Thus, the joint PDF of (X, Y) is a two-dimensional Gaussian PDF. The average location of the hits is at the origin of the target, and the standard deviation in each direction is the same and is denoted as σ. Assuming X and Y are independent, find the probability that the hits will be located within an annular ring of width $\mathrm{d}r$ located a distance r from the origin; that is, find the probability density function for hits as a function of the radius from the origin.

5.49 Let X and Y be independent and both exponentially distributed with

$$f_X(v) = f_Y(v) = b e^{-bv} u(v).$$

Find the PDF of $Z = X - Y$.

5.50 Let X and Y be jointly Gaussian random variables. Show that $Z = aX + bY$ is also a Gaussian random variable. Hence, any linear transformation of two Gaussian random variables produces a Gaussian random variable.

5.51 Let X and Y be jointly Gaussian random variables with $E[X] = 1$, $E[Y] = -2$, $\mathrm{Var}(X) = 4$, $\mathrm{Var}(Y) = 9$, and $\rho_{X,Y} = 1/3$. Find the PDF of $Z = 2X - 3Y - 5$. *Hint:* To simplify this problem, use the result of Exercise 5.50.

5.52 Let X and Y be independent Rayleigh random variables such that

$$f_X(v) = f_Y(v) = v\exp\left(-\frac{v^2}{2}\right)u(v).$$

(a) Find the PDF of $Z = \max(X, Y)$.
(b) Find the PDF of $W = \min(X, Y)$.

5.53 Suppose X is a Rayleigh random variable and Y is an arcsine random variable, so that

$$f_X(x) = \frac{x}{\sigma^2}\exp\left(-\frac{x^2}{2\sigma^2}\right)u(x) \quad \text{and} \quad f_Y(y) = \frac{1}{\pi\sqrt{1-y^2}} , \quad |y| < 1.$$

Furthermore, assume X and Y are independent. Find the PDF of $Z = XY$.

5.54 Let X and Y be independent and both uniformly distributed over $(0, 2\pi)$. Find the PDF of $Z = (X + Y) \bmod 2\pi$.

5.55 Let X be a Gaussian random variable and let Y be a Bernoulli random variable with $\Pr(Y=1) = p$ and $\Pr(Y=-1) = 1 - p$. If X and Y are independent, find the PDF of $Z = XY$. Under what conditions is Z a Gaussian random variable?

5.56 Let X and Y be independent zero-mean, unit-variance Gaussian random variables. Consider forming the new random variable U, V according to

$$U = X\cos(\theta) - Y\sin(\theta),$$
$$V = X\sin(\theta) + Y\cos(\theta).$$

Note that this transformation produces a coordinate rotation through an angle of θ. Find the joint PDF of U and V. *Hint:* The result of Example 5.25 will be helpful here.

5.57 Let X and Y be zero-mean, unit-variance Gaussian random variables with correlation coefficient, ρ. Suppose we form two new random variables using a linear transformation:

$$U = aX + bY ,$$
$$V = cX + dY.$$

Find constraints on the constants a, b, c, and d such that U and V are independent.

5.58 Suppose X and Y are independent and Gaussian with means of μ_X and μ_Y, respectively, and equal variances of σ^2. The polar variables are formed according to $R = \sqrt{X^2 + Y^2}$ and $\Theta = \tan^{-1}(Y/X)$.
(a) Find the joint PDF of R and Θ.
(b) Show that the marginal PDF of R follows a Rician distribution.

5.59 Suppose X and Y are independent, zero-mean Gaussian random variables with variances of σ_X^2 and σ_Y^2 respectively. Find the joint PDF of

$$Z = X^2 + Y^2 \text{ and } W = X^2 - Y^2.$$

5.60 Suppose X and Y are independent, Cauchy random variables with PDFs specified by

$$f_X(u) = f_y(u) = \frac{1/\pi}{1+u^2}.$$

Find the joint PDF of

$$Z = X^2 + Y^2 \text{ and } W = XY.$$

5.61 Suppose M and N are independent discrete random variables. Find the PMF of $L = M + N$ for each of the following cases:

(a) M and N both follow a uniform distribution,

$$P_M(k) = P_N(k) = 1/K, \; k = 0, 1, 2, \ldots, K-1.$$

(b) M and N follow different geometric distributions,

$$P_M(m) = (1-p)p^m, \; m = 0, 1, 2, \ldots,$$

$$P_N(n) = (1-q)q^n, \; n = 0, 1, 2, \ldots.$$

(c) M and N both follow the same geometric distribution,

$$P_M(m) = P_N(m) = (1-p)p^m, \; m = 0, 1, 2, \ldots.$$

5.62 Suppose M and N are independent discrete random variables with identical Poisson distributions,

$$P_M(k) = P_N(k) = \frac{\alpha^k}{k!}e^{-\alpha}, \; k = 0, 1, 2, \ldots.$$

Find the PMF of $L = M - N$. *Hint:* For this problem, you may find the series expansion for the modified Bessel function helpful:

$$I_p(x) = \sum_{k=0}^{\infty} \frac{(x/2)^{2k+p}}{k!(k+p)!}.$$

Section 5.10: Complex Random Variables

5.63 A complex random variable is defined by $Z = Ae^{j\Theta}$, where A and Θ are independent and Θ is uniformly distributed over $(0, 2\pi)$.

(a) Find $E[Z]$.

(b) Find $\mathrm{Var}(Z)$. For this part, leave your answer in terms of the moments of A.

5.64 Suppose $Z = X + jY$ is a circular Gaussian random variable whose PDF is described by Equation (5.70),

$$f_Z(z) = \frac{1}{2\pi\sigma^2}\exp\left(-\frac{|z - \mu_Z|^2}{2\sigma^2}\right).$$

Find the characteristic function associated with this complex Gaussian random variable, $\Phi_Z(\omega) = E[\exp(j\omega Z)]$. Do you get the same (or different) results as with a real Gaussian random variable.

5.65 Suppose $Z = X + jY$ is a circular Gaussian random variable whose PDF is described by Equation (5.70),

$$f_Z(z) = \frac{1}{2\pi\sigma^2}\exp\left(-\frac{|z - \mu_Z|^2}{2\sigma^2}\right).$$

(a) Find the PDF of the magnitude, $R = |Z|$, and phase angle, $\Theta = \angle Z$, for the special case when $\mu_Z = 0$.

(b) Find the PDF of the magnitude, $R = |Z|$, and phase angle, $\Theta = \angle Z$, for the general case when $\mu_Z \neq 0$. *Hint:* In this case, you will have to leave the PDF of the phase angle in terms of a Q-function.

(c) For the case when $\mu_Z \gg \sigma$, show that the PDF of the phase angle is well approximated by a Gaussian PDF. What is the variance of the Gaussian PDF that approximates the PDF of the phase angle?

Section 5.11: Mutual Information, Channel Capacity, and Channel Coding

5.66 Suppose $Q = \begin{bmatrix} 0.8 & 0.1 & 0.1 \\ 0.1 & 0.8 & 0.1 \\ 0.1 & 0.1 & 0.8 \end{bmatrix}$ in Figure 5.7 and $p_i = 1/3$, $i = 1, 2, 3$. Determine the mutual information for this channel.

5.67 Repeat Exercise 5.66 if $Q = \begin{bmatrix} 0.9 & 0.1 & 0 \\ 0 & 0.9 & 0.1 \\ 0 & 0.1 & 0.9 \end{bmatrix}$.

5.68 Repeat Exercise 5.66 if $Q = \begin{bmatrix} 1/3 & 1/3 & 1/3 \\ 1/3 & 1/3 & 1/3 \\ 1/3 & 1/3 & 1/3 \end{bmatrix}$. Can you give an interpretation for your result.

5.69 Find the capacity of the channel described by the transition matrix ,

$$ Q = \begin{bmatrix} 0.8 & 0.2 \\ 0.1 & 0.9 \end{bmatrix} . $$

5.70 For the transition matrix Q given in Exercise 5.66, prove that the equally likely source distribution, $p_i = 1/3$, $i = 1, 2, 3$, is the one that maximizes mutual information and hence the mutual information found in Exercise 5.66 is the capacity associated with the channel described by Q.

Miscellaneous Problems

5.71 Suppose X and Y are independent and exponentially distributed both with unit-mean. Consider the roots of the quadratic equation $z^2 + Xz + Y = 0$.
   (a) Find the probability that the roots are real.
   (b) Find the probability that the roots are complex.
   (c) Find the probability that the roots are equal.

5.72 In this problem, we revisit the light bulb problem of Exercises 3.43. Recall that there were two types of bulbs, long-life (L) and short-life (S) and we were given a box of unmarked bulbs and needed to identify which type of bulbs are in the box. In Exercise 3.43, we chose to run one of the bulbs until it burned out in order to help us identify which type of bulbs are in the box. This time, in order to obtain a more reliable decision, we are going to burn two different bulbs from the box, observe how long it takes each bulb to burn out, and then make a decision as to what type of bulbs are in the box. Let X represent the time that it takes the first bulb to burn out and let Y represent the time it takes the second bulb to burn out. It would seem reasonable to assume that X and Y are independent and since both bulbs are taken from the same box, the PDFs of there lifetimes should be the same. Modeling the conditional PDFs as in Exercise 3.43, we have

$$f_{X|S}(z) = f_{Y|S}(z) = \frac{1}{100}\exp\left(-\frac{z}{100}\right)u(z) \quad \text{and}$$

$$f_{X|L}(z) = f_{Y|L}(z) = \frac{1}{1000}\exp\left(-\frac{z}{1000}\right)u(z).$$

The a priori probability of the bulb types were $\Pr(S) = 0.75$ and $\Pr(L) = 0.25$.

(a) If the two bulbs are tested and it is observed that the first bulb burns out after 200 h and the second bulb burns out after 75 h, which type of bulb was most likely tested?

(b) What is the probability that your decision in part (b) was incorrect?

(c) Determine what decision should be made for each possible observation pair, $\{X=x, Y=y\}$. That is, divide the first quadrant of the (x, y)-plane into two regions, one including all sets of points for which we would decide that the bulbs are S-type and its complement where we decide the bulbs are L-type.

5.73 Once again, we will modify the light bulb problem of Exercise 5.72 in a manner similar to what was done in Exercise 3.44. Suppose we select two light bulbs to turn on when we leave the office for the weekend on Friday at 5 pm. On Monday morning at 8 am we will observe which of the light bulbs have burned out, if any. Let X be the lifetime of the first bulb and Y the lifetime of the second bulb. When we arrive at the office on Monday morning, there are four possible outcomes of the experiment:

(i) both bulbs burned out $\Leftrightarrow \{X<63\} \cap \{Y<63\}$,

(ii) the first bulb burned out while the second did not $\Leftrightarrow \{X<63\} \cap \{Y>63\}$,

(iii) the second bulb burned out while the first did not $\Leftrightarrow \{X>63\} \cap \{Y<63\}$,

(iv) neither bulb burned out $\Leftrightarrow \{X>63\} \cap \{Y>63\}$.

For each of the four cases, determine what decision should be made regarding the type of bulbs that were in the box (i.e., L-type or S-type) and calculate the probability that the decision is wrong. As before, assume a priori probabilities of $\Pr(S) = 0.75$ and $\Pr(L) = 0.25$.

5.74

(a) Repeat Exercise 5.73 if we run the experiment over a 3-day weekend so that the experiment runs for 87 hours instead of 63.

(b) If we could choose the length of the experiment described in Exercise 5.73 to be anything we wanted, how long should we run the experiment in order to maximize our chances of correctly identifying the bulb type?

MATLAB Exercises

5.75 Provide contour plots for the ellipses discussed in Example 5.17. Consider the following cases:

(a) $\sigma_X = \sigma_Y$ and $\rho_{XY} = 0$;

(b) $\sigma_X < \sigma_Y$ and $\rho_{XY} = 0$;

(c) $\sigma_X > \sigma_Y$ and $\rho_{XY} = 0$;

(d) $\sigma_X = \sigma_Y$ and $\rho_{XY} \neq 0$.

Let c^2 be the same for each case. Discuss the effect σ_X, σ_Y and ρ_{XY} have on the shape of the contour. Now select one of the cases and let c^2 increase and decrease. What is the significance of c^2.

5.76 Let X and Y have a joint PDF given by

$$f_{X,\,Y}(x, y) = \frac{1}{2\pi} \exp\left(-\frac{((x-2)^2 + (y-3)^2)}{2} \right)$$

as in Example 5.6. Write a MATLAB program to generate many samples of this pair of random variables. Note that X and Y are independent, Gaussian random variables with unit variances and means of 2 and 3, respectively. After a large number of sample pairs have been generated, compute the relative frequency of the number of pairs that fall within the unit circle, $X^2 + Y^2 < 1$. Compare your answer with that obtained in Example 5.6. How many random samples must you generate in order to get a decent estimate of the probability?

5.77 Let X and Y have a joint PDF given by

$$f_{X,\,Y}(x, y) = \frac{1}{2\pi} \exp\left(-\frac{((x-1)^2 + y^2)}{2} \right).$$

5.78 Write a MATLAB program to evaluate $\Pr((X, Y \in \Re))$, where $\Re$ is the shaded region bounded by the lines $y = x$ and $y = -x$ as shown in the accompanying figure. You should set up the appropriate double integral and use MATLAB to evaluate the integral numerically. Note in this case that one of the limits of integration is infinit How will you deal with this?

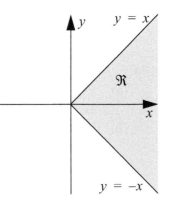

5.79 Write a MATLAB program to generate pairs of random variables that are uniformly distributed over the ellipse $x^2 + 4y^2 < 1$. Use the technique employed in Example 5.16. Also, create a three-dimensional plot of an estimate of the PDF obtained from the random data you generated.

Multiple Random Variables

In many applications, it is necessary to deal with a large numbers of random variables. Often, the number of variables can be arbitrary. In this chapter, the concepts developed previously for single random variables and pairs of random variables are extended to allow for an arbitrary number of random variables. Much of the focus of this chapter is on multidimensional Gaussian random variables, since most non-Gaussian random variables are difficult to deal with in many dimensions. One of the main goals here is to develop a vector/matrix notation which will allow us to represent potentially large sequences of random variables with a compact notation. Many of the concepts developed in Chapter 5 can be extended to multiple dimensions in a very straightforward manner; thus we will devote minimal time to those concepts in our current discussion. Rather, attention is focussed on those ideas that require more than a trivial extension to the work done in previous chapters.

6.1 Joint and Conditional PMFs, CDFs, and PDFs

The concepts of probability mass function (PMF), conditional distribution function (CDF), and probability density function (PDF) are easily extended to an arbitrary number of random variables. Their definitions follow.

Definition 6.1: For a set of N random variables $X_1, X_2, ..., X_N$, the joint PMF, CDF, and PDF are given, respectively, by

$$P_{X_1, X_2, ..., X_N}(x_{k_1}, x_{k_2}, ..., x_{k_N}) = \Pr(X_1 = x_{k_1}, X_2 = x_{k_2}, ..., X_N = x_{k_N});$$ (6.1)

$$F_{X_1, X_2, ..., X_N}(x_1, x_2, ..., x_N) = \Pr(X_1 \leq x_1, X_2 \leq x_2, ..., X_N \leq x_N);$$ (6.2)

$$f_{X_1, X_2, ..., X_N}(x_1, x_2, ..., x_N) = \frac{\partial^N}{\partial x_1 \partial x_2 ... \partial x_N} F_{X_1, X_2, ..., X_N}(x_1, x_2, ..., x_N).$$ (6.3)

When large numbers of random variables are involved, this notation can get cumbersome, so it is convenient to introduce a vector notation to write these quantities in a more compact fashion. Let $\boldsymbol{X} = [X_1, X_2, ..., X_N]^T$ be a column vector[1] consisting of the N random variables

---

[1] We use T to represent the matrix transpose operation so that if $\boldsymbol{v}$ is a row vector, then $\boldsymbol{v}^T$ is a column vector. Also, to avoid confusion throughout the text, we use boldface variables to represent vector and matrix quantities and regular face variables for scalar quantities.

Probability and Random Processes. DOI: 10.116/B978-0-12-386981-4.00005-X

and similarly define $x = [x_1, x_2, ..., x_N]^T$. Then the preceding functions can be expressed, respectively, as $P_X(x)$, $F_X(x)$, and $f_X(x)$.

Marginal CDFs can be found for a subset of the variables by evaluating the joint CDF at infinity for the unwanted variables. For example,

$$F_{X_1, X_2, ..., X_M}(x_1, x_2, ..., x_M) = F_{X_1, X_2, ..., X_N}(x_1, x_2, ..., x_M, \infty, \infty, ..., \infty). \qquad (6.4)$$

Marginal PDFs are found from the joint PDF by integrating out the unwanted variables. Similarly, marginal PMFs are obtained from the joint PMF by summing out the unwanted variables.

$$f_{X_1, X_2, ..., X_M}(x_1, x_2, ..., x_M) \qquad (6.5)$$

$$= \int_{-\infty}^{\infty} \int_{-\infty}^{\infty} \cdots \int_{-\infty}^{\infty} f_{X_1, X_2, ..., X_N}(x_1, x_2, ..., x_N) dx_{M+1} dx_{M+2} \cdots dx_N,$$

$$P_{X_1, X_2, ..., X_M}(x_{k_1}, x_{k_2}, ..., x_{k_M}) = \sum_{k_{M+1}} \sum_{k_{M+2}} \cdots \sum_{k_N} P_{X_1, X_2, ..., X_N}(x_{k_1}, x_{k_2}, ..., x_{k_N}). \qquad (6.6)$$

Similar to that done for pairs of random variables in Chapter 5, we can also establish conditional PMFs and PDFs.

> **Definition 6.2:** For a set of N random variables $X_1, X_2, ..., X_N$, the conditional PMF and PDF of $X_1, X_2, ..., X_M$ conditioned on $X_{M+1}, X_{M+2}, ..., X_N$ are given by

$$P_{X_1, ..., X_M | X_{M+1}, ..., X_N}(x_{k_1}, ..., x_{k_M} | x_{k_{M+1}}, ..., x_{k_N}) = \frac{\Pr(X_1 = x_{k_1}, ..., X_N = x_{k_N})}{\Pr(X_{M+1} = x_{k_{M+1}}, ..., X_N = x_{k_N})}, \qquad (6.7)$$

$$f_{X_1, ..., X_M | X_{M+1}, ..., X_N}(x_1, ..., x_M | x_{M+1}, ..., x_N) = \frac{f_{X_1, ..., X_N}(x_1, ..., x_N)}{f_{X_{M+1}, ..., X_N}(x_{M+1}, ..., x_N)}. \qquad (6.8)$$

Using conditional PDFs, many interesting factorization results can be established for joint PDFs involving multiple random variables. For example, consider four random variables, X_1, X_2, X_3, X_4.

$$f_{X_1, X_2, X_3, X_4}(x_1, x_2, x_3, x_4) = f_{X_1 | X_2, X_3, X_4}(x_1 | x_2, x_3, x_4) f_{X_2, X_3, X_4}(x_2, x_3, x_4)$$

$$= f_{X_1 | X_2, X_3, X_4}(x_1 | x_2, x_3, x_4) f_{X_2 | X_3, X_4}(x_2 | x_3, x_4) f_{X_3, X_4}(x_3, x_4)$$

$$= f_{X_1 | X_2, X_3, X_4}(x_1 | x_2, x_3, x_4) f_{X_2 | X_3, X_4}(x_2 | x_3, x_4) f_{X_3 | X_4}(x_3 | x_4) f_{X_4}(x_4). \qquad (6.9)$$

Almost endless other possibilities exist as well.

Definition 6.3: A set of N random variables are statistically independent if any subset of the random variables are independent of any other disjoint subset. In particular, any joint PDF of $M \le N$ variables should factor into a product of the corresponding marginal PDFs.

As an example, consider three random variables, X, Y, Z. For these three random variables to be independent, we must have each pair independent. This implies that

$$f_{X, Y}(x, y) = f_X(x)f_Y(y), \quad f_{X, Z}(x, z) = f_X(x)f_Z(z), \quad f_{Y, Z}(y, z) = f_Y(y)f_Z(z). \tag{6.10}$$

In addition, the joint PDF of all three must also factor into a product of the marginals,

$$f_{X, Y, Z}(x, y, z) = f_X(x)f_Y(y)f_Z(z). \tag{6.11}$$

Note that all three conditions in Equation (6.10) follow directly from the single condition in Equation (6.11). Hence, Equation (6.11) is a necessary and sufficient condition for three variables to be statistically independent. Naturally, this result can be extended to any number of variables. That is, the elements of a random vector $\mathbf{X} = [X_1, X_2, ..., X_N]^T$ are independent if

$$f_{\mathbf{X}}(\mathbf{x}) = \prod_{n=1}^{N} f_{X_n}(x_n). \tag{6.12}$$

6.2 Expectations Involving Multiple Random Variables

For a vector of random variables $\mathbf{X} = [X_1, X_2, ..., X_N]^T$, we can construct a corresponding mean vector that is a column vector of the same dimension and whose components are the means of the elements of $\mathbf{X}$. Mathematically, we say $\mu = E[\mathbf{X}] = [E[X_1], E[X_2], ..., E[X_N]]^T$. Two other important quantities associated with the random vector are the correlation and covariance matrices.

Definition 6.4: For a random vector $\mathbf{X} = [X_1, X_2, ..., X_N]^T$, the correlation matrix is defined as $\mathbf{R}_{XX} = E[\mathbf{X}\mathbf{X}^T]$. That is, the (i, j) th element of the $N \times N$ matrix $\mathbf{R}_{XX}$ is $E[X_i X_j]$. Similarly, the covariance matrix is defined as $\mathbf{C}_{XX} = E[(\mathbf{X} - \mu)(\mathbf{X} - \mu)^T]$ so that the (i, j) th element of $\mathbf{C}_{XX}$ is $\text{Cov}(X_i, X_j)$.

Theorem 6.1: Correlation matrices and covariance matrices are symmetric and positive definite.

Proof: Recall that a square matrix, R_{XX}, is symmetric if $R_{XX} = R_{XX}^T$. Equivalently, the (i, j) th element must be the same as the (j, i) th element. This is clearly the case here since $E[X_i X_j] = E[X_j X_i]$. Recall that the matrix R_{XX} is positive definite if $z^T R_{XX} z > 0$ for any vector z such that $\|z\| > 0$.

$$z^T R_{XX} z = z^T E[XX^T]z = E[z^T XX^T z] = E[(z^T X)^2]. \tag{6.13}$$

Note that $z^T X$ is a scalar random variable (a linear combination of the components of X). Since the second moment of any random variable is positive (except for the pathological case of a random variable which is identically equal to zero), then the correlation matrix is positive definite. As an aside, this also implies that the eigenvalues of the correlation matrix are all positive. Identical steps can be followed to prove the same properties hold for the covariance matrix. $\square$

Next, consider a linear transformation of a vector random variable. That is, create a new set of M random variables, $Y = [Y_1, Y_2, ..., Y_M]^T$, according to

$$\begin{aligned}
Y_1 &= a_{1,1}X_1 + a_{1,2}X_2 + ... + a_{1,N}X_N + b_1, \\
Y_2 &= a_{2,1}X_1 + a_{2,2}X_2 + ... + a_{2,N}X_N + b_2, \\
&\qquad\qquad ... \\
Y_M &= a_{M,1}X_1 + a_{M,2}X_2 + ... + a_{M,N}X_N + b_M.
\end{aligned} \tag{6.14}$$

The number of new variables, M, does not have to be the same as the number of original variables, N. To write this type of linear transformation in a compact fashion, define a matrix A whose (i, j) th element is the coefficient $a_{i,j}$ and a column vector, $b = [b_1, b_2, ..., b_M]^T$. Then the linear transformation of Equation (6.14) is written in vector/matrix form as $Y = AX + b$. The next theorem describes the relationship between the means of X and Y and the correlation matrices of X and Y.

Theorem 6.2: For a linear transformation of vector random variables of the form $Y = AX + b$, the means of X and Y are related by

$$\mu_Y = A\mu_X + b. \tag{6.15}$$

Also, the correlation matrices of X and Y are related by

$$R_{YY} = AR_{XX}A^T + A\mu_X b^T + b\mu_X^T A^T + bb^T, \tag{6.16}$$

and the covariance matrices of X and Y are related by

$$C_{YY} = AC_{XX}A^{\mathrm{T}}.\tag{6.17}$$

Proof: For the mean vector,

$$\mu_Y = E[Y] = E[AX + b] = AE[X] + b = A\mu_X + b.\tag{6.18}$$

Similarly, for the correlation matrix,

$$\begin{aligned}
R_{YY} &= E[YY^{\mathrm{T}}] = E[(AX + b)(AX + b)^{\mathrm{T}}]\\
&= E[AXX^{\mathrm{T}}A^{\mathrm{T}}] + E[bX^{\mathrm{T}}A^{\mathrm{T}}] + E[AXb^{\mathrm{T}}] + E[bb^{\mathrm{T}}]\\
&= AE[XX^{\mathrm{T}}]A^{\mathrm{T}} + bE[X^{\mathrm{T}}]A^{\mathrm{T}} + AE[X]b^{\mathrm{T}} + bb^{\mathrm{T}}\\
&= AR_{XX}A^{\mathrm{T}} + A\mu_X b^{\mathrm{T}} + b\mu_X^{\mathrm{T}}A^{\mathrm{T}} + bb^{\mathrm{T}}.
\end{aligned}\tag{6.19}$$

To prove the result for the covariance matrix, write $Y - \mu_Y$ as

$$Y - \mu_Y = (AX + b) - (A\mu_X + b) = A(X - \mu_X).\tag{6.20}$$

Then,

$$\begin{aligned}
C_{YY} &= E[(Y - \mu_Y)(Y - \mu_Y)^{\mathrm{T}}] = E[\{A(X - \mu_X)\}\{A(X - \mu_X)\}^{\mathrm{T}}]\\
&= E[A(X - \mu_X)(X - \mu_X)^{\mathrm{T}}A^{\mathrm{T}}] = AE[(X - \mu_X)(X - \mu_X)^{\mathrm{T}}]A^{\mathrm{T}} = AC_{XX}A^{\mathrm{T}}. \ \square
\end{aligned}\tag{6.21}$$

6.3 Gaussian Random Variables in Multiple Dimensions

Recall from the study of two-dimensional random variables in the previous chapter that the functional form of the joint Gaussian PDF was fairly complicated. It would seem that the prospects of forming a joint Gaussian PDF for an arbitrary number of dimensions are grim. However, the vector/matrix notation developed in the previous sections make this task manageable and, in fact, the resulting joint Gaussian PDF is quite simple.

Definition 6.5: The joint Gaussian PDF for a vector of N random variables, X, with mean vector, $\boldsymbol{\mu}_X$, and covariance matrix, C_{XX}, is given by[2]

$$f_X(x) = \frac{1}{\sqrt{(2\pi)^N \det(C_{XX})}} \exp\left(-\frac{1}{2}(x - \boldsymbol{\mu}_X)^{\mathrm{T}} C_{XX}^{-1}(x - \boldsymbol{\mu}_X)\right). \tag{6.22}$$

■ Example 6.1:

To demonstrate the use of this matrix notation, suppose X is a two-element vector and the mean vector and covariance matrix are given by their general forms

$$\boldsymbol{\mu}_X = \begin{bmatrix} \mu_1 \\ \mu_2 \end{bmatrix} \text{ and } C_{XX} = \begin{bmatrix} \sigma_1^2 & \rho\sigma_1\sigma_2 \\ \rho\sigma_1\sigma_2 & \sigma_2^2 \end{bmatrix}.$$

The determinant of the covariance matrix is

$$\det(C_{XX}) = \sigma_1^2\sigma_2^2 - (\rho\sigma_1\sigma_2)^2 = \sigma_1^2\sigma_2^2(1 - \rho^2),$$

while the inverse is

$$C_{XX}^{-1} = \frac{\begin{bmatrix} \sigma_2^2 & -\rho\sigma_1\sigma_2 \\ -\rho\sigma_1\sigma_2 & \sigma_1^2 \end{bmatrix}}{\sigma_1^2\sigma_2^2(1-\rho^2)} = \frac{\begin{bmatrix} \sigma_1^{-2} & -\rho\sigma_1^{-1}\sigma_2^{-1} \\ -\rho\sigma_1^{-1}\sigma_2^{-1} & \sigma_2^{-2} \end{bmatrix}}{(1-\rho^2)}.$$

The quadratic form in the exponent then works out to be

$$(x - \boldsymbol{\mu}_X)^{\mathrm{T}} C_{XX}^{-1}(x - \boldsymbol{\mu}_X) = \begin{bmatrix} x_1 - \mu_1 & x_2 - \mu_2 \end{bmatrix} \frac{\begin{bmatrix} \sigma_1^{-2} & -\rho\sigma_1^{-1}\sigma_2^{-1} \\ -\rho\sigma_1^{-1}\sigma_2^{-1} & \sigma_2^{-2} \end{bmatrix}}{(1-\rho^2)} \begin{bmatrix} x_1 - \mu_1 \\ x_2 - \mu_2 \end{bmatrix}$$

$$= \frac{\left(\dfrac{x_1 - \mu_1}{\sigma_1}\right)^2 - 2\rho\left(\dfrac{x_1 - \mu_1}{\sigma_1}\right)\left(\dfrac{x_2 - \mu_2}{\sigma_2}\right) + \left(\dfrac{x_2 - \mu_2}{\sigma_2}\right)^2}{(1-\rho^2)}.$$

Plugging all these results into the general form for the joint Gaussian PDF gives

$$f_{X_1,X_2}(x_1,x_2) = \frac{1}{\sqrt{(2\mu)^2 \sigma_1^2\sigma_2^2(1-\rho^2)}} \exp\left(-\frac{\left(\dfrac{x_1 - \mu_1}{\sigma_1}\right)^2 - 2\rho\left(\dfrac{x_1 - \mu_1}{\sigma_1}\right)\left(\dfrac{x_2 - \mu}{\sigma_2}\right) + \left(\dfrac{x_2 - \mu_2}{\sigma_2}\right)^2}{2(1-\rho^2)}\right).$$

This is exactly the form of the two-dimensional joint Gaussian PDF given in the Definition 5.10.

■

---

[2] The notation $\det(A)$ refers to the determinant of the matrix A, while A^{-1} is the inverse of A.

■ Example 6.2:

As a special case, suppose a vector of N jointly Gaussian random variables are all mutually uncorrelated. This means that $\text{Cov}(X_i, X_j) = 0$ for all $i \neq j$. A direct result of this is that all of the off-diagonal elements of the covariance matrix of X are zero. In other words, C_{XX} is a diagonal matrix of the general form

$$C_{XX} = \begin{bmatrix} \sigma_1^2 & 0 & \dots & 0 \\ 0 & \sigma_2^2 & \dots & 0 \\ \dots & \dots & & \dots \\ 0 & 0 & \dots & \sigma_N^2 \end{bmatrix}.$$

The determinant of a diagonal matrix is the product of the diagonal entries so that in this case $\det(C_{XX}) = \sigma_1^2 \sigma_2^2 \dots \sigma_N^2$. The inverse is also trivial to compute and takes on the form

$$C_{XX}^{-1} = \begin{bmatrix} \sigma_1^{-2} & 0 & \dots & 0 \\ 0 & \sigma_2^{-2} & \dots & 0 \\ \dots & \dots & & \dots \\ 0 & 0 & \dots & \sigma_N^{-2} \end{bmatrix}.$$

The quadratic form that appears in the exponent of the Gaussian PDF becomes,

$$(x - \mu_X)^{\text{T}} C_{XX}^{-1} (x - \mu_X) = \begin{bmatrix} x_1 - \mu_1 & x_2 - \mu_2 & \dots & x_N - \mu_N \end{bmatrix} \begin{bmatrix} \sigma_1^{-2} & 0 & \dots & 0 \\ 0 & \sigma_2^{-2} & \dots & 0 \\ \dots & \dots & & \dots \\ 0 & 0 & \dots & \sigma_N^{-2} \end{bmatrix} \begin{bmatrix} x_1 - \mu_1 \\ x_2 - \mu_2 \\ \dots \\ x_N - \mu_N \end{bmatrix} = \sum_{n=1}^{N} \left(\frac{x_n - \mu_n}{\sigma_n} \right)^2.$$

The joint Gaussian PDF for a vector of uncorrelated random variables is then

$$f_X(x) = \frac{1}{\sqrt{(2\pi)^N \sigma_1^2 \sigma_2^2 \dots \sigma_N^2}} \exp\left(-\frac{1}{2} \sum_{n=1}^{N} \left(\frac{x_n - \mu_n}{\sigma_n} \right)^2 \right) = \prod_{n=1}^{N} \frac{1}{\sqrt{2\pi\sigma_n^2}} \exp\left(-\frac{(x_n - \mu_n)^2}{2\sigma_n^2} \right).$$

This shows that for any number of uncorrelated Gaussian random variables, the joint PDF factors into the product of marginal PDFs and hence uncorrelated Gaussian random variables are independent. This is a generalization of the same result that was proven in Chapter 5 for two Gaussian random variables. ■

■ Example 6.3:

In this example, we use MATLAB's symbolic capabilities to compute the form of a three-dimensional Gaussian PDF. Suppose we have three jointly Gaussian random variables $[X, Y, Z]^{\text{T}}$ with a mean vector $\mu = [1, 2, 3]^{\text{T}}$ and covariance matrix

(Continued)

$$C = \begin{bmatrix} 9 & 4 & 1 \\ 4 & 9 & 4 \\ 1 & 4 & 9 \end{bmatrix}.$$

The three-dimensional joint PDF, $f_{X,Y,Z}(x,y,z)$, can be found with the following MATLAB code:

```
x=sym('x','real');                    % Define x, y, and z as symbolic.
y=sym('y','real');
z=sym('z','real');
pi=sym('pi');                         % Disable numeric definition of pi.
C=[9 4 1; 4 9 4; 1 4 9];             % Covariance matrix.
mu=[1; 2; 3];                         % mean vector
% Compute PDF symbolically.
v=[x; y; z]-mu;
f=exp(-v'*(inv(C))*v/2)/sqrt((2*pi)^3*det(C));
simplify(f)
```

Executing this program, MATLAB finds the joint PDF to be

$$f_{X,Y,Z}(x,y,z) = \frac{1}{464\sqrt{58}\,\pi^3} \exp\left(-\frac{65}{928}x^2 + \frac{11}{232}x - \frac{125}{232} + \frac{2}{29}xy + \frac{2}{29}y - \frac{7}{464}xz + \frac{69}{232}z - \frac{5}{58}y^2 + \frac{2}{29}yz - \frac{65}{928}z^2\right)$$

The reader is encouraged to try different mean vectors and covariance matrices in the preceding program.

◼

6.4 Transformations Involving Multiple Random Variables

In this section, we discuss various transformations of vector random variables. To exhaustively cover this topic would require much more space than we can devote to it here. Instead, we chose to cover some of the more common transformations encountered in engineering practice. To start with, we extend the formula for 2×2 transformations developed in the previous chapter to the case of NxN transforms. Let $Y = g(X)$ be a vector transformation,

$$
\begin{aligned}
Y_1 &= g_1(X_1, X_2, ..., X_N), \\
Y_2 &= g_2(X_1, X_2, ..., X_N), \\
&\quad\cdots \\
Y_N &= g_N(X_1, X_2, ..., X_N),
\end{aligned}
\tag{6.23}
$$

and let $X = h(Y)$ be the inverse transformation. Given the PDF of X, the PDF of Y is found by

$$f_Y(y) = \frac{f_X(x)}{\left|\det\left[J\left(\begin{matrix} y_1 & y_2 & \cdots & y_N \\ x_1 & x_2 & \cdots & x_N \end{matrix}\right)\right]\right|}\Bigg|_{x=h(y)} = f_X(x)\left|\det\left[J\left(\begin{matrix} x_1 & x_2 & \cdots & x_N \\ y_1 & y_2 & \cdots & y_N \end{matrix}\right)\right]\right|\Bigg|_{x=h(y)} .\tag{6.24}$$

As in Chapter 5, it needs to be understood that if the transformation is not one-to-one, the preceding expression must be evaluated at each root and summed together. This result can be proved using the same sort of derivation that was used for the case of 2×2 transformations.

6.4.1 Linear Transformations

Perhaps the single most important class of transformations is that involving linear transformations of Gaussian random variables. Consider a linear transformation of the general form $Y = AX + b$ when the random vector X has a joint Gaussian PDF as given in Equation (6.22). To begin, consider the case where the dimensionality of X and Y are the same (i.e., both are N element vectors). In that case, the matrix A is a square matrix. Furthermore, it is assumed that the matrix A is invertible ($\det(A) \neq 0$). Then, the Jacobian of the linear transformation is

$$J\left(\begin{matrix} y_1 \; y_2 \; \cdots \; y_N \\ x_1 \; x_2 \; \cdots \; x_N \end{matrix} \right) = \begin{bmatrix} \dfrac{\partial y_1}{\partial x_1} & \dfrac{\partial y_1}{\partial x_2} & \cdots & \dfrac{\partial y_1}{\partial x_N} \\ \dfrac{\partial y_2}{\partial x_1} & \dfrac{\partial y_2}{\partial x_2} & \cdots & \dfrac{\partial y_2}{\partial x_N} \\ \cdots & \cdots & & \cdots \\ \dfrac{\partial y_N}{\partial x_1} & \dfrac{\partial y_N}{\partial x_2} & \cdots & \dfrac{\partial y_N}{\partial x_N} \end{bmatrix} = \begin{bmatrix} a_{1,1} & a_{1,2} & \cdots & a_{1,N} \\ a_{2,1} & a_{2,2} & \cdots & a_{2,N} \\ \cdots & \cdots & & \cdots \\ a_{N,1} & a_{N,2} & \cdots & a_{N,N} \end{bmatrix} = A. \tag{6.25}$$

Also, the inverse transformation is linear and can be written as $X = A^{-1}(Y - b)$. The joint PDF for the vector Y is then

$$f_Y(y) = \left. \frac{f_X(x)}{|\det(A)|} \right|_{x = A^{-1}(y - b)}. \tag{6.26}$$

Plugging in the form of the Gaussian PDF for $f_X(x)$ results in

$$f_Y(y) = \left. \frac{1}{|\det(A)| \sqrt{(2\pi)^N \det(C_{XX})}} \exp\left(-\frac{1}{2}(x - \mu_X)^{\mathrm{T}} C_{XX}^{-1}(x - \mu_X) \right) \right|_{x = A^{-1}(y - b)}. \tag{6.27}$$

To simplify this result, write

$$x - \mu_X \Big|_{x = A^{-1}(y - b)} = A^{-1}(y - b) - \mu_X = A^{-1}(y - (b + A\mu_X)) = A^{-1}(y - \mu_Y). \tag{6.28}$$

The quadratic form in the exponent is then

$$(x - \mu_X)^{\mathrm{T}} C_{XX}^{-1} (x - \mu_X)\Big|_{x = A^{-1}(y - b)} = [A^{-1}(y - \mu_Y)]^{\mathrm{T}} C_{XX}^{-1} [A^{-1}(y - \mu_Y)]$$

$$= (y - \mu_Y)^{\mathrm{T}} (A^{-1})^{T} C_{XX}^{-1} A^{-1} (y - \mu_Y). \tag{6.29}$$

In addition, we can write

$$|\det(A)| \sqrt{\det(C_{XX})} = \sqrt{[\det(A)]^2 \det(C_{XX})}$$

$$= \sqrt{\det(A)\det(C_{XX})\det(A^{\mathrm{T}})} = \sqrt{\det(A C_{XX} A^{\mathrm{T}})}. \tag{6.30}$$

The steps above are carried out using the fact that for a square matrix, $\det(A) = \det(A^{\mathrm{T}})$ and also that the determinant of a product of matrices is equal to the product of the determinants. At this point we have established that

$$f_Y(y) = \frac{1}{\sqrt{(2\pi)^N \det(A C_{XX} A^{\mathrm{T}})}} \exp\left(-\frac{1}{2}(y - \mu_Y)^{\mathrm{T}} (A^{-1})^{\mathrm{T}} C_{XX}^{-1} A^{-1}(y - \mu_Y)\right). \tag{6.31}$$

Finally, recall that for a linear transformation, $C_{YY} = A C_{XX} A^{\mathrm{T}}$. Furthermore, from this relationship, we can also determine that $C_{YY}^{-1} = (A C_{XX} A^{\mathrm{T}})^{-1} = (A^{\mathrm{T}})^{-1} C_{XX}^{-1} A^{-1}$. $= (A^{-1})^{\mathrm{T}} C_{XX}^{-1} A^{-1}$ Hence, the PDF for Y can be written as

$$f_Y(y) = \frac{1}{\sqrt{(2\pi)^N \det(C_{YY})}} \exp\left(-\frac{1}{2}(y - \mu_Y)^{\mathrm{T}} C_{YY}^{-1} (y - \mu_Y)\right). \tag{6.32}$$

This is the general form of a joint Gaussian PDF. Thus, we have shown that any linear transformation of any number of jointly Gaussian random variables produces more jointly Gaussian random variables. Note that this statement applies to more than just $N \times N$ linear transformations. Suppose we wanted to transform N jointly Gaussian random variables to M ($M < N$) new random variables through a linear transformation. We could always form an $N \times N$ transformation producing N new jointly Gaussian random variables. Any subset of M out of N of these random variables will also be jointly Gaussian. In summary, we have proved the following theorem.

> **Theorem 6.3:** Given a vector X of N jointly Gaussian random variables, any linear transformation to a set of M ($M \leq N$) new variables, Y, will produce jointly Gaussian random variables.

Next, suppose we want to create a set of N jointly Gaussian random variables, Y, with a specified covariance matrix, C. We could start with a set of uncorrelated Gaussian random variables (as might be generated by a typical Gaussian random number generator) and then perform a linear transformation to produce a new set of Gaussian random variables with the desired covariance matrix. But, how should the transformation be selected to produce the desired covariance matrix? To answer that question, recall that any covariance matrix is symmetric and any symmetric matrix can be decomposed into

$$C = QDQ^T, \tag{6.33}$$

where D is a diagonal matrix of the eigenvalues of C, and Q is an orthogonal matrix whose columns are the corresponding eigenvectors of C. Note also that C is positive definite and therefore its eigenvalues are all positive. Thus, the matrix D is not only diagonal, but its diagonal elements are all positive, and as a result, the matrix D is a valid covariance matrix. That is, suppose we create a set of N uncorrelated Gaussian random variables, X, with a covariance matrix $C_{XX} = D$. Then, the matrix Q will transform this set of uncorrelated Gaussian random variables to a new set of Gaussian random variables with the desired covariance matrix. If we form $Y = QX$, then according to Theorem 6.2, the covariance matrix of Y will be of the form

$$C_{YY} = QC_{XX}Q^T = QDQ^T = C. \tag{6.34}$$

At this point, the problem has been reduced from creating a set of random variables with an arbitrary covariance matrix to creating a set of random variables with a diagonal covariance matrix. Typical Gaussian random number generators create random variables with a unit variance. To create random variables with unequal variances, simply scale each component by the appropriate value. In particular, suppose[3] $D = \text{diag}(d_1, d_2, ..., d_N)$. Given a set of unit variance uncorrelated Gaussian random variables $Z = [Z_1, Z_2, ..., Z_N]^T$, one could form X with the desired variance according to $X_i = \sqrt{d_i}Z_i$, $i = 1, 2, ..., N$. In matrix notation, we write

$$X = \sqrt{D}Z, \tag{6.35}$$

where $\sqrt{D}$ is understood to mean $\sqrt{D} = \text{diag}(\sqrt{d_1}, \sqrt{d_2}, ..., \sqrt{d_N})$.

In summary, we have a two-step linear transformation. Given a vector of uncorrelated, unit variance Gaussian random variables, we form $X = \sqrt{D}Z$ and then $Y = QX$ to produce the vector of Gaussian random variables with the desired covariance matrix. Naturally, these two consecutive linear transformations can be combined into a single transformation

$$Y = QX = Q\sqrt{D}Z. \tag{6.36}$$

It is common to write the matrix $A = Q\sqrt{D}$ as $\sqrt{C}$ since $AA^T = Q\sqrt{D}\sqrt{D}^TQ^T = QDQ^T = C$.

---

[3] The notation $A = \text{diag}(a_1, a_2, ..., a_N)$ means that A is a diagonal matrix with diagonal elements a_1, a_2, ..., a_N.

Finally, note that if Z is zero-mean, then Y will be zero-mean as well. If it is desired to create Y with a nonzero mean, then a constant term can be added to the transformation to shift the mean to the specified value. This will not alter the covariance matrix of Y. In summary, we have the following result:

> **Theorem 6.4:** Given a vector Z of zero-mean, unit-variance, uncorrelated random variables, then a new set of random variables, Y, with arbitrary mean vector, μ, and covariance matrix, C, can be formed using the linear transformation
>
> $$Y = \sqrt{C}Z + \mu. \tag{6.37}$$
>
> Furthermore, if Z is a Gaussian random vector, then Y will be a Gaussian random vector as well.

If a Gaussian random number generator is not available,[4] one can always use a uniform random number generator together with the Box-Muller transformation described in Example 5.24 to produce Gaussian random variables.

Sometimes it is desired to transform a set of correlated random variables into a new set of uncorrelated random variables. Later in the text, when studying noise, this process will be referred to as "whitening." For now, it is seen that this process is just the opposite of the problem just solved. That is, given a random vector Y with mean, μ, and covariance matrix, C, a vector of zero-mean, unit-variance, uncorrelated random variables can be formed according to

$$Z = (\sqrt{C})^{-1}(Y - \mu). \tag{6.38}$$

Here, the expression $(\sqrt{C})^{-1}$ is interpreted as[5]

$$(\sqrt{C})^{-1} = (Q\sqrt{D})^{-1} = (\sqrt{D})^{-1}Q^{-1} = D^{-1/2}Q^{T}. \tag{6.39}$$

■ Example 6.4:

Let us suppose we desire to create a vector of four random variables with a mean vector of $\mu = [1, 0, 3, -2]^{T}$ and covariance matrix of

$$C = \begin{bmatrix} 30 & -10 & -20 & 4 \\ -10 & 30 & 4 & -20 \\ -20 & 4 & 30 & -10 \\ 4 & -20 & -10 & 30 \end{bmatrix}.$$

---

[4] Many high-level programming languages come with a built-in uniform random number generator, but not a Gaussian random number generator. See Chapter 12 for more details on random number generators.

[5] Since Q is an orthogonal matrix, $Q^{-1} = Q^{T}$. Also, $D^{-1/2} = \mathrm{diag}(d_{1}^{-1/2}, d_{2}^{-1/2}, ..., d_{N}^{-1/2})$.

The eigenvalue matrix and eigenvector matrix are calculated to be (we performed this calculation using MATLAB).

$$D = \begin{bmatrix} 4 & 0 & 0 & 0 \\ 0 & 16 & 0 & 0 \\ 0 & 0 & 36 & 0 \\ 0 & 0 & 0 & 64 \end{bmatrix} \text{ and } Q = \frac{1}{2}\begin{bmatrix} 1 & -1 & 1 & 1 \\ 1 & 1 & 1 & -1 \\ 1 & -1 & -1 & -1 \\ 1 & 1 & -1 & 1 \end{bmatrix}.$$

Thus, the appropriate transformation matrix is

$$A = Q\sqrt{D} = \frac{1}{2}\begin{bmatrix} 1 & -1 & 1 & 1 \\ 1 & 1 & 1 & -1 \\ 1 & -1 & -1 & -1 \\ 1 & 1 & -1 & 1 \end{bmatrix}\begin{bmatrix} 2 & 0 & 0 & 0 \\ 0 & 4 & 0 & 0 \\ 0 & 0 & 6 & 0 \\ 0 & 0 & 0 & 8 \end{bmatrix} = \begin{bmatrix} 1 & -2 & 3 & 4 \\ 1 & 2 & 3 & -4 \\ 1 & -2 & -3 & -4 \\ 1 & 2 & -3 & 4 \end{bmatrix}.$$

Therefore, given a vector of zero-mean, unit-variance, uncorrelated random variables, Z, the required transformation is

$$Y = \begin{bmatrix} 1 & -2 & 3 & 4 \\ 1 & 2 & 3 & -4 \\ 1 & -2 & -3 & -4 \\ 1 & 2 & -3 & 4 \end{bmatrix}Z + \begin{bmatrix} 1 \\ 0 \\ 3 \\ -2 \end{bmatrix}.$$

The MATLAB code to perform the necessary eigendecomposition of this example is very straightforward and is as follows:

```
C=[30 -10 -20 4; -10 30 4 -20; -20 4 30 -10; 4 -20 -10 30]
[Q, D]=eig(C)
A=Q*sqrt(D)
```

∎

6.4.2 *Quadratic Transformations of Gaussian Random Vectors*

In this section, we show how to calculate the PDFs of various quadratic forms of Gaussian random vectors. In particular, given a vector of N zero-mean Gaussian random variables, X, with an arbitrary covariance matrix, C_{XX}, we form a scalar quadratic function of the vector X of the general form

$$Z = X^T B X, \tag{6.40}$$

where B is an arbitrary $N \times N$ matrix. We would then like to find the PDF of the random variable, Z. These types of problem occur frequently in the study of noncoherent communication systems.

One approach to this problem would be to first form the CDF, $F_Z(z) = \Pr(X^T B X \le z)$. This could be accomplished by computing

$$F_Z(z) = \int_{A(z)} f_X(x)dx, \tag{6.41}$$

where $A(z)$ is the region defined by $x^T B x \le z$. While conceptually straightforward, defining the regions and performing the required integration can get quite involved. Instead, we elect to calculate the PDF of Z by first finding its characteristic function. Once the characteristic function is found, the PDF can be found through an inverse transformation.

For the case of Gaussian random vectors, finding the characteristic function of a quadratic form turns out to be surprisingly manageable.

$$\Phi_Z(\omega) = E[e^{j\omega X^T B X}] = \int \frac{1}{\sqrt{(2\pi)^N \det(C_{XX})}} \exp\left(-\frac{1}{2}(x^T[C_{XX}^{-1} - 2j\omega B]x)\right) dx. \tag{6.42}$$

This integral is understood to be over the entire N-dimensional x-plane. To evaluate this integral, we simply manipulate the integrand into the standard form of a N-dimensional Gaussian distribution and then use the normalization integral for Gaussian PDFs. Toward that end, define the matrix F according to $F^{-1} = C_{XX}^{-1} - 2j\omega B$. Then

$$\Phi_Z(\omega) = \int \frac{1}{\sqrt{(2\pi)^N \det(C_{XX})}} \exp\left(-\frac{1}{2}(x^T F^{-1} x)\right) dx$$

$$= \sqrt{\frac{\det(F)}{\det(C_{XX})}} \int \frac{1}{\sqrt{(2\pi)^N \det(F)}} \exp\left(-\frac{1}{2}(x^T F^{-1} x)\right) dx = \sqrt{\frac{\det(F)}{\det(C_{XX})}}. \tag{6.43}$$

The last step is accomplished using the fact that the integral of a multidimensional Gaussian PDF is unity. In addition, using the matrix property that $\det(F^{-1}) = (\det(F))^{-1}$, this can be rewritten in the more convenient form

$$\Phi_Z(\omega) = \sqrt{\frac{\det(F)}{\det(C_{XX})}} = \frac{1}{\sqrt{\det(F^{-1})\det(C_{XX})}}$$

$$= \frac{1}{\sqrt{\det(F^{-1}C_{XX})}} = \frac{1}{\sqrt{\det(I - 2j\omega B C_{XX})}}. \tag{6.44}$$

To get a feel for the functional form of the characteristic function, note that the determinant of a matrix can be written as the product of its eigenvalues. Furthermore, for a matrix of the form

$A = I + cD$, for a constant c, the eigenvalues of A, $\{\lambda_A\}$, can be written in terms of the eigenvalues of the matrix D, $\{\lambda_D\}$, according to $\lambda_A = 1 + c\lambda_D$. Therefore,

$$\Phi_Z(\omega) = \prod_{n=1}^{N} \frac{1}{\sqrt{1 - 2j\omega\lambda_n}}, \tag{6.45}$$

where the λ_ns are the eigenvalues of the matrix BC_{XX}. The particular functional form of the resulting PDF depends on the specific eigenvalues. Two special cases are considered as examples next.

■ Example 6.5:

In this example, we consider the case where the matrix B is an identity so that Z is the sum of the squares of Gaussian random variables, $B = I \Rightarrow Z = \sum_{n=1}^{N} X_n^2$. Further, let us assume that the X_n are uncorrelated and equal variance so that $C_{XX} = \sigma^2 I$. Then, the matrix BC_{XX} has N repeated eigenvalues all equal to σ^2. The resulting characteristic function is

$$\Phi_Z(\omega) = (1 - 2j\omega\sigma^2)^{-N/2} \quad .$$

This is the characteristic function of a chi-square random variable with N degrees of freedom. The corresponding PDF is

$$f_Z(z) = \frac{z^{(N/2)-1}}{(2\sigma^2)^{N/2}\Gamma(N/2)} \exp\left(-\frac{z}{2\sigma^2}\right) u(z).$$

■

■ Example 6.6:

For this example, suppose we need to find the PDF of $Z = X_1 X_2 + X_3 X_4$. In this case, the quantity Z can be expreesed in the general quadratic form of Equation (6.40) if we choose the matrix B according to

$$B = \frac{1}{2} \begin{bmatrix} 0 & 1 & 0 & 0 \\ 1 & 0 & 0 & 0 \\ 0 & 0 & 0 & 1 \\ 0 & 0 & 1 & 0 \end{bmatrix}.$$

Again, we take the X_i to be uncorrelated with equal variance so that $C_{XX} = \sigma^2 I$. In this case, the product matrix BC_{XX} has two pairs of repeated eigenvalues of values $\pm\sigma^2/2$. The resulting characteristic function is

(Continued)

$$\Phi_Z(\omega) = \frac{1}{1+j\omega\sigma^2}\frac{1}{1-j\omega\sigma^2} = \frac{1}{1+(\omega\sigma^2)^2}.$$

This is the characteristic function of a two-sided exponential (Laplace) random variable,

$$f_Z(z) = \frac{1}{2\sigma^2}\exp\left(-\frac{|z|}{\sigma^2}\right).$$

∎

6.4.3 Order Statistics

Suppose a vector of random variables has elements that are independent and identically distributed. In many applications, we need to find the PDF of the largest element in the vector. Or, as a more general problem, we might be interested in the PDF of the mth largest. Let, $X_1, X_2, ..., X_N$ be a sequence of random variables. We create a new set of random variables $Y_1, Y_2, ..., Y_N$ such that Y_1 is the smallest of the X_ns, Y_2 is the second smallest, and so on. The sequence of Y_ns are referred to as *order statistics* of the sequence of X_ns. Given that each of the X_ns follows a common PDF, $f_X(x)$, we seek to find the PDF of the Y_ns.

First, we find the PDF of Y_1 by first finding its CDF.

$$F_{Y_1}(y) = \Pr(Y_1 \le y) = 1 - \Pr(Y_1 \ge y) = 1 - \Pr(X_1 \ge y, X_2 \ge y, ..., X_N \ge y). \qquad (6.46)$$

This expression follows from the observation that if the smallest of a sequence is larger than some threshold, than all elements in the sequence must be above that threshold. Next, using the fact that the X_n are independent and all have the same distribution the previous expression simplifies to

$$F_{Y_1}(y) = 1 - \Pr(X_1 \ge y)\Pr(X_2 \ge y)...\Pr(X_N \ge y) = 1 - (1 - F_X(y))^N. \qquad (6.47)$$

Differentiating with respect to y then produces the desired PDF,

$$f_{Y_1}(y) = Nf_X(y)(1 - F_X(y))^{N-1}. \qquad (6.48)$$

A similar procedure can be followed to determine the PDF of the mth smallest, Y_m. First, we work out an expression for the CDF.

$$F_{Y_m}(y) = \Pr(Y_m \le y) = \Pr(m \text{ or more of the } X_n\text{s are less than } y)$$

$$= \sum_{k=m}^{N} \Pr(k \text{ of the } X_n\text{s are less than } y) \qquad (6.49)$$

To evaluate the probability of the event $\{k$ of the X_ns are less than $y\}$, it is noted that one way for this event to occur is if $X_1 \leq y$, $X_2 \leq y$, ..., $X_k \leq y$, $X_{k+1} > y$, ..., $X_N > y$. The probability of this event is

$$\text{Pr}(X_1 \leq y, X_2 \leq y, ..., X_k \leq y, X_{k+1} > y, ..., X_N > y) = (F_X(y))^k (1 - F_X(y))^{N-k}. \quad (6.50)$$

Of course, we do not have to have the first k elements of the sequence smaller than y. We are looking for the probability that *any* k of the N elements are below y. Thus, we need to count the number of combinations of k out of N variables. This is merely the binomial coefficient. Hence,

$$\text{Pr}(k \text{ of the } X_n\text{s are less than } y) = \binom{N}{k} (F_X(y))^k (1 - F_X(y))^{N-k}. \quad (6.51)$$

Summing over k gives the desired CDF:

$$F_{Y_m}(y) = \sum_{k=m}^{N} \binom{N}{k} (F_X(y))^k (1 - F_X(y))^{N-k}. \quad (6.52)$$

Differentiating with respect to y then gives the expression

$$f_{Y_m}(y) = f_X(y) \sum_{k=m}^{N} \binom{N}{k} k (F_X(y))^{k-1} (1 - F_X(y))^{N-k}$$

$$-f_X(y) \sum_{k=m}^{N} \binom{N}{k} (N-k)(F_X(y))^k (1 - F_X(y))^{N-k-1}. \quad (6.53)$$

It is left as an exercise to the reader (see Exercise 6.22) to show that this expression reduces to the form

$$f_{Y_m}(y) = \frac{N!}{(m-1)!(N-m)!} f_X(y)(F_X(y))^{m-1}(1 - F_X(y))^{N-m}. \quad (6.54)$$

An alternative approach to deriving this expression is outlined in Exercise 6.23. The next example illustrates one possible use of order statistics.

■ Example 6.7:

Suppose we observe a sequence of $N = 2k - 1$ independent and identically distributed random variables, $X_1, ..., X_{2k-1}$, and we wish to estimate the mean of the common distribution. One method to do this would be to use the median (middle) element in the sequence as an estimate of the mean. In terms of order statistics, the median is simply Y_k.

$$f_{Y_k}(y) = \frac{(2k-1)!}{[(k-1)!]^2} f_X(y)[F_X(y)]^{k-1}[1-F_X(y)]^{k-1}.$$

For example, if the X_n are all uniformly distributed over $(0, 1)$, then

$$f_{Y_k}(y) = \frac{(2k+1)!}{[(k-1)!]^2}[y(1-y)]^{k-1}, \ 0 \leq y \leq 1.$$

Some straightforward calculations reveal that this distribution has a mean of $E[Y_k] = 1/2$ and a variance of $\sigma_{Y_k}^2 = 1/(4(2k+1))$. Note that, "on the average," the median is equal to the true mean of the distribution. We say that this estimator is *unbiased*. Furthermore, we also see that as we observe more samples (k gets larger), the variance of the median gets smaller. In other words, as the sample size increases, the median becomes increasingly more precise as an estimator of the mean. In fact, in the limit as $k \to \infty$, the variance goes to zero which means that the median becomes equal to the mean. We will discuss this problem of estimating means of sequences of random variables in the next chapter.

■

6.4.4 Coordinate Systems in Three Dimensions

Coordinate system transformations in three dimensions follow the same procedure as was derived for the two-dimensional problems. Given a random vector X and a corresponding joint PDF $f_X(x)$, the joint PDF of $Y = g(X)$ is given by the general formula expressed in Equation (6.24). An example is included below to illustrate the procedure.

■ Example 6.8: (Cartesian-to-spherical coordinates)

Let the random variables X, Y, and Z in Cartesian coordinates be transformed to spherical coordinates according to

$$R = \sqrt{X^2 + Y^2 + Z^2},$$

$$\Theta = \cos^{-1}\left(\frac{Z}{\sqrt{X^2 + Y^2 + Z^2}}\right),$$

$$\Phi = \tan^{-1}\left(\frac{Y}{X}\right).$$

The inverse transformation is probably more familiar to most readers and is given by

$$X = R\sin(\Theta)\cos(\Phi),$$
$$Y = R\sin(\Theta)\sin(\Phi),$$
$$Z = R\cos(\Theta).$$

The Jacobian of this transformation is

$$J\left(\begin{matrix} x \ x \ z \\ r \ \theta \ \phi \end{matrix}\right) = \begin{bmatrix} \sin(\theta)\cos(\phi) & \sin(\theta)\sin(\phi) & \cos(\theta) \\ r\cos(\theta)\cos(\phi) & r\cos(\theta)\sin(\phi) & -r\sin(\theta) \\ -r\sin(\theta)\sin(\phi) & r\sin(\theta)\cos(\phi) & 0 \end{bmatrix},$$

and the determinant of this matrix works out to be

$$\det\left[J\left(\begin{array}{ccc} x & y & z \\ r & \theta & \phi \end{array}\right)\right] = r^2 \sin(\theta).$$

Suppose X, Y, and Z are jointly Gaussian with a joint PDF given by

$$f_{X,Y,Z}(x,y,z) = \frac{1}{(2\pi\sigma^2)^{3/2}} \exp\left(-\frac{x^2+y^2+z^2}{2\sigma^2}\right).$$

Then, the joint PDF of R, Θ, and Φ is found to be

$$f_{R,\Theta,\Phi}(r,\theta,\phi) = f_{X,Y,Z}(x,y,z) \left|\det\left[J\left(\begin{array}{ccc} x & y & z \\ r & \theta & \phi \end{array}\right)\right]\right|_{\substack{x = r\sin(\theta)\cos(\phi) \\ y = r\sin(\theta)\sin(\phi) \\ z = r\cos(\theta)}} = \frac{r^2 \sin(\theta)}{(2\pi\sigma^2)^{3/2}} \exp\left(-\frac{r^2}{2\sigma^2}\right), \quad \begin{array}{l} r \geq 0 \\ 0 \leq \theta \leq \pi \\ 0 \leq \phi \leq 2\pi \end{array}.$$

The marginal PDFs are found by integrating the unwanted variables out of this joint PDF. In this case, the required integrations are fairly straightforward, resulting in

$$f_R(r) = \sqrt{\frac{2}{\pi}}\frac{r^2}{\sigma^3} \exp\left(-\frac{r^2}{2\sigma^2}\right) u(r),$$

$$f_\Theta(\theta) = \frac{1}{2}\sin(\theta), \quad 0 \leq \theta \leq \pi,$$

$$f_\Phi(\phi) = \frac{1}{2\pi}, \quad 0 \leq \phi \leq 2\pi.$$

Note also that for this example, $f_{R,\Theta,\Phi}(r,\theta,\phi) = f_R(r)f_\Theta(\theta)f_\Phi(\phi)$ so that R, Θ, and Φ are all independent. These PDFs are plotted in Figure 6.1.

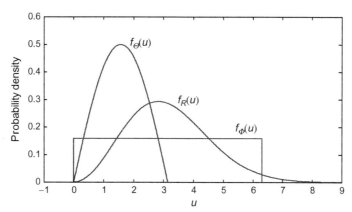

Figure 6.1
PDFs of spherical coordinate variables for Example 6.6. (For color version of this figure, the reader is refered to the web version of this chapter.)

6.5 Estimation and Detection

In this section, we introduce a broad class of problems whereby we are able to observe some random quantities and based on those obervations we must try to infer something about some related quantity. For example, suppose we are able to observe a set of random variables $X = [X_1, X_2, ..., X_n]^T$ and then based on that observation we wish to provide the best estimate that we can of some related random variable, Y. Some examples of these types of problems include the following:

- *Prediction problems:* In this case, the X_i may be samples of some signal and Y represents some future value. We would like to predict the future based on observations of the past.
- *Interpolation problems*: Once again, the X_i may be samples of some signal but in this case, Y may represent some intermediate value of the signal. Thus, given samples of a signal, we wish to interpolate to some in-between point in time.
- *Filtering problems*: In many problems of interest, we are able to observe noisy measurements of some random quantity in which case we may have a model of the form $X_i = Y + N_i$ where Y represents the quantity we are trying to measure and the N_i represents the noise corrupting the i th measurement. In this case, we wish to filter the noise out of the sequence of observations to provide our best estimate of the desired quantity, Y.

In another class of problems, the quantity that we wish to estimate is not a random variable, but rather a deterministic parameter that is somehow associated with the distribution of the observed samples, $X = [X_1, X_2, ..., X_n]^T$. For example, if the X_i are independent observations and all follow some common distribution described by an underlying PDF, $f_X(x)$. We might want to estimate the mean or variance associated with $f_X(x)$. Or, perhaps we might want to estimate a tail probability of the form, $\Pr(X \geq x_0)$.

In the preceding examples, the quantity that we were trying to estimate was continuous (i.e., can take on any of a continuium of values). Sometimes, we are interesting in estimating a quantity that is discrete in nature. One classic example of this type of problem is radar systems where we are merely trying to decide whether or not a target is present based on observing radar returns. Another example is digital communication systems where we are trying to determine whether bits take on values of 0 or 1 based on samples of some receive signal. These types of problems where the quantity to be estimated takes on discrete values are generally referred to as detection problems. One fundamental difference between estimation and detection problems involves how we measure success. In a detection problem, we might ask how often our guess is correct; however, in an estimation problem, such a measure would be meaningless and thus in estimation problems it is more common to measure an error between the true value and the estimated value.

6.5.1 *Maximum a Posteriori Estimation*

The application section at the end of Chapter 2 gave an example of a detection problem where we tried to determine whether or not there was light incident on a photodetector based on observation of the number of photons emitted. In that example, the choice was made according to the a posteriori probabilities of the two options. Following this approach is known as *maximum a posteriori* (MAP) *detection*. In general, suppose we are trying to detect which of a discrete number of possible values $\{y_1, y_2, ..., y_m\}$ the random variable Y has taken on given observation of $\boldsymbol{X} = \boldsymbol{x}$. The MAP detector will choose[6] $\hat{Y} = y_k$ if

$$k = \arg \max_i \quad \Pr(Y{=}y_i|\boldsymbol{X}{=}\boldsymbol{x}). \tag{6.55}$$

That is, given the observation $\boldsymbol{X} = \boldsymbol{x}$, the MAP detector chooses the most probable value of Y. For estimation problems where Y is continuous, the probability in the previous equation is replaced with a PDF and the resulting MAP estimator is $\hat{Y} = y_{\text{MAP}}$, where

$$y_{\text{MAP}} = \arg \quad \max_u f_{Y|\boldsymbol{X}}(u|\boldsymbol{x}) \quad . \tag{6.56}$$

In other words, the MAP estimator for Y will be the value which maximizes the conditional PDF of Y given $\boldsymbol{X}$.

■ Example 6.9:

Suppose X and Y are jointly Gaussian with a joint PDF

$$f_{X, Y}(x, y) = \frac{1}{2\pi\sqrt{1-\rho^2}} \exp\left(-\frac{x^2 - 2\rho xy + y^2}{2(1-\rho^2)}\right).$$

In this case, the conditional PDF works out to be

$$f_{Y|X}(y|x) = \frac{1}{\sqrt{2\pi(1-\rho^2)}} \exp\left(-\frac{(y-\rho x)^2}{2(1-\rho^2)}\right).$$

For a given observation, $X = x$, the value of y that maximizes the conditional PDF is $y = \rho x$. Therefore, the MAP estimator for this example is $\hat{Y} = \rho X$.

■

---

[6] We use the "hat" notation to represent an estimate so that $\hat{Y}$ is an estimate of Y.

■ Example 6.10:

Suppose we wish to accurately measure the temperature in a room. We have available to us a number of thermometers which will provide us with a temperature reading which is equal to the true temperature plus an error which is modeled as a Gaussian random variable with zero-mean and some variance σ_e^2. Each thermometer provides a reading whose error is independent of all other thermometers. If it is known that the true temperature has a Gaussian PDF with mean μ_t and variance σ_t^2, what is the MAP estimate of the temperature given the readings of n thermometers?

In this case, we can model each thermometer reading, X_i, according to

$$X_i = Y + N_i, \quad i = 1, 2, \ldots, n,$$

where Y is the actual temperature and N_i is the error in the ith reading. Conditioned on $Y = y$, each thermometer reading will be an independent Gaussian random variable with mean y and variance σ_e^2. Therefore,

$$f_{X|Y}(x|y) = \frac{1}{(2\pi\sigma_e^2)^{n/2}} \exp\left(-\frac{1}{2\sigma_e^2}\sum_{i=1}^{n}(x_i - y)^2\right).$$

The conditional PDF of Y given X can be found according to Bayes's theorem,

$$f_{Y|X}(y|x) = \frac{f_{X|Y}(x|y)f_Y(y)}{f_X(x)}.$$

The MAP estimator for the temperature will then be the value of y that maximizes the previous conditional PDF. Note that the denominator is not a function of y and so for the purposes of finding the MAP estimator, we can simplify the problem by finding the value of y that maximizes the numerator. The numerator works out to be

$$f_{X|Y}(x|y)f_Y(y) = \left[\frac{1}{(2\pi\sigma_e^2)^{n/2}}\exp\left(-\frac{1}{2\sigma_e^2}\sum_{i=1}^{n}(x_i - y)^2\right)\right]\left[\frac{1}{\sqrt{2\pi\sigma_t^2}}\exp\left(-\frac{(y-\mu_t)^2}{2\sigma_t^2}\right)\right]$$

$$= \frac{1}{(2\pi)^{(n+1)/2}\sigma_e^n\sigma_t}\exp\left(-\frac{1}{2}\left[\left(\frac{n}{\sigma_e^2}+\frac{1}{\sigma_t^2}\right)y^2 - 2\left(\frac{1}{\sigma_e^2}\sum_{i=1}^{n}x_i + \frac{\mu_t}{\sigma_t^2}\right)y + \left(\frac{1}{\sigma_e^2}\sum_{i=1}^{n}x_i^2 + \frac{\mu_t^2}{\sigma_t^2}\right)\right]\right).$$

Differentiating with respect to y and setting the result equal to zero will show that the value of y which optimizes the previous expresion is the solution to the linear equation

$$\left(\frac{n}{\sigma_e^2}+\frac{1}{\sigma_t^2}\right)y - \left(\frac{1}{\sigma_e^2}\sum_{i=1}^{n}x_i + \frac{\mu_t}{\sigma_t^2}\right) = 0.$$

In other words, for this example, the MAP estimator is given by

$$\hat{Y} = \frac{\frac{1}{\sigma_e^2}\left(\sum_{i=1}^{n} X_i\right) + \frac{\mu_t}{\sigma_t^2}}{\frac{n}{\sigma_e^2} + \frac{1}{\sigma_t^2}} = \frac{\left(\frac{1}{n}\sum_{i=1}^{n} X_i\right) + \frac{\mu_t \sigma_e^2}{n \sigma_t^2}}{1 + \frac{\sigma_e^2}{n \sigma_t^2}}.$$

In the previous expression, the term $\frac{1}{n}\sum_{i=1}^{n} X_i$ is the average of all the thermometer readings. It is interesting to note that our MAP estimate of the temperature is not just the average of the individual readings. Instead, the MAP estimate is skewed by our prior knowledge about the temperature as provided by the a priori distribution, $f_Y(y)$.

∎

6.5.2 *Maximum Likelihood Estimation*

In the previous section, we presented the MAP estimator which is chosen to maximize the conditional PDF, $f_{Y|X}(y|x)$. Proceeding as in Example 6.8 and using Bayes's theorem, this conditional PDF can be rewritten as

$$f_{Y|X}(y|x) = \frac{f_{X|Y}(x|y)f_Y(y)}{f_X(x)}. \tag{6.57}$$

As was pointed out in the previous example, the denominator is not a function of y and hence has no bearing on the optimization. It is sufficient to find the value of y which maximizes the numerator. Furthermore, in many problems of interest, the a priori distribution for Y may be modeled using a uniform distribution. In that case, $f_Y(y)$ also does not depend on y (it is constant over all allowable values of y) so that the MAP estimator can be found by maximizing $f_{X|Y}(x|y)$. The estimator that maximizes $f_{X|Y}(x|y)$ is known as a *maximum likelihood* (ML) *estimator*. That is, the ML estimator of Y given observations $X = x$ is $\hat{Y} = y_{ML}$, where

$$y_{ML} = \arg\max_{u} f_{X|Y}(x|u). \tag{6.58}$$

In this case, the same formulation can be used when Y is a discrete random variable.

∎ Example 6.11:

Consider forming the ML estimator for the temperature estimation in Example 6.8. As before, the joint conditional PDF of the set of temperature measurements is given by

$$f_{X|Y}(x|y) = \frac{1}{(2\pi\sigma_e^2)^{n/2}} \exp\left(-\frac{1}{2\sigma_e^2}\sum_{i=1}^{n}(x_i - y)^2\right).$$

(Continued)

The difference in this example is that this time we do not have any prior knowledge about the distribution of the true temperature Y. In that case, we use an ML estimator since we do not have enough information to form a MAP estimator. The ML estimator is simply the value of y that maximizes the previous joint conditional PDF. Differentiating with respect to y and setting equal to zero results in the equation

$$\sum_{i=1}^{n} (x_i - y) = 0.$$

Therefore, the ML estimator is given by

$$\hat{Y} = \frac{1}{n} \sum_{i=1}^{n} X_i.$$

In this case, since we have no prior information about the distribution of the temperature, the best thing to do is to form a temperature estimate which is simply the average of the readings on each thermometer.

6.5.3 Minimum Mean Square Error Estimation

Another common approach to estimation is to construct an estimator that will in some sense minimize the error in the estimator. Define the estimation error, Z, to be the difference between the true value, Y, and the estimated value, $\hat{Y}$,

$$Z = Y - \hat{Y}. \tag{6.59}$$

The error is itself a random variable, so it does not make much sense to try to minimize the error. Instead, we might want to choose our estimator such that certain statistics associated with the error have desirable properties. Consider, for example the mean of the error, $E[Y - \hat{Y}]$. Certainly, we would like this mean to be small, but even if the mean of the error is zero, the estimator may still not be very good. Consider an estimator where the error is large and positive half the time and large and negative the other half of the time. On the average, the error is zero, but the estimator still can have quite large errors. For that reason, it is more common to look at the mean-square value of the error, $E[(Y - \hat{Y})^2]$. An estimator that minimizes this quantity is known as a *minimum mean square error* (MMSE) *estimator*. The following theorem will simplify the problem of finding the MMSE estimator.

Theorem 6.5: Given an observation, $X = x$, the function $g(X)$ which minimizes the mean square error, $E[(Y - g(x))^2 | X = x]$ is the conditional expected value, $g(x) = E[Y | X = x]$.

Proof: Once we condition on $\{X = x\}$, the function $g(x)$ is simply a scalar constant. This optimization problem can then be viewed as finding a value of a constant, c, which minimizes the MSE, $E[(Y-c)^2]$. Differentiating (with respect to c) and setting equal to zero results in $E[2(Y-c)] = 0$ and therefore the optimum value of the constant is $c = E[Y]$. $\square$

■ Example 6.12:

Consider a set of observations, X_i, $i = 1, 2, ..., n$, whose PDFs conditioned on $Y = y$ are each independent and exponentially distributed with means of $1/y$,

$$f_{X|Y}(\mathbf{x}|y) = \prod_{i=1}^{n} y\exp(-yx_i)u(x_i).$$

Furthermore, suppose Y is also an exponential random variable with a mean of 1 so that

$$f_Y(y) = \exp(-y)u(y).$$

In order to find the MMSE estimator of Y given observation of $\{X=x\}$, we need to find the conditional expectation. Toward that end, we find the conditional PDF of Y given $\{X=x\}$,

$$f_{Y|X}(y|\mathbf{x}) = \frac{f_{X|Y}(\mathbf{x}|y)f_Y(y)}{f_X(\mathbf{x})}.$$

The marginal PDF needed in the denominator is found as

$$f_X(\mathbf{x}) = \int_0^\infty f_{X|Y}(\mathbf{x}|y)f_Y(y)\mathrm{d}y = \int_0^\infty y^n\exp\left(-\left(1+\sum_{i=1}^{n}x_i\right)y\right)\mathrm{d}y = \frac{n!}{\left(1+\sum_{i=1}^{n}x_i\right)^{n+1}}.$$

The last step in the preceding equation was acomplished using the normalization integral for an Erlang PDF. The conditional PDF is then

$$f_{Y|X}(y|\mathbf{x}) = \frac{\left(1+\sum_{i=1}^{n}x_i\right)^{n+1}}{n!}y^n\exp\left(-\left(1+\sum_{i=1}^{n}x_i\right)y\right).$$

The mean of this conditional distribution works out to be

$$E[Y|X=\mathbf{x}] = \frac{n+1}{1+\sum_{i=1}^{n}x_i}.$$

Therfore, the MMSE estimator is

$$\hat{Y}_{\mathrm{MMSE}} = \frac{n+1}{1+\sum_{i=1}^{n}X_i}.$$

(Continued)

We could also work out the ML and MAP estimators for this problem. The ML estimator would be the value of y that maximizes $f_{X|Y}(x|y)$, while the MAP estimator maximizes $f_{Y|X}(y|x)$. Since expressions for both of these conditional PDFs have already been given, it is only a matter of some straighforward calculus to produce

$$\hat{Y}_{ML} = \left(\frac{1}{n}\sum_{i=1}^{n} X_i\right)^{-1} \text{ and } \hat{Y}_{MAP} = \frac{n}{1 + \sum_{i=1}^{n} X_i}$$

All three estimators have a similar functional form but yet are all slightly different. ∎

Quite often, the MMSE estimator can be difficult to find. In such cases, it is common to restrict ourselves to linear estimators. Given an observation, $\{X = x\}$, a linear estimator of Y will be of the form $\hat{Y} = g(X) = a_1 X_1 + a_2 X_2 + \cdots + a_n X_n$ for constants a_i, $i = 1, 2, \ldots, n$. With a linear estimator, the MSE becomes

$$E[(Y - \hat{Y})^2] = E[(Y - (a_1 X_1 + a_2 X_2 + \ldots + a_n X_n))^2]. \tag{6.60}$$

Optimizing with respect to the kth of these constants (i.e., differentiating with respect to a_k and setting equal to zero) produces

$$E[X_k(Y - (a_1 X_1 + a_2 X_2 + \ldots + a_n X_n))] = 0. \tag{6.61}$$

This produces a result known as the *orthogonality principle* which states that for linear MMSE (LMMSE) estimators, the estimator must be chosen so that the error in the estimator must be orthogonal to each observation. Noting that the orthogonality principle must hold for each observation (i.e., each X_k) and combining all the results in vector form produces

$$E[X(Y - X^T a)] = 0 \quad, \tag{6.62}$$

where $a = [a_1, a_2, \ldots, a_n]^T$ is the vector of LMMSE coefficients. This is a set of linear equations for the unknown coefficients. Define the vector of correlations between the observations and the value to be estimated as

$$p = E[XY] = \begin{bmatrix} E[X_1 Y] \\ E[X_2 Y] \\ \ldots \\ E[X_n Y] \end{bmatrix}, \tag{6.63}$$

and define the correlation matrix of the observations as

$$R_{XX} = E[XX^T] = \begin{bmatrix} E[X_1^2] & E[X_1X_2] & \dots & E[X_1X_n] \\ E[X_2X_1] & E[X_2^2] & \dots & E[X_2X_n] \\ \dots & \dots & & \dots \\ E[X_nX_1] & E[X_nX_2] & \dots & E[X_n^2] \end{bmatrix}. \tag{6.64}$$

Then the coefficients of the LMMSE estimator are the solution to the linear equations

$$R_{XX}a = p, \tag{6.65}$$

which is given by

$$a = R_{XX}^{-1}p. \tag{6.66}$$

■ Example 6.13:

Suppose we try to find the LMMSE estimator for the temperature measuring problem of Example 6.8. It is interesting to note that, for this example, both the MAP estimator (found in Example 6.8) and the ML estimator (found in Example 6.9) were linear. That is, both estimators formed a linear combination of the obervations, X_i. In order to find the LMMSE coefficients, we will need to evaluate the correlations indicated in Equations (6.63) and (6.64). Fortunately, for this example, these correlations are fairly simple to work out.

$$E[X_iY] = E[(Y+N_i)Y] = E[Y^2] + E[N_iY] = E[Y^2] + E[N_i]E[Y] = E[Y^2] = \sigma_t^2 + \mu_t^2.$$

$$E[X_iX_j] = E[(Y+N_i)(Y+N_j)] = E[Y^2] + E[YN_i] + E[YN_j] + E[N_iN_j] = \begin{cases} \sigma_t^2 + \mu_t^2, & i \neq j, \\ \sigma_t^2 + \mu_t^2 + \sigma_e^2, & i = j. \end{cases}$$

In the previous calculations, we have used the fact that Y is independent of the noise terms, N_i, and that $E[N_i] = 0$. Note that $E[X_iY]$ is the same for all i so that the vector, p, of correlations takes the form of a constant times a column vector of all ones, $p = (\sigma_t^2 + \mu_t^2)1_n$, where we use the notation 1_n to represent an n-element column vector of all ones. Similarly, the correlation matrix, R_{XX}, can be written as a constant times a matrix of all ones plus another constant times an identity matrix, $R_{XX} = (\sigma_t^2 + \mu_t^2)1_n1_n^T + \sigma_e^2I$, where I is an identity matrix. Putting all of these computations into the matrix equation indicated in Equation (6.65) results in

$$[(\sigma_t^2 + \mu_t^2)1_n1_n^T + \sigma_e^2I]a = (\sigma_t^2 + \mu_t^2)1_n.$$

(Continued)

It turns out that the solution to this matrix equation is proportional to the all ones vector. To demonstrate this, assume a solution of the form $a = c1_n$ for some constant c. Plugging this proposed solution into the previous equation produces

$$[(\sigma_t^2 + \mu_t^2)1_n 1_n^T + \sigma_e^2 I]c1_n = (\sigma_t^2 + \mu_t^2)1_n$$

$$[(\sigma_t^2 + \mu_t^2)n + \sigma_e^2]c1_n = (\sigma_t^2 + \mu_t^2)1_n .$$

Here, we have used the fact that the inner product $1_n^T 1_n$ is equal to n and that $I1_n = 1_n$. Therefore, we get equality in the previous equation if the constant is selected according to

$$c = \frac{\sigma_t^2 + \mu_t^2}{(\sigma_t^2 + \mu_t^2)n + \sigma_e^2} .$$

Therefore, the LMMSE estimator for this problem is

$$\hat{Y}_{\text{LMMSE}} = \frac{\sigma_t^2 + \mu_t^2}{(\sigma_t^2 + \mu_t^2)n + \sigma_e^2} \sum_{i=1}^{n} X_i = \frac{\frac{1}{n}\sum_{i=1}^{n} X_i}{1 + \frac{\sigma_e^2}{(\sigma_t^2 + \mu_t^2)n}} .$$

In this case, the LMMSE estimate of the temperature is a scaled version of the average of the individual temperature readings.

∎

In the next sections, an engineering application is presented which demonstrates how LMMSE estimation is used to predict future values of a speech waveform and how that is used in a simple speech coder. In the next chapter, we will turn our attention to problems involving estimation of various parameters of a distribution.

6.6 Engineering Application: Linear Prediction of Speech

In many applications, we are interested in predicting future values of a waveform given current and past samples. This is used extensively in speech coders where the signal-to-quantization noise associated with a quantizer can be greatly increased if only the prediction error is quantized. A fairly simple speech coder which utilizes this idea is illustrated in Figure 6.2. In Section 4.11, we introduced the idea of scalar quantization. The process of sampling (at or above the Nyquist rate), quantizing, and then encoding each quantization level with some binary codeword is known as pulse code modulation (PCM). In Figure 6.2, we consider a slight modification to the basic PCM technique known as *differential PCM* (or DPCM). The basic idea here is that if we can reduce the range of the signal that is being quantized, then we can either reduce the number of quantization levels needed (and hence reduce the bit rate of the speech coder) or reduce the amount of quantization noise and hence increase the SQNR.

Figure 6.2
Block diagram of a simple speech coder using differential pulse code modulation.

A typical speech signal has a frequency content in the range from about 300 to 3500 Hz. In order to be able to recover the signal from its samples, a typical sampling rate of 8 kHz is used which is slightly higher than the Nyquist rate. However, much of the energy content of a speech signal lies in a frequency band below about 1 kHz; thus, when sampled at 8 kHz, a great deal of the speech signal does not change substantially from one sample to the next. Stated another way, when the speech signal is sampled at 8 kHz, we should be able to predict future sample values from current and past samples with pretty good accuracy. The DPCM encoder does exactly that and then only quantizes and encodes the portion of the signal that it is not able to predict.

In Figure 6.2, the X_n represent samples of a speech waveform. These samples are input to the predictor whose job is to make its best estimate of X_n given $X_{n-1}, X_{n-2}, X_{n-3}, \ldots$ as inputs. It is common to use linear prediction, in which case the predictor output is a linear combination of the inputs. That is, assuming the predictor uses the last m samples to form its estimate, the predictor output is of the form

$$Y_n = \sum_{i=1}^{m} a_i X_{n-i},\qquad(6.67)$$

where the a_i are constants that we select to optimize the performance of the predictor. The quantity $Z_n = X_n - Y_n$ is the predictor error, which we want to make as small as possible. This error is quantized with a scalar quantizer which uses 2^b levels and each level is encoded with a b bit codeword. The overall bit rate of the speech coder is $b*f_s$ bits/second, where f_s is the rate (in Hz) at which the speech is sampled. For example, if a 16-level quantizer were used with a speech sampling rate of 8 kHz, the DPCM speech coder would have a bit rate of 32 kbits/second.

An important question is "Can the original samples be recovered from the binary representation of the signal?" Given the encoded bit stream, we can construct the sequence of quantizer outputs, Q_n. As with any quantization scheme, we can never recover the exact quantizer input from the quantizer output, but if we use enough levels in the quantizer, the quantization noise can be kept fairly small. The speech samples are reconstructed according to $X_n = Y_n + Z_n$. Since we do not have Z_n we use Q_n in its place and form

$$\hat{X}_n = Y_n + Q_n = X_n + \varepsilon_n,\qquad(6.68)$$

where $\varepsilon_n = Q_n - Z_n$ is the quantization noise in the nth sample. To complete the process of recovering the sample values, the decoder must also form the Y_n. It can do this by employing an identical predictor as used at the encoder. Unfortunately, the predictor at the decoder does not have access to the same input as the predictor at the encoder. That is, at the decoder we cannot use the true values of the past speech samples, but rather must use the quantized (noisy) versions. This can be problematic since the predictor at the decoder will now form

$$\hat{Y}_n = \sum_{i=1}^{m} a_i \hat{X}_{n-i}. \tag{6.69}$$

If the $\hat{X}_n$ are noisy versions of the X_n, then the $\hat{Y}_n$ will also be noisy. Now, not only do we have quantization noise, but that noise propagates from one sample to the next through the predictor. This leads to the possibility of a snowballing effect, where the noise in our recovered samples gets progressively larger from one sample to the next.

The above problem is circumvented using the modified DPCM encoder shown in Figure 6.3; the corresponding decoder is shown in the figure as well. The difference between this DPCM system and the one in Figure 6.2 is that now the predictor used in the encoder bases its predictions on the quantized samples rather than on the true samples. By doing this, the predicted value may be slightly degraded (but not much if the number of quantization levels is sufficient), but there will be no propagation of errors in the decoder, since the predictor at the decoder now uses the same inputs as the predictor at the encoder.

Now that we have the design of the speech encoder and decoder squared away, we shift our attention to the problem of designing the predictor. Assuming a linear predictor, the problem is essentially to choose the coefficients a_i in Equation (6.67) to minimize the prediction error:

$$Z_n = X_n - Y_n = X_n - \sum_{i=1}^{m} a_i X_{n-i}. \tag{6.70}$$

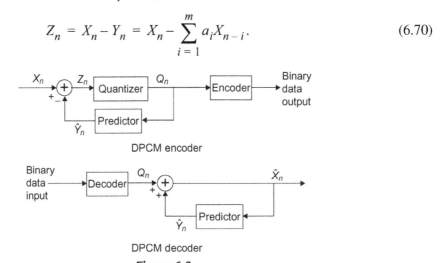

DPCM encoder

DPCM decoder

Figure 6.3
Block diagram of a modified speech coder using differential pulse code modulation.

Following the theory developed in Section 6.5.3, we choose the predictor coefficients to minimize the MSE:

$$E[Z_n^2] = E\left[\left(X_n - \sum_{i=1}^{m} a_i X_{n-i}\right)^2\right]. \tag{6.71}$$

Define the correlation parameter $r_k = E[X_n X_{n-k}]$ to be the correlation between two samples spaced by k sampling intervals. Then the system of equations in Equation (6.65) can be expressed in matrix form as

$$
\begin{bmatrix}
r_0 & r_1 & r_2 & \cdots & r_{m-1} \\
r_1 & r_0 & r_1 & \cdots & r_{m-2} \\
r_2 & r_1 & r_0 & \cdots & r_{m-3} \\
\cdots & \cdots & \cdots & & \cdots \\
r_{m-1} & r_{m-2} & r_{m-3} & \cdots & r_0
\end{bmatrix}
\begin{bmatrix}
a_1 \\ a_2 \\ a_3 \\ \cdots \\ a_m
\end{bmatrix}
=
\begin{bmatrix}
r_1 \\ r_2 \\ r_3 \\ \cdots \\ r_m
\end{bmatrix},
\tag{6.72}
$$

and the predictor coefficients are simply the solution to this set of linear equations.

■ Example 6.14:

Figure 6.4 shows a segment of speech that has a duration of about 2 s, which was sampled at a rate of 8 kHz. From this data, (using MATLAB) we estimated the correlation parameters $r_k = E[X_n X_{n+k}]$; found the linear prediction coefficients, a_i, $i = 1, 2, ..., m$, and then calculated the mean squared estimation error, MSE $= E[(X_n - Y_n)^2]$. The results are shown in Table 6.1. We should note a couple of observations. First, even with a simple one-tap predictor, the size of the error signal is much smaller than the original signal (compare the values of MSE with r_0 in the table). Second, we note that (for this example) there does not seem to be much benefit gained from using more than two previous samples to form the predictor.

Finally, in Figure 6.5 we compare the quality of the encoded speech as measured by the SQNR for PCM and the DPCM scheme of Figure 6.3 using the two-tap predictor specified in Table 6.1. For an equal number of bits per sample, the DPCM scheme improves the SQNR by more than 20 dB. Alternatively, the DPCM scheme can use 3 bits/sample fewer than the PCM scheme and still provide better SQNR.

Table 6.1: Results of linear prediction of speech segment form Figure 6.4

$r_0 = 0.0591$ $r_1 = 0.0568$ $r_2 = 0.0514$ $r_3 = 0.0442$ $r_4 = 0.0360$				
$m = 1$	$a_1 = 0.9615$			MSE $= 0.004473$
$m = 2$	$a_1 = 1.6564$	$a_2 = -0.7228$		MSE $= 0.002144$
$m = 3$	$a_1 = 1.7166$	$a_2 = -0.8492$	$a_3 = 0.0763$	MSE $= 0.002132$
$m = 4$	$a_1 = 1.7272$	$a_2 = -1.0235$	$a_3 = 0.4276$ $a_4 = -0.2052$	MSE $= 0.002044$

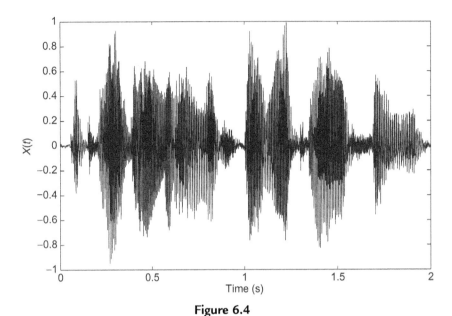

Figure 6.4
Speech segment used in Example 6.7.
(For color version of this figure, the reader is refered to the web version of this chapter.)

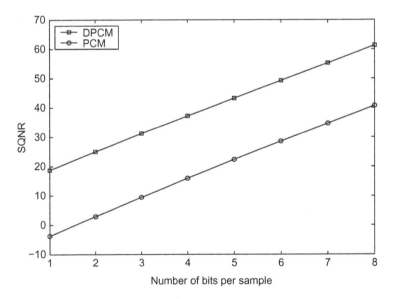

Figure 6.5
SQNR comparison of PCM and DPCM speech coders for the speech segment in Figure 6.4.
(For color version of this figure, the reader is refered to the web version of this chapter.)

Exercises

Section 6.1: Joint and Conditional PMFs, CDFs, and PDFs

6.1 Suppose we flip a coin three times, thereby forming a sequence of heads and tails. Form a random vector by mapping each outcome in the sequence to 0 if a head occurs or to 1 if a tail occurs.

    (a) How many realizations of the vector may be generated? List them.

    (b) Are the realizations independent of one another?

6.2 Let $X = [X_1, X_2, X_3]^T$ represent a three-dimensional vector of random variables that is uniformly distributed over a cubical region

$$f_X(x) = \begin{cases} c, & |x_1| \le 1, |x_2| \le 1, |x_3| \le 1, \\ 0, & \text{otherwise.} \end{cases}$$

    (a) Find the constant c.

    (b) Find the marginal PDF for a subset of two of the three random variables. For example, find $f_{X_1, X_2}(x_1, x_2)$.

    (c) Find the marginal PDF for one of the three random variables. That is, find $f_{X_1}(x_1)$.

    (d) Find the conditional PDFs $f_{X_1|X_2, X_3}(x_1|x_2, x_3)$ and $f_{X_1, X_2|X_3}(x_1, x_2|x_3)$.

    (e) Are the X_i independent?

6.3 Suppose a point in two-dimensional Cartesian space, (X, Y), is equally likely to fall anywhere on the semicircle defined by $X^2 + Y^2 = 1$ and $Y \ge 0$. Find the PDF of Y, $f_Y(y)$.

6.4 Suppose a point in three-dimensional Cartesian space, (X, Y, Z), is equally likely to fall anywhere on the surface of the hemisphere defined by $X^2 + Y^2 + Z^2 = 1$ and $Z \ge 0$.

    (a) Find the PDF of Z, $f_Z(z)$.

    (b) Find the joint PDF of X and Y, $f_{X, Y}(x, y)$.

6.5 Suppose N_1 is a discrete random variable equally likely to take on any integer in the set $\{1, 2, 3\}$. Given that $N_1 = n_1$, the random variable N_2 is equally likely to take on any integer in the set $\{1, 2, ..., n_1\}$. Finally, given that $N_2 = n_2$, the random variable N_3 is equally likely to take on any integer in the set $\{1, 2, ..., n_2\}$.

(a)   Find the two-dimensional joint PMF, $P_{N_1, N_2}(n_1, n_2)$.

(b)   Find the three-dimensional joint PDF, $P_{N_1, N_2, N_3}(n_1, n_2, n_3)$.

(c)   Find the marginal PDFs, $P_{N_2}(n_2)$ and $P_{N_3}(n_3)$.

(d)   What are the chances that none of the three random variables are equal to 1?

6.6   Let $X = [X_1, X_2, X_3]^T$ represent a three-dimensional vector of random variables that is uniformly distributed over the unit sphere. That is,

$$f_X(x) = \begin{cases} c, & \|x\| \le 1, \\ 0, & \|x\| > 1. \end{cases}$$

(a)   Find the constant c.

(b)   Find the marginal PDF for a subset of two of the three random variables. For example, find $f_{X_1, X_2}(x_1, x_2)$.

(c)   Find the marginal PDF for one of the three random variables. That is, find $f_{X_1}(x_1)$.

(d)   Find the conditional PDFs $f_{X_1 | X_2, X_3}(x_1 | x_2, x_3)$ and $f_{X_1, X_2 | X_3}(x_1, x_2 | x_3)$.

Extra: Can you extend this problem to N-dimensions?

6.7   Let $X = [X_1, X_2, ..., X_N]^T$ represent an N-dimensional vector of random variables that is uniformly distributed over the region $x_1 + x_2 + \cdots + x_N \le 1$, $x_i \ge 0$, $i = 1, 2, ..., N$. That is

$$f_X(x) = \begin{cases} c, & \sum_{i=1}^{N} x_i \le 1, x_i \ge 0, \\ 0, & \text{otherwise.} \end{cases}$$

(a)   Find the constant c.

(b)   Find the marginal PDF for a subset of M of the N random variables.

(c)   Are the X_i independent? Are the X_i identically distributed?

Section 6.2: Expectations Involving Multiple Random Variables

6.8 Consider a vector of N random variables, $X = [X_1, X_2, ..., X_N]^T$. Suppose we form a new random variable Z by performing a weighted average of the components of X. That is,

$$Z = \sum_{i=1}^{N} b_i X_i,$$

where $b_i \geq 0$ and $\sum_{i=1}^{N} b_i = 1$. Find the values of the constants b_i such that the variance of Z is minimized.

6.9 A random vector is generated by rolling a die and observing the outcome. The components of the random vector are determined by successive rolls of the die. If the die is rolled two times:

(a) List the possible realizations of the random vector;
(b) Determine the probability of each realization;
(c) Determine the mean vector;
(d) Determine the covariance matrix.

6.10 Repeat parts (c) and (d) of Exercise 6.9 if a three-element vector is formed from three rolls of a die.

6.11 Let $X = [X_1, X_2, ..., X_n]^T$ be a vector of random variables where each component is independent of the others and uniformly distributed over the interval $(0, 1)$.

(a) Find the mean vector, $E[X]$.
(b) Find the correlation matrix, $R_{XX} = E[XX^T]$.
(c) Find the covariance matrix, $C_{XX} = E[(X - E[X])(X - E[X])^T]$.

6.12 A vector random variable, X, has a mean vector and correlation matrix given by

$$E[X] = \begin{bmatrix} 2 \\ -1 \\ 1 \end{bmatrix} \text{ and } R_{XX} = \begin{bmatrix} 13 & 2 & 3 \\ 2 & 10 & 3 \\ 3 & 3 & 10 \end{bmatrix}.$$

A new random vector is formed according to $Y = AX$ where the matrix A is given by

$$A = \begin{bmatrix} 1 & 0 & -1 \\ 0 & -1 & 1 \\ -1 & 1 & 0 \end{bmatrix}.$$

Find the mean vector, correlation matrix and covariance matrix of Y.

6.13 A vector random variable, X, has a covariance matrix and a correlation matrix given by

$$C_{XX} = \begin{bmatrix} 5 & -1 & 2 \\ -1 & 5 & -2 \\ 2 & -2 & 8 \end{bmatrix} \text{ and } R_{XX} = \begin{bmatrix} 6 & 1 & 1 \\ 1 & 9 & -4 \\ 1 & -4 & 9 \end{bmatrix}.$$

Find the mean vector, $E[X]$.

6.14 Three zero-mean random variables $[X, Y, Z]$ have a covariance matrix given by

$$C = E\left[\begin{bmatrix} X \\ Y \\ Z \end{bmatrix} \begin{bmatrix} X & Y & Z \end{bmatrix} \right] = \begin{bmatrix} 1 & 1/2 & 1/4 \\ 1/2 & 1 & 1/2 \\ 1/4 & 1/2 & 1 \end{bmatrix}.$$

Find the value of the constants a and b so that the variance of $Z - aX - bY$ is minimized.

Section 6.3: Gaussian Random Variables in Muliple Dimensions

6.15 Let $\begin{bmatrix} X \\ Y \\ Z \end{bmatrix}$ be a zero-mean Gaussian random vector with covariance matrix,

$$C = \begin{bmatrix} 2 & -1 & 1 \\ -1 & 4 & 1 \\ -1 & 1 & 3 \end{bmatrix}.$$

Write out the joint PDF, $f_{X, Y, Z}(x, y, z)$.

6.16 Let X_1, X_2, and X_3 be a set of three zero-mean Gaussian random variables with a covariance matrix of the form

$$C = \sigma^2 \begin{bmatrix} 1 & \rho & \rho \\ \rho & 1 & \rho \\ \rho & \rho & 1 \end{bmatrix}.$$

Find the following expected values:

(a) $E[X_1|X_2=x_2, X_3=x_3]$,

(b) $E[X_1X_2|X_3=x_3]$,

(c) $E[X_1X_2X_3]$.

6.17 Define the N-dimensional characteristic function for a random vector, $X = [X_1, X_2, ..., X_N]^T$, according to $\Phi_X(\Omega) = E[e^{j\Omega^T X}]$ where $\Omega = [\omega_1, \omega_2, ..., \omega_N]^T$. Show that the N-dimensional characteristic function for a zero-mean Gaussian random vector is given by

$$\Phi_X\Omega = \exp\left(-\frac{\Omega^T C_{XX}\Omega}{2}\right).$$

6.18 For any four zero-mean Gaussian random variables X_1, X_2, X_3, and X_4, show that

$$E[X_1X_2X_3X_4] = E[X_1X_2]E[X_3X_4] + E[X_1X_3]E[X_2X_4] + E[X_1X_4]E[X_2X_3].$$

Hint: You might want to use the result of the previous exercise.

Note: This useful result is referred to as the Gaussian moment-factoring theorem and allows us to decompose fourth-order moments into a series of simpler second-order moments.

Section 6.4: Transformations Involving Multiple Random Variables

6.19 Let X be a two-element zero-mean random vector. Suppose we construct a new random vector Y according to a linear transformation, $Y = TX$. Find the transformation matrix, T, such that Y has a covariance matrix of

$$C_{YY} = \begin{bmatrix} 5 & 1 \\ 1 & 2 \end{bmatrix}.$$

For this problem, assume that the covariance matrix of the vector X is an identity matrix.

6.20 A three-dimensional vector random variable, X, has a covariance matrix of

$$C = \begin{bmatrix} 3 & 1 & -1 \\ 1 & 5 & -1 \\ -1 & -1 & 3 \end{bmatrix}.$$

Find a transformation matrix A such that the new random variables $Y = AX$ will be uncorrelated.

6.21 Suppose X_m, $m = 1, 2, \ldots, n$ are a sequence of independent and exponentially distributed random variables with

$$f_{X_m}(x) = \frac{1}{\mu}\exp(-x/\mu)u(x).$$

Assuming that n is an odd number ($n = 2k - 1$ for some integer k):

(a) Find the PDF of the median of the sequence.

(b) Find the expected value of the median of the sequence. Is the median an unbiased estimate of the mean of the underlying exponential distribution?

(c) Find the variance of the median of the sequence.

6.22 Show that the derivative of

$$F_{Y_m}(y) = \sum_{k=m}^{N} \binom{N}{k}(F_X(y))^k(1 - F_X(y))^{N-k}$$

reduces to the form

$$f_{Y_m}(y) = \frac{N!}{(m-1)!(N-m)!}f_X(y)(F_X(y))^{m-1}(1 - F_X(y))^{N-m}.$$

6.23 In this problem, we formulate an alternative derivation of Equation (6.58) which gives the PDF of the order statistic, Y_m, which is the mth largest of a sequence of N random variables, $X_1, X_2, \ldots, X_N$. Start by writing $f_{Y_m}(y)dy = \Pr(y < Y_m < y + dy)$. Then note that

$\Pr(y < Y_m < y + dy) = \Pr(\{m - 1 \text{ of the } Xs \text{ are less than } y\} \cap$

$\cap \{1 \ X \text{ is between } y \text{ and } y + dy\} \cap \{n - m \text{ of the } Xs \text{ are greater than } y\})$.

Find the probability of the above event and by doing so, prove that the PDF of Y_m is as given by Equation (6.58).

6.24 Suppose X, Y, and Z are independent, zero-mean, unit-variance Gaussian random variables.

(a) Using the techniques outlined in Section 6.4.2, find the characteristic function of $W = XY + XZ + YZ$.

(b) From the characteristic function found in part (a), find the mean and variance of W.

(c) Confirm your answer in part (b) by finding the mean and variance of W directly. In this part, you may want to use the result of the Gaussian moment factoring theorem developed in Exercise 6.18.

6.25 Find the PDF of $Z = X_1X_2 + X_3X_4 + X_5X_6 + X_7X_8$ assuming that all of the X_i are independent zero-mean, unit-variance, Gaussian random variables. *Hint:* Use the result of Special Case #2 in Section 6.4.2.1 to help.

6.26 Let $X_1, X_2, ..., X_5$ be a sequence of five independent discrete random variables, each with a distribution described by:

$$\Pr(X_k=m) = \begin{cases} 0.8, & m = 0, \\ 0.1, & m = \pm 1, \\ 0, & |m| > 1. \end{cases}$$

(a) Find the probability mass function of the median (third largest) of these five samples.

(b) For this random variable, is the median an unbiased estimate of the mean? That is, does the expected value of the median equal the mean of the X_i? Prove your answer.

(c) Find the variance of the median.

6.27 A set of n random variables, $X_1, X_2, X_3, ..., X_n$, are independent and each uniformly distributed over $(0, 1)$.

(a) Find the probability density function of $Z = \max(X_1, X_2, ..., X_n)$.

(b) With Z defined as in part (a) above, let A be the event $\{X_1 = 1/2\}$ and find $f_{Z|A}(z)$. That is, find the conditional PDF of Z given $\{X_1 = 1/2\}$.

Section 6.5: Estimation and Detection

6.28 Let the random variables U and V be as described in Exercise 6.40.

(a) Find the MAP estimator of U given the observation $V = v$.

(b) Find the ML estimator of U given the observation $V = v$.

(c) Find the LMMSE estimator of U given the observation $V = v$.

(d) Find the MSE of each estimator in (a), (b), and (c).

6.29 Repeat Exercise 6.28 assuming we wish to find an estimate of V given the observation $U = u$.

6.30 Let the random variables U, V, and W be as described in Exercise 6.40.

(a) Find the MAP estimator of U given the observation $\{V = v, W=w\}$.

(b) Find the ML estimator of U given the observation $\{V = v, W=w\}$.

    (c)   Find the LMMSE estimator of U given the observation $\{V = v, W=w\}$.

    (d)   Find the MSE of each estimator in (a), (b), and (c).

6.31 Let X be the random vector described in Exercise 6.12.

    (a)   Find the LMMSE estimator of X_1 given observation of $\{X_2=x_2, X_3=x_3\}$.

    (b)   Find the MSE of the estimator in part (a).

    (c)   Explain why we cannot find the MAP or ML estimators in this case.

6.32 Let X, Y, and Z be the random vectors described in Exercise 6.35.

    (a)   Find the LMMSE estimator of X given $\{Y=y, Z=z\}$.

    (b)   Find the LMMSE estimator of Y given $\{X=x, Z=z\}$.

    (c)   Find the LMMSE estimator of Z given $\{X=x, Y=y\}$.

6.33 Repeat Exercise 6.32 using ML estimators.

6.34 Repeat Exercise 6.32 using MAP estimators.

Miscellaneous Exercises

6.35 Suppose X, Y, and Z are jointly Gaussian random variables with mean vector and covariance matrix given by

$$E\begin{bmatrix} X \\ Y \\ Z \end{bmatrix} = \begin{bmatrix} 3 \\ -1 \\ -2 \end{bmatrix} \text{ and } C = E\left[\begin{bmatrix} X \\ Y \\ Z \end{bmatrix} [X\ Y\ Z] \right] = \begin{bmatrix} 4 & -1 & 1 \\ -1 & 4 & 1 \\ 1 & 1 & 4 \end{bmatrix}.$$

    Find $\Pr(X > 2Y - 3X)$.

6.36 The traffic managers of toll roads and toll bridges need specific information to properly staff the toll booths so that the queues are minimized (i.e., the waiting time is minimized).

    (a)   Assume that there is one toll booth on a busy interstate highway and that the number of cars per minute approaching the toll booth follows a Poisson PMF with $\alpha = 10$. The traffic manager wants you to determine the probability that exactly 11 cars approach this toll booth in the minute from noon to 1 min past noon.

    (b)   Now assume that there are N toll booths at a toll plaza and that the number of cars per minute approaching the plaza follows the same Poisson PMF with $\alpha = 30$. The traffic manager wants you to calculate the minimum number of toll booths that need to be staffed if the probability that more than five cars approach any toll booth in one minute is no more than 0.05. For this part, assume the traffic approaching the plaza divides evenly among the N booths such that the traffic approaching each booth is independent and follows a Poisson PMF with $\alpha = 30/N$.

6.37 A sequence of zero mean unit variance independent random variables, X_n, $n = 0, 1, 2, ..., N-1$ are input to a filter that produces an output sequence according to $Y_n = (X_n + X_{n-1})/2$, for $n = 0, 1, 2, ..., N-1$. For initialization purposes, X_{-1} is taken to be zero.

(a) Find the covariance (correlation) matrix of the Y_n.
(b) Now let the variance of the X_n be σ_X^2. Find the covariance (correlation) matrix of the Y_n.

6.38 Repeat Exercise 6.37 with the filter is changed to $Y_n = X_n - X_{n-1}$.

6.39 Suppose a zero mean random sequence X_n has correlation parameters given by $r_k = E[X_n X_{n+k}] = c^{|k|}$. An estimate of a future value of X_n is $\hat{X}_n = a_1 X_{n-1} + a_2 X_{n-2}$ which is a special case of Equation (6.67).

(a) Use Equation (6.72) to find the a_i.
(b) What is the mean squared error, $E[(X_n - \hat{X}_n)^2]$?

6.40 Let X, Y, and Z be a set of independent, zero-mean, unit-variance, Gaussian random variables. Form a new set of random variables according to

$$U = X,$$

$$V = X + Y,$$

$$W = X + Y + Z.$$

(a) Find the three one-dimensional marginal PDFs, $f_U(u)$, $f_V(v)$, and $f_W(w)$.
(b) Find the three two-dimensional joint PDFs, $f_{U,V}(u, v)$, $f_{V,W}(v, w)$, and $f_{U,W}(u, w)$.
(c) Find the three-dimensional joint PDF of U, V, and W, $f_{U,V,W}(u, v, w)$.

6.41 A radio astronomer is attempting to measure radio frequency (RF) emmisions from a certain star. However, these emissions are corrupted by a variety of independent noise sources including thermal noise in his receiving equipment, interference from local RF sources, and galactic noise. The astronomer has studied each of these sources and found them all to be well modeled as zero-mean Gaussian random variables with the following standard deviations:

T, thermal noise, $\sigma_T = 5\mu V$,

I, interference, $\sigma_I = 2\mu V$,

G, galactic noise, $\sigma_G = 1\mu V$.

Let N represent the combined effect of all three noise sources, that is $N = T+I+G$. Suppose the desired emissions from the star are received at a level of 10 μV. What is the probability that the combined noise is larger than the desired emissions? *Hint:* First find the relevant PDF of N and then find the probability that N is bigger than 10 μV.

6.42 A certain system we have designed needs to be powered by a 24-V dc supply. Available to us in our lab are each of the following types of batteries whose statistical characteristics (and quantities available) are as shown in the table. The system can tolerate variations in the supply voltage that are within 1 V of the designed 24 V level. Anything outside that range will damage the system. Your job is to come up with a combination of the batteries that will lead to the highest probability that the combined supply voltage will remain within the range 23-25 V. Assume the voltage of each battery is a Gaussian random variable and is independent of the others and the means and standard deviations of each battery are as shown in the table. State how you would form some combination of the batteries available to produce the desired supply voltage. Also specify what is the probability that your combination of batteries falls within the desired range.

Battery characteristics

"Nominal" Voltage (V)	Number Available (V)	Average Voltage (V)	Standard Deviation (V)
24	1	23.5	1
12	2	11.8	0.5
6	4	5.9	0.3
1.5	12	1.45	0.2
0.5	30	0.475	0.1

6.43 Let X_1, X_2 be a pair of independent random variables with the same exponential PDF,

$$f_{X_i}(x) = \exp(-x)u(x), \quad i = 1, 2.$$

Define Y_1, Y_2 to be the order statistics associated with the X_i. That is,

$$Y_1 = \min(X_1, X_2) \text{ and } Y_2 = \max(X_1, X_2).$$

(a) Find the marginal PDFs of Y_1 and Y_2, $f_{Y_1}(y_1)$ and $f_{Y_2}(y_2)$.
(b) Find the joint PDFs of Y_1 and Y_2, $f_{Y_1, Y_2}(y_1, y_2)$.
(c) Find the MAP estimator of Y_2 given $Y_1 = y_1$.
(d) Find the ML estimator of Y_2 given $Y_1 = y_1$.
(e) Find the LMMSE estimator of Y_2 given $Y_1 = y_1$.
(f) Find the MSE of each estimator in (c), (d), and (e).

MATLAB Exercises

6.44 Three jointly Gaussian random variables $[X, Y, Z]^T$ have a mean vector $\mu = [1, 0, -1]^T$ and covariance matrix

$$C = \begin{bmatrix} 4 & 2 & -1 \\ 2 & 4 & 2 \\ -1 & 2 & 4 \end{bmatrix}.$$

Use MATLAB to help you find the form of the three-dimensional joint PDF, $f_{X, Y, Z}(x, y, z)$.

6.45 For each of the following matrices, determine if the matrix is a valid correlation matrix. In some cases, you may want to use MATLAB to check if the matrix is positive definite.

(a) $C_a = \begin{bmatrix} 3 & -2 & 1 \\ 2 & 6 & 0 \\ -1 & 0 & 2 \end{bmatrix}$,

(b) $C_b = \begin{bmatrix} 3 & -2 & 3 \\ -2 & 6 & 0 \\ 3 & 0 & 2 \end{bmatrix}$,

(c) $C_c = \begin{bmatrix} 3 & -2 & 1 \\ -2 & 6 & 0 \\ 1 & 0 & 2 \end{bmatrix}$,

(d) $C_d = \begin{bmatrix} 1 & -\frac{1}{2} & \frac{1}{4} & -\frac{1}{8} \\ -\frac{1}{2} & 1 & -\frac{1}{2} & \frac{1}{4} \\ \frac{1}{4} & -\frac{1}{2} & 1 & -\frac{1}{2} \\ -\frac{1}{8} & \frac{1}{4} & -\frac{1}{2} & 1 \end{bmatrix}$,

(e) $C_e = \begin{bmatrix} 11 & -3 & 7 & 5 \\ -3 & 11 & 5 & 7 \\ 7 & 5 & 11 & -3 \\ 5 & 7 & -3 & 11 \end{bmatrix}$,

(f) $C_f = \begin{bmatrix} 5 & 1 & 3 & -1 \\ 1 & 5 & -1 & 3 \\ 3 & -1 & 5 & 1 \\ -1 & 3 & 1 & 5 \end{bmatrix}$.

6.46 For each matrix in Exercise 6.45 that is a valid correlation matrix, find a transformation matrix that will transform a set of independent, zero-mean, unit variance random variables into ones with the specified correlation matrix.

6.47 Given a random sequence $X = [X_1, X_2, X_3, X_4]$ with a covariance matrix

$$C_X = \begin{bmatrix} 1 & 0.3 & 0.09 & 0.027 \\ 0.3 & 1 & 0.3 & 0.027 \\ 0.09 & 0.3 & 1 & 0.3 \\ 0.0027 & 0.09 & 0.3 & 1 \end{bmatrix},$$

find a linear transformation that will produce a random sequence $Y = [Y_1, Y_2, Y_3, Y_4]$ with a covariance matrix

$$C_Y = \begin{bmatrix} 1 & 0.1 & 0.2 & 0.3 \\ 0.1 & 1 & 0.1 & 0.2 \\ 0.2 & 0.1 & 1 & 0.1 \\ 0.3 & 0.2 & 0.1 & 1 \end{bmatrix}.$$

Random Sums and Sequences

This chapter forms a bridge between the study of random variables in the previous chapters and the study of random processes to follow. A random process is simply a random function of time. If time is discrete, then such a random function could be viewed as a sequence of random variables. Even when time is continuous, we often choose to sample waveforms (whether they are deterministic or random) in order to work with discrete time sequences rather than continuous time waveforms. Thus, sequences of random variables will naturally occur in the study of random processes. In this chapter, we will develop some basic results regarding both finite and infinite sequences of random variables and random series.

7.1 Independent and Identically Distributed Random Variables

In many applications, we are able to observe an experiment repeatedly. Each new observation can occur with an independent realization of whatever random phenomena control the experiment. This sort of situation gives rise to *independent and identically distributed* (IID or i.i.d.) random variables.

Definition 7.1: A sequence of random variables X_1, X_2, ..., X_n is IID if

$$F_{X_i}(x) = F_X(x) \quad \forall i = 1, 2, ..., n \text{ (identically distributed)}, \tag{7.1}$$

and

$$F_{X_1, X_2, ..., X_n}(x_1, x_2, ..., x_n) = \prod_{i=1}^{n} F_{X_i}(x_i) \text{ (independent)}. \tag{7.2}$$

For continuous random variables, the CDFs can be replaced with PDFs in Equations (7.1) and (7.2), while for discrete random variables, the CDFs can be replaced by PMFs.

Suppose, for example, we wish to measure the voltage produced by a certain sensor. The sensor might be measuring the relative humidity outside. Our sensor converts the humidity

Probability and Random Processes. DOI: 10.116/B978-0-12-386981-4.00007-3

to a voltage level which we can then easily measure. However, as with any measuring equipment, the voltage we measure is random due to noise generated in the sensor as well as in the measuring equipment. Suppose the voltage we measure is represented by a random variable X given by $X = v(h) + N$, where $v(h)$ is the true voltage that should be presented by the sensor when the humidity is h, and N is the noise in the measurement. Assuming that the noise is zero-mean, then $E[X] = v(h)$. That is, on the average, the measurement will be equal to the true voltage $v(h)$. Furthermore, if the variance of the noise is sufficiently small, then the measurement will tend to be close to the true value we are trying to measure. But what if the variance is not small? Then the noise will tend to distort our measurement making our system unreliable. In such a case, we might be able to improve our measurement system by taking several measurements. This will allow us to "average out" the effects of the noise.

Suppose we have the ability to make several measurements and observe a sequence of measurements $X_1, X_2, ..., X_n$. It might be reasonable to expect that the noise that corrupts a given measurement has the same distribution each time (and hence the X_i are identically distributed) and is independent of the noise in any other measurement (so that the X_i are independent). Then the n measurements form a sequence of IID random variables. A fundamental question is then: How do we process an IID sequence to extract the desired information from it? In the preceding case, the parameter of interest, $v(h)$, happens to be the mean of the distribution of the X_i. This turns out to be a fairly common problem and so we start by examining in some detail the problem of estimating the mean from a sequence of IID random variables.

7.1.1 Estimating the Mean of IID Random Variables

Suppose the X_i have some common PDF, $f_X(x)$, which has some mean value, μ_X. Given a set of IID observations, we wish to form some function,

$$\hat{\mu} = g(X_1, X_2, ..., X_n), \tag{7.3}$$

which will serve as an estimate of the mean. But what function should we choose? Even more fundamentally, what criterion should we use to select a function?

There are many criteria that are commonly used. To start with we would like the average value of the estimate of the mean to be equal to the true mean. That is, we want $E[\hat{\mu}] = \mu_X$. If this criterion is met, we say that $\hat{\mu}$ is an *unbiased* estimate of μ_X. Given that the estimate is unbiased, we would also like the error in the estimate to be as small as possible. Define the estimation error to be $\varepsilon = \hat{\mu} - \mu_X$. A common criterion is to choose the estimator which minimizes the second moment of the error (mean-square error), $E[\varepsilon^2] = E[(\hat{\mu} - \mu_X)^2]$. If this criterion is met, we say that $\hat{\mu}$ is an *efficient* estimator of μ_X. To start with a relatively simple approach, suppose we desire to find a linear estimator. That is, we will limit ourselves to estimators of the form

$$\hat{\mu} = a_1 X_1 + a_2 X_2 + \ldots + a_n X_n = \sum_{i=1}^{n} a_i X_i. \tag{7.4}$$

Then, we seek to find the constants, $a_1, a_2, \ldots, a_n$, such that the estimator (1) is unbiased and (2) minimizes the mean-square error. Such an estimator is referred to as the *best linear unbiased estimator* (BLUE).

To simplify notation in this problem, we write $X = [X_1, X_2, \ldots, X_n]^T$ and $a = [a_1, a_2, \ldots, a_n]^T$. The linear estimator $\hat{\mu}$ can then be written as $\hat{\mu} = a^T X$. First, for the estimator to be unbiased, we need

$$\mu_X = E[\hat{\mu}] = E[a^T X] = a^T E[X]. \tag{7.5}$$

Since the X_i are all IID, they all have means equal to μ_X. Hence, the mean vector for X is just $\mu_X \mathbf{1}_n$ where $\mathbf{1}_n$ is an n-element column vector of all ones. The linear estimator will then be unbiased if

$$\sum_{i=1}^{n} a_i = a^T \mathbf{1}_n = 1. \tag{7.6}$$

The mean square error is given by

$$E[\varepsilon^2] = E[(a^T X - \mu_X)^2] = a^T E[XX^T] a - 2\mu_X a^T E[X] + \mu_X^2$$

$$= a^T R a - 2\mu_X^2 a^T \mathbf{1}_n + \mu_X^2 \tag{7.7}$$

In this expression, $R = E[XX^T]$ is the correlation matrix for the vector X. Using the constraint of (7.6), the mean square error simplifies to

$$E[\varepsilon^2] = a^T R a - \mu_X^2. \tag{7.8}$$

The problem then reduces to minimizing the function $a^T R a$ subject to the constraint $a^T \mathbf{1}_n = 1$.

To solve this multidimensional optimization problem, we use standard Lagrange multiplier techniques. Form the auxiliary function

$$h(\lambda) = a^T R a + \lambda a^T \mathbf{1}_n. \tag{7.9}$$

Then solve the equation $\nabla h = 0$. It is not difficult to show that the gradient of the function h works out to be $\nabla h = 2Ra + \lambda \mathbf{1}_n$. Therefore, the optimum vector a will satisfy

$$Ra = \left(-\frac{\lambda}{2}\right) \mathbf{1}_n. \tag{7.10}$$

Solving for a in this equation and then applying the constraint $a^T 1_n = 1$ results in the solution

$$a = \frac{R^{-1} 1_n}{1_n^T R^{-1} 1_n}. \tag{7.11}$$

Due to the fact that the X_i are IID, the form of the correlation matrix can easily be shown to be

$$R = \mu_X^2 1_n 1_n^T + \sigma_X^2 I, \tag{7.12}$$

where I is an identity matrix and σ_X^2 is the variance of the IID random variables. It can be shown using the matrix inversion lemma[1] that the inverse of this correlation matrix is

$$R^{-1} = \sigma_X^{-2} \left[I - \frac{\mu_X^2 / \sigma_X^2}{1 + n \mu_X^2 / \sigma_X^2} 1_n 1_n^T \right]. \tag{7.13}$$

From here, it is easy to demonstrate that $R^{-1} 1_n$ is proportional to 1_n, and therefore the resulting vector of optimum coefficients is

$$a = \frac{1}{n} 1_n. \tag{7.14}$$

In terms of the estimator $\hat{\mu}$, the best linear unbiased estimator of the mean of an IID sequence is

$$\hat{\mu} = \frac{1}{n} 1_n^T X = \frac{1}{n} \sum_{i=1}^{n} X_i. \tag{7.15}$$

This estimator is commonly referred to as the *sample mean*. The preceding derivation proves Theorem 7.1 which follows.

> **Theorem 7.1:** Given a sequence of IID random variables $X_1, X_2, \ldots, X_n$, the sample mean is BLUE.

Another possible approach to estimating various parameters of a distribution is to use the maximum likelihood (ML) approach introduced in Chapter 6 (Section 6.5.2). In the ML

---

[1] The matrix inversion lemma gives a formula to find the inverse of a rank one update of another matrix whose inverse is known. In particular, suppose $A = B + xx^T$ where x is a column vector and the inverse of B is known. Then,.

$$A^{-1} = B^{-1} - \frac{B^{-1} xx^T B^{-1}}{1 + x^T B^{-1} x}.$$

approach, the distribution parameters are chosen to maximize the probability of the observed sample values occurring. Suppose, as in the preceding discussion, we are interested in estimating the mean of a distribution. Given a set of observations, $X_1 = x_1$, $X_2 = x_2$, ..., $X_n = x_n$, the ML estimate of μ_X would be the value of μ_X which maximizes $f_X(x)$. A few examples will clarify this concept.

■ Example 7.1:

Suppose the X_i are jointly Gaussian so that

$$f_X(x) = \frac{1}{(2\pi\sigma^2)^{n/2}}\exp\left(-\frac{1}{2\sigma^2}\sum_{i=1}^{n}(x_i - \mu)^2\right).$$

The value of μ which maximizes this expression will minimize

$$\sum_{i=1}^{n}(x_i - \mu)^2 .$$

Differentiating and setting equal to zero gives the equation

$$-2\sum_{i=1}^{n}(x_i - \mu) = 0 .$$

The solution to this equation works out to be

$$\hat{\mu}_{ML} = \frac{1}{n}\sum_{i=1}^{n}x_i .$$

Hence, the sample mean is also the ML estimate of the mean when the random variables follow a Gaussian distribution.

■

■ Example 7.2:

Now suppose the random variables have an exponential distribution,

$$f_X(x) = \prod_{i=1}^{n}\frac{1}{\mu}\exp\left(-\frac{x_i}{\mu}\right)u(x_i) = \frac{1}{\mu^n}\exp\left(-\frac{1}{\mu}\sum_{i=1}^{n}x_i\right)\prod_{i=1}^{n}u(x_i).$$

Differentiating with respect to μ and setting equal to zero results in

$$-\frac{n}{\mu^{n+1}}\exp\left(-\frac{1}{\mu}\sum_{i=1}^{n}x_i\right) + \frac{1}{\mu^2}\left(\sum_{i=1}^{n}x_i\right)\frac{1}{\mu^n}\exp\left(-\frac{1}{\mu}\sum_{i=1}^{n}x_i\right) = 0.$$

(Continued)

Solving for μ results in

$$\hat{\mu}_{\mathrm{ML}} = \frac{1}{n}\sum_{i=1}^{n} x_i.$$

Once again, the sample mean is the maximum likelihood estimate of the mean of the distribution.

∎

Since the sample mean occurs so frequently, it is beneficial to study this estimator in a little more detail. First, we note that the sample mean is itself a random variable since it is a function of the n IID random variables. We have already seen that the sample mean is an unbiased estimate of the true mean; that is, $E[\hat{\mu}] = \mu_X$. It is instructive to also look at the variance of this random variable.

$$\mathrm{Var}(\hat{\mu}) = E\left[\left(\left(\frac{1}{n}\sum_{i=1}^{n} X_i\right) - \mu_X\right)^2\right] = E\left[\left(\frac{1}{n}\sum_{i=1}^{n} (X_i - \mu_X)\right)^2\right]$$

$$= \frac{1}{n^2}\sum_{i=1}^{n}\sum_{j=1}^{n} E[(X_i - \mu_X)(X_j - \mu_X)]$$

$$= \frac{1}{n^2}\sum_{i=1}^{n}\sum_{j=1}^{n} \mathrm{Cov}(X_i, X_j). \tag{7.16}$$

All terms in the double series in the previous equation are zero except for the ones where $i = j$ since X_i and X_j are uncorrelated for all $i \neq j$. Therefore, the variance of the sample mean is

$$\mathrm{Var}(\hat{\mu}) = \frac{1}{n^2}\sum_{i=1}^{n} \mathrm{Var}(X_i) = \frac{1}{n^2}\sum_{i=1}^{n} \sigma_X^2 = \frac{\sigma_X^2}{n}. \tag{7.17}$$

This means that if we use n samples to estimate the mean, the variance of the resulting estimate is reduced by a factor of n relative to what the variance would be if we used only one sample.

Consider what happens in the limit as $n \to \infty$. As long as the variance of each of the samples is finite, the variance of the sample mean approaches zero. Of course, we never have an infinite number of samples in practice, but this does mean that the sample mean can achieve any level of precision (i.e., arbitrarily small variance) if a sufficient number of samples is taken. We will study this limiting behavior in more detail in a later section. For now, we turn our attention to estimating other parameters of a distribution.

7.1.2 Estimating the Variance of IID Random Variables

Now that we have a handle on how to estimate the mean of IID random variables, suppose we would like to also estimate the variance (or equivalently, the standard deviation). Since the variance is not a linear function of the random variables, it would not make much sense to try to form a linear estimator. That is, to talk about an estimator of the variance being BLUE is meaningless. Hence, we take the ML approach here. As with the problem of estimating the mean, we seek the value of the variance which maximizes the joint PDF of the IID random variables evaluated at their observed values.

■ **Example 7.3:**

Suppose that the random variables are jointly Gaussian so that

$$f_X(x) = \frac{1}{(2\pi\sigma^2)^{n/2}} \exp\left(-\frac{1}{2\sigma^2} \sum_{i=1}^{n} (x_i - \mu)^2\right).$$

Differentiating the joint PDF with respect to σ results in

$$\frac{d}{d\sigma} f_X(x) = \left(-\frac{n}{\sigma} + \frac{1}{\sigma^3} \sum_{i=1}^{n} (x_i - \mu)^2\right) \frac{1}{(2\pi\sigma^2)^{n/2}} \exp\left(-\frac{1}{2\sigma^2} \sum_{i=1}^{n} (x_i - \mu)^2\right).$$

Setting this expression equal to zero and solving results in

$$\hat{\sigma}_{ML}^2 = \frac{1}{n} \sum_{i=1}^{n} (x_i - \mu)^2.$$

■

The result of Example 7.3 seems to make sense. The only problem with this estimate is that it requires knowledge of the mean in order to form the estimate. What if we do not know the mean? One obvious approach would be to replace the true mean in the previous result with the sample mean. That is, one could estimate the variance of an IID sequence using

$$\hat{s}^2 = \frac{1}{n} \sum_{i=1}^{n} (x_i - \hat{\mu})^2, \qquad \hat{\mu} = \frac{1}{n} \sum_{i=1}^{n} x_i. \tag{7.18}$$

This approach, however, can lead to problems. It is left as an exercise for the reader to show that this estimator is biased; that is, in this case, $E[\hat{s}^2] \neq \sigma^2$. To overcome this problem, it is common to adjust the previous form. The following estimator turns out to be an unbiased estimate of the variance:

$$\hat{s}^2 = \frac{1}{n-1}\sum_{i=1}^{n}(x_i - \hat{\mu})^2 .$$

(7.19)

This is known as the *sample variance* and is the most commonly used estimate for the variance of IID random variables. In the previous expression, $\hat{\mu}$ is the usual sample mean.

In summary, given a set of IID random variables, the variance of the distribution is estimated according to:

$$\hat{s}^2 = \begin{cases} \dfrac{1}{n}\sum_{i=1}^{n}(x_i - \mu)^2, & \text{if } \mu \text{ is known,} \\[4mm] \dfrac{1}{n-1}\sum_{i=1}^{n}(x_i - \hat{\mu})^2, & \text{if } \mu \text{ is unknown.} \end{cases}$$

(7.20)

■ Example 7.4:

Suppose we form a random variable Z according to $Z = \sqrt{X^2 + Y^2}$ where X and Y are independent Gaussian random variables with means of μ and variances of σ^2. In this example, we will estimate the mean and variance of Z using the sample mean and sample variance of a large number of MATLAB-generated realizations of the random variable Z. The MATLAB code to accomplish this follows. Upon running this code, we obtained a sample mean of $\hat{\mu} = 5.6336$ and a sample variance of $\hat{s}^2 = 8.5029$. Note that the true mean and variance of the Rician random variable Z can be found analytically (with some effort). For this example, the PDF of the random variable Z is found to take on a Rician form

$$f_Z(z) = \frac{z}{\sigma^2}\exp\left(-\frac{z^2 + 2\mu^2}{2\sigma^2}\right)I_0\left(\frac{\sqrt{2}\mu z}{\sigma^2}\right)u(z) .$$

Using the expressions given in Appendix D (see equations (D.52) and (D.53)) for the mean and variance of a Rician random variable, it is determined that the true mean and variance should be

$$\mu_Z = \sqrt{\frac{\pi\sigma^2}{2}}\exp\left(-\frac{\mu^2}{2\sigma^2}\right)\left[\left(1 + \frac{\mu^2}{\sigma^2}\right)I_0\left(\frac{\mu^2}{2\sigma^2}\right) + \frac{\mu^2}{\sigma^2}I_1\left(\frac{\mu^2}{2\sigma^2}\right)\right],$$

$$\sigma_Z^2 = 2\sigma^2 + 2\mu^2 - \mu_Z^2 .$$

For the values of $\mu = 2$ and $\sigma = 4$ used in the following program, the resulting mean and variance of the Rician random variable should be $\mu_Z = 5.6211$ and $\sigma_Z^2 = 8.4031$.

```
N=10000;
mu=2; sigma=4;                      % set mean and std. dev. of X
                                      and Y.
X=sigma*randn(1,N)+mu;              % generate samples of X
Y=sigma*randn(1,N)+mu;             % generate samples of Y
Z=sqrt(X.^2+Y.^2);                 % Create Z (Rician RVs)
mu_hat=sum(Z)/N                    % sample mean.
s_hat2=sum((Z-mu_hat).^2)/(N-1)  % sample variance
```

7.1.3 Estimating the CDF of IID Random Variables

Suppose instead of estimating the parameters of a distribution, we were interested in estimating the distribution itself. This can be done using some of the previous results. The CDF of the underlying distribution is $F_X(x) = \Pr(X \leq x)$. For any specific value of x, define a set of related variables $Y_1, Y_2, \ldots, Y_n$ such that

$$Y_i = \begin{cases} 1, & \text{if } X_i \leq x, \\ 0, & \text{if } X_i > x. \end{cases} \tag{7.21}$$

It should be fairly evident that if the X_i are IID, then the Y_i must be IID as well. Note that for these Bernoulli random variables, the mean is $E[Y_i] = \Pr(X_i \leq x)$. Hence, estimating the CDF of the X_i is equivalent to estimating the mean of the Y_i which is done using the sample mean

$$\hat{F}_X(x) = \frac{1}{n}\sum_{i=1}^{n} Y_i = \frac{1}{n}\sum_{i=1}^{n}[1 - u(X_i - x)]. \tag{7.22}$$

This estimator is nothing more than the relative frequency interpretation of probability. To estimate $F_X(x)$ from a sequence of n IID observations, we merely count the number of observations that satisfy $X_i \leq x$.

■ Example 7.5:

To illustrate this procedure of estimating the CDF of IID random variables, suppose the X_i are all uniformly distributed over $(0, 1)$. The plot in Figure 7.1 shows the results of one realization of estimating this CDF using n IID random variables for $n = 10$, $n = 100$, and $n = 1000$. Clearly, as n gets larger, the estimate gets better. The MATLAB code that follows can be

(Continued)

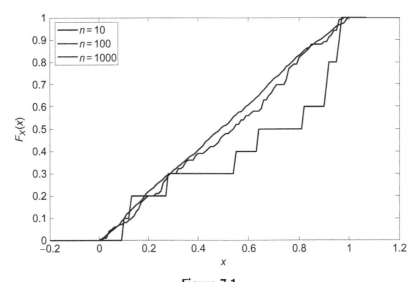

Figure 7.1
Estimate of the CDF of a uniform random variable obtained from n IID random variables,
n =10, 100, and 1000.

used to generate a plot similar to the one in Figure 7.1. The reader is encouraged to try
different types of random variables in this program as well.

```
N=100;                              % Set number of samples
z=[-0.5:0.01:1.5];                  % define variable for horizontal axis
x=rand(1,N);                        % generate uniform random samples
F=zeros(1,length(z));               % initialize CDF estimate
for n=1:N                           % estimate CDF
    F=F+(x(n)<z);
end
F=F/N;
plot(z,F)                           % plot results
xlabel('x'); ylabel('F_X(x)')
```

■

7.2 Convergence Modes of Random Sequences

In many engineering applications, it is common to use various iterative procedures. In such
cases, it is often important to know under what circumstances an iterative algorithm converges
to the desired result. The reader is no doubt familiar with many such applications in the
deterministic world. For example, suppose we wish to solve for the root of some equation
$g(x) = 0$. One could do this with a variety of iterative algorithms (e.g., Newton's method).
The convergence of these algorithms is a quite important topic. That is, suppose x_i is the
estimate of the root at the ith iteration of Newton's method. Does the sequence $x_1, x_2, x_3, \ldots$

converge to the true root of the equation? In this section, we study the topic of random sequences and in particular the issue of convergence of random sequences.

As an example of a random sequence, suppose we started with a set of IID random variables, $X_1, X_2, \ldots, X_n$, and then formed the sample mean according to

$$S_n = \frac{1}{n} \sum_{i=1}^{n} X_i. \tag{7.23}$$

The sequence $S_1, S_2, S_3, \ldots$ is a sequence of random variables. It is desirable that this sequence converges to the true mean of the underlying distribution. An estimator satisfying this condition is called *consistent*. But in what sense can we say that the sequence converges? If a sequence of deterministic numbers $s_1, s_2, s_3, \ldots$ was being considered, the sequence would be convergent to a fixed value s if

$$\lim_{i \to \infty} s_i = s. \tag{7.24}$$

More specifically, if for any $\varepsilon > 0$, there exists an i_ε such that $|s_i - s| < \varepsilon$ for all $i > i_\varepsilon$, then the sequence is said to converge to s.

■ Example 7.6:

Which of the following sequences converge in the limit as $n \to \infty$?

(a) $x_n = \dfrac{1}{1 + n^2}$, $n = 1, 2, 3, \ldots$;

(b) $y_n = \dfrac{n}{n + 1}$, $n = 1, 2, 3, \ldots$;

(c) $z_n = \cos\left(\dfrac{\pi n}{4}\right)$, $n = 1, 2, 3, \ldots$.

For the sequence x_n, $|x_n - 0| < \varepsilon$ will be satisfied provided that

$$\frac{1}{1 + n^2} < \varepsilon \Rightarrow n > \sqrt{\frac{1}{\varepsilon} - 1}.$$

Thus, the sequence x_n converges to $\lim_{n \to \infty} x_n = 0$. Likewise, for the sequence y_n, $|y_n - 1| < \varepsilon$ will be satisfied provided that

$$\left| \frac{n}{n + 1} - 1 \right| < \varepsilon \Rightarrow n > \frac{1}{\varepsilon} - 1.$$

In this case, the sequence y_n converges to $\lim_{n \to \infty} y_n = 1$. Finally, since the sequence z_n oscillates, there is no convergence.

Suppose an experiment, E, is run resulting in a realization, ζ. Each realization is mapped to a particular sequence of numbers. For example, the experiment might be to observe a sequence of IID random variables $X_1, X_2, X_3, \ldots$ and then map them into a sequence of sample means $S_1, S_2, S_3, \ldots$. Each realization, ζ, leads to a specific deterministic sequence, some of which might converge (in the previous sense), while others might not converge. Convergence for a sequence of random variables is not straightforward to define and can occur in a variety of different manners.

7.2.1 Convergence Everywhere

The first and strictest form of convergence is what is referred to as convergence everywhere (a.k.a. sure convergence). A sequence is said to converge everywhere if every realization, ζ, leads to a sequence, $s_n(\zeta)$, which converges to $s(\zeta)$. Note that the limit may depend on the particular realization. That is, the limit of the random sequence may be a random variable.

■ Example 7.7:

Suppose we modify the deterministic sequence in part (a) of Example 7.6 to create a sequence of random variables. Let X be a random variable uniformly distributed over $[0, 1)$. Then define the random sequence

$$X_n = \frac{X}{1+n^2}, \; n = 1, 2, 3, \ldots .$$

In this case, for any realization $X = x$, a deterministic sequence is produced of the form

$$x_n = \frac{x}{1+n^2}$$

which converges to $\lim_{n \to \infty} x_n = 0$. We say that the sequence converges everywhere

(or surely) to $\lim_{n \to \infty} X_n = 0$.

■

■ Example 7.8:

Suppose we tried to modify the deterministic sequence in part (b) of Example 7.6 in a manner similar to what was done in Example 7.7. That is, let Y be a random variable uniformly distributed over $[0, 1)$. Then define the random sequence

$$Y_n = \frac{nY}{n+1}, \; n = 1, 2, 3, \ldots .$$

Now, for any realization $Y = y$, the deterministic sequence

$$y_n = \frac{ny}{n+1}$$

converges to $\lim_{n \to \infty} y_n = y$. In this case, the value that the sequence converges to depends on the particular realization of the random variable Y. In other words, the random sequence converges to a random variable, $\lim_{n \to \infty} Y_n = Y$.

$\blacksquare$

7.2.2 Convergence Almost Everywhere

In many examples, it may be possible to find one or several realizations of the random sequence which do not converge, in which case the sequence (obviously) does not converge everywhere. However, it may be the case that such realizations are so rare that we might not want to concern ourselves with such cases. In particular, suppose that the only realizations that lead to a sequence which does not converge occur with probability zero. Then we say the random sequence converges almost everywhere (a.k.a. almost sure convergence or convergence with probability 1). Mathematically, let A be the set of all realizations that lead to a convergent sequence. Then the sequence converges almost everywhere if $\Pr(A) = 1$.

$\blacksquare$ Example 7.9:

As an example of a sequence that converges almost everywhere, consider the random sequence

$$Z_n = \frac{\sin(\pi n Z)}{\pi n Z}, \quad n = 1, 2, 3, \dots,$$

where Z is a random variable uniformly distributed over $[0, 1)$. For almost every realization $Z = z$, the deterministic sequence

$$z_n = \frac{\sin(\pi n z)}{\pi n z}$$

converges to $\lim_{n \to \infty} z_n = 0$. The one exception is the realization $Z = 0$ in which case the sequence becomes $z_n = 1$ which converges, but not to the same value. Therefore, we say that the sequence Z_n converges almost everywhere (or almost surely) to $\lim_{n \to \infty} Z_n = 0$ since the one exception to this convergence occurs with zero probability; that is, $\Pr(Z = 0) = 0$.

$\blacksquare$

7.2.3 Convergence in Probability

A random sequence $S_1, S_2, S_3, \dots$ converges in probability to a random variable S if for any $\varepsilon > 0$,

$$\lim_{n \to \infty} \Pr(|S_n - S| > \varepsilon) = 0. \tag{7.25}$$

■ Example 7.10:

Let X_k, $k = 1, 2, 3, \ldots,$ be a sequence of IID Gaussian random variables with mean μ and variance σ^2. Suppose we form the sequence of sample means

$$S_n = \frac{1}{n} \sum_{k=1}^{n} X_k, \quad n = 1, 2, 3, \ldots.$$

Since the S_n are linear combinations of Gaussian random variables, then they are also Gaussian with $E[S_n] = \mu$ and $\text{Var}(S_n) = \sigma^2/n$. Therefore, the probability that the sample mean is removed from the true mean by more than ε is

$$\Pr(|S_n - \mu| > \varepsilon) = 2Q\left(\sqrt{\frac{n\varepsilon}{\sigma^2}}\right).$$

As $n \to \infty$, this quantity clearly approaches zero, so that this sequence of sample means converges in probability to the true mean. ■

7.2.4 Convergence in the Mean Square Sense

A random sequence $S_1, S_2, S_3, \ldots$ converges in the mean square (MS) sense to a random variable S if

$$\lim_{n \to \infty} E[|S_n - S|^2] = 0. \tag{7.26}$$

■ Example 7.11:

Consider the sequence of sample means of IID Gaussian random variables described in Example 7.10. This sequence also converges in the MS sense since

$$E[(S_n - \mu)^2] = \text{Var}(S_n) = \frac{\sigma^2}{n}.$$

This sequence of sample variances converges to 0 as $n \to \infty$, thus producing convergence of the random sequence in the MS sense. ■

7.2.5 Convergence in Distribution

Suppose the sequence of random variables $S_1, S_2, S_3, \ldots$ has CDFs given by $F_{S_n}(s)$ and the random variable S has a CDF, $F_S(s)$. Then, the sequence converges in distribution if

$$\lim_{n \to \infty} F_{S_n}(s) = F_S(s) \tag{7.27}$$

for any s which is a point of continuity of $F_S(s)$.

■ Example 7.12:

Consider once again the sequence of sample means of IID Gaussian random variables described in Example 7.10. Since S_n is Gaussian with mean μ and variance σ^2/n, its CDF takes the form

$$F_{S_n}(s) = 1 - Q\left(\frac{s-\mu}{\sigma/\sqrt{n}}\right).$$

For any $s > \mu$, $\lim_{n \to \infty} F_{S_n}(s) = 1$, while for any $s < \mu$, $\lim_{n \to \infty} F_{S_n}(s) = 0$. Thus, the limiting form of the CDF is

$$\lim_{n \to \infty} F_{S_n}(s) = u(s-\mu),$$

where $u(s)$ is the unit step function. Note that the point $s = \mu$ is not a point of continuity of $F_S(s)$ and therefore we do not worry about it for this proof. ■

It should be noted, as was seen in the previous sequence of examples, that some random sequences converge in many of the different senses. In fact, one form of convergence may necessarily imply convergence in several other forms. Table 7.1 illustrates these relationships. For example, convergence in distribution is the weakest form of convergence and does not necessarily imply any of the other forms of convergence. Conversely, if a sequence converges in any of the other modes presented, it will also converge in distribution. The reader will find a number of exercises at the end of the chapter which will illustrate and/or prove some of the relationships in Table 7.1.

Table 7.1: Relationships between convergence modes

Implies this convergence → ↓ This convergence	Everywhere	Almost Everywhere	Probability	Mean Square	Distribution
Everywhere	X	Yes	Yes	No	Yes
Almost Everywhere	No	X	Yes	No	Yes
Probability	No	No	X	No	Yes
Mean Square	No	No	Yes	X	Yes
Distribution	No	No	No	No	X

7.3 The Law of Large Numbers

Having described the various ways in which a random sequence can converge, we return now to the study of sums of random variables. In particular, we look in more detail at the sample mean. The following very well-known result is known as the *weak law of large numbers*.

Theorem 7.2 (The Weak Law of Large Numbers): Let S_n be the sample mean computed from n IID random variables, $X_1, X_2, ..., X_n$. The sequence of sample means, S_n, converges in probability to the true mean of the underlying distribution, $F_X(x)$.

Proof: Recall that if the distribution $F_X(x)$ has a mean of μ and variance σ^2, then the sample mean, S_n, has mean μ and variance σ^2/n. Applying Chebyshev's inequality,

$$\Pr(|S_n - \mu| > \varepsilon) \le \frac{\mathrm{Var}(S_n)}{\varepsilon^2} = \frac{\sigma^2}{n\varepsilon^2}. \tag{7.28}$$

Hence, $\lim_{n \to \infty} \Pr(|S_n - \mu| > \varepsilon) = 0$ for any $\varepsilon > 0$. Thus, the sample mean converges in probability to the true mean. $\square$

The implication of this result is that we can estimate the mean of a random variable with any amount of precision with arbitrary probability if we use a sufficiently large number of samples. A stronger result known as the *strong law of large numbers* shows that the convergence of the sample mean is not just in probability but also almost everywhere. We do not give a proof of this result in this text.

As was demonstrated in Section 7.1.3, the sample mean can be used to estimate more than just means. Suppose we are interested in calculating the probability that some event A results from a given experiment. Assuming that the experiment is repeatable and each time the results of the experiment are independent of all other trials, then $\Pr(A)$ can easily be estimated. Simply define a random variable X_i that is an indicator function for the event A on the ith trial. That is, if the event A occurs on the ith trial, then $X_i = 1$, otherwise $X_i = 0$. Then

$$\Pr(A) = \Pr(X_i = 1) = E[X_i]. \tag{7.29}$$

The sample mean,

$$\hat{\mu}_n = \frac{1}{n} \sum_{i=1}^{n} X_i, \tag{7.30}$$

will give an unbiased estimate of the true probability, $\Pr(A)$. Furthermore, the law of large numbers tells us that as the sample size gets large, the estimate will converge to the true value.

The weak law of large numbers tells us that the convergence is in probability while the strong law of large numbers tells us that the convergence is also almost everywhere.

The technique we have described for estimating the probability of events is known as *Monte Carlo simulation.* It is commonly used, for example, to estimate the bit error probability of a digital communication system. A program is written to simulate transmission and detection of data bits. After a large number of data bits have been simulated, the number of errors are counted and divided by the total number of bits transmitted. This gives an estimate of the true probability of bit error of the system. If a sufficiently large number of bits are simulated, arbitrary precision of the estimate can be obtained.

■ **Example 7.13:**

This example shows how the sample mean and sample variance converges to the true mean for a few different random variables. The results of running the MATLAB code that follows are shown in Figure 7.2. Plot (a) shows the results for a Gaussian distribution, whereas plot (b) shows the same results for an arcsine random variable. In each case, the parameters have been set so that the true mean is $\mu = 3$ and the variance of each sample is 1. Since

Figure 7.2

Convergence of the sample mean for (a) Gaussian and (b) Arcsine random variables.

(Continued)

the variance of the sample mean depends only on the variance of the samples and the number of samples, crudely speaking the "speed" of convergence should be about the same in both cases.

```
N=100;

% Create Gaussian random variables
mu1=3; sigma1=1;

X1=sigma1*randn(1,N)+mu1;
mu_hat1=cumsum(X1)./[1:N];                    % sample means.

% Create Arcsine random variables
mu2=3; b=sqrt(2); sigma2=b^2/2;
X2=b*cos(2*pi*rand(1,N))+mu2;
mu_hat2=cumsum(X2)./[1:N];                     % sample means.

subplot(2,1,1)
plot([1:N],mu_hat1,'-',[1:N], mu1, '--')
xlabel('n'); ylabel('S_n'); title('Gaussian')
axis([0,N,0,2*mu1])
subplot(2,1,2)
plot([1:N],mu_hat2,'-',[1:N], mu2, '--')
xlabel('n'); ylabel('S_n'); title('Arcsine')
axis([0,N,0,2*mu2])
```

7.4 The Central Limit Theorem

Probably the most important result dealing with sums of random variables is the central limit theorem which states that under some mild conditions, these sums converge to a Gaussian random variable in distribution. This result provides the basis for many theoretical models of random phenomena. It also explains why the Gaussian random variable is of such great importance and why it occurs so frequently. In this section, we prove a simple version of the Central Limit Theorem and then discuss some of the generalizations.

Theorem 7.3 (The Central Limit Theorem): Let X_i be a sequence of IID random variables with mean μ_X and variance σ_X^2. Define a new random variable, Z, as a (shifted and scaled) sum of the X_i:

$$Z = \frac{1}{\sqrt{n}} \sum_{i=1}^{n} \frac{X_i - \mu_X}{\sigma_x}.$$

(7.31)

Note that Z has been constructed such that $E[Z] = 0$ and $\text{Var}(Z) = 1$. In the limit as n approaches infinity, the random variable Z converges in distribution to a standard normal random variable.

Proof: The most straightforward approach to prove this important theorem is using characteristic functions. Define the random variable $\tilde{X}_i$ as $\tilde{X}_i = (X_i - \mu_X)/\sigma_X$. The characteristic function of Z is computed as

$$\Phi_Z(\omega) = E[e^{j\omega Z}] = E\left[\exp\left(\frac{j\omega}{\sqrt{n}}\sum_{i=1}^n \tilde{X}_i\right)\right] = E\left[\prod_{i=1}^n \exp\left(\frac{j\omega\tilde{X}_i}{\sqrt{n}}\right)\right]$$

$$= \prod_{i=1}^n E\left[\exp\left(\frac{j\omega\tilde{X}_i}{\sqrt{n}}\right)\right] = \prod_{i=1}^n \phi_{\tilde{X}}\left(\frac{\omega}{\sqrt{n}}\right) = \left[\phi_{\tilde{X}}\left(\frac{\omega}{\sqrt{n}}\right)\right]^n. \tag{7.32}$$

Next, recall Taylor's theorem[2] which states that any function $g(x)$ can be expanded in a power series of the form

$$g(x) = g(x_o) + \frac{dg}{dx}\bigg|_{x=x_o}(x-x_o) + \dots + \frac{1}{k!}\frac{d^k g}{dx^k}\bigg|_{x=x_o}(x-x_o)^k + r_k(x,x_o), \tag{7.33}$$

where the remainder $r_k(x,x_o)$ is small compared to $(x-x_o)^k$ as $x\to x_o$. Applying the Taylor series expansion about the point $\omega = 0$ to the characteristic function of $\tilde{X}$ results in

$$\phi_{\tilde{X}}(\omega) = \phi_{\tilde{X}}(0) + \phi_{\tilde{X}}'(0)\omega + \frac{1}{2}\phi_{\tilde{X}}''(0)\omega^2 + r_3(\omega), \tag{7.34}$$

where $r_3(\omega)$ is small compared to ω^2 as $\omega\to 0$. Furthermore, we note that $\phi_{\tilde{X}}(0) = 1$, $\phi_{\tilde{X}}'(0) = jX[\tilde{X}] = 0$, and $\phi_{\tilde{X}}''(0) = -E[\tilde{X}^2] = -1$. Therefore, Equation (7.34) reduces to

$$\phi_{\tilde{X}}(\omega) = 1 - \frac{\omega^2}{2} + r_3(\omega). \tag{7.35}$$

The characteristic function of Z is then

$$\Phi_Z(\omega) = \left(1 - \frac{\omega^2}{2n} + r_3\left(\frac{\omega}{\sqrt{n}}\right)\right)^n. \tag{7.36}$$

---

[2] See for example Marsden, Tromba, Vector Calculus, 2nd ed., 1976, W. H. Freeman and Co,.

Note that as $n \to \infty$, the argument of $r_3(\ )$ goes to zero for any finite ω. Thus, as $n \to \infty$, $r_3\left(\frac{\omega}{\sqrt{n}}\right)$ becomes negligible compared to ω^2/n. Therefore, in the limit, the characteristic function of Z approaches[3]

$$\lim_{n \to \infty} \Phi_Z(\omega) = \lim_{n \to \infty} \left(1 - \frac{\omega^2}{2n}\right)^n = \exp\left(-\frac{\omega^2}{2}\right). \tag{7.37}$$

This is the characteristic function of a standard normal random variable. $\square$

Several remarks about this theorem are in order at this point. First, no restrictions were put on the distribution of the X_i. The preceding proof applies to any infinite sum of IID random variables, regardless of the distribution. Also, the central limit theorem guarantees that the sum converges in *distribution* to Gaussian, but this does not necessarily imply convergence in *density*. As a counter example, suppose that the X_i are discrete random variables, then the sum must also be a discrete random variable. Strictly speaking, the density of Z would then not exist, and it would not be meaningful to say that the density of Z is Gaussian. From a practical standpoint, the probability density of Z would be a series of impulses. While the envelope of these impulses would have a Gaussian shape to it, the density is clearly not Gaussian. If the X_i are continuous random variables, the convergence in density generally occurs as well.

The proof of the central limit theorem given above assumes that the X_i are IID. This assumption is not needed in many cases. The central limit theorem also applies to independent random variables that are not necessarily identically distributed. Loosely speaking[4], all that is required is that no term (or small number of terms) dominates the sum, and the resulting infinite sum of independent random variables will approach a Gaussian distribution in the limit as the number of terms in the sum goes to infinity. The central limit theorem also applies to some cases of dependent random variables, but we will not consider such cases here.

From a practical standpoint, the central limit theorem implies that for the sum of a sufficiently large (but finite) number of random variables, the sum is *approximately* Gaussian distributed. Of course, the goodness of this approximation depends on how many terms are in the sum and also the distribution of the individual terms in the sum. The next examples show some illustrations to give the reader a feel for the Gaussian approximation.

---

[3] Here we have used the well-known fact that $\lim_{n \to \infty} \left(1 + \frac{x}{n}\right)^n = e^x$. To establish this result, the interested reader is encouraged to expand both sides in a Taylor series and show that in the limit, the two expansions become equivalent.

[4] Formal conditions can be found in *Papoulis, Probability, Random Variables, and Stochastic Processes, 3rd ed.*, 1991, McGraw-Hill.

■ Example 7.14:

Suppose the X_i are all independent and uniformly distributed over $(-1/2, 1/2)$. Consider the sum

$$Z = \sqrt{\frac{12}{n}} \sum_{i=1}^{n} X_i \ .$$

The sum has been normalized so that Z has zero-mean and unit variance. It was shown previously that the PDF of the sum of independent random variables is just the convolution of the individual PDFs. Hence, if we define $Y = X_1 + X_2 + \ldots + X_n$ then

$$f_Y(z) = f_X(z) * f_X(z) * \ldots * f_X(z), \text{ and } f_Z(z) = \sqrt{\frac{n}{12}} f_Y\left(z\sqrt{\frac{n}{12}}\right).$$

The results of performing this n-fold convolution are shown in Figure 7.3 for several values of n. Note that for as few as $n = 4$ or $n = 5$ terms in the series, the resulting PDF of the sum looks very much like the Gaussian PDF.

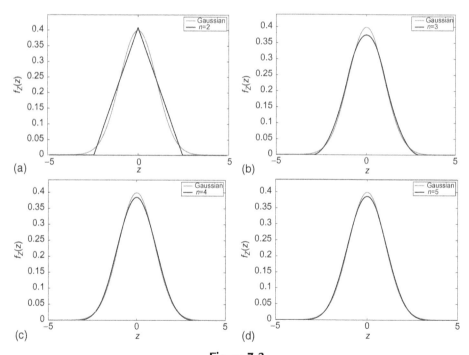

Figure 7.3
PDF of the sum of independent uniform random variables: (a) $n = 2$, (b) $n = 3$, (c) $n = 4$, and (d) $n = 5$.

■

■ Example 7.15:

In this example, suppose the X_i are now discrete Bernoulli distributed random variables such that $Pr(X_i = 1) = Pr(X_i = 0) = 0.5$. In this case, the sum $Y = X_1 + X_2 + ... + X_n$ is a binomial random variable with PMF given by

$$Pr(Y=k) = \binom{n}{k}\left(\frac{1}{2}\right)^n, \quad k = 0, 1, 2, ..., n.$$

The corresponding CDF is

$$F_Y(y) = \sum_{k=0}^{n} \binom{n}{k}\left(\frac{1}{2}\right)^n u(y-k).$$

The random variable Y has a mean of $E[Y] = n/2$ and variance of $Var(Y) = n/4$. In Figure 7.4, this binomial distribution is compared to a Gaussian distribution with the same mean and variance. It is seen that for this discrete random variable, many more terms are needed in the sum before good convergence to a Gaussian distribution is achieved.

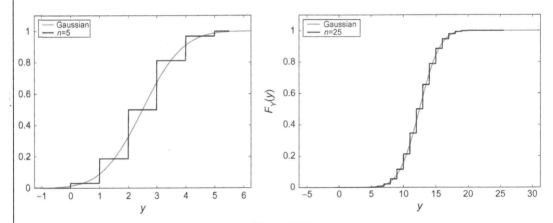

Figure 7.4
CDF of the sum of independent Bernoulli random variables; $n = 5, 25$.

■

7.5 Confidence Intervals

Consider once again the problem of estimating the mean of a distribution from n IID observations. When the sample mean $\hat{\mu}$ is formed, what have we actually learned? Loosely speaking, we might say that our best guess of the true mean is $\hat{\mu}$. However, in most cases, we know that the event $\{\hat{\mu} = \mu\}$ occurs with zero probability (since if $\hat{\mu}$ is a continuous random variable, the probability of it taking on any point value is zero). Alternatively, it could be said that (hopefully) the true mean is "close" to the sample mean. While this is a vague statement, with the help of the central limit theorem, we can make the statement mathematically precise.

If a sufficient number of samples are taken, the sample mean can be well approximated by a Gaussian random variable with a mean of $E[\hat{\mu}] = \mu_x$ and $\text{Var}(\hat{\mu}) = \sigma_X^2/n$. Using the Gaussian distribution, the probability of the sample mean being within some amount ε of the true mean can be easily calculated,

$$\Pr(|\hat{\mu} - \mu_X| < \varepsilon) = \Pr(\mu_X - \varepsilon < \hat{\mu} < \mu_X + \varepsilon) = 1 - 2Q(\varepsilon\sqrt{n}/\sigma_X). \tag{7.38}$$

Stated another way, let ε_α be the value of ε such that the right hand side of the above equation is $1 - \alpha$; that is,

$$\varepsilon_\alpha = \frac{\sigma_X}{\sqrt{n}} Q^{-1}\left(\frac{\alpha}{2}\right), \tag{7.39}$$

where $Q^{-1}(\ )$ is the inverse of the Q-function. Then, given n samples which lead to a sample mean $\hat{\mu}$, the true mean will fall in the interval $(\hat{\mu} - \varepsilon_\alpha, \hat{\mu} + \varepsilon_\alpha)$ with probability $1 - \alpha$. The interval $(\hat{\mu} - \varepsilon_\alpha, \hat{\mu} + \varepsilon_\alpha)$ is referred to as the *confidence interval* while the probability $1 - \alpha$ is the *confidence level* or, alternatively, α is the *level of significance*. The confidence level and level of significance are usually expressed as percentages. The corresponding values of the quantity $c_\alpha = Q^{-1}(\alpha/2)$ are provided in Table 7.2 for several typical values of α. Other values not included in the table can be found from tables of the Q-function (such as provided in Appendix E).

Table 7.2: Constants used to calculate confidence intervals

Percentage of Confidence Level $(1-\alpha)*100\%$	Percentage of Level of Significance $\alpha*100\%$	$c_\alpha = Q^{-1}\left(\frac{\alpha}{2}\right)$
90	10	1.64
95	5	1.96
99	1	2.58
99.9	0.1	3.29
99.99	0.01	3.89

■ Example 7.16:

Suppose the IID random variables each have a variance of $\sigma_X^2 = 4$. A sample of $n = 100$ values is taken and the sample mean is found to be $\hat{\mu} = 10.2$. Determine the 95% confidence interval for the true mean μ_X. In this case, $\sigma_X/\sqrt{n} = 0.2$ and the appropriate value of c_α is $c_{0.05} = 1.96$ from Table 7.2. The 95% confidence interval is then

$$\left(\hat{\mu} - \frac{\sigma_X}{\sqrt{n}} c_{0.05}, \hat{\mu} + \frac{\sigma_X}{\sqrt{n}} c_{0.05}\right) = (9.808, 10.592) \ .$$

■

■ Example 7.17:

Looking again at Example 7.16, suppose we want to be 99 % confident that the true mean falls within a factor of ± 0.5 of the sample mean. How many samples need to be taken in forming the sample mean? To ensure this level of confidence, it is required that

$$\frac{\sigma_X}{\sqrt{n}} c_{0.01} = 0.5$$

and therefore

$$n = \left(\frac{c_{0.01} \sigma_X}{0.5}\right)^2 = \left(\frac{2.58 * 2}{0.5}\right)^2 = 106.5 \ .$$

Since n must be an integer, it is concluded that at least 107 samples must be taken.

■

In summary, to achieve a level of significance specified by α, we note that by virtue of the central limit theorem, the sum

$$\hat{Z}_n = \frac{\hat{\mu} - \mu_X}{\sigma_X/\sqrt{n}} \ , \tag{7.40}$$

approximately follows a standard normal distribution. We can then easily specify a symmetric interval about zero in which a standard normal random variable will fall with probability $1 - \alpha$. As long as n is sufficiently large, the original distribution of the IID random variables does not matter.

Note that in order to form the confidence interval as specified, the standard deviation of the X_i must be known. While in some cases, this may be a reasonable assumption, in many applications, the standard deviation is also unknown. The most obvious thing to do in that case would be to replace the true standard deviation in Equation (7.40) with the sample standard deviation. That is, we form a statistic

$$\hat{T}_n = \frac{\hat{\mu} - \mu_X}{\hat{s}/\sqrt{n}}, \tag{7.41}$$

and then seek a symmetric interval about zero $(-t_\alpha, t_\alpha)$ such that the probability that $\hat{T}_n$ falls in that interval is $1 - \alpha$. For very large n, the sample standard deviation will converge to the true standard deviation and hence $\hat{T}_n$ will approach $\hat{Z}_n$. Hence, in the limit as $n \to \infty$, $\hat{T}_n$ can be treated as having a standard normal distribution, and the confidence interval is found in the same manner we have described. That is, as $n \to \infty$, $t_\alpha \to c_\alpha$. For values of n that are not very large, the actual distribution of the statistic $\hat{T}_n$ must be calculated in order to form the appropriate confidence interval.

Naturally, the distribution of $\hat{T}_n$ will depend on the distribution of the X_i. One case where this distribution has been calculated for finite n is when the X_i are Gaussian random variables. In this case, the statistic $\hat{T}_n$ follows the so-called *Student's t-distribution*[5] with $n - 1$ degrees of freedom:

$$f_{\hat{T}_n}(t) = \frac{(1 + t^2/n)^{-(n+1)/2} \Gamma((n+1)/2)}{\sqrt{n\pi}\Gamma(n/2)}, \tag{7.42}$$

where $\Gamma(\ )$ is the gamma function (see Chapter 3, Equation (3.22) or Appendix E, Equation (E.39)).

From this PDF, one can easily find the appropriate confidence interval for a given level of significance, α, and sample size, n. Tables of the appropriate confidence interval, t_α, can be found in any text on statistics. It is common to use the t-distribution to form confidence intervals even if the samples are not Gaussian distributed. Hence, the t-distribution is very commonly used for statistical calculations.

Many other statistics associated with related parameter estimation problems are encountered and have been carefully expounded in the statistics literature. We believe that with the probability theory developed to this point, the motivated student can now easily understand the motivation and justification for the variety of statistical tests that appear in the literature. Several exercises at the end of this chapter walk the reader through some of the more commonly used statistical distributions including the t-distribution of Equation (7.42), the chi-square distribution (see Chapter 3, Section 3.4.6), and the F-distribution (see Appendix D).

---

[5] The Student's t-distribution was developed by the English mathematician W. S. Gossett who published under the pseudonym "A. Student."

▪ Example 7.18:

Suppose we wish to estimate the failure probability of some system. We might design a simulator for our system and count the number of times the system fails during a long sequence of operations of the system. Examples might include bit errors in a communication system, defective products in an assembly line, etc. The failure probability can then be estimated as discussed at the end of Section 7.3. Suppose the true failure probability is p (which of course is unknown to us). We simulate operation of the system n times and count the number of errors observed, N_e. The estimate of the true failure probability is then just the relative frequency,

$$\hat{p} = \frac{N_e}{n}.$$

If errors occur independently, then the number of errors we observe in n trials is a binomial random variable with parameters n and p. That is,

$$P_{N_e}(k) = \binom{n}{k} p^k (1-p)^{n-k} , \quad k = 0, 1, 2, \ldots, n.$$

From this, we infer that the mean and variance of the estimated failure probability is $E[\hat{p}] = p$ and $\mathrm{Var}(\hat{p}) = n^{-1}p(1-p)$. From this, we can develop confidence intervals for our failure probability estimates. The MATLAB code that follows creates estimates as described and plots the results, along with error bars indicating the confidence intervals associated with each estimate. The plot resulting from running this code is shown in Figure 7.5.

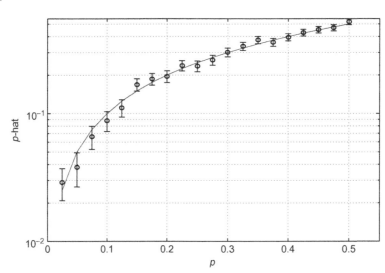

Figure 7.5

Estimates of failure probabilities along with confidence intervals. The solid line is the true probability while the circles represent the estimates.

```
    N=1000;                                  % Number of samples generated.
    c=1.64;                                  % For 90% confidence level.
    points=[0.025:0.025:0.5];                % values of p

    for k=1:length(points)
        p=points(k);
        X=rand(1,N)<p;                       % 1=error, 0=no error
        p_hat(k)=sum(X)/N;                   % relative frequency
        sigma=sqrt(p*(1-p)/N);
        eps(k)=sigma*c;                      % compute confidence interval
    end

    % plot results
    semilogy(points,points,'-')             % true values
    axis([0 0.55 0.01 0.55])
    grid on
    xlabel('p')
    ylabel('p-hat')
    hold on
    errorbar(points,p_hat,eps,'o')          % estimated values with
    hold off                                 % confidence intervals.
```

■

7.6 Random Sums of Random Variables

The sums of random variables considered up to this point have always had a fixed number of terms. Occasionally, one also encounters sums of random variables where the number of terms in the sum is also random. For example, a node in a communication network may queue packets of variable length while they are waiting to be transmitted. The number of bytes in each packet, X_i, might be random as well as the number of packets in the queue at any given time, N. The total number of bytes stored in the queue would then be a random sum of the form

$$S = \sum_{i=1}^{N} X_i. \tag{7.43}$$

Theorem 7.4: Given a set of IID random variables X_i with mean μ_X and variance σ_X^2 and an independent random integer N, the mean and variance of the random sum of the form given in Equation (7.43) are given by

$$E[S] = \mu_X E[N], \tag{7.44}$$

$$\text{Var}(S) = E[N]\sigma_X^2 + \text{Var}(N)\mu_X^2. \tag{7.45}$$

Proof: To calculate the statistics of S, it is easier to first condition on N and then average the resulting conditional statistics with respect to N. To start with, consider the mean:

$$E[S] = E_N[E[S|N]] = E_N\left[E\left[\sum_{i=1}^{N} X_i \middle| N\right]\right] = E_N[N\mu_X] = \mu_X E[N]. \quad (7.46)$$

The variance is found following a similar procedure. The second moment of S is found according to

$$E[S^2] = E_N[E[S^2|N]] = E_N\left[E\left[\sum_{i=1}^{N}\sum_{j=1}^{N} X_i X_j \middle| N\right]\right] = E_N\left[\sum_{i=1}^{N}\sum_{j=1}^{N} E[X_i X_j]\right] (7.47)$$

Note that X_i and X_j are uncorrelated unless $i = j$. Therefore, this expected value works out to be

$$E[S^2] = E_N[(N^2 - N)\mu_X^2 + NE[X_i^2]] = E[N^2]\mu_X^2 + E[N]\sigma_X^2. \quad (7.48)$$

Finally, using $\text{Var}(S) = E[S^2] - (E[S])^2$ results in

$$\text{Var}(S) = E[N^2]\mu_X^2 + E[N]\sigma_X^2 - (E[N])^2\mu_X^2 = \text{Var}(N)\mu_X^2 + E[N]\sigma_X^2. \quad (7.49)$$

One could also derive formulas for higher order moments in a similar manner. $\square$

Theorem 7.5: Given a set of IID random variables X_i with a characteristic function $\Phi_X(\omega)$ and an independent random integer N with a probability generating function $H_N(z)$, the characteristic function of the random sum of the form given in Equation (7.43) is given by

$$\Phi_S(\omega) = H_N(\Phi_X(\omega)). \quad (7.50)$$

Proof: Following a derivation similar to the last theorem,

$$\Phi_S(\omega) = E[e^{j\omega S}] = E_N[E[e^{j\omega S}|N]] = E_N\left[E\left[\exp\left(j\omega\sum_{i=1}^{N} X_i\right)\middle| N\right]\right]$$

$$= E_N\left[E\left[\prod_{i=1}^{N} \exp(j\omega X_i)\middle| N\right]\right] = E_N[(\Phi_X(\omega))^N]$$

$$= \sum_k \text{Pr}(N{=}k)(\Phi_X(\omega))^k = H_N(\Phi_X(\omega)). \quad \square \qquad (7.51)$$

■ Example 7.19:

Suppose the X_i are Gaussian random variables with zero mean and unit variance and N is a binomial random variable with a PMF,

$$\Pr(N=k) = \binom{n}{k} p^k (1-p)^{n-k}.$$

The mean and variance of this discrete distribution are $E[N] = np$ and $\mathrm{Var}(N) = np(1-p)$, respectively. From the results of Theorem 7.4, it is found that

$$E[S] = \mu_X E[N] = 0 \text{ and } \mathrm{Var}(S) = \mathrm{Var}(N)\mu_X^2 + E[N]\sigma_X^2 = np.$$

The corresponding characteristic function of X_i and probability-generating function of N are given by

$$\Phi_X(\omega) = \exp\left(-\frac{\omega^2}{2}\right) \text{ and } H_N(z) = (1-p+pz)^n.$$

The characteristic function of the random sum is then

$$\Phi_S(\omega) = \left(1 - p + p\exp\left(-\frac{\omega^2}{2}\right)\right)^n.$$

It is interesting to note that the sum of any (fixed) number of Gaussian random variables produces a Gaussian random variable. Yet, the preceding characteristic function is clearly not that of a Gaussian random variable, and hence, a random sum of Gaussian random variables is not Gaussian.

■

All of the results presented thus far in this section have made the assumption that the IID variables, X_i, and the number of terms in the series, N, are statistically independent. Quite often, these two quantities are dependent. For example, one might be interested in accumulating terms in the sum until the sum exhibits a specified characteristic (e.g., until the sample standard deviation falls below some threshold). Then, the number of terms in the sum would clearly be dependent on the values of the terms themselves. In such a case, the preceding results would not apply, and similar results for dependent variables would have to be developed. The following application section considers such a situation.

7.7 Engineering Application: A Radar System

In this section, we consider a simple radar system like that depicted in Figure 1.3. At known instants of time, the system transmits a known pulse and then waits for a reflection. Suppose the system is looking for a target at a known range so the system can determine exactly when the reflection should appear at the radar receiver. To make this discussion as simple as possible, suppose that the system has the ability to "sample" the received signal at the appropriate time instant and further that each sample is a random variable X_j that is modeled as a Gaussian random variable with variance σ^2. Let A_1 be the event that there is indeed a target present in which case X_j is taken to have a mean of μ, whereas A_0 is the event that there is no target present and the resulting mean is zero. That is, our received sample consists of a signal part (if it is present) that is some fixed voltage, μ, plus a noise part which we model as Gaussian and zero-mean. As with many radar systems, we assume that the reflection is fairly weak (μ is not large compared to σ), and hence, if we try to decide whether or not a target is present based on a single observation, we will likely end up with a very unreliable decision. As a result, our system is designed to transmit several pulses (at nonoverlapping time instants) and observe several returns, X_j, $j = 1, 2, ..., n$, that we take to be IID. The problem is to determine how to process these returns in order to make the best decision and also to determine how many returns we need to collect in order to have our decisions attain a prescribed reliability.

We consider two possible approaches. In the first approach, we decide ahead of time how many returns to collect and call that fixed number, n. We then process that random vector $\boldsymbol{X} = (X_1, X_2, ..., X_n)^T$ and form a decision. While there are many ways to process the returns, we will use what is known as a probability ratio test. That is, given $\boldsymbol{X} = \boldsymbol{x}$, we want to determine if the ratio $\Pr(A_1|\boldsymbol{x})/\Pr(A_0|\boldsymbol{x})$ is greater or less than 1. Recall that

$$\Pr(A_i|\boldsymbol{x}) = \frac{f_{\boldsymbol{X}}(\boldsymbol{x}|A_i)\Pr(A_i)}{f_{\boldsymbol{X}}(\boldsymbol{x})}, \quad i = 0, 1. \tag{7.52}$$

Thus, the probability ratio test makes the following comparison

$$\frac{\Pr(A_1|\boldsymbol{x})}{\Pr(A_0|\boldsymbol{x})} = \frac{f_{\boldsymbol{X}}(\boldsymbol{x}|A_1)\Pr(A_1)}{f_{\boldsymbol{X}}(\boldsymbol{x}|A_0)\Pr(A_0)} \underset{?}{\overset{?}{\gtrless}} 1. \tag{7.53}$$

This can be written in terms of an equivalent likelihood ratio test:

$$\Lambda(\boldsymbol{x}) \underset{?}{\overset{?}{\gtrless}} \Lambda, \tag{7.54}$$

where $\Lambda(\boldsymbol{x}) = f_{\boldsymbol{X}}(\boldsymbol{x}|A_1)/f_{\boldsymbol{X}}(\boldsymbol{x}|A_0)$ is the likelihood ratio and the threshold $\Lambda = \Pr(A_0)/\Pr(A_1)$ depends on the a priori probabilities. In practice, we may have no idea about the a priori probabilities of whether or not a target is present. However, we can still proceed by choosing the threshold for the likelihood ratio test to provide some prescribed level of performance.

Let the false alarm probability be defined as $P_{\text{fa}} = \Pr(\Lambda(\boldsymbol{X}) > \Lambda|A_0)$. This is the probability that the system declares a target is present when in fact there is none. Similarly, define the correct detection probability as $P_{\text{d}} = \Pr(\Lambda(\boldsymbol{X}) > \Lambda|A_1)$. This is the probability that the system correctly identifies a target as being present. These two quantities, P_{fa} and P_{d}, will specify the performance of our radar system. Given that the X_j are IID Gaussian as described above, the likelihood ratio works out to be

$$\Lambda(x) = \frac{f_X(x|A_1)}{f_X(x|A_0)} = \frac{(2\pi\sigma^2)^{-n/2}\exp\left(-\frac{1}{2\sigma^2}\sum_{j=1}^{n}(x_j - \mu)^2\right)}{(2\pi\sigma^2)^{-n/2}\exp\left(-\frac{1}{2\sigma^2}\sum_{j=1}^{n}x_j^2\right)}$$

$$= \exp\left(\frac{n\mu}{\sigma^2}\left[\left(\frac{1}{n}\sum_{j=1}^{n}x_j\right) - \frac{\mu}{2}\right]\right). \tag{7.55}$$

Clearly, comparing this with a threshold is equivalent to comparing the sample mean with a threshold. That is, for IID Gaussian returns, the likelihood ratio test simplifies to

$$\hat{\mu} = \frac{1}{n}\sum_{j=1}^{n}x_j \overset{?}{\underset{?}{\gtrless}} \mu_0 \,, \tag{7.56}$$

where the threshold μ_0 is set to produce the desired system performance. Since the X_j are Gaussian, the sample mean is also Gaussian. Hence, when there is no target present $\hat{\mu} \sim N(0, \sigma^2/n)$ and when there is a target present $\hat{\mu} \sim N(\mu, \sigma^2/n)$. With these distributions, the false alarm and detection probabilities work out to be

$$P_{\text{fa}} = Q\left(\frac{\sqrt{n}\mu_0}{\sigma}\right) \text{ and } 1 - P_{\text{d}} = Q\left(\frac{\sqrt{n}(\mu - \mu_0)}{\sigma}\right). \tag{7.57}$$

By adjusting the threshold, we can trade off false alarms for missed detections. Since the two probabilities are related, it is common to write the detection probability in terms of the false alarm probability as

$$1 - P_{\text{d}} = Q\left(\frac{\sqrt{n}\mu}{\sigma} - Q^{-1}(P_{\text{fa}})\right). \tag{7.58}$$

From this equation, we can determine how many returns need to be collected in order to attain a prescribed system performance specified by (P_{fa}, P_d). In particular,

$$n = \frac{[Q^{-1}(1 - P_{\text{d}}) + Q^{-1}(P_{\text{fa}})]^2}{\mu^2/\sigma^2}. \tag{7.59}$$

The quantity μ^2 / σ^2 has the physical interpretation of the strength of the signal (when it is present) divided by the strength of the noise, or simply the signal-to-noise ratio.

Since the radar system must search at many different ranges and many different angles of azimuth, we would like to minimize the amount of time it has to spend collecting returns at each point. Presumably, the amount of time we spend observing each point in space depends on the number of returns we need to collect. We can often reduce the number of returns needed by noting that the number of returns required to attain a prescribed reliability as specified by (P_{fa}, P_d) will depend on the particular realization of returns encountered. For example, if the first few returns come back such that the sample mean is very large, we may be very certain that a target is present and hence there is no real need to collect more returns. In other instances, the first few returns may produce a sample mean near $\mu/2$. This inconclusive data would lead us to wait and collect more data before making a decision. Using a variable number of returns whose number depends on the data themselves is known as *sequential detection*.

The second approach we consider will use a sequential detection procedure whereby after collecting n returns, we compare the likelihood ratio with two thresholds, Λ_0 and Λ_1, and decide according to

$$\Lambda(x) \begin{cases} \geq \Lambda_1, & \text{decide a target is present,} \\ \in (\Lambda_0, \Lambda_1), & \text{collect another return,} \\ \leq \Lambda_0, & \text{decide no target is present.} \end{cases} \qquad (7.60)$$

The performance of a sequential detection scheme can be determined as follows. Define the region $R_1^{(n)}$ to be the set of data points $x^{(n)} = (x_1, x_2, ..., x_n)$ which lead to a decision in favor of A_1 after collecting exactly n data points. That is, $\Lambda(x^{(n)}) > \Lambda_1$ and $\Lambda_0 < \Lambda(x^{(j)}) < \Lambda_1$ for $j = 1, 2, ..., n-1$. Similarly define the region $R_0^{(n)}$ to be the set of data points $x^{(n)}$ which lead to a decision in favor of A_0 after collecting exactly n data points. Let $P_{fa}^{(n)}$ be the probability of a false alarm occurring after collecting exactly n returns and $P_d^{(n)}$ the probability of making a correct detection after collecting exactly n returns. The overall false alarm and detection probabilities are then

$$P_{fa} = \sum_{n=1}^{\infty} P_{fa}^{(n)} \text{ and } P_d = \sum_{n=1}^{\infty} P_d^{(n)}. \qquad (7.61)$$

We are now in a position to establish the following fundamental result which will instruct us in how to set the decision thresholds in order to obtain the desired performance.

Theorem 7.6 (Wald's Inequalities): For a sequential detection strategy, the false alarm and detection strategies satisfy

$$P_d \geq \Lambda_1 P_{fa}, \tag{7.62}$$

$$(1 - P_d) \leq \Lambda_0 (1 - P_{fa}). \tag{7.63}$$

Proof: First note that

$$P_{fa}^{(n)} = \Pr(x^{(n)} \in R_1^{(n)} | A_0) = \int_{R_1^{(n)}} f_{X^{(n)}}(x^{(n)} | A_0) dx^{(n)} \tag{7.64}$$

and similarly

$$P_d^{(n)} = \Pr(x^{(n)} \in R_1^{(n)} | A_1) = \int_{R_1^{(n)}} f_{X^{(n)}}(x^{(n)} | A_1) dx^{(n)}. \tag{7.65}$$

For all $x^{(n)} \in R_1^{(n)}$, $f_{X^{(n)}}(x^{(n)} | A_1) \geq \Lambda_1 f_{X^{(n)}}(x^{(n)} | A_0)$ and therefore

$$P_d^{(n)} \geq \Lambda_1 \int_{R_1^{(n)}} f_{X^{(n)}}(x^{(n)} | A_0) dx^{(n)} = \Lambda_1 P_{fa}^{(n)}. \tag{7.66}$$

Summing over all n then produces Equation (7.62). Equation (7.63) is derived in a similar manner. $\square$

Since the likelihood ratio is often exponential in form, it is common to work with the log of the likelihood ratio, $\lambda(x) = \ln(\Lambda(x))$. For the case of Gaussian IID data, we get

$$\lambda(x^{(n)}) = \frac{n\mu}{\sigma^2}\left[\left(\frac{1}{n}\sum_{j=1}^{n} x_j\right) - \frac{\mu}{2}\right] = \lambda(x^{(n-1)}) + \frac{\mu}{\sigma^2}x_n - \frac{\mu^2}{2\sigma^2}. \tag{7.67}$$

In terms of log-likelihood ratios, the sequential decision mechanism is

$$\lambda(x) \begin{cases} \geq \lambda_1, & \text{decide a target is present,} \\ \in (\lambda_0, \lambda_1), & \text{collect another return,} \\ \leq \lambda_0, & \text{decide no target is present,} \end{cases} \tag{7.68}$$

where $\lambda_j = \ln(\Lambda_j)$, $j = 0, 1$. The corresponding versions of Wald's Inequalities are then

$$\ln(P_d) \geq \lambda_1 + \ln(P_{fa}), \tag{7.69a}$$

$$\ln(1 - P_d) \leq \lambda_0 + \ln(1 - P_{fa}). \tag{7.70b}$$

For the case when the signal-to-noise ratio is small, each new datum collected adds a small amount to the sum in Equation (7.67), and it will typically take a large number of terms before the sum will cross one of the thresholds, λ_0 or λ_1. As a result, when the requisite number of data are collected so that the log-likelihood ratio crosses a threshold, it will usually be only incrementally above the threshold. Hence, Wald's inequalities will be approximate equalities and the decision thresholds that lead to (approximately) the desired performance can be found according to

$$\lambda_1 = \ln\left(\frac{P_d}{P_{fa}}\right) \text{ and } \lambda_0 = \ln\left(\frac{1 - P_d}{1 - P_{fa}}\right). \tag{7.71}$$

Now that the sequential detection strategy can be designed to give any desired performance, we are interested in determining how much effort we save relative to the fixed sample size test. Let N be the instant at which the test terminates, and to simplify notation, define

$$S_N = \lambda(\boldsymbol{X}^{(N)}) = \sum_{i=1}^{N} Z_i, \tag{7.72}$$

where from Equation (7.67) $Z_i = \mu(X_i - \mu/2)/\sigma^2$. Note that S_N is a random sum of IID random variables as studied in section 7.6, except that now the random variable N is not independent of the Z_i. Even with this dependence, for this example it is still true that

$$E[S_N] = E[N]E[Z_i]. \tag{7.73}$$

The reader is led through a proof of this in Exercise 7.39. Note that when the test terminates:

$$S_N \cong \begin{cases} \lambda_1, & \text{if the test terminates in } A_1, \\ \lambda_0, & \text{if the test terminates in } A_0, \end{cases} \tag{7.74}$$

and therefore,

$$E[S_N] \cong \lambda_1 \Pr(\text{test terminates in } A_1) + \lambda_0 \Pr(\text{test terminates in } A_0). \tag{7.75}$$

Suppose that A_0 is true. Then the event that the test terminates in A_1 is simply a false alarm. Combining Equations (7.73) and (7.75) results in

$$E[N|A_0] \cong \frac{E[S_N|A_0]}{E[Z_i|A_0]} = -\frac{\lambda_1 P_{\mathrm{fa}} + \lambda_0(1 - P_{\mathrm{fa}})}{\mu^2/2\sigma^2}. \tag{7.76}$$

Similarly, when A_1 is true:

$$E[N|A_1] \cong \frac{E[S_N|A_1]}{E[Z_i|A_1]} = \frac{\lambda_1 P_{\mathrm{d}} + \lambda_0(1 - P_{\mathrm{d}})}{\mu^2/2\sigma^2}. \tag{7.77}$$

It is noted that not only is the number of returns collected a random variable, but the statistics of this random variable depend on whether or not a target is present. This may work to our advantage in that the average number of returns we need to observe might be significantly smaller in the more common case when there is no target present.

■ Example 7.20:

Suppose we want to achieve a system performance specified by $P_{\mathrm{fa}} = 10^{-6}$ and $P_{\mathrm{d}} = 0.99$. Furthermore, suppose the signal-to-noise ratio for each return is $\mu^2/\sigma^2 = 0.1 = -10\mathrm{dB}$. Then the fixed sample size test will use a number of returns given by

$$n = \frac{[Q^{-1}(1-P_{\mathrm{d}}) + Q^{-1}(P_{\mathrm{fa}})]^2}{\mu^2/\sigma^2} = \frac{[2.326 + 4.755]^2}{0.1} = 501.4.$$

Since n must be an integer, 502 samples must be taken to attain the desired performance. For the sequential test, the two thresholds for the log-likelihood test are set according to Wald's inequalities,

$$\lambda_0 = \ln\left(\frac{1-P_{\mathrm{d}}}{1-P_{\mathrm{fa}}}\right) = -4.6 \text{ and } \lambda_1 = \ln\left(\frac{P_{\mathrm{d}}}{P_{\mathrm{fa}}}\right) = 13.8.$$

With these thresholds set, the average number of samples needed for the test to terminate is

$$E[N|A_0] = -\frac{\lambda_1 P_{\mathrm{fa}} + \lambda_0(1-P_{\mathrm{fa}})}{\mu^2/2\sigma^2} = 92.1$$

when there is no target present, and

$$E[N|A_1] = \frac{\lambda_1 P_{\mathrm{d}} + \lambda_0(1-P_{\mathrm{d}})}{\mu^2/2\sigma^2} = 272.4$$

when a target is present. Clearly, for this example, the sequential test saves us significantly in terms of the amount of data that needs to be collected to make a reliable decision.

■

Section 7.1: IID Random Variables

7.1 A random variable, X, has a Gaussian PDF with mean 5 and unit variance. We measure 10 independent samples of the random variable.
   (a) Determine the expected value of the sample mean.
   (b) Determine the variance of the sample mean.
   (c) Determine the expected value of the unbiased sample variance.

7.2 Two independent samples of a random variable X are taken. Determine the expected value and variance of the sample mean estimate of μ_X if the PDF is exponential, (i.e., $f_X(x) = \exp(-x)u(x)$).

7.3 The noise level in a room is measured n times. The error ε for each measurement is independent of the others and is normally distributed with zero-mean and standard deviation $\sigma_\varepsilon = 0.1$. In terms of the true mean, μ, determine the PDF of the sample mean, $\hat{\mu}$, for $n = 100$.

7.4 Suppose X is a vector of N IID random variables where each element has some PDF, $f_X(x)$. Find an example PDF such that the median is a better estimate of the mean than the sample mean.

7.5 Suppose the variance of an IID sequence of random variables is formed according to

$$\hat{\sigma}^2 = \frac{1}{n} \sum_{m=1}^{n} (X_m - \hat{\mu})^2,$$

where $\hat{\mu}$ is the sample mean. Find the expected value of this estimate and show that it is biased.

7.6 Find the variance of the sample standard deviation,

$$\hat{s}^2 = \frac{1}{n-1} \sum_{m=1}^{n} (X_m - \hat{\mu})^2,$$

assuming that the X_i are IID Gaussian random variables with mean μ and variance σ^2.

7.7 Show that if X_n, $n = 1, 2, 3, \dots$, is a sequence of IID Gaussian random variables, the sample mean and sample variance are statistically independent.

7.8 A sequence of random variables, X_n, is to be approximated by a straight line using the estimate, $\hat{X}_n = a + bn$. Determine the least squares (i.e., minimum mean squared error) estimates for a and b if N samples of the sequence are observed.

Section 7.2: Convergence Modes of Random Sequences

7.9 (a) Prove that any sequence that converges in the mean square sense must also converge in probability. *Hint:* Use Markov's inequality.

(b) Prove by counterexample that convergence in probability does not necessarily imply convergence in the mean square sense.

7.10 Consider a sequence of IID random variables, X_n, $n = 1, 2, 3, \dots$, each with CDF $F_{X_n}(x) = F_X(x) = 1 - Q\left(\frac{x - \mu}{\sigma}\right)$. This sequence clearly converges in distribution since $F_{X_n}(x)$ is equal to $F_X(x)$ for all n. Show that this sequence does not converge in any other sense and therefore convergence in distribution does not imply convergence in any other form.

7.11 (a) Show by counterexample that convergence almost everywhere does not imply convergence in the MS sense.

(b) Show by counterexample that convergence in the MS sense does not imply convergence almost everywhere.

7.12 Prove that convergence almost everywhere implies convergence in probability.

7.13 Consider the random sequence $X_n = X/(1 + n^2)$, where X is a Cauchy random variable with PDF,

$$f_X(x) = \frac{1/\pi}{1 + x^2}.$$

Determine which forms of convergence apply to this random sequence.

7.14 Let X_n be a sequence of IID Gaussian random variables. Form a new sequence according to

$$Y_n = \frac{1}{2}X_{n-1} - X_n + \frac{1}{2}X_{n+1}.$$

Determine which forms of convergence apply to the random sequence, Y_n.

7.15 Let X_n be a sequence of IID Gaussian random variables. Form a new sequence according to

$$Z_n = \sum_{i=1}^{n} \left(\frac{1}{2}\right)^{n-i} X_i.$$

Determine which forms of convergence apply to the random sequence, Z_n.

7.16 Let X_k, $k = 1, 2, 3, \ldots$, be a sequence of IID random variables with finite mean and variance. Show that the sequence of sample means

$$S_n = \sum_{k=1}^{n} X_k$$

converges in the MS sense.

7.17 Suppose X_k is a sequence of zero-mean Gaussian random variables with covariances described by $\text{Cov}(X_k, X_m) = \rho^{|k-m|}$ for some $|\rho| < 1$. Form the sequence of sample means

$$S_n = \sum_{k=1}^{n} X_k, \quad n = 1, 2, 3, \ldots.$$

Note that in this case we are forming the sequence of sample means of *dependent* random variables.
(a) Determine if the sequence S_n converges in distribution.
(b) Determine if the sequence S_n converges in probability.
(c) Determine if the sequence S_n converges in the MS sense.

Section 7.3: The Law of Large Numbers

7.18 Suppose $X_1, X_2, \ldots, X_n$ is a sequence of IID positive random variables. Define

$$Y_n = \prod_{i=1}^{n} X_i.$$

Show that as $n \to \infty$, Y_n converges in distribution, and find the distribution to which it converges.

7.19 Let X_k, $k = 1, 2, 3, \ldots$, be a sequence of IID random variables with finite mean, μ, and let S_n be the sequence of sample means,

$$S_n = \sum_{k=1}^{n} X_k, \quad n = 1, 2, 3, \ldots.$$

(a)   Show that the characteristic function of S_n can be written as

$$\Phi_{S_n}(\omega) = \left(\Phi_X\!\left(\frac{\omega}{n}\right) \right)^n.$$

(b)   Use Taylor's theorem to write the characteristic function of the X_k as

$$\Phi_X(\omega) = c_0 + c_1 \omega + r_2(\omega),$$

where the remainder term $r_2(\omega)$ is small compared to ω as $\omega \to 0$. Find the constants c_0 and c_1.

(c)   Writing the characteristic function of the sample mean as

$$\Phi_{S_n}(\omega) = \left(c_0 + c_1 \frac{\omega}{n} + r_2\!\left(\frac{\omega}{n}\right) \right)^n,$$

show that as $n \to \infty$

$$\lim_{n \to \infty} \Phi_{S_n}(\omega) = \exp(j\mu\omega).$$

In so doing, you have proved that the distribution of the sample mean is that of a constant in the limit as $n \to \infty$. Thus, the sample mean converges in distribution.

7.20 Prove that if a sequence converges in distribution to a constant value, then it also converges in probability. *Note:* The results of Exercise 7.19 and this one together constitute an alternative proof to the weak law of large numbers.

7.21 Prove that the sequence of sample means of IID random variables converges in the MS sense. What conditions are required on the IID random variables for this convergence to occur?

7.22 Let X_k, $k = 1, 2, 3, \ldots$, be a sequence of IID Cauchy random variables with

$$f_X(x) = \frac{1/\pi}{1 + x^2},$$

and let S_n be the sequence of sample means,

$$S_n = \sum_{k=1}^{n} X_k, \qquad n = 1, 2, 3, \ldots.$$

(a)   Show that S_n also follows a Cauchy distribution.

(b)   Prove that in this case, the sample mean does not converge in probability and therefore the weak law of large numbers does not apply. What assumption has been violated in this case that makes the weak law of large numbers not applicable?

Section 7.4: The Central Limit Theorem

7.23 Independent samples are taken of a random variable X. If the PDF of X is uniform over the interval $[-1/\sqrt{12}, 1/\sqrt{12})$ and zero elsewhere, then approximate the density of the sample mean with a normal density, assuming the number of samples is large. Write the approximation as an equation.

7.24 Consider the lottery described in Exercise 2.61.

    (a) Assuming six million tickets are sold and that each player selects his/her number independent of all others, find the exact probability of fewer than 3 players winning the lottery.

    (b) Approximate the probability in part (a) using the Poisson approximation to the binomial distribution.

    (c) Approximate the probability in part (a) using the central limit theorem. In this example, which approximation is more accurate?

7.25 A communication system transmits bits over a channel such that the probability of being received in error is $p = 0.02$. Bits are transmitted in blocks of length 1023 bits and an error correction scheme is used such that bit errors can be corrected provided that no more than 30 errors occur in a 1023 bit block. Use the central limit theorem to approximate the probability that no more than 30 errors occur in a 1023 bit block.

7.26 A certain class of students takes a standardized test where each student's score is modeled as a random variable with mean, $\mu = 85$, and standard deviation, $\sigma = 5$. The school will be put on probationary status if the class average falls below 75. If there are 100 students in the class, use the central limit theorem to approximate the probability that the class average falls below 75.

7.27 Let X_k be a sequence of IID exponential random variables with mean of 1. We wish to compute $\Pr\left(\sum_{k=1}^{n} X_k > y \right)$ for some constant y (such that $y > 25$).

    (a) Find a bound to the probability using Markov's inequality.

    (b) Find a bound to the probability using Chebyshev's inequality.

    (c) Find a bound to the probability using the Chernoff bound.

    (d) Find an approximation to the probability using the central limit theorem.

    (e) Find the exact probability.

    (f) Plot all five results from (a) through (e) for $y > 25$ and determine for what range of y the central limit theorem gives the most accurate approximation compared with the 3 bounds.

Section 7.5: Confidence Intervals

7.28 Suppose we wish to estimate the probability, p_A, of some event, A. We do so by repeating an experiment n times and observing whether or not the event A occurs during each experiment. In particular, let

$$X_i = \begin{cases} 1, & A \text{ occured during } i\text{th experiment,} \\ 0, & \text{otherwise.} \end{cases}$$

We then estimate p_A using the sample mean of the X_i,

$$\hat{p}_A = \frac{1}{n} \sum_{i=1}^{n} X_i.$$

(a)   Assuming n is large enough so that the central limit theorem applies, find an expression for $\Pr(|\hat{p}_A - p_A| < \varepsilon)$.

(b)   Suppose we want to be 95% certain that our estimate is within $\pm 10\%$ of the true value. That is, we want $\Pr(|\hat{p}_A - p_A| < 0.1 p_A) = 0.95$. How large does n need to be? In other words, how many time do we need to run the experiment?

(c)   Let Y_n be the number of times that we observe the event A during our n repetitions of the experiment. That is, let $Y_n = X_1 + X_2 + \ldots + X_n$. Assuming that n is chosen according to the results of part (b), find an expression for the average number of times the event A is observed, $E[Y_n]$. Show that for rare events (i.e., $p_A \ll 1$) $E[Y_n]$ is essential independent of p_A. Thus, even if we have no idea about the true value of p_A, we can run the experiment until we observe the event A for a predetermined number of times and be assured of a certain degree of accuracy in our estimate of p_A.

7.29 Suppose we wish to estimate the probability, p_A, of some event A as outlined in Exercise 7.28. As motivated by the result of part (c) of Exercise 7.28, suppose we repeat our experiment for a random number of trials, N. In particular, we run the experiment until we observe the event A exactly m times and then form the estimate of p_A according to

$$\hat{p}_A = \frac{m-1}{N-1}.$$

Here, the random variable N represents the number of trials until the mth occurrence of A.

(a)   Find $E[\hat{p}_A]$. Is this estimate unbiased?

(b)   Would it be better to use $\hat{p}_A = \frac{m}{N}$ as an estimate?

7.30 A company manufactures five-volt power supplies. However, since there are manufacturing tolerances, there are variations in the voltage design. The standard deviation in the design voltage is 5%. Using a 99% confidence level, determine whether or not the following samples fall within the confidence interval:
(a)   100 samples, the estimate of $\mu_X = 4.7$
(b)   100 samples, the estimate of $\mu_X = 4.9$
(c)   100 samples, the estimate of $\mu_X = 5.4$
Hint: refer to Equation (7.39)

7.31 You collect a sample size N_1 of data and find that a 90% confidence level has width, w. What should the sample size N_2 be to increase the confidence level to 99.9% and yet maintain the same interval width, w?

7.32 Company A manufactures computer applications boards. They are concerned with the mean time before failures (MTBF), which they regularly measure. Denote the sample MTBF as $\hat{\mu}_M$ and the true MTBF as μ_M. Determine the number of failures that must be measured before $\hat{\mu}_M$ lies within 20 % of the true μ_M with a 90% probability. Assume the PDF is exponential, i.e., $f_M(x) = (1/\mu_M)\exp(-x/\mu_M)u(x)$.

7.33 A political polling firm is conducting a poll in order to determine which candidate is likely to win an upcoming election. The polling firm interviews n likely voters and asks each whether or not they will vote for the republican (R) or the democrat (D) candidate. They then tabulate the percentage that respond R and D.
(a)   How many voters should the firm poll in order to correctly estimate the correct proportion of R and D respondents in the general population to within ±3 percentage points with 90% probability?
(b)   Repeat part (a) if we want 95% confidence in our polling data.

Section 7.6: Random Sums of Random Variables

7.34 A node in a communication network receives data packets of variable length. Each packet has a random number of bits that is uniformly distributed over the integers $\{100, 101, 102, ..., 999\}$. The number of packet arrivals per minute is a Poisson random variable with a mean of 50.
(a)   What is the average number of data bits per minute arriving at the node?
(b)   What is the variance of the number of data bits per minute arriving at the node?

7.35 The number of cars approaching a toll booth in a minute follows a geometric random variable with a mean of 2 cars/minute. The time it takes the toll collector to serve each car is an exponential random variable with a mean of 20 seconds.
(a)   Find the mean time that the toll collector will require to serve cars that arrive in a one-minute interval.

(b)   Find the PDF of the time that the toll collector will require to serve cars that arrive in a one-minute interval.

(c)   What is the probability that the toll collector will require more than one minute to serve the cars that arrive during a one-minute interval, thereby causing a queue to form?

7.36 Let $S = \sum_{k=1}^{N} X_k$ be a random sum of discrete IID random variables. Further, let $H_N(z)$ and $H_X(z)$ be the probability-generating functions of N and X, respectively. Find the probability-generating function of S assuming that N is independent of the X_k.

7.37 A gambler plays a game of chance where he wins \$1 with probability p and loses \$1 with probability $1 - p$ each time he plays. The number of games he plays in an hour, N, is a random variable with a geometric PMF, $P_N(n) = (1-q)q^{n-1}$, $n = 1, 2, 3, \ldots$.
(a)   What is the PGF of the gambler's total winnings after playing for an hour?
(b)   What is the probability that the gambler has not lost any money after an hour if $p = 0.48$ and $q = 7/8$?

Miscellaneous Exercises

7.38 In this exercise, a proof of equation (7.73) is constructed. Write the random sum as

$$S_N = \sum_{i=1}^{N} Z_i = \sum_{i=1}^{\infty} Y_i Z_i,$$

where Y_i is a Bernoulli random variable in which $Y_i = 1$ if $N \geq i$ and $Y_i = 0$ if $N < i$.
(a)   Prove that Y_i and Z_i are independent and hence

$$E[S_N] = \sum_{i=1}^{\infty} E[Y_i]E[Z_i].$$

(b)   Prove that the equation of part (a) simplifies to

$$E[S_N] = E[Z_i]E[N].$$

7.39 Suppose that X_k is a sequence of IID Gaussian random variables. Recall that the sample variance is given by

$$\hat{s}^2 = \frac{1}{n-1} \sum_{k=1}^{n} (X_k - \hat{\mu})^2 \quad \text{where } \hat{\mu} = \frac{1}{n} \sum_{k=1}^{n} X_k.$$

(a) Show that the sample variance can be written as a quadratic form $\hat{s}^2 = \mathbf{X}^\mathsf{T}\mathbf{B}\mathbf{X}$ and find the corresponding form of the matrix $\mathbf{B}$.

(b) Use the techniques outlined in Section 6.4.2 to show that the characteristic function of $\hat{s}^2$ is

$$\Phi_{\hat{s}^2}(\omega) = \frac{1}{\left(1 - 2j\omega\dfrac{\sigma^2}{n-1}\right)^{\frac{n-1}{2}}}.$$

(c) Show that the PDF of $\hat{s}^2$ is that of a chi-square random variable.

7.40 Let X be a zero-mean, unit-variance, Gaussian random variable and let Y be a chi-square random variable with $n - 1$ degrees of freedom (see Appendix D, section D.1.4). If X and Y are independent, find the PDF of

$$T = \frac{X}{\sqrt{Y/n}}.$$

Hint: One way to accomplish this is to define an auxiliary random variable, $U = Y$, and then find the joint PDF of T and U using the 2×2 transformation techniques outlined in Section 5.9. Once the joint PDF is found, the marginal PDF of T can be found by integrating out the unwanted variable U.

Note: This is the form of the statistic

$$\hat{T} = \frac{\hat{\mu} - \mu_X}{\hat{s}/\sqrt{n}}$$

of Equation (7.41) where the sample mean is Gaussian and the sample variance is chi-square (by virtue of the results of Exercise 7.39) assuming that the underlying X_k are Gaussian.

7.41 Suppose we form a sample variance $\hat{s}_1^2 = \dfrac{1}{n-1}\displaystyle\sum_{k=1}^{n}(X_k - \hat{\mu}_X)^2$ from a sequence of IID Gaussian random variables and then form another sample variance

$\hat{s}_s^2 = \dfrac{1}{m-1}\displaystyle\sum_{k=1}^{m}(Y_k - \hat{\mu}_Y)^2$ from a different sequence of IID Gaussian random variables

that are independent from the first set. We wish to determine if the true variances of the two sets of Gaussian random variables are the same or if they are significantly different, so we form the ratio of the sample variances

$$F = \frac{\hat{s}_1^2}{\hat{s}_2^2}$$

to see if this quantity is either large or small compared to 1. Assuming that the X_k and the Y_k are both standard normal, find the PDF of the statistic F and show that it follows an F distribution (see Appendix D, Section D.1.7). *Hint:* Use the conditional PDF approach outlined in Section 5.9.

MATLAB Exercises

7.42 Let X, Y, and Z be independent Gaussian random variables with equal means of $\mu = 3$ and variances of $\sigma^2 = 4$. Estimate the mean and variance of $W = \sqrt{X^2 + Y^2 + Z^2}$ by constructing a large number of realizations of this random variable in MATLAB and then computing the sample mean and sample variance. How many samples of the random variable were needed before the sample mean and sample variance seemed to converge to a fairly accurate estimate. (To answer this, you must define what you mean by "fairly accurate.")

7.43 For the random variable W described in Exercise 7.42, form an estimate of the CDF by following the procedure outlined in Example 7.5. Also, form an estimate of the PDF of this random variable. Explain the procedure you used to estimate the PDF.

7.44 A player engages in the following dice tossing game ("craps"). Two dice are rolled. If the player rolls the dice such that the sum is either 7 or 11, he immediately wins the game. If the sum is 2, 3, or 12, he immediately loses. If he rolls a 4, 5, 6, 8, 9, or 10, this number is called the "point" and the player continues to roll the dice. If he is able to roll the point again before he rolls a 7, he wins. If he rolls a 7 before he rolls the point again, he loses. Write a MATLAB program to simulate this dice game and estimate the probability of winning.

7.45 Let X_i, $i = 1, 2, ..., n$, be a sequence of IID random variables uniformly distributed over $(0, 1)$. Suppose we form the sum $Z = \sum_{i=1}^{n} X_i$. First, find the mean and variance of Z. Then write a MATLAB program to estimate the PDF of Z. Compare the estimated PDF with a Gaussian PDF of the same mean and variance. Over what range of Z is the Gaussian approximation of the PDF within 1% of the true PDF? Repeat this problem for $n = 5, 10, 20, 50,$ and 100.

7.46 Suppose you are given an observation of sample values of a sequence of random variables, x_n, $n = 1, 2, 3, ..., m$. Write a MATLAB program to plot these data points along with a least squares curve fit to the data (see the results of Exercise 7.8). Run your program using the following sequence:

$$(0, 1, 0, -1, 2, -3, 5, 0, -7, 8).$$

Random Processes

This chapter introduces the concept of a random process. Most of the treatment in this text views a random process as a random function of time. However, time need not be the independent variable. We can also talk about a random function of position, in which case there may be two or even three independent variables and the function is more commonly referred to as a random field. The concept of a random process allows us to study systems involving signals which are not entirely predictable. These random signals play fundamental roles in the fields of communications, signal processing, and control systems, and many other engineering disciplines. This and the following chapters will extend the study of signal and system theory to include randomness. In this chapter, we introduce some basic concepts, terminologies, notations, and tools for studying random processes and present several important examples of random processes as well.

8.1 Definition and Classification of Processes

In the study of deterministic signals, we often encounter four types or classes of signals:

- *Continuous time and continuous amplitude* signals are a function of a continuous independent variable, time. The amplitude of the function is also continuous.
- *Continuous time and discrete amplitude* signals are a function of a continuous independent variable, time—but the amplitude is discrete.
- *Discrete time and continuous amplitude* signals are functions of a quantized or discrete independent time variable, while amplitude is continuous.
- *Discrete time and discrete amplitude* signals are functions where both the independent time variable and the amplitude are discrete.

In this text, we write a continuous function of time as $x(t)$, where t is the continuous time variable. For discrete-time signals, the time variable is typically limited to regularly spaced discrete points in time, $t = nt_o$. In this case, we use the notation $x[n] = x(nt_o)$ to represent the discrete sequence of numbers. Most of the discussion that follows is presented in terms of continuous time signals, but the conversion to the discrete-time case will be straightforward in most cases.

Recall from Chapter 3 that a random variable, X, is a function of the possible outcomes, ζ, of an experiment. Now, we would like to extend this concept so that a function of time $x(t)$ (or $x[n]$ in the discrete-time case) is assigned to every outcome, ζ, of an experiment. The

Probability and Random Processes. DOI: 10.1016/B978-0-12-386981-4-00008-5 function, $x(t)$, may be real or complex and it can be discrete or continuous in amplitude. Strictly speaking, the function is really a function of two variables, $x(t, \zeta)$, but to keep the notation simple, we typically do not explicitly show the dependence on the outcome, just as we have not in the case of random variables. The function $x(t)$ may have the same general dependence on time for every outcome of the experiment or each outcome could produce a completely different waveform. In general, the function $x(t)$ is a member of an *ensemble* (family, set, collection) of functions. Just as we did for random variables, an ensemble of member functions, $X(t)$, is denoted with an upper case letter. Thus, $X(t)$ represents the random process, while $x(t)$ is one particular member or *realization* of the random process. In summary, we have the following definition of a random process:

> **Definition 8.1:** A random process is a function of the elements of a sample space, S, as well as another independent variable, t. Given an experiment, E, with sample space, S, the random process, $X(t)$, maps each possible outcome, $\zeta \in S$, to a function of t, $x(t, \zeta)$, as specified by some rule.

■ Example 8.1:

Suppose an experiment consists of flipping a coin. If the outcome is heads, $\zeta = \mathrm{H}$, the random process takes on the functional form $x_{\mathrm{H}}(t) = \sin(\omega_0 t)$; whereas, if the outcome is tails, $\zeta = \mathrm{T}$, the realization $x_{\mathrm{T}}(t) = \sin(2\omega_0 t)$ occurs, where ω_0 is some fixed frequency. The two realizations of this random process are illustrated in Figure 8.1.

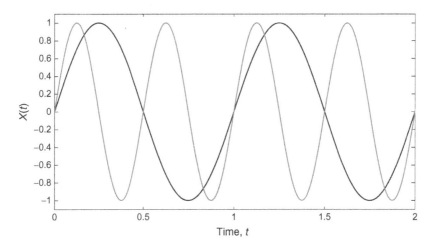

Figure 8.1
Member functions for the random process of Example 8.1.

■

The random process in Example 8.1 actually has very little randomness. There are only two possible realizations of the random process. Furthermore, given an observation of the realization of the random process at one point in time, $X(t_1)$, one could determine the rest of the realization (as long as $\omega_0 t_1 \neq n\pi$). The next example shows that a random process could have this last property, even if the number of realizations were infinite.

■ Example 8.2:

Now suppose that an experiment results in a random variable A that is uniformly distributed over $[0, 1)$. A random process is then constructed according to $X(t) = A\sin(\omega_0 t)$. Since the random variable is continuous, there are an uncountably infinite number of realizations of the random process. A few are shown in Figure 8.2. As with the previous example, given an observation of the realization of the random process at one point in time, $X(t_1)$, one could determine the rest of the realization (as long as $\omega_0 t_1 \neq n\pi$).

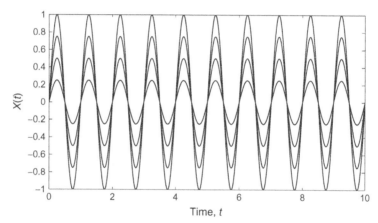

Figure 8.2
Some member functions for the random process of Example 8.2.

■ Example 8.3:

This example is a generalization of that given in Example 8.1. Suppose now the experiment consists of flipping a coin repeatedly and observing the sequence of outcomes. The random process $X(t)$ is then constructed as $X(t) = \sin(\Omega_i t)$, $(i-1)T \leq t < iT$, where $\Omega_i = \omega_0$ if the ith flip of the coin results in "heads" and $\Omega_i = 2\omega_0$ if the ith flip of the coin results in "tails." One possible realization of this random process is illustrated in Figure 8.3. This is the sort of signal that might be produced by a frequency shift keying modem. In that application, the frequencies are not determined by coin tosses, but by random data bits instead.

(Continued)

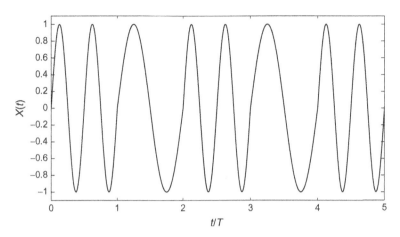

Figure 8.3
One possible realization for the random process of Example 8.3.

■ Example 8.4:

As an example of a random process that is discrete in amplitude but continuous in time, we present the so-called "random telegraph" process. Let $T_1, T_2, T_3, \ldots$ be a sequence of IID random variables, each with an exponential distribution,

$$f_T(s) = \lambda e^{-\lambda s} u(s).$$

At any time instant, the random telegraph signal, $X(t)$, takes on one of two possible states, $X(t) = 0$ or $X(t) = 1$. Suppose the process starts (at time $t = 0$) in the zero state. It then remains in that state for a time interval equal to T_1 at which point it switches to the state $X(t) = 1$. The process remains in that state for another interval of time equal in length to T_2 and then switches states again. The process then continues to switch after waiting for time intervals specified by the sequence of exponential random variables. One possible realization is shown in Figure 8.4. The MATLAB code for generating such a process follows.

```
N=10;                        % number of switches in realization
Fs=100;                      % Sample rate (samples per second)
lambda=1/2;                  % switching rate (switches per second)
X=[];
S=rand(1,N);                 % uniform random variables.
T=-log(S)/lambda;            % transform to exponential RVs.
V=cumsum(T);                 % switching times.
state=0; Nsold=1;
```

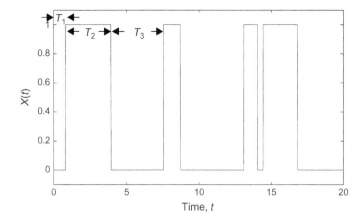

Figure 8.4
One possible realization for the random telegraph signal of Example 8.4.

```
for k=1:N
    Nsnew=ceil(V(k)*Fs);              % new switching time
    Ns=Nsnew-Nsold;                   % number of samples in current
                                      % switching interval

    X=[X state*ones(1,Ns)];
    state=1-state;                    % switch state
    Nsold=Nsnew;
end
t=[1:length(X)]/Fs;                   % time axis
plot(t,X)                             % plot results
xlabel('time, t'); ylabel('X(t)')
axis([0 max(t) -0.1 1.1])             % manual scale of axes
```

■ Example 8.5:

As an example of a discrete-time random process, suppose each outcome of an experiment produces a sequence of IID, zero-mean Gaussian random variables, $W_1, W_2, W_3, \ldots$. A discrete-time random process $X[n]$ could be constructed according to:

$$X[n] = X[n-1] + W_n,$$

with the initial condition $X[0] = 0$. The value of the process at each point in time is equal to the value of the process at the previous point in time plus a random increase (decrease) which follows a Gaussian distribution. Note also that $X[n]$ is just the sum of the first n terms in the sequence W_i. A sample realization of this random process is shown in Figure 8.5. The MATLAB code for generating this process is provided here.

(Continued)

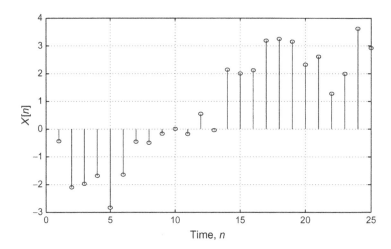

Figure 8.5
One possible realization for the discrete-time random process of Example 8.5.

The reader is encouraged to run this program several times to see several different realizations of the same random process.

```
N=25;                               % number of time instants in process.
W=randn(1,N);                       % Gaussian random variables.
X=[0 cumsum(W)];                    % Samples of X[n]
stem([0:N],X,'o')                   % plot realization of X[n]
xlabel('time, n'); ylabel('X[n]');
```

■

8.2   Mathematical Tools for Studying Random Processes

As with random variables, we can mathematically describe a random process in terms of a cumulative distribution function, probability density function, or a probability mass function. In fact, given a random process, $X(t)$, which is sampled at some specified point in time, $t = t_k$, the result is a random variable, $X_k = X(t_k)$. This random variable can then be described in terms of its PDF, $f_X(x_k;t_k)$. Note that an additional time variable has been added to the PDF. This is necessary due to the fact that the PDF of the sample of the random process may depend on when the process is sampled. If desired, the CDF or PMF can be used rather than the PDF to describe the sample of the random process.

■ Example 8.6:

Consider the random telegraph signal of Example 8.4. Since this process is binary valued, any sample will be a Bernoulli random variable. The only question is, what is the probability that $X_k = X(t_k)$ is equal to 1 (or 0)? Suppose that there are exactly n switches in the time interval $[0, t_k)$. Then $X(t_k) = n \bmod 2$. Stated another way, define $S_n = T_1 + T_2 + \ldots + T_n$. There will be exactly n switches in the time interval $[0, t_k)$ provided that $S_n < t_k < S_{n+1}$. Therefore,

$$\Pr(n \text{ switches in } [0, t_k)) = \Pr(S_n < t_k < S_{n+1}) = \int \Pr(S_n < t_k < S_{n+1}|S_n=s)f_{S_n}(s)\mathrm{d}s$$

$$= \int_0^{t_k} \Pr(t_k < S_{n+1}|S_n=s)f_{S_n}(s)\mathrm{d}s = \int_0^{t_k} \Pr(T_{n+1} > t_k - s)f_{S_n}(s)\mathrm{d}s \, .$$

Since the T_i are IID and exponential, S_n will follow a Gamma distribution. Using the Gamma PDF for $f_{S_n}(s)$ and the fact that $\Pr(T_{n+1} > t_k - s) = \exp(-\lambda(t_k - s))$ results in

$$\Pr(n \text{ switches in } [0, t_k)) = \int_0^{t_k} \mathrm{e}^{-\lambda(t_k - s)}\frac{\lambda^n s^{n-1}}{(n-1)!}\mathrm{e}^{-\lambda s}\mathrm{d}s = \frac{\lambda^n \mathrm{e}^{-\lambda t_k}}{(n-1)!}\int_0^{t_k} s^{n-1}\mathrm{d}s = \frac{(\lambda t_k)^n}{n!}\mathrm{e}^{-\lambda t_k} \, .$$

So, it is seen that the number of switches in the interval $[0, t_k)$ follows a Poisson distribution. The sample of the random process will be equal to 0 if the number of switches is even. Thus,

$$\Pr(X(t_k)=0) = \sum_{n \text{ even}} \Pr(n \text{ switches in } [0, t_k)) = \sum_{n \text{ even}} \frac{(\lambda t_k)^n}{n!}\mathrm{e}^{-\lambda t_k} = \mathrm{e}^{-\lambda t_k}\cosh(\lambda t_k) = \frac{1}{2} + \frac{1}{2}\mathrm{e}^{-2\lambda t_k} \, .$$

Likewise,

$$\Pr(X(t_k)=1) = \sum_{n \text{ odd}} \Pr(n \text{ switches in } [0, t_k)) = \sum_{n \text{ odd}} \frac{(\lambda t_k)^n}{n!}\mathrm{e}^{-\lambda t_k} = \mathrm{e}^{-\lambda t_k}\sinh(\lambda t_k) = \frac{1}{2} - \frac{1}{2}\mathrm{e}^{-2\lambda t_k} \, .$$

The behavior of this distribution as it depends on time is shown in Figure 8.6 and should make intuitive sense. For very small values of t_k , it is most likely that there are no switches in the interval $[0, t_k)$, in which case $\Pr(X(t_k)=0)$ should be close to one. On the other hand, for large t_k , many switches will likely occur and it should be almost equally likely that the process take on the values of 0 or 1.

■

■ Example 8.7:

Now consider the PDF of a sample of the discrete-time process of Example 8.5. Note that since $X[n]$ is formed by summing n IID Gaussian random variables, $X[n]$ will itself be a Gaussian random variable with mean of $E[X[n]] = n\mu_W$ and variance $\mathrm{Var}(X[n]) = n\sigma_W^2$. In this case, since the W_i's were taken to be zero-mean, the PDF of $X[n]$ is

$$f_X(x;n) = \frac{1}{\sqrt{2\pi n\sigma_W^2}}\exp\left(-\frac{x^2}{2n\sigma_W^2}\right) \, .$$

Once again, we see that for this example, the form of the PDF does indeed depend on when the sample is taken.

■

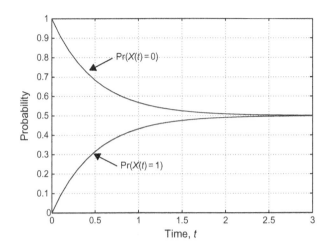

Figure 8.6
Time dependence of the PMF for the random telegraph signal.

The PDF (or CDF or PMF) of a sample of a random process taken at an arbitrary point in time goes a long way toward describing the random process, but it is not a complete description. To see this, consider two samples, $X_1 = X(t_1)$ and $X_2 = X(t_2)$, taken at two arbitrary points in time. The PDF, $f_X(x;t)$, describes both X_1 and X_2, but it does not describe the relationship between X_1 and X_2. For some random processes, it might be reasonable to expect that X_1 and X_2 would be highly correlated if t_1 is near t_2, while X_1 and X_2 might be virtually uncorrelated if t_1 and t_2 are far apart. To characterize relationships of this sort, a joint PDF of the two samples would be needed. That is, it would be necessary to construct a joint PDF of the form $f_{X_1, X_2}(x_1, x_2; t_1, t_2)$. This is referred to as a *second-order PDF* of the random process $X(t)$.

Continuing with this reasoning, in order to completely describe the random process, it is necessary to specify an nth order PDF for an arbitrary n. That is, suppose the random process is sampled at time instants t_1, t_2, ..., t_n, producing the random variables $X_1 = X(t_1)$, $X_2 = X(t_2)$, ..., $X_n = X(t_n)$. The joint PDF of the n samples, $f_{X_1, X_2, ..., X_n}(x_1, x_2, ..., x_n; t_1, t_2, ..., t_n)$, for an arbitrary n and arbitrary sampling times will give a complete description of the random process. In order to make this notation more compact, the vectors $\boldsymbol{X} = (X_1, X_2, ..., X_n)^T$, $\boldsymbol{x} = (x_1, x_2, ..., x_n)^T$, and $\boldsymbol{t} = (t_1, t_2, ..., t_n)^T$ are introduced and the nth order joint PDF is written as $f_{\boldsymbol{X}}(\boldsymbol{x};\boldsymbol{t})$.

Unfortunately, for many realistic random processes, the prospects of writing down an n-th order PDF are rather daunting. One notable exception is the Gaussian random process which will be described in detail in Section 8.5. However, for most other cases, specifying a joint

PDF of *n* samples may be exceedingly difficult, and hence it is necessary to resort to a simpler but less complete description of the random process. The simplest is the mean function of the process.

> **Definition 8.2:** The *mean function* of a random process is simply the expected value of the process. For continuous time processes, this is written as

$$\mu_X(t) = E[X(t)] = \int x f_X(x;t) dx, \tag{8.1}$$

> while for discrete-time processes, the following notation is used:

$$\mu_X[n] = E[X[n]] = \int x f_X(x;n) dx. \tag{8.2}$$

In general, the mean of a random process may change with time, but in many cases, this function is constant. Also, it is noted that only the first-order PDF of the process is needed to compute the mean function.

Example 8.8:

Consider the random telegraph process of Example 8.4. It was shown in Example 8.6 that the first-order PMF of this process was described by a Bernoulli distribution with

$$\Pr(X(t) = 1) = \frac{1}{2} - \frac{1}{2}\exp(-\lambda t).$$

The mean function then follows as

$$\mu_X(t) = E[X(t)] = 1 * \Pr(X(t) = 1) + 0 * \Pr(X(t) = 0) = \frac{1}{2} - \frac{1}{2}\exp(-\lambda t).$$

Example 8.9:

Next, consider the sinusoidal random process of Example 8.2 where $X(t) = A\sin(\omega_0 t)$ and A was a uniform random variable over $[0, 1)$. In this case,

$$u_X(t) = E[X(t)] = E[A\sin(\omega_0 t)] = E[A]\sin(\omega_0 t) = \frac{1}{2}\sin(\omega_0 t).$$

This example illustrates a very important concept in that quite often it is not necessary to explicitly evaluate the first-order PDF of a random process in order to evaluate its mean function.

■ Example 8.10:

Now suppose the random process of the previous example is slightly modified. In particular, consider a sine-wave process where the random variable is the phase, Θ, which is uniformly distributed over $[0, 2\pi)$. That is, $X(t) = a\sin(\omega_0 t + \Theta)$. For this example, the amplitude of the sine wave, a, is taken to be fixed (not random). The mean function is then

$$\mu_X(t) = E[X(t)] = E[a\sin(\omega_0 t + \Theta)] = a\int f_\Theta(\theta)\sin(\omega_0 t + \theta)d\theta = \frac{a}{2\pi}\int_0^{2\pi}\sin(\omega_0 t + \theta)d\theta = 0,$$

which is a constant. Why is the mean function of the previous example a function of time and this one is not? Consider the member functions of the respective ensembles for the two random processes.

■

■ Example 8.11:

Now consider a sinusoid with a random frequency $X(t) = \cos(2\pi Ft)$, where F is a random variable uniformly distributed over some interval $(0, f_0)$. The mean function can be readily determined to be

$$\mu_X(t) = E[\cos(2\pi Ft)] = \frac{1}{f_0}\int_0^{f_0}\cos(2\pi ft)df = \frac{\sin(2\pi f_0 t)}{2\pi f_0 t} = \mathrm{sinc}(2f_0 t).$$

We can also estimate the mean function through simulation. Below we provide some MATLAB code to produce many realizations of this random process. The mean function is then found by taking the sample mean of all the realizations created. The sample mean and the ensemble mean are shown in Figure 8.7. Naturally, more or less accuracy in the sample mean can be obtained by varying the number of realizations generated.

```
fo=2;                                   % max frequency
N=1000;                                 % number of realizations
t=[-4.995:0.01:4.995];                  % time axis
F=fo*rand(N,1);                         % uniform frequencies
x=cos(2*pi*F*t);                        % each row is a
                                        % realization of process
sample_mean=sum(x)/N;                   % compute sample mean
true_mean=sin(2*pi*fo*t)./(2*pi*fo*t);  % compute ensemble mean
plot(t,sample_mean,'-',t,true_mean,'--')% plot results
xlabel('t (seconds)'); ylabel('mu(t)');
```

■

To partially describe the second-order characteristics of a random process, the autocorrelation function is introduced.

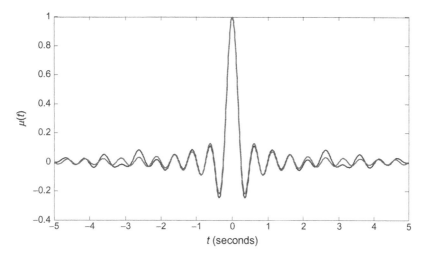

Figure 8.7
Comparison of the sample mean and ensemble mean for the sinusoid with random frequency of
Example 8.11. The solid line is the sample mean while the dashed line is the ensemble mean.

Definition 8.3: The *autocorrelation function*, $R_{XX}(t_1, t_2)$, of a continuous-time
random process, $X(t)$, is defined as the expected value of the product $X(t_1)X(t_2)$:

$$R_{XX}(t_1, t_2) = E[X(t_1)X(t_2)] = \int_{-\infty}^{\infty} \int_{-\infty}^{\infty} x_1 x_2 f_{X_1, X_2}(x_1, x_2; t_1, t_2) dx_1 dx_2. \qquad (8.3)$$

For discrete-time processes, the autocorrelation function is

$$R_{XX}[n_1, n_2] = E[X[n_1]X[n_2]] = \int_{-\infty}^{\infty} \int_{-\infty}^{\infty} x_1 x_2 f_{X_1, X_2}(x_1, x_2; n_1, n_2) dx_1 dx_2. \qquad (8.4)$$

Naturally, the autocorrelation function describes the relationship (correlation) between
two samples of a random process. This correlation will depend on when the samples are
taken; thus, the autocorrelation function is, in general, a function of two time variables.
Quite often we are interested in how the correlation between two samples depends on
how far apart the samples are spaced. To explicitly draw out this relationship, define a
time difference variable, $\tau = t_2 - t_1$, and the autocorrelation function can then be
expressed as

$$R_{XX}(t, t + \tau) = E[X(t)X(t + \tau)] , \qquad (8.5)$$

where we have replaced t_1 with t to simplify the notation even further.

■ **Example 8.12:**

Consider the sine wave process with a uniformly distributed amplitude as described in Examples 8.2 and 8.9, where $X(t) = A\sin(\omega_0 t)$. The autocorrelation function is found as

$$R_{XX}(t_1, t_2) = E[X(t_1)X(t_2)] = E[A^2\sin(\omega_0 t_1)\sin(\omega_0 t_2)] = \frac{1}{3}\sin(\omega_0 t_1)\sin(\omega_0 t_2)$$

or

$$R_{XX}(t, t + \tau) = \frac{1}{3}\sin(\omega_0 t)\sin(\omega_0 (t + \tau)).$$

■

■ **Example 8.13:**

Now consider the sine wave process with random phase of Example 8.10 where $X(t) = a\sin(\omega_0 t + \Theta)$. Then

$$R_{XX}(t_1, t_2) = E[X(t_1)X(t_2)] = E[a^2\sin(\omega_0 t_1 + \theta)\sin(\omega_0 t_2 + \theta)].$$

To aid in calculating this expected value, we use the trigonometric identity

$$\sin(x)\sin(y) = \frac{1}{2}\cos(x - y) - \frac{1}{2}\cos(x + y).$$

The autocorrelation then simplifies to

$$R_{XX}(t_1, t_2) = \frac{a^2}{2}E[\cos(\omega_0(t_2 - t_1))] + \frac{a^2}{2}E[\cos(\omega_0(t_1 + t_2 + 2\theta))] = \frac{a^2}{2}\cos(\omega_0(t_2 - t_1))$$

or

$$R_{XX}(t, t + \tau) = \frac{a^2}{2}\cos(\omega_0 \tau).$$

Note that in this case, the autocorrelation function is only a function of the difference between the two sampling times. That is, it does not matter where the samples are taken, only how far apart they are.

■

■ **Example 8.14:**

Recall the random process of Example 8.5 where $X[n] = X[n-1] + W_n$, $X[0] = 0$ and the W_n were a sequence of IID, zero-mean Gaussian random variables. In this case, it is easier to calculate the autocorrelation function using the alternative expression,

$$X[n] = \sum_{i=1}^{n} W_i$$

Then,

$$R_{XX}[n_1, n_2] = E[X[n_1]X[n_2]] = E\left[\sum_{i=1}^{n_1} W_i \sum_{j=1}^{n_2} W_j\right] = \sum_{i=1}^{n_1} \sum_{j=1}^{n_2} E[W_i W_j] .$$

Since the W_i are IID and zero-mean, $E[W_i W_j] = 0$ unless $i = j$. Therefore,

$$R_{XX}[n_1, n_2] = \min(n_1, n_2)\sigma_W^2 .$$

■

Definition 8.4: The autocovariance function, $C_{XX}(t_1, t_2)$, of a continuous time random process, $X(t)$, is defined as the covariance of $X(t_1)$ and $X(t_2)$:

$$C_{XX}(t_1, t_2) = \text{Cov}(X(t_1), X(t_2)) = E[(X(t_1) - \mu_X(t_1))(X(t_2) - \mu_X(t_2))] . \qquad (8.6)$$

The definition is easily extended to discrete-time random processes.

As with the covariance function for random variables, the autocovariance function can be written in terms of the autocorrelation function and the mean function:

$$C_{XX}(t_1, t_2) = R_{XX}(t_1, t_2) - \mu_X(t_1)\mu_X(t_2). \qquad (8.7)$$

Once the mean and autocorrelation functions of a random process have been computed, the autocovariance function is trivial to find.

The autocovariance function is helpful when studying random processes which can be represented as the sum of a deterministic signal, $s(t)$, plus a zero-mean noise process, $N(t)$. If $X(t) = s(t) + N(t)$, then the autocorrelation function of $X(t)$ is

$$R_{XX}(t_1, t_2) = E[(s(t_1) + N(t_1))(s(t_2) + N(t_2))] = s(t_1)s(t_2) + R_{NN}(t_1, t_2), \qquad (8.8)$$

using the fact that $\mu_N(t) = 0$. If the signal is strong compared to the noise, the deterministic part will dominate the autocorrelation function, and thus $R_{XX}(t_1, t_2)$ will not tell us much about the randomness in the process $X(t)$. On the other hand, the autocovariance function is

$$C_{XX}(t_1, t_2) = R_{XX}(t_1, t_2) - s(t_1)s(t_2) = R_{NN}(t_1, t_2) = C_{NN}(t_1, t_2). \qquad (8.9)$$

Therefore, the autocovariance function allows us to isolate the noise which is the source of randomness in the process.

Definition 8.5: For a pair of random processes $X(t)$ and $Y(t)$, the cross-correlation function is defined as

$$R_{XY}(t_1, t_2) = E[X(t_1)Y(t_2)] . \qquad (8.10)$$

Likewise, the cross-covariance function is

$$C_{XY}(t_1, t_2) = E[(X(t_1) - \mu_X(t_1))(Y(t_2) - \mu_Y(t_2))]. \tag{8.11}$$

■ Example 8.15:

Suppose $X(t)$ is a zero-mean random process with autocorrelation function $R_{XX}(t_1, t_2)$. A new process $Y(t)$ is formed by delaying $X(t)$ by some amount t_d. That is, $Y(t) = X(t - t_d)$. Then the cross-correlation function is

$$R_{XY}(t_1, t_2) = E[X(t_1)Y(t_2)] = E[X(t_1)X(t_2 - t_d)] = R_{XX}(t_1, t_2 - t_d).$$

In a similar fashion, it is seen that $R_{YX}(t_1, t_2) = R_{XX}(t_1 - t_d, t_2)$ and $R_{YY}(t_1, t_2) = R_{XX}(t_1 - t_d, t_2 - t_d)$.

■

8.3 Stationary and Ergodic Random Processes

From the few simple examples given in the preceding section, we conclude that the mean function and the autocorrelation (or autocovariance) function can provide information about the temporal structure of a random process. We will delve into the properties of the autocorrelation function in more detail later in this chapter, but first the concepts of stationarity and ergodicity must be introduced.

Definition 8.6: A continuous time random process $X(t)$ is *strict sense stationary* if the statistics of the process are invariant to a time shift. Specifically, for any time shift τ and any integer $n \geq 1$,

$$f_{X_1, X_2, \ldots, X_n}(x_1, x_2, \ldots, x_n; t_1, t_2, \ldots, t_n) =$$

$$f_{X_1, X_2, \ldots, X_n}(x_1, x_2, \ldots, x_n; t_1 + \tau, t_2 + \tau, \ldots, t_n + \tau). \tag{8.12}$$

In general, it is quite difficult to show that a random process is strict sense stationary since to do so, one needs to be able to express the general nth order PDF. On the other hand, to show that a process is not strict sense stationary, one needs to show only that one PDF of any order is not invariant to a time shift. One example of a process that can be shown to be stationary in the strict sense is an IID process. That is, suppose $X(t)$ is a random process that has the property that $X(t)$ has an identical distribution for any t and that $X(t_1)$ and $X(t_2)$ are independent for any $t_1 \neq t_2$. In this case,

$$f_{X_1, X_2, \ldots, X_n}(x_1, x_2, \ldots, x_n; t_1, t_2, \ldots, t_n) = \prod_{i=1}^{n} f_{X_i}(x_i; t_i). \tag{8.13}$$

Since the nth order PDF is the product of first order PDFs and the first-order PDF is invariant to a time shift, then the nth order PDF must be invariant to a time shift.

■ Example 8.16:

Consider the sinusoidal process with random amplitude from Example 8.2, where $X(t) = A\sin(\omega_0 t)$. This process is clearly not stationary since if we take any realization of the process, $x(t) = a\sin(\omega_0 t)$, then a time shift $x(t + \tau) = a\sin(\omega_0(t + \tau))$ would not be a realization in the original ensemble. Now suppose the process has a random phase rather than a random amplitude as in Example 8.10, resulting in $X(t) = a\sin(\omega_0 t + \Theta)$. It was already shown in Example 8.10 that $\mu_X(t) = 0$ for this process, and therefore the mean function is invariant to a time shift. Furthermore, in Example 8.13, it was shown that $R_{XX}(t, t + \tau) = (a^2/2)\cos(\omega_0 \tau)$, and thus the autocorrelation function is also invariant to a time shift. It is not difficult to show that the first-order PDF follows an arcsine distribution

$$f_X(x;t) = \frac{1}{\pi\sqrt{1 - x^2}}, \qquad -1 < x < 1,$$

and it is also independent of time and thus invariant to a time shift. It seems that this process might be stationary in the strict sense, but it would be rather cumbersome to prove it because the nth order PDF is difficult to specify.

■

As was seen in Example 8.16, it may be possible in some examples to determine that some of the statistics of a random process are invariant to time shifts, but determining stationarity in the strict sense may be too big of a burden. In those cases, we often settle for a looser form of stationarity.

Definition 8.7: A random process is *wide sense stationary* (WSS) if the mean function and autocorrelation function are invariant to a time shift. In particular, this implies that

$$\mu_X(t) = \mu_X = \text{constant}, \tag{8.14}$$

$$R_{XX}(t, t + \tau) = R_{XX}(\tau) \ \text{(function only of } \tau\text{)}. \tag{8.15}$$

All strict sense stationary random processes are also WSS, provided that the mean and autocorrelation function exist. The converse is not true. A WSS process does not necessarily need to be stationary in the strict sense. We refer to a process which is not WSS as *non-stationary*.

■ Example 8.17:

Suppose we form a random process $Y(t)$ by modulating a carrier with another random process, $X(t)$. That is, let $Y(t) = X(t)\cos(\omega_0 t + \Theta)$ where Θ is uniformly distributed over $[0, 2\pi)$ and independent of $X(t)$. Under what conditions is $Y(t)$ WSS? To answer this, we calculate the mean and autocorrelation function of $Y(t)$.

$$\mu_Y(t) = E[X(t)\cos(\omega_0 t + \Theta)] = E[X(t)]E[\cos(\omega_0 t + \Theta)] = 0,$$

$$R_{YY}(t, t+\tau) = E[X(t)X(t+\tau)\cos(\omega_0 t + \Theta)\cos(\omega_0(t+\tau) + \Theta)]$$

$$= E[X(t)X(t+\tau)]\left\{\frac{1}{2}\cos(\omega_0 \tau) + \frac{1}{2}E[\cos(\omega_0(2t+\tau) + 2\Theta)]\right\}$$

$$= \frac{1}{2}R_{XX}(t, t+\tau)\cos(\omega_0 \tau).$$

While the mean function is a constant, the autocorrelation is not necessarily only a function of τ. The process $Y(t)$ will be WSS provided that $R_{XX}(t, t+\tau) = R_{XX}(\tau)$. Certainly if $X(t)$ is WSS, then $Y(t)$ will be as well.

■

■ Example 8.18:

Let $X(t) = At + B$ where A and B are independent random variables, both uniformly distributed over the interval $(-1, 1)$. To determine whether this process is WSS, calculate the mean and autocorrelation functions:

$$\mu_X(t) = E[At + B] = E[A]t + E[B] = 0,$$

$$R_{XX}(t, t+\tau) = E[(At + B)(A(t+\tau) + B)] = E[A^2]t(t+\tau) + E[B^2] + E[AB](2t+\tau) = \frac{1}{3}t(t+\tau) + \frac{1}{3}.$$

Clearly, this process is not WSS.

■

Many of the processes we deal with are WSS and hence have a constant mean function and an autocorrelation function that depends only on a single time variable. Hence, in the remainder of the text, when a process is known to be WSS or if we are assuming it to be WSS, then we will represent its autocorrelation function by $R_{XX}(\tau)$. If a process is non-stationary or if we do not know if the process is WSS, then we will explicitly write the autocorrelation function as a function of two variables, $R_{XX}(t, t+\tau)$. For example, if we say that a process has a mean function of $\mu_X = 1$, and an autocorrelation function, $R_{XX}(\tau) = \exp(-|\tau|)$, then the reader can infer that the process is WSS, even if it is not explicitly stated.

In order to calculate the mean or autocorrelation function of a random process, it is necessary to perform an ensemble average. In many cases, this may not be possible as we may not be able to observe all realizations (or a large number of realizations) of a random process. In fact, quite often we may be able to observe only a single realization. This would occur in situations where the conditions of an experiment cannot be duplicated and therefore the experiment is not repeatable. Is it possible to calculate the mean and/or autocorrelation function from a single realization of a random process? The answer is sometimes, depending on the nature of the process.

To start with, consider the mean. Suppose a WSS random process $X(t)$ has a mean μ_X. We are able to observe one realization of the random process, $x(t)$, and wish to try to determine μ_X from this realization. One obvious approach would be to calculate the time average[1] of the realization:

$$\langle x(t) \rangle = \lim_{t_o \to \infty} \frac{1}{2t_o} \int_{-t_o}^{t_o} x(t)\mathrm{d}t. \tag{8.16}$$

However, it is not obvious if the time average of one realization is necessarily equal to the ensemble average. If the two averages are the same, then we say that the random process is *ergodic in the mean*.

One could take the same approach for the autocorrelation function. Given a single realization, $x(t)$, form the time-average autocorrelation function:

$$\Re_{xx}(\tau) = \langle x(t)x(t+\tau) \rangle = \lim_{t_o \to \infty} \frac{1}{2t_o} \int_{-t_o}^{t_o} x(t)x(t+\tau)\mathrm{d}t. \tag{8.17}$$

If $\Re_{xx}(\tau) = R_{XX}(\tau)$ for any realization, $x(t)$, then the random process is said to be *ergodic in the autocorrelation*. In summary, we have the following definition of ergodicity:

> **Definition 8.8:** A WSS random process is *ergodic* if ensemble averages involving the process can be calculated using time averages of any realization of the process. Two limited forms of ergodicity are:
>
> • Ergodic in the mean - $\langle x(t) \rangle = E[X(t)]$,
> • Ergodic in the autocorrelation - $\langle x(t)x(t+\tau) \rangle = E[X(t)X(t+\tau)]$.

■ Example 8.19:

As a simple example, suppose $X(t) = A$ where A is a random variable with some arbitrary PDF $f_A(a)$. Note that this process is stationary in the strict sense since for any realization, $x(t) = x(t+\tau)$. That is, not only are the statistics of the process invariant to

(Continued)

---

[1] Throughout the text, angular brackets $\langle \ \rangle$ are used as a shorthand notation to represent the time-average operator.

time shifts, but every realization is also invariant to any time shift. If we take the time average of a single realization, $x(t) = a$, we get $\langle x(t) \rangle = a$. Hence, each different realization will lead to a different time average and will not necessarily give the ensemble mean, μ_A. Although this process is stationary in the strict sense, it is not ergodic in any sense.

■

■ Example 8.20:

Now consider the sinusoid with random phase $X(t) = a\sin(\omega_0 t + \Theta)$, where Θ is uniform over $[0, 2\pi)$. It was demonstrated in Example 8.13 that this process is WSS. But is it ergodic? Given any realization $x(t) = a\sin(\omega_0 t + \theta)$, the time average is $\langle x(t) \rangle = \langle a\sin(\omega_0 t + \theta) \rangle = 0$. That is, the average value of any sinusoid is zero. So, this process is ergodic in the mean since the ensemble average of this process was also zero. Next, consider the sample autocorrelation function:

$$
\langle x(t)x(t + \tau) \rangle = a^2 \langle \sin(\omega_0 t + \theta)\sin(\omega_0 t + \omega_0 \tau + \theta) \rangle
$$

$$
= \frac{a^2}{2} \langle \cos(\omega_0 \tau) \rangle - \frac{a^2}{2} \langle \cos(2\omega_0 t + \omega_0 \tau + 2\theta) \rangle
$$

$$
= \frac{a^2}{2} \cos(\omega_0 \tau).
$$

This also is exactly the same expression obtained for the ensemble averaged autocorrelation function. Therefore, this process is also ergodic in the autocorrelation.

■

■ Example 8.21:

For a process that is known to be ergodic, the autocorrelation function can be estimated by taking a sufficiently long time average of the autocorrelation function of a single realization. We demonstrate this via MATLAB for a process that consists of a sum of sinusoids of fixed frequencies and random phases,

$$
X(t) = \sum_{k=1}^{n} \cos(2\pi f_k t + \theta_k),
$$

where the θ_K are IID and uniform over $(0, 2\pi)$. For an arbitrary signal $x(t)$, we note the similarity between the time-averaged autocorrelation function and the convolution of $x(t)$ and $x(-t)$. If we are given a single realization, $x(t)$, which lasts only for the time interval, $(-t_0, t_0)$, then these two expressions are given by

$$
\langle x(t)x(t + \tau) \rangle = \frac{1}{2t_0 - \tau} \int_{-t_0}^{t_0 - \tau} x(t)x(t + \tau)dt,
$$

$$x(t)*x(-t) = \int_{-t_0}^{t_0-\tau} x(t)x(t+\tau)dt,$$

for $\tau > 0$. In general, we have the relationship

$$\langle x(t)x(t+\tau) \rangle = \frac{x(t)*x(-t)}{2t_0 - |\tau|}.$$

By using the MATLAB convolution function, conv, the time-averaged autocorrelation can easily be computed. This is demonstrated in the code that follows. Figure 8.8 shows a comparison between the ensemble averaged autocorrelation and the time-averaged autocorrelation taken from a single realization. From the figure, it is noted that the agreement between the two is good for $\tau \ll t_0$, but not good when $\tau \sim t_0$. This is due to the fact that when τ approaches t_0, the time window over which the time average is computed gets too small to produce an accurate estimate.

```
N=4;                            % number of sinusoids
to=5;                           % time duration
Ts=0.01;                        % sample interval
t=[-to:Ts:to];                  % time axis
tau=[-2*to:Ts:2*to];            % tau axis
theta=rand(1,N);                % random phases
f=1./[1:N];                     % frequencies (not random)
x=zeros(size(t));
```

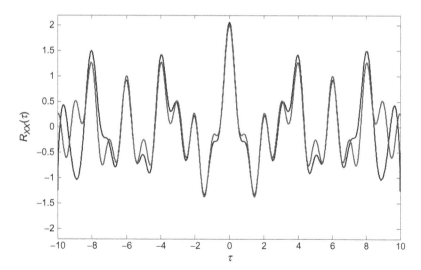

Figure 8.8

Comparison of the time-average autocorrelation and the ensemble-average autocorrelation for the sum of sinusoids process of Example 8.21. The solid line is the time-average autocorrelation while the dashed line is the ensemble-average autocorrelation.

(Continued)

```
      True_Rxx=zeros(size(tau));
      for k=1:N
          x=x+cos(2*pi*f(k)*t+2*pi*theta(k));      % construct process
          True_Rxx=True_Rxx+cos(2*pi*f(k)*tau)/2;   % compute Rxx(tau)
      end
      z=conv(x,fliplr(x));                          % x(t)*x(-t)
      Rxx=Ts*z./(2*to-abs(tau));                    % time averaged Rxx
      plot(tau,Rxx,'-',tau,True_Rxx,'--')           % plot results
      xlabel('tau'); ylabel('R_X_X(tau)')
      axis([-2*to 2*to -1.1*N/2 1.1*N/2])
```

∎

The previous examples show two different random processes, one that is ergodic and one that is not. What characteristics of a random process make it ergodic? To get some better insight toward answering this question, consider a discrete-time process, $X[n]$, where each random variable in the sequence is IID and consider forming the time average,

$$\langle X[n] \rangle = \lim_{m \to \infty} \frac{1}{m} \sum_{n=1}^{m} X[n]. \tag{8.18}$$

The right hand side of the previous equation is nothing more than the sample mean. By virtue of the law of large numbers, the limit will indeed converge to the ensemble mean of the random process, μ_X, and thus this process is ergodic in the mean. In this case, the time average converges to the ensemble average because each time sample of the random process gives an independent observation of the underlying randomness. If the samples were highly correlated, then taking more samples would not necessarily cause the sample mean to converge to the ensemble mean. So, it seems that the form of the autocorrelation function will play some role in determining if a random process is ergodic in the mean.

To formalize this concept, consider the time average of an arbitrary WSS random process.

$$\langle X(t) \rangle = \lim_{t_o \to \infty} X_{t_o} \quad \text{where} \quad X_{t_o} = \frac{1}{2t_o} \int_{-t_o}^{t_o} X(t)dt. \tag{8.19}$$

Note that X_{t_o} is a random variable with an expected value given by[2]

$$E[X_{t_o}] = E\left[\frac{1}{2t_o} \int_{-t_o}^{t_o} X(t)dt\right] = \frac{1}{2t_o} \int_{-t_o}^{t_o} E[X(t)]dt = \frac{1}{2t_o} \int_{-t_o}^{t_o} \mu_X dt = \mu_X. \tag{8.20}$$

Therefore, X_{t_o} is an unbiased estimate of the true mean, but for the process to be ergodic in the mean, it is required that X_{t_o} converges to μ_X as $t_o \to \infty$. This convergence will occur (in the

---

[2] We exchange the order of expectation and integration since they are both linear operators.

mean square sense) if the variance of X_{t_0} goes to zero in the limit as $t_0 \to \infty$. To see under what conditions this occurs, we calculate the variance.

$$\text{Var}(X_{t_0}) = E[(X_{t_0} - \mu_X)^2] = E\left[\left(\frac{1}{2t_0}\int_{-t_0}^{t_0}(X(t) - \mu_X)dt\right)^2\right]$$

$$= \frac{1}{4t_0^2}\int_{-t_0}^{t_0}\int_{-t_0}^{t_0} E[(X(t) - \mu_X)(X(s) - \mu_X)]dtds = \frac{1}{4t_0^2}\int_{-t_0}^{t_0}\int_{-t_0}^{t_0} C_{XX}(t,s)dtds. \qquad (8.21)$$

Since the random process $X(t)$ is WSS, the autocovariance is only a function of $\tau = t - s$. As a result, the double integral can be converted to a single integral.[3] The result is

$$\text{Var}(X_{t_0}) = \frac{1}{2t_0}\int_{-2t_0}^{2t_0}\left(1 - \frac{|\tau|}{2t_0}\right)C_{XX}(\tau)d\tau = \frac{1}{t_0}\int_0^{2t_0}\left(1 - \frac{\tau}{2t_0}\right)C_{XX}(\tau)d\tau. \qquad (8.22)$$

Thus, the random process will be ergodic in the mean if this expression goes to zero in the limit as $t_0 \to \infty$. This proves the following theorem.

Theorem 8.1: A continuous WSS random process $X(t)$ will be ergodic in the mean if

$$\lim_{t_0 \to \infty}\frac{1}{t_0}\int_0^{2t_0}\left(1 - \frac{\tau}{2t_0}\right)C_{XX}(\tau)d\tau = 0. \qquad (8.23)$$

One implication of Theorem 8.1 is that if $C_{XX}(\tau)$ tends to a constant as $\tau \to \infty$, then that constant must be zero for the process to be ergodic. Stated in terms of the autocorrelation function, a sufficient condition for a process to be ergodic in the mean is that

$$\lim_{\tau \to \infty} R_{XX}(\tau) = \mu_X^2. \qquad (8.24)$$

Similar relationships can be developed to determine when a process is ergodic in the autocorrelation, but that topic is beyond the intended scope of this text.

■ Example 8.22:

Consider the process $X(t) = A$ of Example 8.19. It is easily found that the autocovariance function of this process is $C_{XX}(\tau) = \sigma_A^2$ for all τ. Plugging this into the left-hand side of Equation (8.23) results in

$$\lim_{t_0 \to \infty}\frac{1}{t_0}\int_0^{2t_0}\left(1 - \frac{\tau}{2t_0}\right)C_{XX}(\tau)d\tau = \lim_{t_0 \to \infty}\frac{\sigma_A^2}{t_0}\int_0^{2t_0}\left(1 - \frac{\tau}{2t_0}\right)d\tau = \lim_{t_0 \to \infty}\frac{\sigma_A^2}{t_0}t_0 = \sigma_A^2.$$

(Continued)

---

[3] The procedure for doing this conversion will be described in detail in Chapter 10, where a similar integral will be encountered in the proof of the Wiener-Khintchine-Einstein theorem.

Since this limit is not equal to zero, the process clearly does not meet the condition for ergodicity. Next, consider the sinusoidal process of Example 8.20. In that case, the autocovariance function is $C_{XX}(\tau) = (a^2/2)\cos(\omega_o \tau)$ and the left hand side of Equation (8.23) produces

$$\lim_{t_o \to \infty}\frac{1}{t_o}\int_0^{2t_o}\left(1-\frac{\tau}{2t_o}\right)C_{XX}(\tau)d\tau = \lim_{t_o \to \infty}\frac{a^2}{2t_o}\int_0^{2t_o}\left(1-\frac{\tau}{2t_o}\right)\cos(\omega_o\tau)d\tau = \lim_{t_o \to \infty}a^2\left(\frac{1-\cos(2\omega_o t_o)}{(2\omega_o t_o)^2}\right) = 0 \,.$$

So, even though this autocorrelation function does not approach zero in the limit as $\tau \to 0$ (it oscillates), it still meets the condition for ergodicity.

■

8.4 Properties of the Autocorrelation Function

Since the autocorrelation function, along with the mean, is considered to be a principal statistical descriptor of a WSS random process, we will now consider some properties of the autocorrelation function. It should quickly become apparent that not just any function of τ can be a valid autocorrelation function.

Property 8.4.1: The autocorrelation function evaluated at $\tau = 0$, $R_{XX}(0)$, is the *average normalized power* in the random process, $X(t)$.

To clarify this, note that $R_{XX}(0) = E[X^2(t)]$. Now suppose the random process $X(t)$ was a voltage measured at some point in a system. For a particular realization, $x(t)$, the instantaneous power would be $p(t) = x^2(t)/r$, where r is the impedance in Ohms (Ω). The average power (averaged over all realizations in the ensemble) would then be $P_{avg} = E[X^2(t)]/r = R_{XX}(0)/r$. If, on the other hand, $X(t)$ were a current rather than a voltage, then the average power would be $P_{avg} = R_{XX}(0)r$. From a systems level, it is often desirable not to concern ourselves with whether a signal is a voltage or a current. Accordingly, it is common to speak of a normalized power, which is the power measured using a 1Ω impedance. With $r = 1$, the two expression for average power are the same and equal to the autocorrelation function evaluated at zero.

Property 8.4.2: The autocorrelation function of a WSS random process is an even function; that is, $R_{XX}(\tau) = R_{XX}(-\tau)$.

This property can easily be established from the definition of autocorrelation. Note that $R_{XX}(-\tau) = E[X(t)X(t-\tau)]$. Since $X(t)$ is WSS, this expression is the same for any value of t. In particular, replace t in the previous expression with $t+\tau$ so that $R_{XX}(-\tau) = E[X(t+\tau)X(t)] = R_{XX}(\tau)$. As a result of this property, any function of τ which is not even cannot be a valid autocorrelation function.

Property 8.4.3: The autocorrelation function of a WSS random process is maximum at the origin; that is, $|R_{XX}(\tau)| \leq R_{XX}(0)$ for all τ.

This property is established using the fact that for any two random variables, X and Y,

$$(E[XY])^2 \leq E[X^2]E[Y^2]. \tag{8.25}$$

This fact was previously demonstrated in the proof of Theorem 5.4. Letting $X = X(t)$ and $Y = X(t+\tau)$ results in

$$R_{XX}^2(\tau) = \{E[X(t)X(t+\tau)]\}^2 \leq E[X^2(t)]E[X^2(t+\tau)] = R_{XX}^2(0). \tag{8.26}$$

Taking square roots of both sides results in Property 8.4.3.

Property 8.4.4: If $X(t)$ is ergodic and has no periodic components, then
$$\lim_{\tau \to \infty} R_{XX}(\tau) = \mu_X^2.$$

Property 8.4.5: If $X(t)$ has a periodic component, then $R_{XX}(\tau)$ will have a periodic component with the same period.

From these properties, it is seen that an autocorrelation function can oscillate, can decay slowly or rapidly, and can have a non-zero constant component. As the name implies, the autocorrelation function is intended to measure the extent of correlation of samples of a random process as a function of how far apart the samples are taken.

8.5 Gaussian Random Processes

One of the most important classes of random processes is the Gaussian random process which is defined as follows.

Definition 8.9: A random process, $X(t)$, for which any n samples, $X_1 = X(t_1)$, $X_2 = X(t_2), \ldots, X_n = X(t_n)$, taken at arbitrary points in time $t_1, t_2, \ldots, t_n$, form a set of jointly Gaussian random variables for any $n = 1, 2, 3, \ldots$ is a Gaussian random process.

In vector notation, the vector of n samples, $\boldsymbol{X} = [X_1, X_2, \ldots, X_n]^T$, will have a joint PDF given by

$$f_{\boldsymbol{X}}(\boldsymbol{x}) = \frac{1}{\sqrt{(2\pi)^n \det(\boldsymbol{C_{XX}})}} \exp\left(-\frac{1}{2}(\boldsymbol{x} - \mu_{\boldsymbol{X}})^T \boldsymbol{C_{XX}^{-1}}(\boldsymbol{x} - \mu_{\boldsymbol{X}})\right). \tag{8.27}$$

As with any joint Gaussian PDF, all that is needed to specify the PDF is the mean vector and the covariance matrix. When the vector of random variables consists of samples of a random process, to specify the mean vector, all that is needed is the mean function of the random process, $\mu_X(t)$, since that will give the mean for any sample time. Similarly, all that is needed to specify the elements of the covariance matrix, $C_{i,j} = \text{Cov}(X(t_i), X(t_j))$, would be the autocovariance function of the random process, $C_{XX}(t_1, t_2)$, or equivalently the autocorrelation function, $R_{XX}(t_1, t_2)$, together with the mean function. Therefore, the mean and autocorrelation functions provide sufficient information to specify the joint PDF for any number of samples of a Gaussian random process. Note that since any n th order PDF is completely specified by $\mu_X(t)$ and $R_{XX}(t_1, t_2)$, if a Gaussian random process is WSS, then the mean and autocorrelation functions will be invariant to a time shift and therefore any PDF will be invariant to a time shift. Hence, any WSS Gaussian random process is also stationary in the strict sense.

■ Example 8.23:

Consider the random process $X(t) = A\cos(\omega_0 t) + B\sin(\omega_0 t)$, where A and B are independent, zero-mean Gaussian random variables with equal variances of σ^2. This random process is formed as a linear combination of two Gaussian random variables, and therefore samples of this process are also Gaussian random variables. The mean and autocorrelation functions of this process are found as

$$\mu_X(t) = E[A\cos(\omega_0 t) + B\sin(\omega_0 t)] = E[A]\cos(\omega_0 t) + E[B]\sin(\omega_0 t) = 0,$$

$$R_{XX}(t_1, t_2) = E[(A\cos(\omega_0 t_1) + B\sin(\omega_0 t_1))(A\cos(\omega_0 t_2) + B\sin(\omega_0 t_2))]$$

$$= E[A^2]\cos(\omega_0 t_1)\cos(\omega_0 t_2) + E[B^2]\sin(\omega_0 t_1)\sin(\omega_0 t_2)$$
$$+ E[AB]\{\cos(\omega_0 t_1)\sin(\omega_0 t_2) + \sin(\omega_0 t_1)\cos(\omega_0 t_2)\}$$

$$= \sigma^2\{\cos(\omega_0 t_1)\cos(\omega_0 t_2) + \sin(\omega_0 t_1)\sin(\omega_0 t_2)\} = \sigma^2\cos(\omega_0(t_2 - t_1)).$$

Note that this process is WSS since the mean is constant and the autocorrelation function depends only on the time difference. Since the process is zero-mean, the first-order PDF is that of a zero-mean Gaussian random variable:

$$f_X(x;t) = \frac{1}{\sqrt{2\pi\sigma^2}}\exp\left(-\frac{x^2}{2\sigma^2}\right).$$

This PDF is independent of time as would be expected for a stationary random process. Now consider the joint PDF of two samples, $X_1 = X(t)$ and $X_2 = X(t + \tau)$. Since the process is zero-mean, the mean vector is simply the all-zeros vector. The covariance matrix is then of the form

$$C_{XX} = \begin{bmatrix} R_{XX}(0) & R_{XX}(\tau) \\ R_{XX}(\tau) & R_{XX}(0) \end{bmatrix} = \sigma^2 \begin{bmatrix} 1 & \cos(\omega_0 \tau) \\ \cos(\omega_0 \tau) & 1 \end{bmatrix}.$$

The joint PDF of the two samples would then be

$$f_{X_1, X_2}(x_1, x_2; t, t+\tau) = \frac{1}{2\pi\sigma^2|\sin(\omega_0\tau)|}\exp\left(-\frac{x_1^2 - 2x_1x_2\cos(\omega_0\tau) + x_2^2}{2\sigma^2\sin^2(\omega_0\tau)}\right).$$

Note once again that this joint PDF is dependent only on time difference, τ, and not on absolute time t. Higher order joint PDFs could be worked out in a similar manner.

■

Suppose we create a random process that jumps by an amount of $\pm\delta$ units every Δt seconds. One possible realization of such a random process is illustrated in Figure 8.9. This proces is often referred to as a random walk. Mathematically, we can express this process as

$$X(t) = \delta\sum_{k=1}^{n} W_k, \tag{8.28}$$

where the W_k are IID Bernoulli random variables with $\Pr(W_k=1) = \Pr(W_k=-1) = 1/2$ and $n = \lfloor t/\Delta t \rfloor$. Note that the mean of this process is

$$\mu_X(t) = E[X(t)] = E\left[\delta\sum_{k=1}^{n} W_k\right] = \delta\sum_{k=1}^{n} E[W_k] = 0. \tag{8.29}$$

Next, consider a limiting form of the random walk where we let both the time between jumps, Δt, and the size of the jumps, δ, simultaneously go to zero. Then the process consists of an infinite number of infinitesimal jumps. As $\Delta t \to 0$, the number of terms in the series in Equation (8.28) becomes infinite, and by virtue of the central limit theorem, $X(t)$ will follow a Gaussian distribution. Thus, we have created a zero-mean Gaussian random process.

Since the process is Gaussain, to complete the statistical characterization of this process, we merely need to find the covariance (or correlation) function. Assuming $t_2 > t_1$,

$$R_{X, X}(t_1, t_2) = E[X(t_1)X(t_2)] = E[X(t_1)(X(t_1) + (X(t_2) - X(t_1)))]$$
$$= E[X^2(t_1)] + E[X(t_1)(X(t_2) - X(t_1))]. \tag{8.30}$$

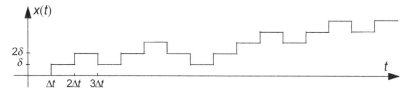

Figure 8.9
One possible realization of a random walk process.

To simplify the second term in the preceding equation, note that $X(t_1)$ represents the accumulation of jumps in the time interval $[0, t_1)$ while $X(t_2) - X(t_1)$ represents the accumulation of jumps in the time interval $[t_1, t_2)$. Due to the manner in which the process is constructed, these terms are independent, thus

$$E[X(t_1)(X(t_2) - X(t_1))] = E[X(t_1)]E[X(t_2) - X(t_1)] = 0. \qquad (8.31)$$

Therefore, $R_{X,X}(t_1, t_2) = E[X^2(t_1)]$ for $t_2 > t_1$. Similarly, if $t_1 > t_2$, we would find that $R_{X,X}(t_1, t_2) = E[X^2(t_2)]$. From Equation (8.28), the variance of the process works out to be

$$E[X^2(t)] = \mathrm{Var}(X(t)) = \delta^2 \sum_{k=1}^{n} \mathrm{Var}(W_k) = \delta^2 n = \frac{\delta^2 t}{\Delta t}. \qquad (8.32)$$

If we let $\delta \to 0$ and $\Delta t \to 0$ in such a way that $\delta = \sqrt{\lambda \Delta t}$ for some constant λ, then

$$E[X^2(t)] = \lambda t, \qquad (8.33)$$

and the covariance function is

$$C_{X,X}(t_1, t_2) = R_{X,X}(t_1, t_2) = \lambda \min(t_1, t_2). \qquad (8.34)$$

In principle, knowing the mean function and the covariance function of this Gaussian random process will allow us to specify the joint PDF of any number of samples.

This mathematical process is known as a Wiener process and finds applications in many different fields. It is perhaps most famously used to model Brownian motion which is the random fluctuation of minute particles suspended in fluids due to impact with neighboring atomic particles. It has also been used in the fields of finance, quantum mechanics, physical cosmology, and as will be seen in future chapters; in electrical engineering, the Weiner process is closely related to the commonly used white noise model of noise in electronic equipment.

8.6 Poisson Processes

Consider a process $X(t)$ which counts the number of occurrences of some event in the time interval $[0, t)$. The event might be the telephone calls arriving at a certain switch in a public telephone network, customers entering a certain store, or the birth of a certain species of animal under study. Since the random process is discrete (in amplitude), we will describe it in terms of a probability mass function, $P_X(i; t) = \Pr(X(t) = i)$. Each occurrence of the event being counted is referred to as an *arrival* or a *point*. These types of processes are referred to as *counting processes* or *birth processes*. Suppose this random process has the following general properties:

- *Independent Increments* The number of arrivals in two non-overlapping intervals are independent. That is, for two intervals $[t_1, t_2)$ and $[t_3, t_4)$ such that $t_1 \le t_2 \le t_3 \le t_4$, the number of arrivals in $[t_1, t_2)$ is statistically independent of the number of arrivals in $[t_3, t_4)$.
- *Stationary Increments* The number of arrivals in an interval $[t, t + \tau)$ depends only on the length of the interval τ and not on where the interval occurs, t.
- *Distribution of Infinitesimal Increments* For an interval of infinitesimal length, $[t, t + \Delta t)$, the probability of a single arrival is proportional to Δt, and the probability of having more than one arrival in the interval is negligible compared to Δt. Mathematically, we say that for some arbitrary constant λ:[4]

$$\Pr(\text{no arrivals in } [t, t + \Delta t)) = 1 - \lambda \Delta t + o(\Delta t), \tag{8.35}$$

$$\Pr(\text{one arrival in } [t, t + \Delta t)) = \lambda \Delta t + o(\Delta t), \tag{8.36}$$

$$\Pr(\text{more than one arrival in } [t, t + \Delta t)) = o(\Delta t). \tag{8.37}$$

Surprisingly enough, these rather general properties are enough to exactly specify the distribution of the counting process as shown next.

Consider the PMF of the counting process at time $t + \Delta t$. In particular, consider finding the probability of the event $\{X(t + \Delta t) = 0\}$.

$$
\begin{aligned}
P_X(0; t + \Delta t) &= \Pr(\text{no arrivals in } [0, t + \Delta t)) \\
&= \Pr(\text{no arrivals in } [0, t))\Pr(\text{no arrivals in } [t, t + \Delta t)) \\
&= P_X(0; t)[1 - \lambda \Delta t + o(\Delta t)]. \tag{8.38}
\end{aligned}
$$

Subtracting $P_X(0; t)$ from both sides and dividing by Δt results in

$$\frac{P_X(0; t + \Delta t) - P_X(0; t)}{\Delta t} = -\lambda P_X(0; t) + \frac{o(\Delta t)}{\Delta t} P_X(0; t). \tag{8.39}$$

Passing to the limit as $\Delta t \to 0$ gives the first-order differential equation

$$\frac{\mathrm{d}}{\mathrm{d}t} P_X(0; t) = -\lambda P_X(0; t). \tag{8.40}$$

---

[4] The notation $o(x)$ refers to an arbitrary function of x which goes to zero as $x \to 0$ in a faster than linear fashion. That is, some function $g(x)$ is said to be $o(x)$ if $\lim_{x \to 0} \frac{g(x)}{x} = 0$.

The solution to this equation is of the general form

$$P_X(0;t) = c\exp(-\lambda t)u(t) \tag{8.41}$$

for some constant c. The constant c is found to be equal to unity by using the fact that at time zero, the number of arrivals must be zero. That is $P_X(0;0) = 1$. Therefore,

$$P_X(0;t) = \exp(-\lambda t)u(t). \tag{8.42}$$

The rest of the PMF for the random process $X(t)$ can be specified in a similar manner. We find the probability of the general event $\{X(t+\Delta t) = i\}$ for some integer $i > 0$.

$$
\begin{aligned}
P_X(i;t+\Delta t) &= \Pr(i \text{ arrivals in } [0,t))\Pr(\text{no arrivals in } [t,t+\Delta t)) \\
&+ \Pr(i-1 \text{ arrivals in } [0,t))\Pr(\text{one arrival in } [t,t+\Delta t)) \\
&+ \Pr(\text{less than } i-1 \text{ arrivals in } [0,t))\Pr(\text{more than one arrival in } [t,t+\Delta t))
\end{aligned}
$$

$$= P_X(i;t)[1 - \lambda\Delta t + o(\Delta t)] + P_X(i-1;t)[\lambda\Delta t + o(\Delta t)] + \sum_{j=0}^{i-2} P_X(i;t)o(\Delta t) \tag{8.43}$$

As before, subtracting $P_X(i;t)$ from both sides and dividing by Δt results in

$$\frac{P_X(i;t+\Delta t) - P_X(i;t)}{\Delta t} = -\lambda P_X(i;t) + \lambda P_X(i-1;t) + \sum_{j=0}^{i} P_X(i;t)\frac{o(\Delta t)}{\Delta t}. \tag{8.44}$$

Passing to the limit as $\Delta t \to 0$ gives another first-order differential equation,

$$\frac{\mathrm{d}}{\mathrm{d}t}P_X(i;t) + \lambda P_X(i;t) = \lambda P_X(i-1;t). \tag{8.45}$$

It is fairly straightforward to solve this set of differential equations. For example, for $i = 1$ we have

$$\frac{\mathrm{d}}{\mathrm{d}t}P_X(1;t) + \lambda P_X(1;t) = \lambda P_X(0;t) = \lambda e^{-\lambda t}u(t), \tag{8.46}$$

together with the initial condition that $P_X(1;0) = 0$. The solution to this equation can be shown to be

$$P_X(1;t) = \lambda t e^{-\lambda t}u(t). \tag{8.47}$$

It is left as an exercise for the reader (see Exercise 8.32) to verify that the general solution to the family of differential equations specified in (8.45) is

$$P_X(i;t) = \frac{(\lambda t)^i}{i!} e^{-\lambda t} u(t). \tag{8.48}$$

Starting with the three mild assumptions made about the nature of this counting process at the start of this section, we have demonstrated that $X(t)$ follows a Poisson distribution, hence this process is referred to as a *Poisson counting process*. Starting with the PMF for the Poisson counting process specified in Equation (8.48), one can easily find the mean and autocorrelation functions for this process. First, the mean function is given by

$$\mu_X(t) = E[X(t)] = \sum_{i=0}^{\infty} i \frac{(\lambda t)^i}{i!} e^{-\lambda t} u(t) = \lambda t u(t). \tag{8.49}$$

In other words, the average number of arrivals in the interval $[0, t)$ is λt. This gives the parameter λ the physical interpretation of the average rate of arrivals, or as it is more commonly referred to the *arrival rate* of the Poisson process. Another observation we can make from the mean process is that the Poisson counting process is not stationary.

The autocorrelation function can be calculated as follows:

$$R_{XX}(t_1, t_2) = E[X(t_1)X(t_2)] = E[X(t_1)\{X(t_1) + (X(t_2) - X(t_1))\}]$$

$$= E[X^2(t_1)] + E[X(t_1)\{X(t_2) - X(t_1)\}]. \tag{8.50}$$

To simplify the second expression, we use the independent increments property of the Poisson counting process. Assuming that $t_1 < t_2$, then $X(t_1)$ represents the number of arrivals in the interval $[0, t_1)$, while $X(t_2) - X(t_1)$ is the number of arrivals in the interval $[t_1, t_2)$. Since these two intervals are non-overlapping, the number of arrivals in the two intervals are independent. Therefore,

$$R_{XX}(t_1, t_2) = E[X^2(t_1)] + E[X(t_1)]E[X(t_2) - X(t_1)]$$
$$= \text{Var}(X(t_1)) + \mu_X(t_1)\mu_X(t_2) = \lambda t_1 + \lambda^2 t_1 t_2. \tag{8.51}$$

This can be written more concisely in terms of the autocovariance function,

$$C_{XX}(t_1, t_2) = \text{Var}(X(t_1)) = \lambda t_1. \tag{8.52}$$

If $t_2 < t_1$, then the roles of t_1 and t_2 need to be reversed. In general for the Poisson counting process, we have

$$C_{XX}(t_1, t_2) = \lambda \min(t_1, t_2). \tag{8.53}$$

Another feature that can be extracted from the PMF of the Poisson counting process is the distribution of the inter-arrival time. That is, let T be the time at which the first arrival occurs. We seek the distribution of the random variable T. The CDF of T can be found as

$$F_T(t) = \Pr(T \le t) = \Pr(\text{at least one arrival in } [0, t))$$

$$= 1 - \Pr(\text{no arrivals in } [0, t)) = [1 - e^{-\lambda t}]u(t) \,. \tag{8.54}$$

Therefore, it follows that the arrival time is an exponential random variable with a mean value of $E[T] = 1/\lambda$. The PDF of T is

$$f_T(t) = \lambda e^{-\lambda t}u(t). \tag{8.55}$$

We could get the same result starting from any point in time. That is, we do not need to measure the time to the next arrival starting from time zero. Picking any arbitrary point in time t_o, we could define T to be the time until the first arrival after time t_o. Using the same reasoning as above we would arrive at the same exponential distribution. If we pick t_o to be the time of a specific arrival, and then define T to be the time to the next arrival, then T is interpreted as an inter-arrival time. Hence, we conclude that the time between successive arrivals in the Poisson counting process follows an exponential distribution with a mean of $1/\lambda$.

The Poisson counting process can be represented as a sum of randomly shifted unit step functions. That is, let S_i be the time of the ith arrival. Then,

$$X(t) = \sum_{i=1}^{\infty} u(t - S_i)\,. \tag{8.56}$$

The random variables, S_i, are sometimes referred to as *points* of the Poisson process. Many other related random processes can be constructed by replacing the unit step functions with alternative functions. For example, if the step function is replaced by a delta function, the *Poisson impulse process* results, which is expressed as

$$X(t) = \sum_{i=1}^{\infty} \delta(t - S_i)\,. \tag{8.57}$$

A sample realization of this process is shown in Figure 8.10.

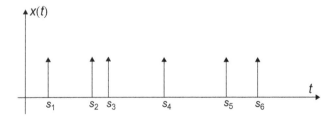

Figure 8.10
A sample realization of the Poisson impulse process.

8.7 Engineering Application—Shot Noise in a *p–n* Junction Diode

Both the Poisson counting process and the Poisson impulse process can be viewed as special cases of a general class of processes referred to as *shot noise processes*. Given an arbitrary waveform $h(t)$ and a set of Poisson points, S_i, the shot noise process is constructed as

$$X(t) = \sum_{i=1}^{\infty} h(t - S_i). \tag{8.58}$$

As an example of the physical origin of such a process, consider the operation of a *p–n* junction diode. When a forward bias voltage is applied to the junction, a current is generated. This current is not constant, but actually consists of discrete holes from the *p* region and electrons from the *n* region which have sufficient energy to overcome the potential barrier at the junction. Carriers do not cross the junction in a steady deterministic fashion; rather, each passage is a random event which might be modeled as a Poisson point process. The arrival rate of that Poisson process would be dependent on the bias voltage across the junction. As a carrier crosses the junction, it produces a current pulse, which we represent with some pulse shape $h(t)$, such that the total area under $h(t)$ is equal to the charge in an electron, q. Thus, the total current produced by the *p–n* junction diode can be modeled as a shot noise process.

To start with, we compute the mean function of a shot noise process. However, upon examining Equation (8.58), it is not immediately obvious how to take the expected value for an arbitrary pulse shape $h(t)$. There are several ways to achieve the goal. One approach is to divide the time axis into infinitesimal intervals of length Δt. Then, define a sequence of Bernoulli random variables V_n such that $V_n = 1$ if a point occurred within the interval $[n\Delta t, (n+1)\Delta t)$ and $V_n = 0$ if no points occurred in the same interval. Since the intervals are taken to be infinitesimal, the probability of having more than one point in a single interval is negligible. Furthermore, from the initial assumptions that led to the Poisson process, the distribution of the V_n is given by

$$\Pr(V_n{=}1) = \lambda \Delta t \text{ and } \Pr(V_n{=}0) = 1 - \lambda \Delta t. \tag{8.59}$$

The shot noise process can be approximated by

$$X(t) \cong \sum_{n=0}^{\infty} V_n h(t - n\Delta t). \tag{8.60}$$

In the limit as $\Delta t \to 0$, the approximation becomes exact. Using this alternative representation of the shot noise process, calculation of the mean function is straightforward.

$$E[X(t)] \cong \sum_{n=0}^{\infty} E[V_n]h(t-n\Delta t) = \lambda \sum_{n=0}^{\infty} h(t-n\Delta t)\Delta t. \tag{8.61}$$

Note that in this calculation, the fact that $E[V_n] = \lambda\Delta t$ was used. Passing to the limit as $\Delta t \to 0$ results in

$$\mu_X(t) = \lambda \int_0^{\infty} h(t-u)du = \lambda \int_0^t h(v)dv. \tag{8.62}$$

Strictly speaking, the mean function of the shot noise process is not a constant, and therefore the process is not stationary. However, in practice, the current pulse will be time limited. Suppose the current pulse, $h(t)$, has a time duration of t_h. That is, for $t > t_h$, $h(t)$ is essentially equal to zero. For the example of the p–n junction diode, the time duration of the current pulse is the time it takes the carrier to pass through the depletion region. For most devices, this number may be a small fraction of a nanosecond. Then for any $t > t_h$,

$$\mu_X(t) = \lambda \int_0^{\infty} h(v)dv = \lambda q = \text{constant}. \tag{8.63}$$

For example, using the fact that the charge on an electron is $1.6 \times 10^{-19}\text{C}$, if carriers made transitions at an average rate of 10^{15} per second (1 per femtosecond), then the average current produced in the diode would be 0.16 mA.

Next, we seek the autocorrelation (or autocovariance) function of the shot noise process. The same procedure used to calculate the mean function can also be used here.

$$R_{XX}(t, t+\tau) = E[X(t)X(t+\tau)] = \sum_{n=0}^{\infty}\sum_{m=0}^{\infty} E[V_n V_m]h(t-n\Delta t)h(t+\tau-m\Delta t)$$

$$= \sum_{n=0}^{\infty} E[V_n^2]h(t-n\Delta t)h(t+\tau-n\Delta t) + \sum_{n=0}^{\infty} E[V_n]h(t-n\Delta t)\sum_{m \neq n}^{\infty} E[V_m]h(t+\tau-m\Delta t)$$

$$= \lambda \sum_{n=0}^{\infty} h(t-n\Delta t)h(t+\tau-n\Delta t)\Delta t + \lambda^2 \sum_{n=0}^{\infty} h(t-n\Delta t)\Delta t \sum_{m \neq n}^{\infty} h(t+\tau-m\Delta t)\Delta t$$

$$= \sum_{n=0}^{\infty} h(t-n\Delta t)h(t+\tau-n\Delta t)[\lambda\Delta t - (\lambda\Delta t)^2]$$

$$+ \lambda^2 \sum_{n=0}^{\infty} h(t-n\Delta t)\Delta t \sum_{m=0}^{\infty} h(t+\tau-m\Delta t)\Delta t \tag{8.64}$$

Passing to the limit as $\Delta t \rightarrow 0$, we note that the term involving $(\lambda \Delta t)^2$ is negligible compared to $\lambda \Delta t$. The resulting limit then takes the form

$$R_{XX}(t, t + \tau) = \lambda \int_0^\infty h(t - u)h(t + \tau - u)\mathrm{d}u + \lambda^2 \int_0^\infty h(t - u)\mathrm{d}u \int_0^\infty h(t + \tau - u)\mathrm{d}u$$

$$= \lambda \int_0^t h(v)h(v + \tau)\mathrm{d}v + \lambda^2 \int_0^t h(v)\mathrm{d}v \int_0^{t+\tau} h(v)\mathrm{d}v. \tag{8.65}$$

Note that the last term (involving the product of integrals) is just the product of the mean function evaluated at time t and the mean function evaluated at time $t + \tau$. Thus, we have,

$$R_{XX}(t, t + \tau) = \lambda \int_0^t h(v)h(v + \tau)\mathrm{d}v + \mu_X(t)\mu_X(t + \tau), \tag{8.66}$$

or equivalently, in terms of the autocovariance function,

$$C_{XX}(t, t + \tau) = \lambda \int_0^t h(v)h(v + \tau)\mathrm{d}v. \tag{8.67}$$

As with the mean function, it is seen that the autocovariance function is a function of not only τ but also t. Again, for sufficiently large t, the upper limit in the preceding integral will be much longer than the time duration of the pulse, $h(t)$. Hence, for $t > t_h$,

$$C_{XX}(t, t + \tau) = C_{XX}(\tau) = \lambda \int_0^\infty h(v)h(v + \tau)\mathrm{d}v, \tag{8.68}$$

or

$$R_{XX}(t, t + \tau) = R_{XX}(\tau) = \lambda \int_0^\infty h(v)h(v + \tau)\mathrm{d}v + \mu_X^2. \tag{8.69}$$

We say that the shot noise process is asymptotically WSS. That is, after waiting a sufficiently long period of time, the mean and autocorrelation functions will be invariant to time shifts. In this case, the phrase "sufficiently long time" may mean a small fraction of a nanosecond! So for all practical purposes, the process is WSS. Also, it is noted that the width of the autocovariance function is t_h. That is, if $h(t)$ is time limited to a duration of t_h, then $C_{XX}(\tau)$ is zero for $|\tau| > t_h$. This relationship is illustrated in Figure 8.11, assuming $h(t)$ is a square pulse, and implies that any samples of the shot noise process that are separated by more than t_h will be uncorrelated.

Finally, in order to characterize the PDF of the shot noise process, consider the approximation to the shot noise process given in Equation (8.60). At any fixed point in time, the process

$X(t)$ can be viewed as the linear combination of a large number of independent Bernoulli random variables. By virtue of the central limit theorem, this sum can be very well approximated by a Gaussian random variable. Since the shot noise process is WSS (at least in the asymptotic sense) and is a Gaussian random process, then the process is also stationary in the strict sense. Also, samples spaced by more than t_h are independent.

■ Example 8.24:

Consider a shot noise process in a p–n junction diode where the pulse shape is square as illustrated in Figure 8.11. The mean current is $\mu_X = \lambda q$, which is presumably the desired signal we are trying to measure. The fluctuation of the shot noise process about the mean, we view as the unwanted disturbance, or noise. It would be interesting to measure the ratio of the power in the desired part of the signal to the power in the noise part of the signal. The desired part has a time-average power of $\mu_X^2 = (\lambda q)^2$, while the noise part has a power of $\sigma_X^2 = C_{XX}(0) = \lambda q^2/t_h$. The signal-to-noise ratio (SNR) is then

$$\text{SNR} = \frac{\mu_X^2}{\sigma_X^2} = \frac{(\lambda q)^2}{\lambda q^2/t_h} = \lambda t_h .$$

We write this in a slightly different form,

$$\text{SNR} = \lambda t_h = \lambda q\left(\frac{t_h}{q}\right) = \mu_x\left(\frac{t_h}{q}\right).$$

For example, if the pulse duration were $t_h = 10\text{picoseconds}$, the SNR as it depends on the strength of the desired part of the signal would be as illustrated in Figure 8.12. It is noted that the SNR is fairly strong until we try to measure signals which are below a microamp .

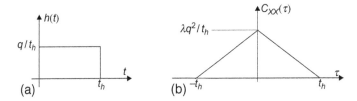

Figure 8.11
(a) A square current pulse and (b) the corresponding autocovariance function.

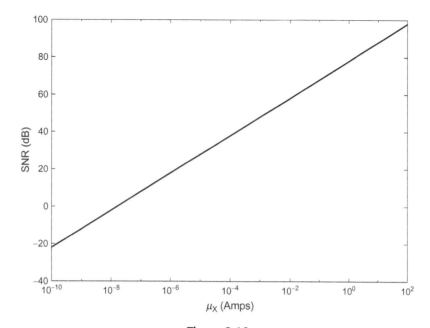

Figure 8.12
Signal-to-noise ratio in a shot noise process for an example p–n junction diode.

In Chapter 10, we will view random processes in the frequency domain. Using the frequency domain tools we will develop in that chapter, it will become apparent that the noise power in the shot noise process is distributed over a very wide bandwidth (about 100 GHz for the previous example). Typically, our measuring equipment would not respond to that wide of a frequency range, and so the amount of noise power we actually see would be much less than that presented in Example 8.24 and would be limited by the bandwidth of our equipment.

■ Example 8.25:

In this example, we provide some MATLAB code to generate a sample realization of a shot noise process. We chose to use a current pulse shape of the form, $h(t) = t\exp(-t^2)$, but the reader could easily modify this to use other pulse shapes as well. A typical realization is shown in Figure 8.13. Note that after a short initial transient period, the process settles into a steady-state behavior.

(Continued)

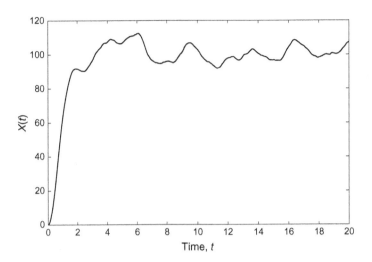

Figure 8.13
A typical realization of a shot noise process.

```
dt=0.001;                           % time sample interval
t=[0:dt:20];                        % time axis
v=rand(size(t))<0.2;                % impulse process
h=t.*exp(-t.^2);                    % pulse shape
x=conv(v,h);                        % shot noise process
plot(t,x(1:length(t)))              % plot results
xlabel('time, t'); ylabel('X(t)')
```

Section 8.2: Mathematical Tools for Studying Random Processes

8.1 A discrete random process, $X[n]$, is generated by repeated tosses of a coin. Let the occurrence of a head be denoted by 1 and that of a tail by -1. A new discrete random process is generated by $Y[2n] = X[n]$ for $n = 0, \pm 1, \pm 2, \ldots$ and $Y[n] = X[n+1]$ for n odd (either positive or negative). Find the autocorrelation function for $Y[n]$.

8.2 Let W_n be an IID sequence of zero-mean Gaussian random variables with variance σ_W^2. Define a discrete-time random process $X[n] = pX[n-1] + W_n$, $n = 1, 2, 3, \ldots$, where $X[0] = W_0$ and p is a constant.
   (a) Find the mean function, $\mu_X[n]$.
   (b) Find the autocorrelation function, $R_{X,X}[n_1, n_2]$.

8.3 Let X_k, $k = 1, 2, 3, \ldots$, be a sequence of IID random variables with mean μ_X and variance σ_X^2. Form the sample mean process

$$S[n] = \frac{1}{n} \sum_{k=1}^{n} X_k, \quad n = 1, 2, 3, \ldots .$$

   (a) Find the mean function, $\mu_S[n] = E[S[n]]$.
   (b) Find the autocorrelation function, $R_{S,S}[k, n] = E[S[k]S[n]]$.

8.4 Define a random process according to

$$X[n] = X[n-1] + W_n, \quad n = 1, 2, 3, \ldots,$$

where $X[0] = 0$ and W_n is a sequence of IID Bernoulli random variables with $\Pr(W_n=1) = p$ and $\Pr(W_n=0) = 1-p$.
   (a) Find the PMF, $P_X(k;n) = \Pr(X[k]=n)$.
   (b) Find the joint PMF, $P_{X_1, X_2}(k_1, k_2; n_1, n_2) = \Pr(X[k_1]=n_1, X[k_2]=n_2)$.
   (c) Find the mean function, $\mu_X[n] = E[X[n]]$.
   (d) Find the autocorrelation function, $R_{X,X}[k, n] = E[X[k]X[n]]$.

Probability and Random Processes.
© 2012 by Elsevier Inc. All rights reserved.

8.5   Consider the random process defined in Example 8.5. The PDF, $f_X(x;n)$, and the mean function, $\mu_X[n]$, were found in Example 8.7.
   (a)   Find the joint PDF, $f_{X_1, X_2}(x_1, x_2; n_1, n_2)$.
   (b)   Find the autocorrelation function, $R_{X, X}(k, n) = E[X[k]X[n]]$.

Section 8.3: Stationary and Ergodic Random Processes

8.6   A random process $X(t)$ consists of three-member functions: $x_1(t) = 1$, $x_2(t) = -3$, and $x_3(t) = \sin(2\pi t)$. Each member function occurs with equal probability.
   (a)   Find the mean function, $\mu_X(t)$.
   (b)   Find the autocorrelation function, $R_{X, X}(t_1, t_2)$.
   (c)   Is the process WSS? Is it stationary in the strict sense?

8.7   A random process $X(t)$ has the following member functions: $x_1(t) = -2\cos(t)$, $x_2(t) = -2\sin(t)$, $x_3(t) = 2[\cos(t) + \sin(t)]$, $x_4(t) = [\cos(t) - \sin(t)]$, $x_5(t) = [\sin(t) - \cos(t)]$. Each member function occurs with equal probability.
   (a)   Find the mean function, $\mu_X(t)$.
   (b)   Find the autocorrelation function, $R_{X, X}(t_1, t_2)$.
   (c)   Is the process WSS? Is it stationary in the strict sense?

8.8   Let a discrete random process $X[n]$ be generated by repeated tosses of a fair die. Let the values of the random process be equal to the results of each toss.
   (a)   Find the mean function, $\mu_X[n]$.
   (b)   Find the autocorrelation function, $R_{X, X}[k_1, k_2]$.
   (c)   Is the process WSS? Is it stationary in the strict sense?

8.9   Let $X[n]$ be a wide sense stationary, discrete random process with autocorrelation function $R_{XX}[n]$, and let c be a constant.
   (a)   Find the autocorrelation function for the discrete random process $Y[n] = X[n] + c$.
   (b)   Are $X[n]$ and $Y[n]$ independent? Uncorrelated? Orthogonal?

8.10 A wide sense stationary, discrete random process, $X[n]$, has an autocorrelation function of $R_{XX}[k]$. Find the expected value of $Y[n] = (X[n+m] - X[n-m])^2$, where m is an arbitrary integer.

8.11 A random process is given by $X(t) = A\cos(\omega t) + B\sin(\omega t)$, where A and B are independent zero-mean random variables.
   (a)   Find the mean function, $\mu_X(t)$.
   (b)   Find the autocorrelation function, $R_{X, X}(t_1, t_2)$.
   (c)   Under what conditons (on the variances of A and B) is $X(t)$ WSS?

8.12 Show by example that the random process $Z(t) = X(t) + Y(t)$ may be a wide sense stationary process even though the random processes $X(t)$ and $Y(t)$ are not. *Hint:* Let $A(t)$ and $B(t)$ be independent, wide sense stationary random processes with zero-means and identical autocorrelation functions. Then let $X(t) = A(t)\sin(t)$ and $Y(t) = B(t)\cos(t)$. Show that $X(t)$ and $Y(t)$ are not wide sense stationary. Then show that $Z(t)$ is wide sense stationary.

8.13 Let $X(t) = A(t)\cos(\omega_0 t + \Theta)$, where $A(t)$ is a wide sense stationary random process independent of Θ and let Θ be a random variable distributed uniformly over $[0, 2\pi)$. Define a related process $Y(t) = A(t)\cos((\omega_0 + \omega_1)t + \Theta)$. Show that $X(t)$ and $Y(t)$ are stationary in the wide sense but that the cross-correlation $R_{XY}(t, t + \tau)$, between $X(t)$ and $Y(t)$, is not a function of τ only and, therefore, $Z(t) = X(t) + Y(t)$ is not stationary in the wide sense.

8.14 Let $X(t)$ be a modified version of the random telegraph process. The process switches between the two states $X(t) = 1$ and $X(t) = -1$ with the time between switches following exponential distributions, $f_T(s) = \lambda\exp(-\lambda s)u(s)$. Also, the starting state is determined by flipping a biased coin so that $\Pr(X(0) = 1) = p$ and $\Pr(X(0) = -1) = 1 - p$.
  (a) Find $\Pr(X(t) = 1)$ and $\Pr(X(t) = -1)$.
  (b) Find the mean function, $\mu_X(t)$.
  (c) Find the autocorrelation function, $R_{X, X}(t_1, t_2)$.
  (d) Is this process WSS?

8.15 Let $s(t)$ be a periodic square wave as illustrated in the accompanying figure. Suppose a random process is created according to $X(t) = s(t - T)$, where T is a random variable uniformly distributed over $(0, 1)$.

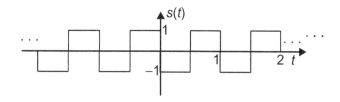

  (a) Find the probability mass function of $X(t)$.
  (b) Find the mean function, $\mu_X(t)$.
  (c) Find the autocorrelation function, $R_{X, X}(t_1, t_2)$.
  (d) Is this process WSS?

8.16 Let $s(t)$ be a periodic triangle wave as illustrated in the accompanying figure. Suppose a random process is created according to $X(t) = s(t-T)$, where T is a random variable uniformly distributed over $(0, 1)$.

(a) Find the probability mass function of $X(t)$.
(b) Find the mean function, $\mu_X(t)$.
(c) Find the autocorrelation function, $R_{X,X}(t_1, t_2)$.
(d) Is this process WSS?

8.17 Let a random process consist of a sequence of pulses with the following properties: (i) the pulses are rectangular of equal duration, Δ (with no "dead" space in between pulses), (ii) the pulse amplitudes are equally likely to be ± 1, (iii) all pulses amplitudes are statistically independent, and (iv) the various members of the ensemble are not synchronized.

(a) Find the mean function, $\mu_X(t)$.
(b) Find the autocorrelation function, $R_{X,X}(t_1, t_2)$.
(c) Is this process WSS?

8.18 A random process is defined by $X(t) = \exp(-At)u(t)$ where A is a random variable with PDF, $f_A(a)$.

(a) Find the PDF of $X(t)$ in terms of $f_A(a)$.
(b) If A is an exponential random variable, with $f_A(a) = e^{-a}u(a)$, find $\mu_X(t)$ and $R_{X,X}(t_1, t_2)$. Is the process WSS?

8.19 Two zero-mean discrete-time random processes, $X[n]$ and $Y[n]$, are statistically independent. Let a new random process be $Z[n] = X[n] + Y[n]$. Let the autocorrelation functions for $X[n]$ and $Y[n]$ be

$$R_{XX}[k] = \left(\frac{1}{2}\right)^{|k|}, \qquad R_{YY}[k] = \left(\frac{1}{3}\right)^{|k|}.$$

Find $R_{ZZ}[k]$. Plot all three autocorrelation functions (you may want to use MATLAB to help).

8.20 Consider a discrete-time wide sense stationary random processes whose autocorrelation function is of the form

$$R_{XX}[k] = a^{|k|}, \quad \text{where } |a| < 1.$$

Assume this process has zero-mean. Is the process ergodic in the mean?

8.21 Let $X(t)$ be a wide sense stationary random process that is ergodic in the mean and the autocorrelation. However, $X(t)$ is not zero-mean. Let $Y(t) = CX(t)$, where C is a random variable independent of $X(t)$ and C is not zero-mean. Show that $Y(t)$ is not ergodic in the mean or the autocorrelation.

8.22 Let $X(t)$ be a WSS random process with mean μ_X and autocorrelation function $R_{XX}(\tau)$. Consider forming a new process according to

$$Y(t) = \frac{X(t + t_o) - X(t)}{t_o}.$$

(a)   Find the mean function of $Y(t)$.

(b)   Find the autocorrelation function of $Y(t)$. Is $Y(t)$ WSS?

Section 8.4: Properties of the Autocorrelation Function

8.23 Which of the following could be the correlation function of a stationary random process?

(a) $R_a(\tau) = \exp(-|\tau|)$.

(d) $R_d(\tau) = \sin(\tau)$.

(b) $R_b(\tau) = \exp(-\tau)u(\tau)$.

(e) $R_e(\tau) = \cos(\tau)$.

(c) $R_c(\tau) = \exp(|\tau|)$.

(f) $R_f(\tau) = \text{sinc}(\tau)$.

8.24 A stationary random process, $X(t)$, has a mean of μ_X and correlation function, $R_{X,X}(\tau)$. A new process is formed according to $Y(t) = aX(t) + b$ for constants a and b. Find the correlation function $R_{Y,Y}(\tau)$ in terms of μ_X and $R_{X,X}(\tau)$.

8.25 An ergodic random process has a correlation function given by

$$R(\tau) = \frac{3 + \tau^2}{2 + 2\tau^2}.$$

What is the mean of this process?

8.26 For each of the functions in Exercise 8.23 that represents a valid correlation function, construct a random process that possesses that function as its correlation function.

Section 8.5: Gaussian Random Processes

8.27 Let $X(t)$ and $Y(t)$ be two jointly wide sense stationary Gaussian random processes with zero-means and with autocorrelation and cross-correlation functions denoted as $R_{XX}(\tau)$, $R_{YY}(\tau)$, and $R_{XY}(\tau)$. Determine the cross-correlation function between $X^2(t)$ and $Y^2(t)$.

8.28 If $X(t)$ is a wide sense stationary Gaussian random process, find the cross-correlation between $X(t)$ and $X^3(t)$ in terms of the autocorrelation function $R_{XX}(\tau)$.

8.29 Suppose $X(t)$ is a Weiner process with diffusion parameter $\lambda = 1$ as described in Section 8.5.

(a) Write the joint PDF of $X_1 = X(t_1)$ and $X_2 = X(t_2)$ for $t_2 > t_1$ by evaluating the covariance matrix of $X = [X_1, X_2]^T$ and using the general form of the joint Gaussian PDF in Equation 6.22.

(b) Evaluate the joint PDF of $X_1 = X(t_1)$ and $X_2 = X(t_2)$ for $t_2 > t_1$ indirectly by defining the related random variables $Y_1 = X_1$ and $Y_2 = X_2 - X_1$. Noting that Y_1 and Y_2 are independent and Gaussian, write down the joint PDF of Y_1 and Y_2 and then form the joint PDF of X_1 and X_2 be performing the appropriate 2×2 transformation.

(c) Using the technique outlined in part (b), find the joint PDF of three samples of a Wiener process, $X_1 = X(t_1)$, $X_2 = X(t_2)$, and $X_3 = X(t_3)$ for $t_1 < t_2 < t_3$.

8.30 Let $X(t)$ be a wide sense stationary Gaussian random process and form a new process according to $Y(t) = X(t)\cos(\omega t + \theta)$ where ω and θ are constants.

(a) Is $Y(t)$ wide sense stationary?

(b) Is $Y(t)$ a Gaussian random process?

8.31 Let $X(t)$ be a wide sense stationary Gaussian random process and form a new process according to $Y(t) = X(t)\cos(\omega t + \Theta)$ where ω is a constant and Θ is a random variable uniformly distributed over $[0, 2\pi)$ and independent of $X(t)$.

(a) Is $Y(t)$ wide sense stationary?

(b) Is $Y(t)$ a Gaussian random process?

Section 8.6: Poisson Processes

8.32 Prove that the family of differential equations,

$$\frac{d}{dt}P_X(0;t) + \lambda P_X(0;t) = 0,$$

$$\frac{d}{dt}P_X(i;t) + \lambda P_X(i;t) = \lambda P_X(i-1;t), \quad i = 1, 2, 3, \ldots,$$

leads to the Poisson distribution,

$$P_X(i;t) = \frac{(\lambda t)^i}{i!}e^{-\lambda t}.$$

8.33 Consider a Poisson counting process with arrival rate λ.

    (a) Suppose it is observed that there is exactly one arrival in the time interval $[0, t_o)$. Find the PDF of that arrival time.

    (b) Now suppose there were exactly two arrivals in the time interval $[0, t_o)$. Find the joint PDF of those two arrival times.

    (c) Extend these results to an arbitrary number, n, of arrivals?

8.34 Let $N(t)$ be a Poisson counting process with arrival rate λ. Find $\Pr(N(t)=k \mid N(t+\tau)=m)$ where $\tau > 0$ and $m \geq k$.

8.35 Let $X_i(t)$, $i = 1, 2, \ldots, n$, be a sequence of independent Poisson counting processes with arrival rates, λ_i. Show that the sum of all of these Poisson processes,

$$X(t) = \sum_{i=1}^{n} X_i(t) ,$$

is itself a Poisson process. What is the arrival rate of the sum process?

8.36 A workstation is used until it fails and then it is sent out for repair. The time between failures, or the length of time the workstation functions until it needs repair, is a random variable T. Assume the times between failures, T_1, T_2, ..., T_n of the workstations available are independent random variables that are identically distributed. For $t > 0$, let the number of workstations that have failed be $N(t)$.

    (a) If the time between failures of each workstation has an exponential PDF, then what type of process is $N(t)$?

    (b) Assume that you have just purchased 10 new workstations and that each has a 90-day warranty. If the mean time between failures (MTBF) is 250 days, what is the probability that at least one workstation will fail before the end of the warranty period?

8.37 Suppose the arrival of calls at a switchboard is modeled as a Poisson process with the rate of calls per minute being $\lambda_a = 0.1$.

    (a) What is the probability that the number of calls arriving in a 10-minute interval is less than 10?

    (b) What is the probability that the number of calls arriving in a 10-minute interval is less than 10 if $\lambda_a = 10$?

    (c) Assuming $\lambda_a = 0.1$, what is the probability that one call arrives during the first 10-minute interval and two calls arrive during the second 10-minute interval?

8.38 Let $X(t)$ be a Poisson counting process with arrival rate, λ. We form two related counting processes, $Y_1(t)$ and $Y_2(t)$, by deterministically splitting the Poisson process,

$X(t)$. Each arrival associated with $X(t)$ is alternately assigned to one of the two new processes. That is, if S_i is the ith arrival time of $X(t)$, then

$$Y_1(t) = \sum_{i \text{ odd}} u(t - S_i),$$

$$Y_2(t) = \sum_{i \text{ even}} u(t - S_i).$$

Find the PMFs of the two split processes, $P_{Y_1}(k; t) = \Pr(Y_1(t) = k)$ and $P_{Y_2}(k; t) = \Pr(Y_2(t) = k)$. Are the split processes also Poisson processes?

8.39 Let $X(t)$ be a Poisson counting process with arrival rate, λ. We form two related counting processes, $Y_1(t)$ and $Y_2(t)$, by randomly splitting the Poisson process, $X(t)$. In random splitting, the ith arrival associated with $X(t)$ will become an arrival in process $Y_1(t)$ with probability p and will become an arrival in process $Y_2(t)$ with probability $1 - p$. That is, let S_i be the ith arrival time of $X(t)$ and define W_i to be a sequence of IID Bernoulli random variables with $\Pr(W_i = 1) = p$ and $\Pr(W_i = 0) = 1 - p$. Then the split processes are formed according to

$$Y_1(t) = \sum_{i=1}^{\infty} W_i u(t - S_i),$$

$$Y_2(t) = \sum_{i=1}^{\infty} (1 - W_i) u(t - S_i).$$

Find the PMFs of the two split processes, $P_{Y_1}(k;t) = \Pr(Y_1(t) = k)$ and $P_{Y_2}(k;t) = \Pr(Y_2(t) = k)$. Are the split processes also Poisson processes?

8.40 Consider a Poisson counting process with arrival rate, λ. Suppose it is observed that there have been exactly n arrivals in $[0, t)$ and let $S_1, S_2, \ldots, S_n$ be the times of those n arrivals. Next, define $X_1, X_2, \ldots, X_n$ to be a sequence of IID random variables uniformly distributed over $[0, t)$ and let $Y_1, Y_2, \ldots, Y_n$ be the order statistics associated with the X_i. Show that the joint PDF of the order statistics, $f_Y(y)$, is identical to the joint PDF of the Poisson arrival times, $f_S(s)$. Hence, the order statistics are statistically identical to the arrival times.

Section 8.7: Shot Noise in a *p–n* Junction Diode

8.41 Model lightning strikes to a power line during a thunderstorm as a Poisson impulse process. Suppose the number of lightning strikes in time interval t has a mean rate of arrival given by s, which is one strike per 3 minutes.
    (a)   What is the expected number of lightning strikes in 1 minute? in 10 minutes?
    (b)   What is the average time between lightning strikes?

8.42 Suppose the power line in the previous problem has an impulse response that may be approximated by $h(t) = te^{-at}u(t)$, where $a = 10\text{s}^{-1}$.
   (a) What does the shot noise on the power line look like? Sketch a possible member function of the shot noise process.
   (b) Find the mean function of the shot noise process.
   (c) Find the autocorrelation function of the shot noise process.

8.43 A shot noise process with random amplitudes is defined by

$$X(t) = \sum_{i=1}^{\infty} A_i h(t - S_i),$$

where the S_i are a sequence of points from a Poisson process and the A_i are IID random variables which are also indpendent of the Poisson points.
   (a) Find the mean function of $X(t)$.
   (b) Find the autocorrelation function of $X(t)$.

8.44 In this problem, we develop an alternative derivation for the mean function of the shot noise process described in Section 8.7,

$$X(t) = \sum_{i=1}^{\infty} h(t - S_i),$$

where the S_i are the arrival times of a Poisson process with arrival rate, λ, and $h(t)$ is an arbitrary pulse shape which we take to be causal. That is, $h(t) = 0$ for $t < 0$. In order to find the mean function, $\mu_X(t) = E[X(t)]$, we condition on the event that there were exactly n arrivals in $[0, t)$. Then, the conditional mean function is

$$E[X(t)|n \text{ arrivals in } [0, t)] = \sum_{i=1}^{n} E[h(t - S_i)|n \text{ arrivals in } [0, t)].$$

   (a) Use the results of Exercise 8.40 to justify that

$$E[X(t)|n \text{ arrivals in } [0, t)] = \sum_{i=1}^{n} E[h(t - X_i)],$$

   where the X_i are a sequence of IID random variables uniformly distributed over $[0, t)$.
   (b) Show that the expectation in part (a) reduces to

$$E[X(t)|n \text{ arrivals in } [0, t)] = \frac{n}{t} \int_0^t h(t - u) du.$$

   (c) Finally, average over the Poisson distribution of the number of arrivals to show that

$$E[X(t)] = \lambda \int_0^t h(t) dt.$$

8.45 Use the technique outlined in Exercise 8.44 to derive the autocovariance function of the shot noise process.

Miscellaneous Exercises

8.46 A random process $X(t)$ is said to be mean square continuous at some point in time t, if

$$\lim_{t_o \to 0} E[(X(t+t_o) - X(t))^2] = 0.$$

(a) Prove that $X(t)$ is mean square continuous at time t if its correlation function, $R_{X,X}(t_1, t_2)$ is continuous at the point $t_1 = t$, $t_2 = t$.

(b) Prove that if $X(t)$ is mean square continuous at time t, then the mean function $\mu_X(t)$ must be continuous at time t. *Hint:* Consider $\text{Var}(X(t+t_o) - X(t))$ and note that any variance must be non-negative.

(c) Prove that for a WSS process $X(t)$, if $R_{X,X}(\tau)$ is continuous at $\tau = 0$, then $X(t)$ is mean square continuous at all points in time.

8.47 Let $N(t)$ be a Poisson counting process with arrival rate, λ. Determine whether or not $N(t)$ is mean square continuous. *Hint:* See the definition of mean square continuity and the associated results in Exercise 8.46.

8.48 Let $X(t)$ be a Weiner process with diffusion parameter λ as described in Section 8.5. Determine whether or not $X(t)$ is mean square continuous. *Hint:* See the definition of mean square continuity and the associated results in Exercise 8.46.

8.49 Let W_k be a sequence of IID Bernoulli random variables with $\Pr(W_k = 1) = \Pr(W_k = -1) = 1/2$ and form a random process according to

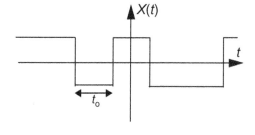

$$X(t) = \sum_{k=-\infty}^{\infty} W_k p(t - kt_o)$$

where

$$p(t) = \begin{cases} 1, & -\dfrac{t_o}{2} \le t < \dfrac{t_o}{2}, \\ 0, & \text{otherwise.} \end{cases}$$

A sample realization of this process is shown in the accompanying figure. Is this process mean square continuous? *Hint:* See the definition of mean square continuity and the associated results in Exercise 8.46.

MATLAB Exercises

8.50 You are given a member function of a random process as $y(t) = 10\sin(2\pi t + \pi/2)$ where the amplitude is in volts. Quantize the amplitude of $y(t)$ into 21 levels with the intervals ranging from -10.5 to 10.5 in 1-volt steps. Consider 100 periods of $y(t)$ and let t take on discrete values given by nt_s where $t_s = 5\text{ms}$. Construct a histogram of $y(t)$.

8.51 Write a MATLAB program to generate a Bernoulli process $X[n]$ for which each time instant of the process is a Bernoulli random variable with $\Pr(X[n] = 1) = 0.1$ and $\Pr(X[n] = 0) = 0.9$. Also, the process is IID (i.e., $X[n]$ is independent of $X[m]$ for all $m \neq n$). Once you have created the program to simulate $X[n]$, then create a counting process $Y[n]$ which counts the number of occurrences of $X[m] = 1$ in the interval $m \in [0, n]$. Plot member functions of each of the two processes.

8.52 Let W_n, $n = 0, 1, 2, 3, \ldots$, be a sequence of IID zero-mean Gaussian random variables with variance $\sigma_W^2 = 1$.
   (a) Write a MATLAB program to generate the process

   $$X[n] = \frac{1}{2}X[n-1] - \frac{1}{4}X[n-2] - \frac{1}{4}X[n-3] + W_n,$$

   where $X[0] = W_0$, and $X[n] = 0$ for $n < 0$.
   (b) Estimate the mean function of this process by generating a large number of realizations of the random process and computing the sample mean.
   (c) Compute the time-averaged mean of the process from a single realization. Does this seem to give the same result as the ensemble mean estimated in part (b)?

8.53 A certain random process is created as a sum of a large number, n, of sinusoids with random frequencies and random phases,

$$X(t) = \sum_{k=1}^{n} \cos(2\pi F_k t + \theta_k) ,$$

where the random phases θ_k are IID and uniformly distributed over $(0, 2\pi)$ and the random frequencies are given by $F_k = f_o + f_d\cos(\beta_k)$, where the β_k are IID and uniformly distributed over $(0, 1)$. (*Note:* These types of processes occur in the study of wireless commuincation systems.) For this exercise, we will take the constants f_o and

f_d to be $f_o = 25$Hz and $f_d = 10$ Hz, while we will let the number of terms in the sum be $n = 32$.

(a)   Write a MATLAB program to generate realizations of this random process.

(b)   Assuming the process is stationary and ergodic, use a single realization to estimate the first-order PDF of the process, $f_X(x)$.

(c)   Assuming the process is stationary and ergodic, use a single realization to estimate the mean of the process, μ_X.

(d)   Assuming the process is stationary and ergodic, use a single realization to estimate the autocorrelation function of the process, $R_{XX}(\tau)$.

8.54   Write a MATLAB program to generate a shot noise process with

$$h(t) = b(at)\exp(-at)u(t),$$

where $a = 10^{12}$s^{-1} and the constant b is chosen so that $q = \int h(t)dt$. For this program, assume that carriers cross the depletion region at a rate of 10^{13} per second. Plot a member function of this random process.

Markov Processes

In this chapter, we study a class of random processes that possess a certain characteristic that could crudely be described as memoryless. These processes appear in numerous applications including queuing systems, computer communication networks, biological systems, and a wide variety of other applications. As a result of their frequent occurrence, these processes have been studied extensively, and a wealth of theory exists to solve problems related to these processes. We make no attempt to give an exhaustive treatment here, but rather present some of the fundamental concepts involving Markov processes.

9.1 Definition and Examples of Markov Processes

Definition 9.1: A random process, $X(t)$, is said to be a Markov process if for any time instants, $t_1 < t_2 < \ldots < t_n < t_{n+1}$, the random process satisfies

$$F_X(X(t_{n+1}) \leq x_{n+1} | X(t_n) = x_n, X(t_{n-1}) = x_{n-1}, \ldots, X(t_1) = x_1)$$

$$= F_X(X(t_{n+1}) \leq x_{n+1} | X(t_n) = x_n). \tag{9.1}$$

To understand this definition, we interpret t_n as the present time so that t_{n+1} represents some point in the future and $t_1, t_2, \ldots, t_{n-1}$ represent various points in the past. The Markovian property then states that given the present, the future is independent of the past. Or, in other words, the future of the random process only depends on where it is now and not on how it got there.

■ Example 9.1:

A classical example of a continuous time Markov process is the Poisson counting process studied in the previous chapter. Let $X(t)$ be a Poisson counting process with rate λ. Then its probability mass function satisfies

$$\Pr(X(t_{n+1}) = x_{n+1} | X(t_n) = x_n, X(t_{n-1}) = x_{n-1}, \ldots, X(t_1) = x_1)$$

Probability and Random Processes. DOI: 10.1016/B978-0-12-386981-4.00009-7

$$
= \begin{cases} 0, & x_{n+1} < x_n, \\ \dfrac{(\lambda(t_{n+1} - t_n))^{x_{n+1} - x_n}}{(x_{n+1} - x_n)!} e^{-\lambda(t_{n+1} - t_n)}, & x_{n+1} \ge x_n. \end{cases}
$$

Clearly this is independent of $\{X(t_{n-1}) = x_{n-1}, \ldots, X(t_1) = x_1\}$. In fact, the Markovian property must be satisfied because of the independent increments assumption of the Poisson process. ∎

To start with, we will focus our attention on discrete-valued Markov processes in discrete time, better known as Markov chains. Let $X[k]$ be the value of the process at time instant k. Since the process is discrete-valued, $X[k] \in \{x_1, x_2, x_3, \ldots\}$ and we say that if $X[k] = x_n$ then the process is in state n at time k. A Markov chain is described statistically by its transition probabilities which are defined as follows.

Definition 9.2: Let $X[k]$ be a Markov chain with states $\{x_1, x_2, x_3, \ldots\}$. Then, the probability of transitioning from state i to state j in one time instant is

$$p_{i,j} = \Pr(X[k+1] = j | X[k] = i). \tag{9.2}$$

If the Markov chain has a finite number of states, n, then it is convenient to define a *transition probability matrix*,

$$
\boldsymbol{P} = \begin{bmatrix} p_{1,1} & p_{1,2} & \cdots & p_{1,n} \\ p_{2,1} & p_{2,2} & \cdots & p_{2,n} \\ & & & \\ p_{n,1} & p_{n,2} & \cdots & p_{n,n} \end{bmatrix}. \tag{9.3}
$$

One can encounter processes where the transition probabilities vary with time and therefore need to be explicitly written as a function of k (e.g., $p_{i,j,k}$) but we do not consider such processes in this text and henceforth it is assumed that transition probabilities are independent of time.

■ Example 9.2:

Suppose every time a child buys a kid's meal at his favorite fast food restaurant he receives one of four superhero action figures. Naturally, the child wants to collect all four action figures and so he regularly eats lunch at this restaurant in order to complete the collection. This process can be described by a Markov chain. In this case, let $X[k] \in \{0, 1, 2, 3, 4\}$ be the number of different action figures that the child has collected after purchasing k meals. Assuming each meal contains one of the four superheroes with equal probability and that the action figure in any meal is independent of what is contained in any previous or future meals, then the transition probability matrix easily works out to be

$$P = \begin{bmatrix} 0 & 1 & 0 & 0 & 0 \\ 0 & 1/4 & 3/4 & 0 & 0 \\ 0 & 0 & 1/2 & 1/2 & 0 \\ 0 & 0 & 0 & 3/4 & 1/4 \\ 0 & 0 & 0 & 0 & 1 \end{bmatrix}.$$

Initially (before any meals are bought), the process starts in state 0 (the child has no action figures). When the first meal is bought, the Markov chain must move to state 1 since no matter which action figure is contained in the meal, the child will now have one superhero. Therefore, $p_{0,1} = 1$ and $p_{0,j} = 0$ for all $j \neq 1$. If the child has one distinct action figure, when he buys the next meal he has a 25 % chance of receiving a duplicate and a 75% chance of getting a new action figure. Thus, $p_{1,1} = 1/4$, $p_{1,2} = 3/4$, and $p_{1,j} = 0$ for $j \neq 1, 2$. Similar logic is used to complete the rest of the matrix. The child might be interested in knowing the average number of lunches he needs to buy until his collection is completed. Or, maybe the child has saved up only enough money to buy 10 lunches and wants to know what are his chances of completing the set before running out of money. We will develop the theory needed to answer such questions.

■

The transition process of a Markov chain can also be illustrated graphically using a state diagram. Such a diagram is illustrated in Figure 9.1 for the Markov chain in Example 9.2. In the figure, each directed arrow represents a possible transition and the label on each arrow represents the probability of making that transition. Note that for this Markov chain, once we reach state 4, we remain there forever. This type a state is referred to as an *absorbing state*.

Figure 9.1
State diagram for the Markov chain of Example 9.2.

■ Example 9.3 (The Gambler's Ruin Problem):

Suppose a gambler plays a certain game of chance (e.g., blackjack) against the "house." Every time the gambler wins the game, he increases his fortune by one unit (say, a dollar) and every time he loses, his fortune decreases by one unit. Suppose the gambler wins each game with probability p and loses with probability $q = 1-p$. Let X_n represent the amount of the gambler's fortune after playing the game n times. If the gambler ever reaches the state $X_n = 0$, the gambler is said to be "ruined" (he has lost all of his money). Assuming that the outcome of each game is independent of all others, the sequence x_n, $n = 0, 1, 2, \dots$, forms a Markov chain. The state transition matrix is of the form

$$\mathbf{P} = \begin{bmatrix} 1 & 0 & 0 & 0 & 0 & \dots \\ q & 0 & p & 0 & 0 & \dots \\ 0 & q & 0 & p & 0 & \dots \\ 0 & 0 & q & 0 & p & \dots \\ \dots & & & \dots & & \dots \end{bmatrix}.$$

The state transition diagram for the gambler's ruin problem is shown in Figure 9.2a. One might then be interested in determining how long it might take before the gambler is ruined (enters the zero state). Is ruin inevitable for any p, or if the gambler is sufficiently proficient at the game, can he avoid ruin indefinitely? A more realistic alternative to this model is one where the house also has a finite amount of money. Suppose the gambler initially starts with d dollars and the house has $b - d$ dollars so that between the two competitors there is a total of b dollars in the game. Now if the gambler ever gets to the state 0, he is ruined, while if he gets to the state b, he has "broken the bank" (i.e., the house is ruined). Now the Markov chain has two absorbing states as shown in Figure 9.2b. It would seem that sooner or later the gambler must have a run of bad luck sufficient to send him to the 0 state (i.e., ruin) or a run of good luck which will cause him to enter the state b (i.e., break the bank). It would be interesting to find the probabilities of each of these events.

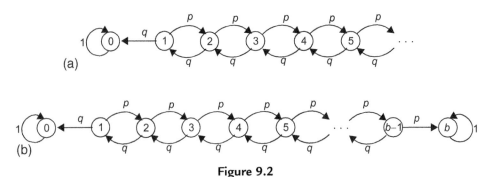

Figure 9.2
State transition diagram for Example 9.3 (The Gamblers Ruin Problem) with
(a) one absorbing state and (b) with two absorbing states.

■

The previous example is one of a class of Markov chains known as *random walks*. Random walks are often used to describe the motion of a particle. There are many applications that can be described by a random walk that do not involve the movement of a particle, but it is helpful to think of such a particle when describing random walks. In one dimension, a random walk is a Markov chain whose states are the integers and whose transition probabilities satisfy $p_{i,j} = 0$ for any $j \neq i-1, i, i+1$. In other words, at each time instant, the state of the Markov chain can either increase by one, stay the same, or decrease by one. If $p_{i, i+1} = p_{i, i-1}$, then the random walk is said to be symmetric, whereas, if $p_{i, i+1} \neq p_{i, i-1}$, the random walk is said to have drift. Often the state space of the random walk will be a finite range of integers, $n, n+1, n+2, \ldots, m-1, m$ (for $m > n$), in which case the states n and m are said to be boundaries or barriers. The gamblers ruin problem is an example of a random walk with absorbing boundaries, where $p_{n,n} = p_{m,m} = 1$. Once the particle reaches the boundary, it is absorbed and remains there forever. We could also construct a random walk with reflecting boundaries, in which case $p_{n, n+1} = p_{m, m-1} = 1$. That is, whenever the particle reaches the boundary, it is always reflected back to the adjacent state.

■ Example 9.4 (A Queueing System):

A common example of Markov chains (and Markov processes in general) is that of queueing systems. Consider, for example, a taxi stand at a busy airport. A line of taxis, which for all practical purposes can be taken to be infinitely long, is available to serve travelers. Customers wanting a taxi enter a queue and are given a taxi on a first come, first serve basis. Suppose it takes one unit of time (say, a minute) for the customer at the head of the queue to load himself and his luggage into a taxi. Hence, during each unit of time, one customer in the queue receives service and leaves the queue while some random number of new customers enter the end of the queue. Suppose at each time instant, the number of new customers arriving for service is described by a discrete distribution $(p_0, p_1, p_2, \ldots)$, where p_k is the probability of k new customers. For such a system, the transition probability matrix of the Markov chain would look like

$$P = \begin{bmatrix} p_0 & p_1 & p_2 & p_3 & \cdots \\ p_0 & p_1 & p_2 & p_3 & \cdots \\ 0 & p_0 & p_1 & p_2 & \cdots \\ 0 & 0 & p_0 & p_1 & \cdots \\ \cdots & & \cdots & & \cdots \end{bmatrix}.$$

The manager of the taxi stand might be interested in knowing the probability distribution of the queue length. If customers have to wait too long, they may get dissatisfied and seek other forms of transportation.

■ Example 9.5 (A Branching Process):

Branching processes are commonly used in biological studies. Suppose a certain species of organism has a fixed lifespan (one time unit). At the end of its lifespan, the nth organism in the species produces a number of offspring described by some random variable, Y_n, whose sample space is the set of non-negative integers. Also, assume that the number of offspring produced by each organism is independent and identically distributed (IID). Then, X_k, the number of species in the organism during the kth generation is a random variable which depends only on the number of organisms in the previous generation. In particular, if $X_k = i$, then $X_{k+1} = \sum_{n=1}^{i} Y_n$. The transition probability is then

$$p_{i,j} = \Pr(X_{k+1} = j | X_k = i) = \Pr\left(\sum_{n=1}^{i} Y_n = j\right).$$

Let $H_Y(z)$ be the probability-generating function of the random variable Y_n (i.e., $H_Y(z) = \sum_{i=0}^{\infty} \Pr(Y_n = i) z^i$). Then, due to the IID nature of the Y_n, $\sum_{n=1}^{i} Y_n$ will have a probability-generating function given by $[H_Y(z)]^i$. Hence, the transition probability, $p_{i,j}$, will be given by the coefficient of z^j in the power series expansion of $[H_Y(z)]^i$.

■

■ Example 9.6 (A Genetic Model):

Suppose a gene contains n units. Of these units, i are mutant and $n-i$ are normal. Every time a cell doubles, the n units double and each of the two cells receives a gene composed of n units. After doubling, there is a pool of $2n$ units of which $2i$ are mutant. These $2n$ units are grouped into two sets of n units randomly. As we trace a single line of descent, the number of mutant units in each gene forms a Markov chain. Define the kth gene to be in state i ($X_k = i$) if it is composed of i mutant and $n-i$ normal units. It is not difficult to show that, given $X_k = i$,

$$p_{i,j} = \Pr(X_{k+1} = j | X_k = i) = \frac{\binom{2i}{j}\binom{2n-2i}{n-j}}{\binom{2n}{n}}.$$

■

9.2 Calculating Transition and State Probabilities in Markov Chains

The state transition probability matrix of a Markov chain gives the probabilities of transitioning from one state to another in a single time unit. It will be useful to extend this concept to longer time intervals.

Definition 9.3: The n-step transition probability for a Markov chain is

$$p_{i,j}^{(n)} = \Pr(X_{k+n}=j|X_k=i).$$ (9.4)

Also, define an n-step transition probability matrix $\boldsymbol{P}^{(n)}$ whose elements are the n-step transition probabilities in Equation (9.4).

Given the one-step transition probabilities, it is straightforward to calculate higher order transition probabilities using the following result.

Theorem 9.1 (Chapman–Kolmogorov Equation):

$$p_{i,j}^{(n)} = \sum_k p_{i,k}^{(m)} p_{k,j}^{(n-m)}, \quad \text{for any } m = 0, 1, 2, \ldots, n.$$ (9.5)

Proof: First, condition on the event that in the process of transitioning from state i to state j, the Markov chain passes through state k at some intermediate point in time. Then using the principle of total probability

$$\Pr(X_{l+n}=j|X_l=i) = \sum_k \Pr(X_{l+n}=j|X_l=i, X_{l+m}=k)\Pr(X_{l+m}=k|X_k=i).$$ (9.6)

Using the Markov property, the expression reduces to the desired form:

$$\Pr(X_{l+n}=j|X_l=i) = \sum_k \Pr(X_{l+n}=j|X_{l+m}=k)\Pr(X_{l+m}=k|X_k=i). \quad \Box$$ (9.7)

This result can be written in a more compact form using transition probability matrices. It is easily seen that the Chapman–Kolmogorov equations can be written in terms of the n-step transition probability matrices as

$$\boldsymbol{P}^{(n)} = \boldsymbol{P}^{(m)}\boldsymbol{P}^{(n-m)}.$$ (9.8)

Then, starting with the fact that $\boldsymbol{P}^{(1)} = \boldsymbol{P}$, it follows that $\boldsymbol{P}^{(2)} = \boldsymbol{P}^{(1)}\boldsymbol{P}^{(1)} = \boldsymbol{P}^2$, and using induction, it is established that

$$\boldsymbol{P}^{(n)} = \boldsymbol{P}^n.$$ (9.9)

Hence, we can find the n-step transition probability matrix through matrix multiplication. If n is large, it may be more convenient to compute $\boldsymbol{P}^n$ via eigendecomposition. In many cases[1] the

---

[1] It is noted that not all transition matrices are diagonalizable. For those that are not, this approach will not work.

matrix P can be expanded as $P = UAU^{-1}$, where A is the diagonal matrix of eigenvalues and U is the matrix whose columns are the corresponding eigenvectors. Then,

$$P^n = UA^nU^{-1}. \tag{9.10}$$

Another quantity of interest is the probability distribution of the Markov chain at some time instant k. If the initial probability distribution of the Markov chain is known, then the distribution at some later point in time can easily be found. Let $\pi_j(k) = \Pr(X_k=j)$ and $\pi(k)$ be the row vector whose jth element is $\pi_j(k)$. Then

$$\pi_j(k) = \Pr(X_k=j) = \sum_i \Pr(X_k=j|X_0=i)\Pr(X_0=i) = \sum_i p_{i,j}^{(k)}\pi_i(0), \tag{9.11}$$

or in vector form,

$$\pi(k) = \pi(0)P^k. \tag{9.12}$$

■ Example 9.7 (continuation of Example 9.2):

Recall in Example 9.2, the child who purchased kid's meals at his favorite restaurant in order to collect a set of four superhero action figures. Initially, before any meals are purchased, the child has no action figures and so the initial probability distribution is $\pi(0) = (1, 0, 0, 0, 0)$. Repeated application of Equation (9.12) with the probability transition matrix given in Example 9.2 results in

$$\pi(1) = (0, 1, 0, 0, 0),$$

$$\pi(2) = (0, 1/4, 3/4, 0, 0),$$

$$\pi(3) = (0, 1/16, 9/16, 3/8, 0),$$

$$\pi(4) = (0, 1/64, 21/64, 9/16, 3/32),$$

$$\pi(5) = (0, 1/256, 45/256, 75/128, 15/64),$$

$$\pi(6) = (0, 1/1024, 93/1024, 135/256, 195/512),$$

and so on. It is to be expected that if the child buys enough meals, he will eventually complete the collection (i.e., get to state 4) with probability approaching unity. This can easily be verified analytically by calculating the limiting form of P^k as $k \to \infty$. Recall that for this example, P is a triangular matrix and hence its eigenvalues are simply the diagonal entries. Hence, the diagonal matrix of eigenvalues is

$$\Lambda = \begin{bmatrix} 0 & 0 & 0 & 0 & 0 \\ 0 & 1/4 & 0 & 0 & 0 \\ 0 & 0 & 1/2 & 0 & 0 \\ 0 & 0 & 0 & 3/4 & 0 \\ 0 & 0 & 0 & 0 & 1 \end{bmatrix}.$$

It should be clear that $\lim_{k \to \infty} \Lambda^k$ is a matrix with all zero entries except the one in the lower right corner, which is equal to one. Using MATLAB (or some other math package) to calculate the corresponding matrix of eigenvectors, it is found that

$$\lim_{k \to \infty} \boldsymbol{P}^k = \boldsymbol{U}(\lim_{k \to \infty} \Lambda^k)\boldsymbol{U}^{-1} = \begin{bmatrix} 0 & 0 & 0 & 0 & 1 \\ 0 & 0 & 0 & 0 & 1 \\ 0 & 0 & 0 & 0 & 1 \\ 0 & 0 & 0 & 0 & 1 \\ 0 & 0 & 0 & 0 & 1 \end{bmatrix}.$$

Then, using the initial distribution of $\boldsymbol{\pi}(0) = (1, 0, 0, 0, 0)$, the state distribution as $k \to \infty$ works out to be

$$\boldsymbol{\pi} = \lim_{k \to \infty} \pi(k) = \lim_{k \to \infty} \boldsymbol{\pi}(0)\boldsymbol{P}^k = \begin{bmatrix} 0 & 0 & 0 & 0 & 1 \end{bmatrix}.$$

∎

In Example 9.6, it was seen that as $k \to \infty$, the k-step transition probability matrix approached that of a matrix whose rows were all identical. In that case, the limiting product $\lim_{k \to \infty} \boldsymbol{\pi}(0)\boldsymbol{P}^k$ is the same regardless of the initial distribution $\boldsymbol{\pi}(0)$. Such a Markov chain is said to have a unique steady-state distribution, $\boldsymbol{\pi}$. It should be emphasized that not all Markov chains have a steady-state distribution. For example, the Poisson counting process of Example 9.1 clearly does not, since any counting process is a monotonic non-decreasing function of time and, therefore, it is expected that the distribution should skew toward larger values as time progresses.

This concept of a steady-state distribution can be viewed from the perspective of stationarity. Suppose at time k, the process has some distribution, $\boldsymbol{\pi}(k)$. The distribution at the next time instant is then $\boldsymbol{\pi}(k+1) = \boldsymbol{\pi}(k)\boldsymbol{P}$. If $\boldsymbol{\pi}(k) = \boldsymbol{\pi}(k+1)$, then the process has reached a point where the distribution is stationary (independent of time). This stationary distribution, $\boldsymbol{\pi}$, must satisfy the relationship

$$\boldsymbol{\pi} = \boldsymbol{\pi P}. \tag{9.13}$$

In other words, $\boldsymbol{\pi}$ (if it exists) is the left eigenvector of the transition probability matrix $\boldsymbol{P}$, that corresponds to the eigenvalue $\lambda = 1$. The next example shows that this eigenvector is not always unique.

■ **Example 9.8 (The Gambler's Ruin revisited):**

Suppose a certain gambler has $5 and plays against another player (the house). The gambler decides that he will play until he either doubles his money or loses it all. Suppose the house has designed this game of chance so that the gambler will win with probability $p = 0.45$ and the house will win with probability $q = 0.55$. Let X_k be the amount of money the gambler has after playing the game k times. The transition probability matrix for this Markov chain is

$$
P = \begin{bmatrix}
1 & 0 & 0 & 0 & 0 & 0 & 0 & 0 & 0 & 0 & 0 \\
0.55 & 0 & 0.45 & 0 & 0 & 0 & 0 & 0 & 0 & 0 & 0 \\
0 & 0.55 & 0 & 0.45 & 0 & 0 & 0 & 0 & 0 & 0 & 0 \\
0 & 0 & 0.55 & 0 & 0.45 & 0 & 0 & 0 & 0 & 0 & 0 \\
0 & 0 & 0 & 0.55 & 0 & 0.45 & 0 & 0 & 0 & 0 & 0 \\
0 & 0 & 0 & 0 & 0.55 & 0 & 0.45 & 0 & 0 & 0 & 0 \\
0 & 0 & 0 & 0 & 0 & 0.55 & 0 & 0.45 & 0 & 0 & 0 \\
0 & 0 & 0 & 0 & 0 & 0 & 0.55 & 0 & 0.45 & 0 & 0 \\
0 & 0 & 0 & 0 & 0 & 0 & 0 & 0.55 & 0 & 0.45 & 0 \\
0 & 0 & 0 & 0 & 0 & 0 & 0 & 0 & 0.55 & 0 & 0.45 \\
0 & 0 & 0 & 0 & 0 & 0 & 0 & 0 & 0 & 0 & 1
\end{bmatrix}.
$$

This matrix has (two) repeated eigenvalues of $\lambda = 1$, and the corresponding eigenvectors are $[1\ 0\ 0\ 0\ 0\ 0\ 0\ 0\ 0\ 0\ 0]$ and $[0\ 0\ 0\ 0\ 0\ 0\ 0\ 0\ 0\ 0\ 1]$. Note that any linear combination of these will also be an eigenvector. Therefore, any vector of the form $[p\ 0\ 0\ 0\ 0\ 0\ 0\ 0\ 0\ 0\ 1-p]$ is a left eigenvector of P and hence there is no unique stationary distribution for this Markov chain. For this example, the limiting form of the state distribution of the Markov chain depends on the initial distribution. The limiting form of P^k can easily be found to be

$$
\lim_{k \to \infty} P^k = \begin{bmatrix}
1 & 0\ 0\ 0\ 0\ 0\ 0\ 0\ 0\ & 0 \\
0.9655 & 0\ 0\ 0\ 0\ 0\ 0\ 0\ 0\ 0\ & 0.0345 \\
0.9233 & 0\ 0\ 0\ 0\ 0\ 0\ 0\ 0\ 0\ & 0.0767 \\
0.8717 & 0\ 0\ 0\ 0\ 0\ 0\ 0\ 0\ 0\ & 0.1283 \\
0.8087 & 0\ 0\ 0\ 0\ 0\ 0\ 0\ 0\ 0\ & 0.1913 \\
0.7317 & 0\ 0\ 0\ 0\ 0\ 0\ 0\ 0\ 0\ & 0.2683 \\
0.6376 & 0\ 0\ 0\ 0\ 0\ 0\ 0\ 0\ 0\ & 0.3624 \\
0.5225 & 0\ 0\ 0\ 0\ 0\ 0\ 0\ 0\ 0\ & 0.4775 \\
0.3819 & 0\ 0\ 0\ 0\ 0\ 0\ 0\ 0\ 0\ & 0.6181 \\
0.2101 & 0\ 0\ 0\ 0\ 0\ 0\ 0\ 0\ 0\ & 0.7899 \\
0 & 0\ 0\ 0\ 0\ 0\ 0\ 0\ 0\ & 1
\end{bmatrix}
$$

Using the initial distribution $\pi(0) = [0\ 0\ 0\ 0\ 0\ 1\ 0\ 0\ 0\ 0\ 0]$ (that is, the gambler starts off in state 5), then it is seen that the steady-state distribution is $\lim_{k \to \infty} \pi(k) = [0.7317\ 0\ 0\ 0\ 0\ 0\ 0\ 0\ 0\ 0\ 0.2683]$. So, when the gambler starts with $5, he has about a 73% chance of losing all of his money and about a 27% chance of doubling his money.

■

As seen in Example 9.7, with some Markov chains, the limiting form of P^k (as $k \rightarrow \infty$) does not necessarily converge to a matrix whose rows are all identical. In that case, the limiting form of the state distribution will depend on the starting distribution. In the case of the gambler's ruin problem, we probably could have guessed this behavior. If the gambler had started with very little money, we would expect him to end up in the state of ruin with very high probability; whereas, if the gambler was very wealthy and the house had very little money, we would expect a much greater chance of the gambler eventually breaking the house. Accordingly, our intuition tells us that the probability distribution of the gamblers ultimate state should depend on the starting state.

In general, there are several different manners in which a Markov chain's state distribution can behave as $k \rightarrow \infty$. In some cases, $\lim_{k \to \infty} \pi(k)$ does not exist. Such would be the case when the process tends to oscillate between two or more states. A second possibility, as in Example 9.7, is that $\lim_{k \to \infty} \pi(k)$ does in fact converge to a fixed distribution, but the form of this limiting distribution depends on the starting distribution. The last case is when $\lim_{k \to \infty} \pi(k) = \pi$. That is the state distribution converges to some fixed distribution, π, and the form of π is independent of the starting distribution. Here, the transition probability matrix, P, will have a single (not repeated) eigenvalue at $\lambda = 1$, and the corresponding eigenvector (properly normalized) will be the steady-state distribution, π. Furthermore, the limiting form of P^k will be one whose rows are all identical and equal to the steady-state distribution, π. In the next section, we look at some conditions that must be satisfied for the Markov chain to achieve a unique steady-state distribution.

■ Example 9.9:

In this example, we provide the MATLAB code to simulate the distribution of the queue length of the taxi stand described in Example 9.4. For this example, we take the number of arrivals per time unit, X, to be a Poisson random variable whose PMF is

$$P_X(k) = \frac{\lambda^k e^{-\lambda}}{k!}.$$

Recall that in the taxi stand, one customer is served per time unit (assuming there is a least one customer in the queue waiting to be served). The following code can be used to estimate and plot the PMF of the queue length. The average queue length was also calculalted to be 3.36 customers for an arrival rate of $\lambda = 0.85$ customers per time unit. Figure 9.3 shows a histogram of the PMF of the queue length for the same arrival rate.

```
N=10000;                          % Length of simulation.
a=0.85;                           % arrival rate.
k=[0:10];
Poisson=zeros(size(k));           % Calculate Poisson PMF
for m=k
```

(Continued)

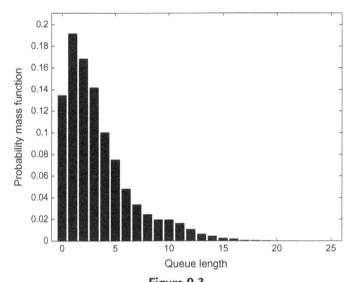

Figure 9.3

Histogram of the queue length for the taxi stand of Example 9.4 assuming a Poisson arrival process with an average arrival rate of 0.85 arrivals per time unit.

```
    Poisson(m+1)=a.^m*exp(-a)./factorial(m);
end
queue(1)=0;                              % Initial queue size.
for n=1:N
    x=rand(1);
    arrivals=sum(x>cumsum(Poisson));     % Poisson RV
    departures=queue(n)>0;
    queue(n+1)=queue(n)+arrivals-departures;  % current queue length
end
mean_queue_length=sum(queue)/length(queue)  % compute avg. queue length
bins=[0:25]
y=hist(queue,bins);
PMF=y/N;                                 % estimate PMF
bar(bins,PMF)                            % plot results
axis([min(bins)-1 max(bins)+1 0 1.1*max(PMF)])
```

9.3 Characterization of Markov Chains

Using the methods presented in the previous sections, calculating the steady-state distribution of a Markov chain requires performing an eigendecomposition of the transition probability matrix, P. If the number of states in the Markov chain is large (or infinite), performing the required linear algebra may be difficult (or impossible). Thus, it would be useful to seek alternative methods to determine if a steady-state distribution exists, and if so, to calculate it. To develop the necessary theory, we must first proceed through a sequence of definitions and classifications of the states of a Markov chain.

> **Definition 9.4:** State j is *accessible* from state i if for some finite n, $p_{i,j}^{(n)} > 0$. This simply means that if the process is in state i, it is possible for the process to get to state j in a finite amount of time. Furthermore, if state j is accessible from state i and state i is accessible from state j, then the states i and j are said to *communicate*. It is common to use the shorthand notation $i \leftrightarrow j$ to represent the relationship "state i communicates with state j."

The states of any Markov chain can be divided into sets or classes of states where all the states in a given class communicate with each other. It is possible for the process to move from one communicating class of states to another, but once that transition is made, the process can never return to the original class. If it did, the two classes would communicate with each other and hence would be a part of a single class.

> **Definition 9.5:** A Markov chain for which all of the states are part of a single communicating class is called an *irreducible* Markov chain. Also, the corresponding transition probability matrix is called an irreducible matrix.

The examples in Section 9.1 can be used to help illustrate these concepts. For both processes in Examples 9.1 and 9.2, none of the states communicate with any other states. This is a result of the fact that both processes are counting processes and as a result, it is impossible to go backward in the chain of states. Therefore, if $j > i$, state j is accessible from state i but state i is not accessible from state j. As a result, for any counting process, all states form a communication class of their own. That is, the number of classes is identical to the number of states. In the gambler's ruin problem of Example 9.3, the two absorbing states do not communicate with any other state since it is impossible to ever leave an absorbing state, while all the states in between communicate with each other. Therefore, this Markov chain has three communicating classes; the two absorbing states each form of class to themselves, while the third class consists of all the states in between.

The queueing system (taxi stand) of Example 9.4 represents a Markov chain where all states communicate with each other, and therefore that Markov chain is irreducible. For the branching process of Example 9.5, all states communicate with each other except the state 0, which represents the extinction of the species, which presumably is an absorbing state. The genetic model of Example 9.6 is similar to the gambler's ruin problem in that the Markov chain has two absorbing states at the end points, while everything in between forms a single communicating class.

Definition 9.6: The *period* of state i, $d(i)$, is the greatest common divisor of all integers $n \geq 1$ such that $p_{i,i}^{(n)} > 0$. Stated another way, $d(i)$ is the period of state i if any transition from state i to itself must occur in a number of steps that is a multiple of $d(i)$. Furthermore, a Markov chain for which all states have a period of $d(i) = 1$ is called an *aperiodic* Markov chain.

Most Markov chains are aperiodic. The class of random walks is an exception. Suppose a Markov chain defined on the set of integers has transition probabilities that satisfy

$$p_{i,j} = \begin{cases} p, & j = i+1, \\ 1-p, & j = i-1, \\ 0, & \text{otherwise.} \end{cases} \tag{9.14}$$

Then each state will have a period of 2. If we add absorbing boundaries to the random walk, then the absorbing states are not periodic because for the absorbing states, $p_{i,i}^{(n)} = 1$ for all n and thus the period of an absorbing state is 1. It is left as an exercise for the reader (see Exercise 9.26) to establish the fact that the period is a property of a class of communicating states. That is, if $i \leftrightarrow j$, then $d(i) = d(j)$ and therefore all states in the same class must have the same period.

Definition 9.7: Let $f_{i,i}^{(n)}$ be the probability that given a process is in state i, the first return to state i will occur in exactly n steps. Mathematically,

$$f_{i,i}^{(n)} = \Pr(X_{k+n} = i, X_{k+m} \neq i \text{ for } m = 1, 2, \ldots, n-1 | X_k = i). \tag{9.15}$$

Also, define $f_{i,i}$ to be the probability that the process will eventually return to state i. The probability of eventual return is related to the first return probabilities by

$$f_{i,i} = \sum_{n=1}^{\infty} f_{i,i}^{(n)}. \tag{9.16}$$

It should be noted that the first return probability $f_{i,i}^{(n)}$ is not the same thing as the n-step transition probability, but the two quantities are related. To develop this relationship, it is observed that

$$p_{i,i}^{(n)} = \sum_{m=0}^{n} \Pr(\{X_{k+n}=i\}, \{\text{first return to state } i \text{ occurs in } m \text{ steps}\} | X_k=i)$$

$$= \sum_{m=0}^{n} \Pr(X_{k+n}=i | X_{k+m}=i)\Pr(\text{first return to state } i \text{ occurs in } m \text{ steps} | X_{k+m}=i)$$

$$= \sum_{m=0}^{n} p_{i,i}^{(n-m)} f_{i,i}^{(m)}, \quad n = 1, 2, 3, \ldots. \tag{9.17}$$

In Equation (9.17), $p_{i,i}^{(0)}$ is taken to be equal to 1 and $f_{i,i}^{(0)}$ is taken to be equal to 0. Given the n-step transition probabilities, $p_{ii}^{(n)}$, one could solve the preceding system of equations for the first return probabilities. However, since the previous equation is a convolution, this set of equations may be easier to solve using frequency domain techniques. Define the generating functions[1],

$$P_{i,i}(z) = \sum_{n=0}^{\infty} p_{i,i}^{(n)} z^n, \tag{9.18}$$

$$F_{i,i}(z) = \sum_{n=0}^{\infty} f_{i,i}^{(n)} z^n. \tag{9.19}$$

It is left as an exercise for the reader (see Exercise 9.27) to demonstrate that these two generating functions are related by

$$P_{i,i}(z) - 1 = P_{i,i}(z) F_{i,i}(z). \tag{9.20}$$

This relationship provides an easy way to compute the first return probabilities from the n-step transition probabilities. Note that if the transition probability matrix is known, then calculating the generating function, $P_{i,i}(z)$, is straightforward. Recall that $\boldsymbol{P}^n = \boldsymbol{U}\boldsymbol{\Lambda}^n\boldsymbol{U}^{-1}$ and therefore if $[\boldsymbol{U}]_{i,j}$ is the element in the ith row and jth column of $\boldsymbol{U}$, then

$$p_{i,i}^{(n)} = \sum_{j} [\boldsymbol{U}]_{i,j} \boldsymbol{\Lambda}^n [\boldsymbol{U}^{-1}]_{j,i}. \tag{9.21}$$

---

[1] Note that these generating functions are not necessarily probability-generating functions as the sequences involved are not necessarily probability distributions.

Forming the generating function from this equation results in

$$P_{i,i}(z) = \sum_{n=0}^{\infty} p_{i,i}^{(n)} z^n = \sum_{n=0}^{\infty} \sum_{j} [U]_{i,j} (\Lambda z)^n [U^{-1}]_{j,i}$$

$$= \sum_{j} [U]_{i,j} \left(\sum_{n=0}^{\infty} (\Lambda z)^n \right) [U^{-1}]_{j,i} \quad . \qquad (9.22)$$

$$= \sum_{j} [U]_{i,j} (I - \Lambda z)^{-1} [U^{-1}]_{j,i} .$$

In other words, $P_{i,i}(z)$ is the element in the ith row and ith column of the matrix $U(I - \Lambda z)^{-1} U^{-1}$.

Definition 9.8: The ith state of a Markov chain is *transient* if $f_{i,i} < 1$ and *recurrent* if $f_{i,i} = 1$. Since $f_{i,i}$ represents the probability of the process eventually returning to state i given that it is in state i, the state is transient if there is some non-zero probability that it will never return, and the state is recurrent if the process must eventually return with probability 1.

Theorem 9.2: State i of a Markov chain is recurrent if and only if

$$\sum_{n=1}^{\infty} p_{i,i}^{(n)} = \infty . \qquad (9.23)$$

Since this sum represents the expected number of returns to state i, it follows that a state is recurrent if and only if the expected number of returns is infinite.

Proof: First, note that $f_{i,i} = \sum_{n=1}^{\infty} f_{i,i}^{(n)} = \lim_{z \to 1} F_{i,i}(z)$. From Equation (9.20), this would imply that

$$\lim_{z \to 1} P_{i,i}(z) - 1 = \lim_{z \to 1} P_{i,i}(z) f_{i,i} . \qquad (9.24)$$

As a result,

$$\lim_{z \to 1} P_{i,i}(z) = \sum_{n=1}^{\infty} p_{i,i}^{(n)} = \frac{1}{1 - f_{i,i}} \; . \tag{9.25}$$

If state i is transient, then $f_{i,i} < 1$ and hence, $\sum_{n=1}^{\infty} p_{i,i}^{(n)} < \infty$, whereas if state i is

recurrent, $f_{i,i} = 1$ and $\sum_{n=1}^{\infty} p_{i,i}^{(n)} = \infty$. $\square$

We leave it to the reader (see Exercise 9.29) to verify that recurrence is a class property. That is, if one state in a communicating class is recurrent, then all are recurrent, and if one is transient, then all are transient.

■ Example 9.10:

Consider a random walk on the integers (both positive and negative) that initially starts at the origin ($X_0 = 0$). At each time instant, the process either increases by 1 with probability p or decreases by 1 with probability $1 - p$:

$$p_{i,j} = \begin{cases} p, & \text{if } j = i + 1, \\ 1 - p, & \text{if } j = i - 1, \\ 0, & \text{otherwise.} \end{cases}$$

First note that this Markov chain is periodic with period 2 and hence $p_{i,i}^{(n)} = 0$ for any odd n. For even n, given the process is in state i, the process will be in state i again n time instants later if during those time instants the process increases $n/2$ times and decreases $n/2$ times. This probability follows a binomial distribution so that

$$p_{i,i}^{(n)} = \binom{n}{n/2} p^{n/2} (1-p)^{n/2}, \text{ for even } n.$$

To determine if the states of this random walk are recurrent or transient, we must determine whether or not the series

$$\sum_{n=1}^{\infty} p_{i,i}^{(2n)} = \sum_{n=1}^{\infty} \binom{2n}{n} (p(1-p))^n$$

converges. To help make this determination, the identity

$$\sum_{n=1}^{\infty} \binom{2n}{n} x^n = \frac{1}{\sqrt{1 - 4x}} - 1, \quad |x| < 1/4,$$

(Continued)

is used. This identity can easily be verified by expanding the binomial on the right hand side in powers of x and confirming (after a little algebra) that the coefficients of the power series expansion do take on the desired form. Applying this identity results in

$$\sum_{n=1}^{\infty} p_{i,i}^{(2n)} = \frac{1}{\sqrt{1 - 4p(1-p)}} - 1 .$$

Note that for a probability p, $4p(1-p) \leq 1$ with equality if and only if $p = 1/2$. Therefore, the series converges and all states are transient if $p \neq 1/2$, while if $p = 1/2$, the series diverges and all states are recurrent.

■

Definition 9.9: The *mean time to first return* for a recurrent state i of a Markov chain is

$$\mu_i = \sum_{n=1}^{\infty} n f_{i,i}^{(n)} . \tag{9.26}$$

If the state is transient, then the mean time to first return must be infinite, since with some non-zero probability, the process will never return.

Definition 9.10: A recurrent state is referred to as *null recurrent* if $\mu_i = \infty$, while the state is *positive recurrent* if $\mu_i < \infty$.

The mean time to first return of a recurrent state is related to the steady-state probability of the process being in that state. To see this, define a sequence of random variables T_1, T_2, T_3, ... where T_m represents the time between the $m-1$ th and m th returns to the state i. That is, suppose that $X_k = i$ for some time instant k which is sufficiently large so that the process has pretty much reached steady state. The process then returns to state i at time instants $k + T_1$, $k + T_1 + T_2$, $k + T_1 + T_2 + T_3$, and so on. Over some period of time where the process visits state i exactly n times, the fraction of time the process spends in state i can be written as

$$\text{fraction of time process is in state } i = \frac{n}{\sum_{j=1}^{n} T_j} = \frac{1}{\frac{1}{n}\sum_{j=1}^{n} T_j} . \tag{9.27}$$

As $n \to \infty$ (assuming the process is ergodic), the left hand side of the previous equation becomes the steady-state probability that the process is in state i, π_i. Furthermore, due to the law of large numbers, the denominator of the right hand side converges to μ_i. This proves the following key result.

Theorem 9.3: For an irreducible, aperiodic, recurrent Markov chain, the steady-state distribution is unique and is given by

$$\pi_i = \frac{1}{\mu_i}. \tag{9.28}$$

Note that if a state is positive recurrent, then $\pi_i > 0$, while if a state is null recurrent, then $\pi_i = 0$. Note that for any transient state, $\mu_i = \infty$ and as a result, $\pi_i = 0$.

■ Example 9.11:

Continuing with the random walk from the previous example, the generating function for the n-step transition probabilities is found to be

$$P_{i,i}(z) = \sum_{n=0}^{\infty} p_{i,i}^{(n)} z^n = \sum_{n=0}^{\infty} \binom{2n}{n} (p(1-p))^n z^{2n} = \frac{1}{\sqrt{1 - 4p(1-p)z^2}}.$$

Using the relationship $P_{i,i}(z) - 1 = P_{i,i}(z)F_{i,i}(z)$, the generating function for the first return probabilities is

$$F_{i,i}(z) = 1 - \sqrt{1 - 4p(1-p)z^2}.$$

Since the random walk is only recurrent for $p = 1/2$, we consider only that case so that $F_{i,i}(z) = 1 - \sqrt{1 - z^2}$. The mean time to first return can be found directly from the generating function.

$$\mu_i = \sum_{n=1}^{\infty} n f_{i,i}^{(n)} = \lim_{z \to 1} \frac{d}{dz} F_{i,i}(z) = \lim_{z \to 1} \frac{z}{\sqrt{1 - z^2}} = \infty.$$

Thus, when the transition probabilities of the random walk are balanced so that all states are recurrent, then the mean time to first return is infinite and, in fact, all states are null recurrent.

■

9.4 Continuous Time Markov Processes

In this section, we investigate Markov processes where the time variable is continuous. In particular, most of our attention will be devoted to the so-called birth–death processes which are a generalization of the Poisson counting process studied in the previous chapter. To start with, consider a random process $X(t)$ whose state space is either finite or countably infinite so that we can represent the states of the process by the set of integers, $X(t) \in \{\ldots, -3, -2, -1, 0, 1, 2, 3, \ldots\}$. Any process of this sort that is a Markov process has the interesting property that the time between any change of states is an exponential random variable. To see this, define T_i to be the time between the ith and the $i+1$th change of state and let $h_i(t)$ be the complement to its CDF, $h_i(t) = \Pr(T_i > t)$. Then, for $t > 0, s > 0$

$$h_i(t+s) = \Pr(T_i > t+s) = \Pr(T_i > t+s, T_i > s) = \Pr(T_i > t+s \mid T_i > s)\Pr(T_i > s). \quad (9.29)$$

Due to the Markovian nature of the process, $\Pr(T_i > t+s \mid T_i > s) = \Pr(T_i > t)$ and hence the previous equation simplifies to

$$h_i(t+s) = h_i(t)h_i(s). \qquad (9.30)$$

The only function which satisfies this type of relationship for arbitrary t and s is an exponential function of the form $h_i(t) = e^{-\rho_i t}$ for some constant ρ_i. Furthermore, for this function to be a valid probability, the constant ρ_i must not be negative. From this, the PDF of the time between change of states is easily found to be $f_{T_i}(t) = \rho_i e^{-\rho_i t} u(t)$.

As with discrete-time Markov chains, the continuous-time Markov process can be described by its transition probabilities.

Definition 9.11: Define $p_{i,j}(t) = \Pr(X(t_0 + t) = j \mid X(t_0) = i)$ to be the transition probability for a continuous time Markov process. If this probability does not depend on t_0, then the process is said to be a *homogeneous* Markov process.

Unless otherwise stated, we assume for the rest of this chapter that all continuous time Markov processes are homogeneous. The transition probabilities, $p_{i,j}(t)$, are somewhat analogous to the n-step transition probabilities used in the study of discrete-time processes and as a result, these probabilities satisfy a continuous time version of the Chapman–Kolmogorov equations:

$$p_{i,j}(t+s) = \sum_k p_{i,k}(t)p_{k,j}(s), \quad \text{for } t, s > 0. \qquad (9.31)$$

One of the most commonly studied class of continuous time Markov processes is the birth–death process. These processes get their name from applications in the study of biological systems, but they are also commonly used in the study of queueing theory, and many other applications. The birth–death process is similar to the discrete-time random walk studied in the previous section in that when the process changes states, it either increases by 1 or decreases by 1. As with the Poisson counting process, the general class of birth–death processes can be described by the transition probabilities over an infinitesimal period of time, Δt. For a birth–death process,

$$p_{i,j}(\Delta t) = \begin{cases} \lambda_i \Delta t + o(\Delta t), & \text{if } j = i+1, \\ \mu_i \Delta t + o(\Delta t), & \text{if } j = i-1, \\ 1 - (\lambda_i + \mu_i)\Delta t + o(\Delta t), & \text{if } j = i, \\ o(\Delta t), & \text{if } j \neq i-1, i, i+1. \end{cases} \qquad (9.32)$$

The parameter λ_i is called the birth rate while μ_i is the death rate when the process is in state i. In the context of queueing theory, λ_i and μ_i are referred to as the arrival and departure rates, respectively.

Similar to what was done with the Poisson counting process, by letting $s = \Delta t$ in Equation (9.31) and then applying the infinitesimal transition probabilities, a set of differential equations can be developed that will allow us to solve for the general transition probabilities. From Equation (9.31),

$$p_{i,j}(t + \Delta t) = \sum_k p_{i,k}(t) p_{k,j}(\Delta t)$$

$$= (\lambda_{j-1}\Delta t) p_{i,j-1}(t) + (1 - (\lambda_j + \mu_j)\Delta t) p_{i,j}(t) + (\mu_{j+1}\Delta t) p_{i,j+1}(t) + o(\Delta t) . \quad (9.33)$$

Rearranging terms and dividing by Δt produces

$$\frac{p_{i,j}(t + \Delta t) - p_{i,j}(t)}{\Delta t} = \lambda_{j-1} p_{i,j-1}(t) - (\lambda_j + \mu_j) p_{i,j}(t) + \mu_{j+1} p_{i,j+1}(t) + \frac{o(\Delta t)}{\Delta t} . \quad (9.34)$$

Finally, passing to the limit as $\Delta t \to 0$ results in

$$\frac{\mathrm{d}}{\mathrm{d}t} p_{i,j}(t) = \lambda_{j-1} p_{i,j-1}(t) - (\lambda_j + \mu_j) p_{i,j}(t) + \mu_{j+1} p_{i,j+1}(t). \quad (9.35)$$

This set of equations is referred to as the *forward Kolmogorov equations*. One can follow a similar procedure (see Exercise 9.32) to develop a slightly different set of equations known as the *backward Kolmogorov equations*,

$$\frac{\mathrm{d}}{\mathrm{d}t} p_{i,j}(t) = \lambda_i p_{i+1,j}(t) - (\lambda_i + \mu_i) p_{i,j}(t) + \mu_i p_{i-1,j}(t). \quad (9.36)$$

For all but the simplest examples, it is very difficult to find a closed-form solution for this system of equations. However, the Kolmogorov equations can lend some insight into the behavior of the system. For example, consider the steady-state distribution of the Markov process. If a steady state exists, we would expect that as $t \to \infty$, $p_{i,j}(t) \to \pi_j$ independent of i and also that $\mathrm{d}p_{i,j}(t)/\mathrm{d}t \to 0$. Plugging these simplifications into the forward Kolmogorov equations leads to

$$\lambda_{j-1} \pi_{j-1} - (\lambda_j + \mu_j) \pi_j + \mu_{j+1} \pi_{j+1} = 0. \quad (9.37)$$

These equations are known as the *global balance equations*. From them, the steady-state distribution can be found (if it exists). The solution to the balance equations is surprisingly easy to obtain. First, we rewrite the difference equation in the more symmetric form

$$\lambda_j \pi_j - \mu_{j+1} \pi_{j+1} = \lambda_{j-1} \pi_{j-1} - \mu_j \pi_j. \quad (9.38)$$

Next, assume that the Markov process is defined on the states $j = 0, 1, 2, \ldots$. Then the previous equation must be adjusted for the end point $j = 0$ according to (assuming $\mu_0 = 0$ which merely states that there can be no deaths when the population size is zero)

$$\lambda_0 \pi_0 - \mu_1 \pi_1 = 0. \tag{9.39}$$

Combining Equations (9.38) and (9.39) results in

$$\lambda_j \pi_j - \mu_{j+1} \pi_{j+1} = 0, \quad j = 0, 1, 2, \ldots, \tag{9.40}$$

which leads to the simple recursion

$$\pi_{j+1} = \frac{\lambda_j}{\mu_{j+1}} \pi_j, \quad j = 0, 1, 2, \ldots, \tag{9.41}$$

whose solution is given by

$$\pi_j = \pi_0 \prod_{i=1}^{j} \frac{\lambda_{i-1}}{\mu_i}, \quad j = 1, 2, 3, \ldots. \tag{9.42}$$

This gives the π_j in terms of π_0. In order to determine π_0, the constraint that the π_j must form a distribution is imposed.

$$\sum_{j=0}^{\infty} \pi_j = 1 \Rightarrow \pi_0 = \frac{1}{1 + \sum\limits_{j=1}^{\infty} \prod\limits_{i=1}^{j} \frac{\lambda_{i-1}}{\mu_i}}. \tag{9.43}$$

This completes the proof of the following theorem.

Theorem 9.4: For a Markov birth–death process with birth rate λ_n, $n = 0, 1, 2, \ldots$, and death rate μ_n, $n = 1, 2, 3, \ldots$, the steady-state distribution is given by

$$\pi_k = \lim_{t \to \infty} p_{i,k}(t) = \frac{\prod\limits_{i=1}^{k} \frac{\lambda_{i-1}}{\mu_i}}{1 + \sum\limits_{j=1}^{\infty} \prod\limits_{i=1}^{j} \frac{\lambda_{i-1}}{\mu_i}}. \tag{9.44}$$

If the series in the denominator diverges, then $\pi_k = 0$ for any finite k. This indicates that a steady-state distribution does not exist. Likewise, if the series converges, the π_k will be non-zero resulting in a well-behaved steady-state distribution.

■ Example 9.12 (The M/M/1 Queue):

In this example, we consider the birth–death process with constant birth rate and constant death rate. In particular, we take

$$\lambda_n = \lambda, \quad n = 0, 1, 2, \dots \quad \text{and } \mu_0 = 0, \ \mu_n = \mu, \ n = 1, 2, 3, \dots \ .$$

This model is commonly used in the study of queueing systems and, in that context, is referred to as the M/M/1 queue. In this nomenclature, the first "M" refers to the arrival process as being Markovian, the second "M" refers to the departure process as being Markovian, and the "1" is the number of servers. So this is a single server queue, where the interarrival time of new customers is an exponential random variable with mean $1/\lambda$ and the service time for each customer is exponential with mean $1/\mu$. For the M/M/1 queueing system, $\lambda_{i-1}/\mu_i = \lambda/\mu$ for all i so that

$$1 + \sum_{j=1}^{\infty} \prod_{i=1}^{j} \frac{\lambda_{i-1}}{\mu_i} = \sum_{j=0}^{\infty} \left(\frac{\lambda}{\mu}\right)^j = \frac{1}{1 - \lambda/\mu} \quad \text{for } \lambda < \mu.$$

The resulting steady-state distribution of the queue size is then

$$\pi_k = \frac{(\lambda/\mu)^k}{\dfrac{1}{1 - \lambda/\mu}} = (1 - \lambda/\mu)(\lambda/\mu)^k, \quad k = 0, 1, 2, \dots, \ \text{for } \lambda < \mu.$$

Hence, if the arrival rate is less than the departure rate, the queue size will have a steady state. It makes sense that if the arrival rate is greater than the departure rate, then the queue size will tend to grow without bound. ■

■ Example 9.13 (The M/M/∞ Queue):

Next suppose the last example is modified so that there are an infinite number of servers available to simultaneously provide service to all customers in the system. In that case, there are no customers ever waiting in line, and the process $X(t)$ now counts the number of customers in the system (receiving service) at time t. As before, we take the arrival rate to be constant $\lambda_n = \lambda$, but now the departure rate needs to be proportional to the number of customers in service, $\mu_n = n\mu$. In this case, $\lambda_{i-1}/\mu_i = \lambda/(i\mu)$ and

$$1 + \sum_{j=1}^{\infty} \prod_{i=1}^{j} \frac{\lambda_{i-1}}{\mu_i} = 1 + \sum_{j=1}^{\infty} \prod_{i=1}^{j} \frac{\lambda}{i\mu} = 1 + \sum_{j=1}^{\infty} \frac{(\lambda/\mu)^j}{j!} = e^{\lambda/\mu}.$$

Note that the series converges for any λ and μ, and hence the M/M/∞ queue will always have a steady-state distribution given by

$$\pi_k = \frac{(\lambda/\mu)^k}{k!} e^{-\lambda/\mu}.$$

■

■ Example 9.14:

This example demonstrates one way to simulate the M/M/1 queueing system of Example 9.12. One realization of this process as produced by the code that follows is illustrated in Figure 9.4. In generating the figure, we use an average arrival rate of $\lambda = 20$ customers per hour and an average service time of $1/\mu = 2$ minutes. This leads to the condition $\lambda < \mu$ and the M/M/1 queue exhibits stable behavior. The reader is encouraged to run the program for the case when $\lambda > \mu$ to observe the unstable behavior (the queue size will tend to grow continuously over time).

```
a=20;                               % arrival rate (customers/hour)
b=30;                               % departure rate (1/b=avg service time)
N=25;                               % no. of arrivals in simulation
X=-log(rand(1,N))/a;                % random interarrival times
X=cumsum(X);                        % random arrival times
Y=-log(rand(1,N))/b;                % service times for each customer
serv_start=X(1);                    % first customer starts service
                                    % immediately upon arrival.

Z(1)=serv_start+Y(1);               % departure time of first customer
for k=2:N                           % kth customer
```

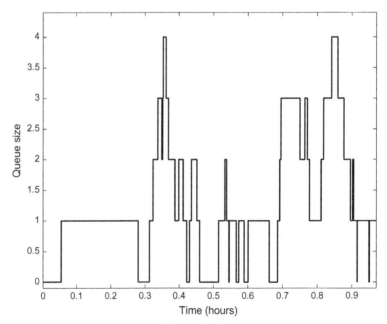

Figure 9.4
Simulated realization of the birth/death process for an M/M/1 queueing system
of Example 9.12.

```
        serv_start=max([Z(k-1), X(k)]);   % beginning of service time
        Z(k)=serv_start+Y(k);             % end of service time
    end
    % Construct data to plot graph of queue size vs time
    xaxis=[0, X(1)];                      % vector of points for the MM1
                                          % birth/death process
    yaxis=[0, 0];                         % vector of queue sizes at points
                                          % in above vector
    qs=1;                                 % current queue size
    X=X(2:length(X));
    while length(X)>0
        if X(1)<Z(1)                      % next point is arrival
            qs=qs+1;                      % increase queue size
            xaxis=[xaxis xaxis(length(xaxis)) X(1)];
            yaxis=[yaxis qs qs];
            X=X(2:length(X));
        else                              % next point is departure
            qs=qs-1; % decrease queue size
            xaxis=[xaxis xaxis(length(xaxis)) Z(1)];
            yaxis=[yaxis qs qs];
            Z=Z(2:length(Z));
        end
    end
    plot(xaxis,yaxis)                     % plot realization of birth/death
                                          % process
    xlabel('time (hours)');
    ylabel('queue size')
```

If the birth–death process is truly modeling the size of a population of some organism, then it would be reasonable to consider the case when $\lambda_0 = 0$. That is, when the population size reaches zero, no further births can occur. In that case, the species is extinct and the state $X(t) = 0$ is an absorbing state. A fundamental question would then be, is extinction a certain event and if not what is the probability of the process being absorbed into the state of extinction? Naturally, the answer to these questions would depend on the starting population size. Let q_i be the probability that the process eventually enters the absorbing state, given that it is initially in state i. Note that if the process is currently in state i, after the next transition, the birth–death process must be either in state $i-1$ or state $i+1$. The time to the next birth, B_i, is a random variable with an exponential distribution with a mean of $1/\lambda_i$, while the time to the next death is an exponential random variable, D_i, with a mean of $1/\mu_i$. Thus, the process will transition to state $i+1$ if $B_i < D_i$, otherwise it will transition to state $i-1$. The reader can easily verify that $\Pr(B_i < D_i) = \lambda_i/(\lambda_i + \mu_i)$. The absorption probability can then be written as

$$q_i = \Pr(\text{absorbtion}|\text{in state } i)$$

$$= \Pr(\text{absorbtion, next state is } i+1|\text{in state } i)$$

$$+ \Pr(\text{absorbtion, next state is } i-1|\text{in state } i) \tag{9.45}$$

$$= \Pr(\text{absorbtion}|\text{in state } i+1)\Pr(\text{next state is } i+1|\text{in state } i)$$

$$+ \Pr(\text{absorbtion}|\text{in state } i-1)\Pr(\text{next state is } i-1|\text{in state } i)$$

$$= q_{i+1}\frac{\lambda_i}{\lambda_i + \mu_i} + q_{i-1}\frac{\mu_i}{\lambda_i + \mu_i}, \quad i = 1, 2, 3, \dots .$$

This provides a recursive set of equations that can be solved to find the absorption probabilities. To solve this set of equations, we rewrite them as

$$q_{i+1} - q_i = \frac{\mu_i}{\lambda_i}(q_i - q_{i-1}), \, i = 1, 2, 3, \dots . \tag{9.46}$$

After applying this recursion repeatedly and using the fact that $q_0 = 1$,

$$q_{i+1} - q_i = (q_1 - 1)\prod_{j=1}^{i}\frac{\mu_j}{\lambda_j}. \tag{9.47}$$

Summing this equation from $i = 1, 2, \dots, n$ results in

$$q_{n+1} - q_1 = (q_1 - 1)\sum_{i=1}^{n}\prod_{j=1}^{i}\frac{\mu_j}{\lambda_j}. \tag{9.48}$$

Next, suppose that the series on the right hand side of the previous equation diverges as $n \to \infty$. Since the q_i are probabilities, the left hand side of the equation must be bounded, which implies that $q_1 = 1$. Then from Equation (9.47), it is determined that q_n must be equal to 1 for all n. That is, if

$$\sum_{i=1}^{\infty}\prod_{j=1}^{i}\frac{\mu_j}{\lambda_j} = \infty, \tag{9.49}$$

then absorption will eventually occur with probability 1 regardless of the starting state. If $q_1 < 1$ (absorption is not certain), then the preceding series must converge to a finite number.

It is expected in that case that as $n \to \infty$, $q_n \to 0$. Passing to the limit as $n \to \infty$ in Equation (9.48) then allows a solution for q_1 of the form

$$q_1 = \frac{\sum_{i=1}^{\infty} \prod_{j=1}^{i} \frac{\mu_j}{\lambda_j}}{1 + \sum_{i=1}^{\infty} \prod_{j=1}^{i} \frac{\mu_j}{\lambda_j}}. \tag{9.50}$$

Furthermore, the general solution for the absorption probability is

$$q_n = \frac{\sum_{i=n}^{\infty} \prod_{j=1}^{i} \frac{\mu_j}{\lambda_j}}{1 + \sum_{i=1}^{\infty} \prod_{j=1}^{i} \frac{\mu_j}{\lambda_j}}. \tag{9.51}$$

■ Example 9.15:

Consider a population model where both the birth and death rates are proportional to the population, $\lambda_n = n\lambda$, $\mu_n = n\mu$. For this model,

$$\sum_{i=1}^{\infty} \prod_{j=1}^{i} \frac{\mu_j}{\lambda_j} = \sum_{i=1}^{\infty} \prod_{j=1}^{i} \frac{\mu}{\lambda} = \sum_{i=1}^{\infty} \left(\frac{\mu}{\lambda}\right)^i = \frac{\mu/\lambda}{1 - \mu/\lambda} = \frac{\mu}{\lambda - \mu} \text{ for } \lambda > \mu.$$

Therefore, if $\lambda < \mu$, the series diverges and the species will eventually reach extinction with probability 1. If $\lambda > \mu$,

$$\sum_{i=n}^{\infty} \prod_{j=1}^{i} \frac{\mu_j}{\lambda_j} = \sum_{i=n}^{\infty} \left(\frac{\mu}{\lambda}\right)^i = \frac{(\mu/\lambda)^n}{1 - \mu/\lambda},$$

and the absorption (extinction) probabilities are

$$q_n = \left(\frac{\mu}{\lambda}\right)^n, \quad n = 1, 2, 3, \dots.$$

■

Continuous time Markov processes do not necessarily need to have a discrete amplitude as in the previous examples. In the following, we discuss a class of continuous time, continuous amplitude Markov processes. To start with, it is noted that for any time instants $t_0 < t_1 < t_2$, the conditional PDF of a Markov process must satisfy the Chapman–Kolmogorov equation

$$f(x_2, t_2 | x_0, t_0) = \int_{-\infty}^{\infty} f(x_2, t_2 | x_1, t_1) f(x_1, t_1 | x_0, t_0) \mathrm{d}x_1. \tag{9.52}$$

This is just the continuous amplitude version of Equation (9.31). Here, we use the notation $f(x_2, t_2 | x_1, t_1)$ to represent the conditional probability density of the process $X(t_2)$ at the point x_2 conditioned on $X(t_1) = x_1$. Next, suppose we interpret these time instants as $t_0 = 0$, $t_1 = t$, and $t_2 = t + \Delta t$. In this case, we interpret $x_2 - x_1 = \Delta x$ as the the infinitesimal change in the process that occurs during the infinitesimal time instant Δt and $f(x_2, t_2 | x_1, t_1)$ is the PDF of that increment.

Define $\Phi_{\Delta x}(\omega)$ to be the characteristic function of $\Delta x = x_2 - x_1$:

$$\Phi_{\Delta x}(\omega) = E[e^{j\omega \Delta x}] = \int_{-\infty}^{\infty} e^{j\omega(x_2 - x_1)} f(x_2, t + \Delta t | x_1, t) dx_2 . \tag{9.53}$$

We note that the characteristic function can be expressed in a Taylor series as

$$\Phi_{\Delta x}(\omega) = \sum_{k=0}^{\infty} \frac{M_k(x_1, t)}{k!} (j\omega)^k , \tag{9.54}$$

where $M_k(x_1, t) = E[(x_2 - x_1)^k | (x_1, t)]$ is the kth moment of the increment Δx. Taking inverse transforms of this expression, the conditional PDF can be expressed as

$$f(x_2, t + \Delta t | x_1, t) = \sum_{k=0}^{\infty} \frac{M_k(x_1, t)}{k!} (-1)^k \frac{\partial^k}{\partial x_2^k} (\delta(x_2 - x_1)) . \tag{9.55}$$

Inserting this result into the Chapman–Kolmogorov equation, Equation (9.52), results in

$$f(x_2, t + \Delta t | x_0, t_0) = \sum_{k=0}^{\infty} \frac{(-1)^k}{k!} \int_{-\infty}^{\infty} M_k(x_1, t) \frac{\partial^k}{\partial x_2^k} \delta(x_2 - x_1) f(x_1, t | x_0, t_0) dx_1$$

$$= \sum_{k=0}^{\infty} \frac{(-1)^k}{k!} \frac{\partial^k}{\partial x_2^k} [M_k(x_2, t) f(x_2, t | x_0, t_0)]$$

$$= f(x_2, t | x_0, t_0) + \sum_{k=1}^{\infty} \frac{(-1)^k}{k!} \frac{\partial^k}{\partial x_2^k} [M_k(x_2, t) f(x_2, t | x_0, t_0)] . \tag{9.56}$$

Subtracting $f(x_2, t | x_0, t_0)$ from both sides of this equation and dividing by Δt results in

$$\frac{f(x_2, t + \Delta t | x_0, t_0) - f(x_2, t | x_0, t_0)}{\Delta t} = \sum_{k=1}^{\infty} \frac{(-1)^k}{k!} \frac{\partial^k}{\partial x_2^k} \left[\frac{M_k(x_2, t)}{\Delta t} f(x_2, t | x_0, t_0) \right] . \tag{9.57}$$

Finally, passing to the limit as $\Delta t \to 0$ results in the partial differential equation

$$\frac{\partial}{\partial t} f(x, t | x_0, t_0) = \sum_{k=1}^{\infty} \frac{(-1)^k}{k!} \frac{\partial^k}{\partial x^k} [K_k(x, t) f(x, t | x_0, t_0)], \qquad (9.58)$$

where the function $K_k(x, t)$ is defined as

$$K_k(x, t) = \lim_{\Delta t \to 0} \frac{E[(X(t + \Delta t) - X(t))^k | X(t)]}{\Delta t}. \qquad (9.59)$$

For many processes of interest, the PDF of an infinitesimal increment can be accurately approximated from its first few moments and hence we take $K_k(x, t) = 0$ for $k > 2$. For such processes, the PDF must satisfy

$$\frac{\partial}{\partial t} f(x, t | x_0, t_0) = -\frac{\partial}{\partial x} (K_1(x, t) f(x, t | x_0, t_0)) + \frac{1}{2} \frac{\partial^2}{\partial x^2} (K_2(x, t) f(x, t | x_0, t_0)). \qquad (9.60)$$

This is known as the (one-dimensional) *Fokker–Planck equation* and is used extensively in diffusion theory to model the dispersion of fumes, smoke, and similar phenomenon.

In general, the Fokker–Planck equation is notoriously difficult to solve and doing such is well beyond the scope of this text. Instead, we consider a simple special case where the functions $K_1(x, t)$ and $K_2(x, t)$ are constants, in which case the Fokker–Planck equation reduces to

$$\frac{\partial}{\partial t} f(x, t | x_0, t_0) = -2c \frac{\partial}{\partial x} f(x, t | x_0, t_0) + D \frac{\partial^2}{\partial x^2} f(x, t | x_0, t_0), \qquad (9.61)$$

where in diffusion theory, D is known as the coefficient of diffusion and c is the drift. This equation is used in models that involve the diffusion of smoke or other pollutants in the atmosphere, the diffusion of electrons in a conductive medium, the diffusion of liquid pollutants in water and soil, and the difussion of plasmas. This equation can be solved in several ways. Perhaps one of the easiest methods is to use Fourier transforms. This is explored further in the exercises where the reader is asked to show that (taking $x_0 = 0$ and $t_0 = 0$) the solution to this diffusion equation is

$$f(x, t | x_0 = 0, t_0 = 0) = \frac{1}{\sqrt{4\pi Dt}} \exp\left(-\frac{(x - 2ct)^2}{4Dt}\right). \qquad (9.62)$$

That is, the PDF is Gaussian with a mean and variance which changes linearly with time. For the case when $c = 0$, this is the Wiener process discussed in Section 8.5. The behavior of this process is explored in the next example.

■ Example 9.16:

In this example, we model the diffusion of smoke from a forest fire that starts in a National Park at time $t = 0$ and location $x = 0$. The smoke from the fire drifts in the positive x directioin due to wind blowing at 10 miles per hour, and the diffusion coefficient is 1 square mile per hour. The probability density function is given in Equation (9.62). We provide a three-dimensional rendition of this function in Figure 9.5 using the following MATLAB program.

```
c=10;                               % drift
D=1;                                % Diffusion coefficient
tpoints=[0.25, 0.5, 1, 1.5, 2];     % time samples
x=[0:0.1:50];                       % x-axis
for k=1:length(tpoints)
    t=tpoints(k);                   % set t
    pdf(k,:)=exp(-(x-2*c*t).^2/(4*D*t))/sqrt(4*pi*D*t);
                                    %f(x,t)
```

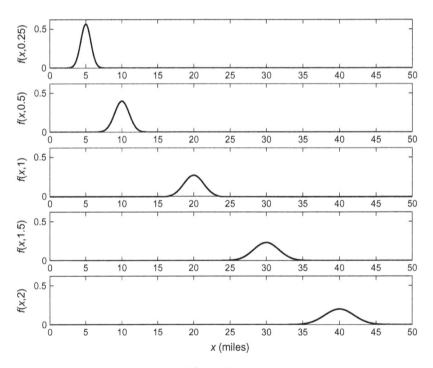

Figure 9.5
Observations of the PDF at different time instants showing the drift and dispersion of smoke for Example 9.16.

```
        subplot(5,1,k)
        plot(x,pdf(k,:))                              % plot PDF
        axis([0 max(x) 0 1.1*max(max(pdf))])
        s=num2str(t);
        leftstr='f(x,';
        rightstr=')';
        txt=[leftstr s rightstr];
        ylabel(txt)
    end
    xlabel('x (miles)')
```

9.5 Engineering Application: A Computer Communication Network

Consider a local area computer network where a cluster of nodes are connected by a common communication line. Suppose for simplicity that these nodes occasionally need to transmit a message of some fixed length (referred to as a packet). Also, assume that the nodes are synchronized so that time is divided into slots, each of which is sufficiently long to support one packet. In this example, we consider a random access protocol known as *slotted Aloha*. Messages (packets) are assumed to arrive at each node according to a Poisson process. Assuming there are a total of n nodes, the packet arrival rate at each node is assumed to be λ/n so that the total arrival rate of packets is fixed at λ packets/slot. In slotted Aloha, every time a new packet arrives at a node, that node attempts to transmit that packet during the next slot. During each slot, one of three events can occur:(1) no node attempts to transmit a packet, in which case the slot is said to be idle; (2) exactly one node attempts to transmit a packet, in which case the transmission is successful; (3) more than one node attempts to transmit a packet, in which case a collision is said to have occurred. All nodes involved in a collision will need to retransmit their packets, but if they all retransmit during the next slot, then they will continue to collide and the packets will never be successfully transmitted. All nodes involved in a collision are said to be backlogged until their packet is successfully transmitted. In the slotted Aloha protocol, each backlogged node chooses to transmit during the next slot with probability p (and hence chooses not to transmit during the next slot with probability $1-p$). Viewed in an alternative manner, every time a collision occurs, each node involved waits a random amount of time until they attempt retransmission, where that random time follows a geometric distribution.

This computer network can be described by a Markov chain, X_k = number of backlogged nodes at the end of the kth slot. To start with, we evaluate the transition probabilities of the Markov chain, $p_{i,j}$. Assuming that there are an infinite number of nodes (or a finite number of nodes each of which could store an arbitrary number of backlogged packets in a buffer), we note that

$$\Pr(m \text{ backlogged nodes attempt to transmit}|X_k{=}n) \ = \ \binom{n}{m}p^m(1-p)^{n-m}, \quad (9.63)$$

$$\Pr(m \text{ new arrivals}|X_k{=}n) = \frac{\lambda^m}{m!}e^{-\lambda}. \tag{9.64}$$

Using these equations, it is straightforward to determine that the transition probabilities are given by

$$p_{i,j} = \begin{cases} 0, & \text{for } j < i-1, \\ ip(i-p)^{i-1}e^{-\lambda}, & \text{for } j=i-1, \\ (1 + \lambda(1-p)^i - ip(1-p)^{i-1})e^{-\lambda}, & \text{for } j=i, \\ (1-(1-p)^i)\lambda e^{-\lambda}, & \text{for } j=i+1, \\ \dfrac{\lambda^{j-i}}{(j-i)!}e^{-\lambda}, & \text{for } j > i+1. \end{cases} \tag{9.65}$$

In order to get a feeling for the steady-state behavior of this Markov chain, we define the drift of the chain in state i as

$$d_i = E[X_{k+1}|X_k = i] - i. \tag{9.66}$$

Given that the chain is in state i, if the drift is positive, then the number of backlogged nodes will tend to increase; whereas, if the drift is negative, the number of backlogged nodes will tend to decrease. Crudely speaking, a drift of zero represents some sort of equilibrium for the Markov chain. Given the preceding transition probabilities, the drift works out to be

$$d_i = \lambda - (1-p)^{i-1}e^{-\lambda}[ip + \lambda(1-p)]. \tag{9.67}$$

Assuming that $p \ll 1$, then we can use the approximations $(1-p) \approx 1$ and $(1-p)^i \approx e^{-ip}$ to simplify the expression for the drift,

$$d_i \approx \lambda - g(i)e^{-g(i)}, \quad \text{where } g(i) = \lambda + ip. \tag{9.68}$$

The parameter $g(i)$ has the physilcal interpretation of the average number of transmissions per slot given there are i backlogged states. To understand the significance of this result, the two terms in the expression for the drift are plotted in Figure 9.6. The first term, λ, has the interpretation of the average number of new arrivals per slot, while the second term, $g\exp(-g)$, is the average number of successful transmissions per slot or the average departure rate. For a very small number of backlogged states, the arrival rate is greater than the departure rate and the number of backlogged states tends to increase. For moderate values of i, the departure rate is greater than the arrival rate and the number of backlogged states tends to decrease. Hence, the drift of the Markov chain is such that the system tends to stabilize

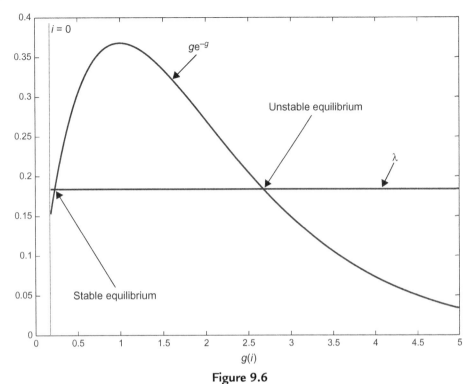

Figure 9.6
Arrival rate and successful transmission rate for a slotted Aloha system.

around the point marked stable equilibrium in Figure 9.6. This is the first point where the two curves cross. Note, however, that for very large i, the drift becomes positive again. If the number of backlogged states ever becomes large enough to push the system to the right of the point marked unstable equilibrium in the figure, then the number of backlogged nodes will tend to grow without bound and the system will become unstable.

Note that the value of λ represents the throughput of the system. If we try to use a value of λ which is greater than the peak value of $g\exp(-g)$, then the drift will always be positive and the system will be unstable from the beginning. This maximum throughput occurs when $g(i) = 1$ and has a value of $\lambda_{max} = 1/e$. By choosing an arrival rate less than λ_{max}, we can get the system to operate near the stable equilibrium, but sooner or later, we will get a string of bad luck and the system will drift into the unstable region. The lower the arrival rate, the longer it will take (on average) for the system to become unstable, but at any arrival rate, the system will eventually reach the unstable region. Thus, slotted Aloha is inherently an unstable protocol. As a result, various modifications have been proposed which exhibit stable behavior.

9.6 Engineering Application: A Telephone Exchange

Consider a base station in a cellular phone system. Suppose calls arrive at the base station according to a Poisson process with some arrival rate λ. These calls are initiated by mobile units within the cell served by that base station. Furthermore, suppose each call has a duration that is an exponential random variable with some mean, $1/\mu$. The base station has some fixed number of channels, m, that can be used to service the demands of the mobiles in its cell. If all m channels are being used, any new call that is initiated cannot be served and the call is said to be blocked. We are interested in calculating the probability that when a mobile initiates a call, the customer is blocked.

Since the arrival process is memoryless and the departure process is memoryless, the number of calls being serviced by the base station at time t, $X(t)$, is a birth–death Markov process. Here, the arrival rate and departure rates (given there are n channels currently being used) are given by

$$\lambda_n = \begin{cases} \lambda, & 0 \le n < m, \\ 0, & n = m, \end{cases} \qquad \mu_n = n\mu, \ 0 \le n \le m. \tag{9.69}$$

The steady-state distribution of this Markov process is given by Equation (9.44). For this example, the distribution is found to be

$$\pi_n = \frac{\prod_{i=1}^{n} \dfrac{\lambda_{i-1}}{\mu_i}}{1 + \sum_{j=1}^{m} \prod_{i=1}^{j} \dfrac{\lambda_{i-1}}{\mu_i}} = \frac{\dfrac{1}{n!}\left(\dfrac{\lambda}{\mu}\right)^n}{\sum_{j=0}^{m} \dfrac{1}{j!}\left(\dfrac{\lambda}{\mu}\right)^j}. \tag{9.70}$$

The blocking probability is just the probability that when a call is initiated, it finds the system in state m. In steady state, this is given by π_m and the resulting blocking probability is the so-called Erlang-B formula,

$$\Pr(\text{blocked call}) = \frac{\dfrac{1}{m!}\left(\dfrac{\lambda}{\mu}\right)^m}{\sum_{j=0}^{m} \dfrac{1}{j!}\left(\dfrac{\lambda}{\mu}\right)^j}. \tag{9.71}$$

This equation is plotted in Figure 9.7 for several values of m. The horizontal axis is the ratio of λ/μ which is referred to in the telephony literature as the traffic intensity. As an example of the use of this equation, suppose a certain base station had 60 channels available to service incoming calls. Furthermore, suppose each user initiated calls at a rate of 1 call per 3 hours

From left to right *m*=1, 2, 3, 4, 5, 6, 8, 10, 12, 15, 20, 25, 30, 40, 50, 60, 70, 80, 90, 100

Figure 9.7
The Erlang-B formula.

and calls had an average duration of 3 minutes (0.05 hours). If a 2% probability of blocking is desired, then from Figure 9.7 we determine that the system can handle a traffic intensity of approximately 50 Erlangs. Note that each user generates an intensity of

$$\frac{\lambda}{\mu} = \frac{1/(3 \text{ hour})}{1/(0.05 \text{ hour})} = \frac{1}{60} \text{ Erlangs}. \tag{9.72}$$

Hence, a total of 50*60 = 3000 mobile users per cell could be supported while still maintaining a 2% blocking probability.

Exercises

Section 9.1: Definition and Examples of Markov Processes

9.1   For a Markov chain, prove or disprove the following statement:

$$\Pr(X_k = i_k | X_{k+1} = i_{k+1}, X_{k+2} = i_{k+2}, ..., X_{k+m} = i_{k+m}) = \Pr(X_k = i_k | X_{k+1} = i_{k+1})$$

9.2   *A Diffusion Model* - Model the diffusion of electrons and holes across a potential barrier in an electronic device as follows. We have n black balls (electrons) in urn A and n whiteballs (holes) in urn B. An experimental outcome selects randomly one ball from each urn. The ball from urn A is placed in urn B and that from urn B is placed in A. Let the state of the process be the number of black balls in urn A. (By knowing the number of black balls in urn A, we know the composition of both urns.) Let k denote the state of the process. Find the transition probabilities, $p_{i,j}$.

9.3   Let $X[n]$ be the sum of n independent rolls of a fair (cubicle) die.
   (a)   Is $X[n]$ a Markov chain?
   (b)   Define a new process according to $Y[n] = X[n] \bmod 3$. That is, $Y[n] \in \{0, 1, 2\}$ is related to $X[n]$ by $X[n] = 3q + Y[n]$ for a non-negative integer q. Find the transition probability matrix for the process $Y[n]$.
   (c)   Now suppose $Z[n] = X[n] \bmod 5$. Find the transition matrix for $Z[n]$.

9.4   Suppose we label the spaces on a monopoly board as $\{0, 1, 2, ..., 39\}$ where,
   0 = Go,
   1 = Mediterranean Ave.,
   2 = Community Chest,
   3 = Baltic Ave.,
   ...
   39 = Boardwalk.

   Let $X[k]$ be the location of a player after k turns. On each turn, a player moves by rolling two (six-sided) dice and moving forward the number of places indicated by the sum of the two dice. Any time the roll of the dice causes the player to land on space 30 (Go to Jail) the player's token is immediately moved to space 10 (Jail). Describe the elements of the transition probability matrix, $p_{i,j}$, for the monopoly Markov chain $X[k]$.

9.5 N balls labeled 1 through N are placed in Box 1 while a Box 2 is initially empty. At each time instant, one of the N balls is chosen (with equally probability) and moved to the other box. Let $X[k]$ be the number of balls in Box 1 at time instant k. Draw a state diagram and find the transition probability matrix for this Markov chain. *Note:* This is known as the Ehrenfest chain and was developed by the dutch Physicist Paul Ehrenfest for the study of molecular dynamics.

9.6   *An Inventory Model* - A hot dog vendor operates a stand where the number of hot dogs he sells each day is modeled as a Poisson random variable with a mean value of 100. Let $X[k]$ represent the number of hot dogs the vendor has at the beginning of each day. At the end of the day, if his inventory of hot dogs on hand falls below some minimum value, α, then the vendor goes out that evening and purchses enough hot dogs to bring the total of his inventory to β. Write an equation to describe the elements of the transition probability matrix, $p_{i,j}$, for this inventory process.

9.7   *A Web Search Engine Model* - Suppose after we enter some keywords into our web search engine it finds five pages that contain those keywords. We will call these pages A, B, C, D, and E. The engine would like to rank the pages according to some measure of importance. To do so, we make note of which pages contain links to which other pages. Suppose we find the following links.

Page	Has links to pages
A	B, C
B	C, D, E
C	A, E
D	A, B, C, E
E	B, D

We then create a random walk where the initial state is equally likely to be any one of the five pages. At each time instant, the state changes with equal probabilty to one of the pages for which a link exists. For example, if we are currently in state A, then at the next time instant we will transition to either state B or state C with equal probability. If we are currently in state B, we will transition to state C, D, or E with equal probability, and so on. Draw a transition diagram and find the probability transition matrix for this Markov chain. *Note:* This process forms the basis for the PageRank algorithm used by Google.

Section 9.2: Calculating Transition and State Probabilities

9.8 Consider a two-state Markov chain with a general transition probability matrix

$$P = \begin{bmatrix} 1-p & p \\ q & 1-q \end{bmatrix},$$

where $0 < p, q < 1$. Find an expression for the n-step transition probability matrix, P^n.

9.9 For the general two-state Markov chain of Exercise 9.8, suppose the states are called 0 and 1. Furthermore, suppose $\Pr(X_0=0) = s$ and $\Pr(X_0=1) = 1-s$.

(a) Find $\Pr(X_1 = 0, X_2 = 1)$.

(b) Find $\Pr(X_1 = 1 | X_0 = 0, X_2 = 0)$.

(c) Find $\Pr(X_2 = X_1)$. Is it the same as $\Pr(X_1 = X_0)$?

9.10 A square matrix P is called a stochastic matrix if all of its elements satisfy $0 \le p_{i,j} \le 1$ and, furthermore, $\sum_j p_{i,j} = 1$ for all i. Every stochastic matrix is the transition probability matrix for some Markov chain; however, not every stochastic matrix is a valid two-step transition probability matrix. Prove that a 2×2 stochastic matrix is a valid two-step transition probability matrix for a two-state Markov chain if and only if the sum of the diagonal elements is greater than or equal to 1.

9.11 A PCM waveform has the two states +1 and 0. Suppose the transition matrix is

$$P = \begin{bmatrix} 0.5 & 0.5 \\ 0.25 & 0.75 \end{bmatrix}.$$

The initial value of the waveform is determined by the flip of a coin, with the outcome of a head corresponding to +1 and a tail to 0.

(a) What is the probability that the waveform will be at +1 after one step if the coin is a fair coin?

(b) Find the same probability if the coin is biased such that a head occurs with probability $1/3$.

(c) Repeat the problem for two steps.

9.12 A three-state Markov chain has the following transition matrix:

$$P = \begin{bmatrix} 0.25 & 0.5 & 0.25 \\ 0.4 & 0.6 & 0 \\ 1 & 0 & 0 \end{bmatrix}.$$

(a)   Does this Markov chain have a unique steady-state probability vector? If so, find it.
(b)   What is the approximate value of $p_{1,3}^{(100)}$? What interpretation do you give to this result?
(c)   What is the probability that after the third step you are in state 3 if the initial state probability vector is (1/3 1/3 1/3)?

9.13 The three letters C, A, and T represent the states of a word-generating system. Let the initial state probability vector be (1/3 1/3 1/3) for the three letters, respectively. The transition matrix is given as

$$\begin{array}{cc} & \begin{array}{ccc} \text{C} & \text{A} & \text{T} \end{array} \\ P = \begin{array}{c} \text{C} \\ \text{A} \\ \text{T} \end{array} & \begin{bmatrix} 0.1 & 0.7 & 0.2 \\ 0.6 & 0.1 & 0.3 \\ 0.1 & 0.8 & 0.1 \end{bmatrix}. \end{array}$$

What is the probability of generating a proper three-letter English dictionary word after two transitions from the initial state?

9.14 Two students play the following game. Two dice are tossed. If the sum of the numbers showing is less than 7, student A collects a dollar from student B. If the total is greater than 7, then student B collects a dollar from student A. If a 7 appears, then the student with the fewest dollars collects a dollar from the other. If the students have the same amount, then no dollars are exchanged. The game continues until one student runs out of dollars. Let student A's number of dollars represent the states. Let each student start with 3 dollars.

(a)   What is the transition matrix, P?
(b)   If student A reaches state 0 or 6, then he stays there with probability 1. What is the probability that student B loses in 3 tosses of the dice?
(c)   What is the probability that student A loses in 5 or fewer tosses ?

9.15 A biologist would like to estimate the size of a certain population of fish. A sequential approach is proposed whereby a member of the population is sampled at random, tagged and then returned. This process is repeated until a member is drawn that has been previously tagged. If desired, we could then begin tagging again with a new kind of tag. Let M be the trial at which the first previously tagged fish is sampled and N be the total population size. This process can be described in terms of a Markov chain where X_k is the number of successive untagged members observed. That is, $X_k = k$ for $k = 1, 2, \ldots, M-1$ and $X_M = 0$.

(a) For a fixed $N = n$, find the form of the transition probability matrix.

(b) Find $\Pr(M = m | X_0 = 0)$ for $m = 2, 3, 4, \ldots, n$.

9.16 A person with a contagious disease enters the population. Every day he either infects a new person (which occurs with probability p) or his symptoms appear and he is discovered by health officials (which occurs with probability $1 - p$). Assuming all infected persons behave in the same manner, compute the probability distribution of the number of infected but undiscovered people in the population at the time of first discovery of the disease.

9.17 A certain three-state Markov chain has a transition probability matrix given by

$$P = \begin{bmatrix} 0.4 & 0.5 & 0.1 \\ 0.05 & 0.7 & 0.25 \\ 0.05 & 0.5 & 0.45 \end{bmatrix}.$$

Determine if the Markov chain has a unique steady-state distribution or not. If it does, find that distribution.

9.18 Suppose a process can be considered to be in one of two states (let's call them state A and state B), but the next state of the process depends not only on the current state but also on the previous state as well. We can still describe this process using a Markov chain, but we will now need four states. The chain will be in state (X, Y), $X, Y \in \{A, B\}$ if the process is currently in state X and was previously in state Y.

(a) Show that the transition probability matrix of such a four-state Markov chain must have zeros in at least half of its entries.

(b) Suppose that the transition probability matrix is given by

$$\begin{array}{ccccc} & (A, A) & (A, B) & (B, A) & (B, B) \\ P = \begin{array}{c} (A, A) \\ (A, B) \\ (B, A) \\ (B, B) \end{array} & \begin{bmatrix} 0.8 & 0.2 & 0 & 0 \\ 0 & 0 & 0.4 & 0.6 \\ 0.6 & 0.4 & 0 & 0 \\ 0 & 0 & 0.1 & 0.9 \end{bmatrix}. \end{array}$$

Find the steady-state distribution of the Markov chain.

(c) What is the steady-state probability that the underlying process is in state A?

9.19 A communication system sends data in the form of packets of fixed length. Noise in the communication channel may cause a packet to be received incorrectly. If this happens, then the packet is retransmitted. Let the probability that a packet is received incorrectly be q.

(a) Determine the average number of transmissions that are necessary before a packet is received correctly. Draw a state diagram for this problem.

(b) Let the the transmission time be T_t seconds for a packet. If the packet is received incorrectly, then a message is sent back to the transmitter stating that the message was received incorrectly. Let the time for sending such a message be T_a. Assume that if the packet is received correctly that we do not send an acknowledgment. What is the average time for a successful transmission ? Draw a state diagram for this problem.

(c) Now suppose there are three nodes. The packet is to be sent from node 1 to node 2 to node 3 without an error. The probability of the packets being received incorrectly at each node is the same and is q. The transmission time is T_t and the time to acknowledge that a packet is received incorrectly is T_a. Draw a state diagram for this problem. Determine the average time for the packet to reach node 3 correctly.

9.20 Consider the scenario of Example 9.2 where a child buys kid's meals at a local restaurant in order to complete his collection of superhero action figures. Recall the states were $X[k] \in \{0, 1, 2, 3, 4\}$ where $X[k]$ represents the number of distinct action figures collected after k meals are purchased.

(a) Find an expression for $p_{0,4}^{(n)}$, the probability of having a complete set if n meals are purchased.

(b) Find the probability that the set is first completed after purchasing n meals.

(c) Find the average number of meals that must be purchased to complete the set.

9.21 Find the steady-state probability distribution for the web search engine model of Exercise 9.7. It is this distibution that is used as the ranking for the each web page and ultimately determines which pages show up on the top of your list when your search results are displayed.

Section 9.3: Characterization of Markov Chains

9.22 A random waveform is generated as follows. The waveform starts at 0 voltage. Every t_s seconds, the waveform switches to a new voltage level. If the waveform is at a voltage level of 0 volts, it may move to +1 volt with probability p or it may move to −1 volt with probability $q = 1 - p$. Once the waveform is at +1 (or −1), the waveform will return (with probability 1) to 0 volts at the next switching instant.

(a) Model this process as a Markov chain. Describe the states of the system and give the transition probability matrix.

(b)   Determine whether each state is periodic or aperiodic. If periodic, determine the period of each state.

(c)   For each instant of time, determine the PMF for the value of the waveform.

9.23   A student takes this course at period 1 on Monday, Wednesday, and Friday. Period 1 starts at 7:25 A.M. Consequently, the student sometimes misses class. The student's attendance behavior is such that she attends class depending only on whether or not she went to the last class. If she attended class on one day, then she will go to class the next time it meets with probability 1/2. If she did not go to one class, then she will go to the next class with probability 3/4.

(a)   Find the transition matrix P.

(b)   Find the probability that if she went to class on Wednesday that she will attend class on Friday.

(c)   Find the probability that if she went to class on Monday that she will attend class on Friday.

(d)   Does the Markov chain described by this transition matrix have a steady-state distribution? If so, find that distribution.

9.24   Let X_n be the sum of n independent rolls of a fair (cubicle) die.

(a)   Find $\lim_{n \to \infty} \Pr(X_n \text{ is a multiple of } 3)$.

(b)   Find $\lim_{n \to \infty} \Pr(X_n \text{ is a multiple of } 5)$.

9.25   For a Markov chain with each of the transition probability matrices in (a)–(c), find the communicating classes and the periodicity of the various states.

(a) $\begin{bmatrix} 0 & 0 & 1 & 0 \\ 1 & 0 & 0 & 0 \\ \frac{1}{2} & \frac{1}{2} & 0 & 0 \\ \frac{1}{3} & \frac{1}{3} & \frac{1}{3} & 0 \end{bmatrix}$,   (b) $\begin{bmatrix} 0 & 1 & 0 & 0 \\ 0 & 0 & 0 & 1 \\ 0 & 1 & 0 & 0 \\ \frac{1}{3} & 0 & \frac{2}{3} & 0 \end{bmatrix}$,   (c) $\begin{bmatrix} 0 & 1 & 0 & 0 \\ \frac{1}{2} & 0 & 0 & \frac{1}{2} \\ 0 & 0 & 0 & 1 \\ 0 & \frac{1}{2} & \frac{1}{2} & 0 \end{bmatrix}$.

9.26   Prove that if $i \leftrightarrow j$, then $d(i) = d(j)$ and hence all states in the same class must have the same period.

9.27   Demonstrate that the two generating functions defined in Equations (9.18) and (9.19) are related by

$$P_{i,\,i}(z) - 1 = P_{i,\,i}(z) F_{i,\,i}(z) .$$

9.28 Define the generating functions

$$P_{i,j}(z) = \sum_{n=0}^{\infty} p_{i,j}^{(n)} z^n \text{ and } F_{i,j}(z) = \sum_{n=0}^{\infty} f_{i,j}^{(n)} z^n .$$

(a)   Show that $P_{i,j}(z) = F_{i,j}(z) P_{jj}(z)$.

(b)   Prove that if state j is a transient state, then for all i,

$$\sum_{n=1}^{\infty} p_{i,j}^{(n)} < \infty .$$

9.29 Verify that recurrence is a class property. That is, if one state in a communicating class is recurrent then all are recurrent, and if one is transient then all are transient.

9.30 Suppose a Bernoulli trial results in a success with probability p and a failure with probability $1 - p$. Suppose the Bernoulli trial is repeated indefinitely with each repitition independent of all others. Let X_n be a "success runs" Markov chain where X_n represents the number of most recent consecutive successes that have been observed at the nth trial. That is, $X_n = m$ if trial numbers $n, n - 1, n - 2, ..., n - m + 1$ were all successes but trial number $n - m$ was a failure. Note that $X_n = 0$ if the nth trial was a failure.

(a)   Find an expression for the one-step transition probabilities, $p_{i,j}$.

(b)   Find an expression for the n-step first return probabilities for state 0, $f_{0,0}^{(n)}$.

(c)   Prove that state 0 is recurrent for any $0 < p < 1$. Note that since all states communicate with one another, this result together with the result of Exercise 9.29 is sufficient to show that all states are recurrent.

9.31 Find the steady-state distribution of the success runs Markov chain described in Exercise 9.30.

Section 9.4: Continuous Time Markov Processes

9.32 Derive the *backward Kolmogorov equations,*

$$\frac{\mathrm{d}}{\mathrm{d}t} p_{i,j}(t) = \lambda_i p_{i+1,j}(t) - (\lambda_i + \mu_i) p_{i,j}(t) + \mu_i p_{i-1,j}(t) .$$

9.33 In this problem, you will demonstrate that the Gaussian PDF in Equation (9.64) is in fact the solution to the diffusion Equation (9.63). To do this, we will use frequency domain methods. Define $\Phi(\omega, t) = E[e^{j\omega X(t)}] = \int_{-\infty}^{\infty} f(x, t | x_0 = 0, t_0 = 0) e^{j\omega x} dx$ to be the time-varying characteristic function of the random process $X(t)$.

(a) Starting from the diffusion Equation (9.63), show that the characteristic function must satisfy

$$\frac{\partial}{\partial t} \Phi(\omega, t) = (2cj\omega - D\omega^2) \Phi(\omega, t).$$

Also, determine the appropriate initial condition for this differential equation. That is, find $\Phi(\omega, 0)$.

(b) Solve the first-order differential equation in part (a) and show that the characteristic function is of the form

$$\Phi(\omega, t) = \exp(-D\omega^2 + 2cj\omega).$$

(c) From the characteristic function, find the resulting PDF given by Equation (9.62).

MATLAB Exercises

9.34 On the first day of the new year it is cloudy. What is the probability that it is sunny on July 4 if the following transition matrix applies?

$$
\begin{array}{c}
\phantom{\text{sunny}} \quad \text{sunny} \quad \text{cloudy} \quad \text{rainy} \\
\begin{array}{c} \text{sunny} \\ \text{cloudy} \\ \text{rainy} \end{array}
\begin{bmatrix}
0.7 & 0.2 & 0.1 \\
0.3 & 0.2 & 0.5 \\
0.3 & 0.3 & 0.4
\end{bmatrix}.
\end{array}
$$

How much does your answer change if it is a leap year?

9.35 Determine which of the following transition matrices (a)–(g) represents a regular Markov chain. Find the steady-state distribution for the regular matrices. Note a Markov chain is regular if some power of the transition matrix has only positive (non-zero) entries. This implies that a regular chain has no periodic states.

(a) $\begin{bmatrix} 1/3 & 2/3 \\ 5/6 & 1/6 \end{bmatrix}$, (b) $\begin{bmatrix} 0 & 1 \\ 1/4 & 3/4 \end{bmatrix}$, (c) $\begin{bmatrix} 0 & 1 \\ 1 & 0 \end{bmatrix}$,

(d) $\begin{bmatrix} 1/5 & 4/5 \\ 1 & 0 \end{bmatrix}$, (e) $\begin{bmatrix} 1/2 & 1/2 & 0 \\ 0 & 1/2 & 1/2 \\ 1/3 & 1/3 & 1/3 \end{bmatrix}$, (f) $\begin{bmatrix} 1/3 & 0 & 2/3 \\ 0 & 1 & 0 \\ 0 & 1/5 & 4/5 \end{bmatrix}$,

(g) $\begin{bmatrix} 1/2 & 1/4 & 1/4 \\ 1/3 & 2/3 & 0 \\ 0 & 1/4 & 3/4 \end{bmatrix}$.

9.36 Write a MATLAB program to simulate a three-state Markov chain with the following transition probability matrix

$$P = \begin{bmatrix} 1/2 & 1/4 & 1/4 \\ 1/2 & 0 & 1/2 \\ 1/4 & 1/4 & 1/2 \end{bmatrix}.$$

Assuming that the process starts in the third state, generate a sequence of 500 states. Estimate the steady-state probability distribution, π, using the sequence you generated. Does it agree with the theoretical answer? Does the steady-state distribution depend on the starting state of the process?

9.37 Write a MATLAB program to simulate the M/M/1 queueing system. If you like, you may use the program provided in Example 9.14. Use your program to estimate the average amount of time a customer spends waiting in line for service. Assume an arrival rate of $\lambda = 15$ customers/hour and an average service time of $1/\mu = 3$ min. Note, that if a customer arrives to find no others in the system, the waiting time is zero.

9.38 Modify the program of Example 9.14 to simulate the M/M/∞ queue of Example 9.13. Based on your simulation results, estimate the PMF of the number of customers in the system. Compare your results with the analytical results found in Example 9.13.

9.39 For the monopoly Markov chain described in Exercise 9.4, write a MATLAB program to construct the transition probability matrix. Find the steady-state probability of being on each of the following spaces:
    (a)   Go—space 0,
    (b)   Jail—Space 10,
    (c)   New York Ave.—Space 19,
    (d)   Park Place—Space 37.

9.40 For the monopoly Markov chain described in Exercise 9.4, write a MATLAB program to simulate the movement of a token around the board keeping track of each space visited. Using your program estimate the relative frequency of visiting each of the spaces listed in Exercise 9.39. Do your simulation results agree with the analytical results of Exercise 9.39.

Power Spectral Density

In the study of deterministic signals and systems, frequency domain techniques (e.g., Fourier transforms) provide a valuable tool that allows the engineer to gain significant insights into a variety of problems. In this chapter, we develop frequency domain tools for studying random processes. This will prepare us for the study of random processes in linear systems in the next chapter.

For a deterministic continuous signal, $x(t)$, the Fourier transform is used to describe its spectral content. In this text, we write the Fourier transform as[1]

$$X(f) = \mathcal{F}[x(t)] = \int_{-\infty}^{\infty} x(t)e^{-j2\pi ft}dt, \tag{10.1}$$

and the corresponding inverse transform is

$$x(t) = \mathcal{F}^{-1}[X(f)] = \int_{-\infty}^{\infty} X(f)e^{j2\pi ft}df. \tag{10.2}$$

For discrete-time signals, we could use a discrete Fourier transform or a z-transform. The Fourier transform, $X(f)$, is referred to as the spectrum of $x(t)$ since it describes the spectral contents of the signal. In general, $X(f)$ is a complex function of frequency and hence we also speak of an amplitude (magnitude) spectrum, $|X(f)|$, and a phase spectrum, $\angle X(f)$. In order to study random processes in the frequency domain, we seek a similar quantity which will describe the spectral characteristics of a random process.

The most obvious thing to do would be to try to define the Fourier transform of a random process as perhaps

---

[1] Even though we use an upper case letter to represent a Fourier transform, it is not necessarily random. Clearly, the Fourier transform of a non-random signal is also not random. While this is inconsistent with our previous notation of using upper case letters to represent random quantities, this notation of using upper case letters to represent Fourier Transforms is so common in the literature, we felt it necessary to retain this convention. The context should make it clear whether a function of frequency is random or not.

Probability and Random Processes. DOI: 10.1016/B978-0-12-386981-4-00010-3

$$X(f) = \int_{-\infty}^{\infty} X(t)e^{-j2\pi ft}dt = \mathscr{F}[X(t)] \; ; \tag{10.3}$$

however, this leads to several problems. First of all, there are problems with existence. Since $X(t)$ is a random process, there is not necessarily any guarantee that the integral exists for every possible realization, $x(t)$. That is, not every realization of the random process may have a Fourier transform. Even for processes that are well-behaved in the sense that every realization has a well-defined Fourier transform, we are still left with the problem that $X(f)$ is itself a random process. In Chapter 8, we described the temporal characteristics of random processes in terms of deterministic functions such as the mean function and the autocorrelation function. In a similar way, we seek a deterministic description of the spectral characteristics of a random process. The power spectral density (PSD) function, which is defined in the next section, will play that role.

10.1 Definition of PSD

To start with, for a random process $X(t)$, define a truncated version of the random process as

$$X_{t_o}(t) = \begin{cases} X(t), & |t| \leq t_o, \\ 0, & |t| > t_o. \end{cases} \tag{10.4}$$

The energy of this random process is

$$E_{X_{t_o}} = \int_{-t_o}^{t_o} X^2(t)dt = \int_{-\infty}^{\infty} X_{t_o}^2(t)dt, \tag{10.5}$$

and hence the time-averaged power is

$$P_{X_{t_o}} = \frac{1}{2t_o}\int_{-\infty}^{\infty} X_{t_o}^2(t)dt = \frac{1}{2t_o}\int_{-\infty}^{\infty} |X_{t_o}(f)|^2 df. \tag{10.6}$$

The last equality is obtained using Parseval's theorem. The quantity $X_{t_o}(f)$ is the Fourier transform of $X_{t_o}(t)$. Since the random process has been truncated to a finite time interval, there will generally not be any problem with the existence of the Fourier transform. Note that $P_{X_{t_o}}$ is a random variable and so to get the ensemble averaged power, we must take an expectation,

$$\overline{P_{X_{t_o}}} = E[P_{X_{t_o}}] = \frac{1}{2t_o}\int_{-\infty}^{\infty} E[|X_{t_o}(f)|^2]df. \tag{10.7}$$

The power in the (untruncated) random process $X(t)$ is then found by passing to the limit as $t_o \to \infty$,

$$\overline{P}_X = \lim_{t_o \to \infty} \frac{1}{2t_o} \int_{-\infty}^{\infty} E[|X_{t_o}(f)|^2] df = \int_{-\infty}^{\infty} \lim_{t_o \to \infty} \frac{E[|X_{t_o}(f)|^2]}{2t_o} df. \tag{10.8}$$

Define $S_{XX}(f)$ to be the integrand in equation (10.8). That is, let

$$S_{XX}(f) = \lim_{t_o \to \infty} \frac{E[|X_{t_o}(f)|^2]}{2t_o}. \tag{10.9}$$

Then, the average power in the process can be expressed as

$$\overline{P}_X = \int_{-\infty}^{\infty} S_{XX}(f) df. \tag{10.10}$$

Therefore, this function of frequency which we have simply referred to as $S_{XX}(f)$ has the property that when integrated over all frequency, the total power in the process is obtained. In other words, $S_{XX}(f)$ has the units of power per unit frequency and so it is the power density function of the random process in the frequency domain. Hence, the quantity $S_{XX}(f)$ is given the name PSD. In summary, we have the following definition of PSD.

Definition 10.1: For a random process $X(t)$, the power spectral density (PSD) is defined as

$$S_{XX}(f) = \lim_{t_o \to \infty} \frac{E[|X_{t_o}(f)|^2]}{2t_o}, \tag{10.11}$$

where $X_{t_o}(f)$ is the Fourier transform of the truncated version of the process as described in equation (10.4)

Several properties of the PSD function should be evident from Definition 10.1 and from the development that lead to that definition:

(1) $S_{XX}(f)$ is a real function. (10.12a)

(2) $S_{XX}(f)$ is a non-negative function. (10.12b)

(3) $S_{XX}(f)$ is an even function. (10.12c)

(4) The average power in a random process is given by $\overline{P}_X = \int_{-\infty}^{\infty} S_{XX}(f) df.$ (10.12d)

■ Example 10.1:

As a simple example, consider a sinusoidal process $X(t) = A\sin(\omega_0 t + \Theta)$ with random amplitude and phase. Assume the phase is uniform over $[0, 2\pi)$ and independent of the amplitude which we take to have an arbitrary distribution. Since each realization of this process is a sinusoid at frequency f_0, we would expect that all of the power in this process should be located at $f = f_0$ (and $f = -f_0$). Mathematically we have

$$X_{t_0}(t) = A\sin(\omega_0 t + \Theta)\text{rect}\left(\frac{t}{2t_0}\right),$$

where $\text{rect}(t)$ is a square pulse of unit height and unit width and centered at $t = 0$. The Fourier transform of this truncated sinusoid works out to be

$$X_{t_0}(f) = -jt_0 A e^{j\Theta}\text{sinc}(2(f-f_0)t_0) + jt_0 A e^{-j\Theta}\text{sinc}(2(f+f_0)t_0),$$

where the "sinc" function is $\text{sinc}(x) = \sin(\pi x)/(\pi x)$. We next calculate the expected value of the magnitude squared of this function.

$$E[|X_{t_0}(f)|^2] = E[A^2]t_0^2\{\text{sinc}^2(2(f-f_0)t_0) + \text{sinc}^2(2(f+f_0)t_0)\}.$$

The PSD function for this random process is then

$$S_{XX}(f) = \lim_{t_0 \to \infty}\frac{E[|X_{t_0}(f)|^2]}{2t_0} = \lim_{t_0 \to \infty}\frac{E[A^2]t_0}{2}\{\text{sinc}^2(2(f-f_0)t_0) + \text{sinc}^2(2(f+f_0)t_0)\}.$$

To calculate this limit, we observe that as t_0 gets large, the function $g(f) = t_0\text{sinc}^2(2ft_0)$ becomes increasingly narrower and taller. Thus, we could view the limit as an infinitely tall, infinitely narrow pulse. This is one way to define a delta function. One property of a delta function that is not necessarily shared by the function under consideration is that $\int \delta(f)df = 1$. Therefore, the limiting form of $g(t)$ will have to be a scaled (in amplitude) delta function. To figure out what the scale factor needs to be, the integral of $g(f)$ is calculated:

$$\int_{-\infty}^{\infty} g(f)df = \int_{-\infty}^{\infty} t_0\text{sinc}^2(2ft_0)df = \frac{1}{2}\int_{-\infty}^{\infty}\text{sinc}^2(u)du = \frac{1}{2}.$$

Therefore,

$$\lim_{t_0 \to \infty} t_0\text{sinc}^2(2ft_0) = \frac{1}{2}\delta(f).$$

The resulting PSD is then simplified to

$$S_{XX}(f) = \frac{E[A^2]}{4}\{\delta(f-f_0) + \delta(f+f_0)\}.$$

This is consistent with our intuition. The power in a sinusoid with amplitude A is $A^2/2$. Thus, the average power in the sinusoidal process is $E[A^2]/2$. This power is evenly split between the two points $f = f_0$ and $f = -f_0$.

■

One important lesson to learn from the previous example is that even for very simplistic random processes, it can be quite complicated to evaluate the PSD using the definition given in Example 10.11. The next section presents a very important result that allows us to greatly simplify the process of finding the PSD of many random processes.

10.2 The Wiener–Khintchine–Einstein Theorem

Theorem 10.1 (Wiener–Khintchine–Einstein): For a wide sense stationary (WSS) random process $X(t)$ whose autocorrelation function is given by $R_{XX}(\tau)$, the PSD of the process is

$$S_{XX}(f) = \mathscr{F}[R_{XX}(\tau)] = \int_{-\infty}^{\infty} R_{XX}(\tau)e^{-j2\pi f\tau}\,d\tau. \tag{10.13}$$

In other words, the autocorrelation function and PSD form a Fourier transform pair.

Proof: Starting from the definition of PSD,

$$E[|X_{t_o}(f)|^2] = E\left[\int_{-t_o}^{t_o}\int_{-t_o}^{t_o} X(t)X(s)e^{-j2\pi f(t-s)}\,dt\,ds\right] \tag{10.14}$$

$$= \int_{-t_o}^{t_o}\int_{-t_o}^{t_o} E[X(t)X(s)]e^{-j2\pi f(t-s)}\,dt\,ds = \int_{-t_o}^{t_o}\int_{-t_o}^{t_o} R_{XX}(t,s)e^{-j2\pi f(t-s)}\,dt\,ds\,.$$

Using the assumption that the process is WSS, the autocorrelation function is only a function of a single time variable, $t-s$. Hence, the expression above is rewritten as

$$E[|X_{t_o}(f)|^2] = \int_{-t_o}^{t_o}\int_{-t_o}^{t_o} R_{XX}(t-s)e^{-j2\pi f(t-s)}\,dt\,ds\,. \tag{10.15}$$

It is noted that the preceding integrand is only a function of a single variable; therefore, with the appropriate change of variables, the double integral can be reduced to a single integral. The details are given in the following.

The region of integration is a square in the s-t plane of width $2t_o$ centered at the origin. Consider an infinitesimal strip bounded by the lines $t-s = \tau$ and $t-s = \tau+d\tau$. This strip is illustrated in Figure 10.1. Let $a(\tau)$ be the area of that

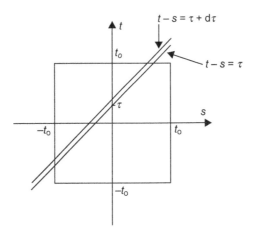

Figure 10.1
Illustration of the change of variables for the double integral in Equation (10.15).

strip which falls within the square region of integration. A little elementary geometry reveals that

$$a(\tau) = \begin{cases} 2t_0\left(1 - \dfrac{|\tau|}{2t_0}\right)d\tau, & \text{for } |\tau| < 2t_0, \\[2ex] 0, & \text{for } |\tau| > 2t_0. \end{cases} \qquad (10.16)$$

To obtain the preceding result, one must neglect edge terms that contribute expressions which are quadratic in the infinitesimal $d\tau$. Since the integrand in Equation (10.15) is a function only of $t - s$, it is constant (and equal to $R_{XX}(\tau)e^{-j2\pi\tau}$) over the entire strip. The double integral over the strip can therefore be written as the value of the integrand multiplied by the area of the strip. The double integral over the entire square can be written as a sum of the integrals over all the strips which intersect the square:

$$\int_{-t_0}^{t_0} \int_{-t_0}^{t_0} R_{XX}(t-s)e^{-j2\pi f(t-s)}dt\,ds = \sum_{\text{strips}} R_{XX}(\tau)e^{-j2\pi\tau}a(\tau). \qquad (10.17)$$

Passing to the limit as $d\tau \to 0$, the sum becomes an integral resulting in

$$E[|X_{t_0}(f)|^2] = 2t_0\int_{-2t_0}^{2t_0} R_{XX}(\tau)e^{-j2\pi\tau}\left(1 - \frac{|\tau|}{2t_0}\right)d\tau. \qquad (10.18)$$

The PSD function for the random process $X(t)$ is then

$$S_{XX}(f) = \lim_{t_o \to \infty} \frac{E[|X_{t_o}(f)|^2]}{2t_o} = \lim_{t_o \to \infty} \int_{-2t_o}^{2t_o} \left(1 - \frac{|\tau|}{2t_o}\right) R_{XX}(\tau) e^{-j2\pi f \tau} d\tau. \qquad (10.19)$$

Passing to the limit as $t_o \to \infty$ then gives the desired result in Equation (10.13). $\square$

While most of the random processes we deal with are WSS, for those that are not, Theorem 10.1 needs to be adjusted since the autocorrelation function for a non-stationary process would be a function of two time variables. For non-stationary processes, the Wiener–Khintchine–Einstein theorem is written as

$$S_{XX}(f) = \int_{-\infty}^{\infty} \langle R_{XX}(t, t + \tau) \rangle e^{-j2\pi f \tau} d\tau, \qquad (10.20)$$

where in this case, $\langle\ \rangle$ represents a time average with respect to the time variable t. We leave it as an exercise to the reader to prove this more general version of the theorem.

■ Example 10.2:

Let us revisit the random sinusoidal process, $X(t) = A\sin(\omega_o t + \Theta)$, of Example 10.1. This time the PSD function will be calculated by first finding the autocorrelation function.

$$R_{XX}(t, t + \tau) = E[X(t)X(t + \tau)] = E[A^2 \sin(\omega_o t + \Theta)\sin(\omega_o(t + \tau) + \Theta)]$$

$$= \frac{1}{2}E[A^2]E[\cos(\omega_o \tau) - \cos(\omega_o(2t + \tau) + 2\Theta)] = \frac{1}{2}E[A^2]\cos(\omega_o \tau).$$

The autocorrelation function is only a function of τ and thus the PSD is simply the Fourier transform of the autocorrelation function,

$$S_{XX}(f) = \int_{-\infty}^{\infty} R_{XX}(\tau) e^{-j2\pi f \tau} d\tau = \frac{1}{2}E[A^2]\mathcal{F}[\cos(\omega_o \tau)] = \frac{1}{4}E[A^2]\{\delta(f - f_o) + \delta(f + f_o)\}.$$

This is exactly the same result that was obtained in Example 10.1 using the definition of PSD, but in this case, the result was obtained with much less work.

■

■ Example 10.3:

Now suppose we have a sinusoid with a random amplitude, but a fixed phase, $X(t) = A\sin(\omega_o t + \theta)$. Here, the autocorrelation function is

$$R_{XX}(t, t + \tau) = E[X(t)X(t + \tau)] = E[A^2 \sin(\omega_o t + \theta)\sin(\omega_o(t + \tau) + \theta)]$$

(Continued)

$$= \frac{1}{2}E[A^2]E[\cos(\omega_0\tau) - \cos(\omega_0(2t+\tau) + 2\theta)] = \frac{1}{2}E[A^2]\cos(\omega_0\tau) + \frac{1}{2}E[A^2]\cos(\omega_0(2t+\tau) + 2\theta).$$

In this case, the process is not WSS and so we must take a time average of the autocorrelation before we take the Fourier transform.

$$\langle R_{XX}(t, t+\tau)\rangle = \langle \frac{1}{2}E[A^2]\cos(\omega_0\tau) + \frac{1}{2}E[A^2]\cos(\omega_0(2t+\tau) + 2\theta)\rangle$$

$$= \frac{1}{2}E[A^2]\cos(\omega_0\tau) + \frac{1}{2}E[A^2]\langle\cos(\omega_0(2t+\tau) + 2\theta)\rangle = \frac{1}{2}E[A^2]\cos(\omega_0\tau).$$

The time-averaged autocorrelation is exactly the same as the autocorrelation in the previous example, and hence, the PSD of the sinusoid with random amplitude and fixed phase is exactly the same as the PSD of the sinusoid with random amplitude and random phase. ∎

■ Example 10.4:

Next, consider a modified version of the random telegraph signal of Example 8.4. In this case, the process starts at $X(0) = 1$ and switches back and forth between $X(t) = 1$ and $X(t) = -1$, with the switching times being dictated by a Poisson point process with rate λ. A sample realization is shown in Figure 10.2. To find the PSD, we first find the autocorrelation function.

$$R_{XX}(t, t+\tau) = E[X(t)X(t+\tau)]$$

$$= (1)\Pr(\text{even number of switches in } [t, t+\tau))$$
$$+ (-1)\Pr(\text{odd number of switches in } [t, t+\tau)).$$

The number of switches in a contiguous interval follows a Poisson distribution, and therefore

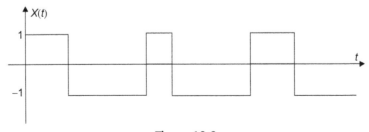

Figure 10.2
A sample realization for the random telegraph signal of Example 10.4.

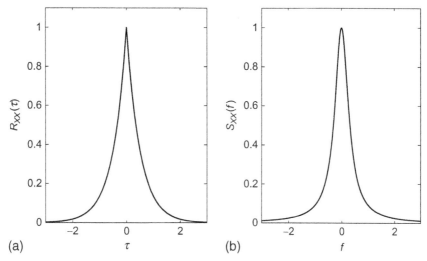

Figure 10.3

(a) Autocorrelation function and (b) PSD for the random telegraph signal of Example 10.4.

$$R_{XX}(t, t+\tau) = \sum_{m\ \text{even}} \frac{(\lambda|\tau|)^m}{m!} e^{-\lambda|\tau|} - \sum_{m\ \text{odd}} \frac{(\lambda|\tau|)^m}{m!} e^{-\lambda|\tau|} = \sum_{m=0}^{\infty} \frac{(-\lambda|\tau|)^m}{m!} e^{-\lambda|\tau|} = e^{-2\lambda|\tau|}.$$

Since this is a function only of τ, we directly take the Fourier transform to find the PSD:

$$S_{XX}(f) = \mathscr{F}[e^{-2\lambda|\tau|}] = \frac{1/\lambda}{1 + (\pi f/\lambda)^2} = \frac{\lambda}{\lambda^2 + (\pi f)^2}.$$

The autocorrelation function and PSD for this random telegraph signal are illustrated in Figure 10.3.

■

■ Example 10.5:

To illustrate how some minor changes in a process can affect its autocorrelation and PSD, let us return to the random telegraph process as it was originally described in Example 8.4. As in the previous example, the process switches back and forth between two values as dictated by an underlying Poisson point process; however, now the two values of the process are $X(t) \in \{0, 1\}$ instead of $X(t) \in \{+1, -1\}$. Also, the process starts at $X(0) = 0$ instead of $X(0) = 1$. Noting that the product $X(t)X(t+\tau)$ is equal to zero unless both $\{X(t)=1\}$ and $\{X(t+\tau)=1\}$ are true, the autocorrelation function is calculated as

$$R_{XX}(t, t+\tau) = E[X(t)X(t+\tau)] = \Pr(\{X(t)=1\} \cap \{X(t+\tau)=1\})$$
$$= \Pr(\text{odd number of switches in } [0, t)) \Pr(\text{even number of switches in } [t, t+\tau))$$

(Continued)

$$= \left(\sum_{m \text{ odd}} \frac{(\lambda t)^m}{m!} e^{-\lambda t} \right) \left(\sum_{m \text{ even}} \frac{(\lambda \tau)^m}{m!} e^{-\lambda \tau} \right) = \left(\frac{1}{2} - \frac{1}{2} e^{-2\lambda t} \right) \left(\frac{1}{2} + \frac{1}{2} e^{-2\lambda \tau} \right) .$$

The last step was accomplished using some of the results obtained in Example 8.6 and assumes that τ is positive. If, on the other hand, τ is negative, then it turns out that

$$R_{XX}(t, t + \tau) = \left(\frac{1}{2} - \frac{1}{2} e^{-2\lambda(t + \tau)} \right) \left(\frac{1}{2} + \frac{1}{2} e^{2\lambda \tau} \right) .$$

Clearly, this process is not stationary since the autocorrelation function is a function of both t and τ. Thus, before the Fourier transform is taken, the time average of the autocorrelation function must be computed.

$$\langle R_{XX}(t, t + \tau) \rangle = \left\langle \frac{1}{2} - \frac{1}{2} e^{-2\lambda t} \right\rangle \left(\frac{1}{2} + \frac{1}{2} e^{-2\lambda \tau} \right) = \frac{1}{4} + \frac{1}{4} e^{-2\lambda \tau}, \text{ for } \tau > 0,$$

$$\langle R_{XX}(t, t + \tau) \rangle = \left\langle \frac{1}{2} - \frac{1}{2} e^{-2\lambda(t + \tau)} \right\rangle \left(\frac{1}{2} + \frac{1}{2} e^{2\lambda \tau} \right) = \frac{1}{4} + \frac{1}{4} e^{2\lambda \tau}, \text{ for } \tau < 0.$$

In summary, the autocorrelation function can be concisely expressed as

$$\langle R_{XX}(t, t + \tau) \rangle = \frac{1}{4} + \frac{1}{4} e^{-2\lambda |\tau|} .$$

The PSD function is then found to be

$$\left(S_{XX}(f) = \mathscr{F}\left[\frac{1}{4} + \frac{1}{4} e^{-2\lambda |\tau|} \right] \right) = \frac{1}{4} \delta(f) + \frac{1}{4} \frac{\lambda}{\lambda^2 + (\pi f)^2} .$$

There are two differences between this result and that of Example 10.4. First, the total power (integral of PSD) in this process is 1/2 the total power in the process of the previous example. This is easy to see since when $X(t) \in \{0, 1\}$, $E[X^2(t)] = 1/2$, while when $X(t) \in \{+1, -1\}$, $E[X^2(t)] = 1$. Second, in this example, there is a delta function in the PSD which was not present in the previous example. This is due to the fact that the mean of the process in this example was (asymptotically) equal to 1/2, whereas in the previous example it was zero. It is left as an exercise for the reader to determine if the initial conditions of the random process would have any effect on the PSD. That is, if the process started at $X(0) = 1$ and everything else remained the same, would the PSD change?

∎

Definition 10.2: The *cross spectral density* between two random processes, $X(t)$ and $Y(t)$, is the Fourier transform of the cross correlation function:

$$S_{XY}(f) = \mathscr{F}[R_{XY}(\tau)] = \int_{-\infty}^{\infty} R_{XY}(\tau) e^{-j2\pi f \tau} d\tau . \tag{10.21}$$

The cross spectral density does not have a physical interpretation nor does it share the same properties of the PSD function. For example, $S_{XY}(f)$ is not necessarily real since $R_{XY}(\tau)$ is not necessarily even. The cross spectral density function does possess a form of symmetry known as *Hermitian symmetry*[2],

$$S_{XY}(f) = S_{YX}(-f) = S_{XY}^*(-f). \tag{10.22}$$

This property follows from the fact that $R_{XY}(\tau) = R_{YX}(-\tau)$. The proof of this property is left to the reader.

10.3 Bandwidth of a Random Process

Now that we have an analytical function which describes the spectral content of a signal, it is appropriate to talk about the bandwidth of a random process. As with deterministic signals, there are many definitions of bandwidth. Which definition is used depends on the application and sometimes on personal preference. Several definitions of bandwidth are given next. To understand these definitions, it is helpful to remember that when measuring the bandwidth of a signal (whether random or deterministic), only positive frequencies are measured. Also, we tend to classify signals according to where their spectral contents lies. Those signals for which most of the power is at or near direct current (d.c.) are referred to as *lowpass* signals, while those signals whose PSD is centered around some non-zero frequency, $f = f_o$ are referred to as *bandpass* processes.

> **Definition 10.3:** For a lowpass process, the absolute bandwidth, B_{abs}, is the largest frequency for which the PSD is non-zero. That is, B_{abs} is the smallest value of B such that $S_{XX}(f) = 0$ for all $f > B$. For a bandpass process, let B_L be the largest value of B such that $S_{XX}(f) = 0$ for all $0 < f < B$ and similarly let B_R be the smallest value of B such that $S_{XX}(f) = 0$ for all $B < f$. Then $B_{abs} = B_R - B_L$. In summary, the absolute bandwidth of a random process is the width of the band which contains all frequency components. The concept of absolute bandwidth is illustrated in Figure 10.4.

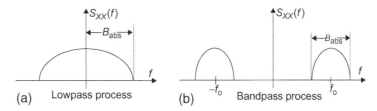

Figure 10.4
Measuring the absolute bandwidth of (a) a lowpass and (b) a bandpass process.

---

[2] Here and throughout the text, the superscript * refers to the complex conjugate.

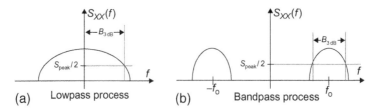

Figure 10.5
Measuring the 3 dB bandwidth of (a) a lowpass and (b) a bandpass process.

Definition 10.4: The *3 dB bandwidth* (or *half-power* bandwidth), $B_{3\,dB}$, is the width of the frequency band where the PSD is within 3 dB of its peak value everywhere within the band. Let S_{peak} be the maximum value of the PSD. Then for a lowpass signal, $B_{3\,dB}$ is the largest value of B for which $S_{XX}(f) > S_{peak}/2$ for all frequencies such that $0 < f < B$. For a bandpass process, $B_{3\,dB} = B_R - B_L$, where $S_{XX}(f) > S_{peak}/2$ for all frequencies such that $B_L < f < B_R$, and it is assumed that the peak value occurs within the band. The concept of 3 dB bandwidth is illustrated in Figure 10.5.

Definition 10.5: The *root-mean-square* (RMS) *bandwidth*, B_{rms}, of a lowpass random process is given by

$$B_{rms}^2 = \frac{\int_0^\infty f^2 S_{XX}(f)\,df}{\int_0^\infty S_{XX}(f)\,df}. \tag{10.23}$$

This measure of bandwidth is analogous to using standard deviation as a measure of the width of a PDF. For bandpass processes, this definition is modified according to

$$B_{rms}^2 = \frac{4\int_0^\infty (f-f_o)^2 S_{XX}(f)\,df}{\int_0^\infty S_{XX}(f)\,df}, \tag{10.24}$$

where

$$f_o = \frac{\int_0^\infty f S_{XX}(f)\,df}{\int_0^\infty S_{XX}(f)\,df}. \tag{10.25}$$

It is left as an exercise to the reader to figure out why the factor of 4 appears in the preceding definition.

■ Example 10.6:

Consider the random telegraph process of Example 10.4 where the PSD was found to be

$$S_{XX}(f) = \frac{\lambda}{\lambda^2 + (\pi f)^2}.$$

The absolute bandwidth of this process is $B_{abs} = \infty$. This can be seen from the picture of the PSD in Figure 10.2. To find the 3 dB bandwidth, it is noted that the peak of the PSD occurs at $f = 0$ and has a value of $S_{peak} = \lambda^{-1}$. The 3 dB bandwidth is then the value of f for which $S_{XX}(f) = 1/(2\lambda)$. This is easily found to be $B_{3dB} = \lambda/\pi$. Finally, the RMS bandwidth of this process is infinite since

$$\int_0^\infty \frac{\lambda f^2}{\lambda^2 + (\pi f)^2} df = \infty.$$

■

10.4 Spectral Estimation

The problem of estimating the PSD of a random process has been the topic of extensive research over the past several decades. Many books are dedicated to this topic alone and hence we cannot hope to give a complete treatment of the subject here; however, some fundamental concepts are introduced in this section that will provide the reader with a basic understanding of the problem and some rudimentary solutions. Spectral estimators are generally grouped into two classes, *parametric* and *non-parametric*. A parametric estimator assumes a certain model for the random process with several unknown parameters and then attempts to estimate the parameters. Given the model parameters, the PSD is then computed analytically from the model. On the other hand, a non-parametric estimator makes no assumptions about the nature of the random process and estimates the PSD directly. Since parametric estimators take advantage of some prior knowledge of the nature of the process, it would be expected that these estimators are more accurate. However, in some cases, prior knowledge may not be available, in which case a non-parametric estimator may be more appropriate. We start with a description of some basic techniques for non-parametric spectral estimation.

10.4.1 Non-parametric Spectral Estimation

Suppose we observe a random process, $X(t)$, over some time interval $(-t_o, t_o)$ (or a discrete-time process $X[n]$ over some time interval $[0, n_o - 1]$) and we wish to estimate its PSD function. Two approaches immediately come to mind. The first method we will refer to as the direct method or the *periodogram*. It is based on the definition of PSD in Equation (10.11). The second method we will refer to as the indirect method or the *correlation method*. The basic idea here is to estimate the autocorrelation function and then take the Fourier transform of the estimated autocorrelation to form an estimate of the PSD. We first describe the correlation method. In all of the discussion on spectral estimation to follow, it is assumed that the random processes are WSS.

An estimate of the autocorrelation function of a continuous time random process can be formed by taking a time average of the particular realization observed:

$$\hat{R}_{XX}(\tau) = \left\langle X\left(t - \frac{\tau}{2}\right) X\left(t + \frac{\tau}{2}\right) \right\rangle = \frac{1}{2t_o - |\tau|} \int_{-t_o + \frac{|\tau|}{2}}^{t_o - \frac{|\tau|}{2}} X\left(t - \frac{\tau}{2}\right) X\left(t + \frac{\tau}{2}\right) dt. \tag{10.26}$$

It is not difficult to show that this estimator is unbiased (i.e., $E[\hat{R}_{XX}(\tau)] = R_{XX}(\tau)$), but at times, it is not a particularly good estimator, especially for large values of τ. The next example illustrates this fact.

■ Example 10.7:

Consider the random telegraph process of Example 10.4. A sample realization of this process is shown in Figure 10.6, along with the estimate of the autocorrelation function. For convenience, the true autocorrelation is shown as well. Note that the estimate matches quite well for small values of τ, but as $|\tau| \to t_o$, the estimate becomes very poor.

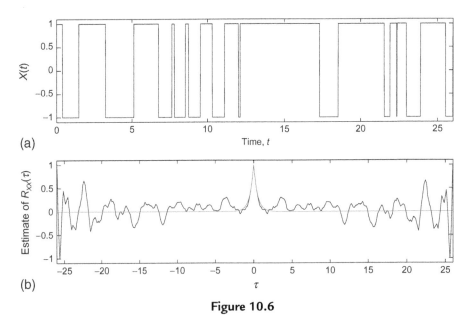

(a)

(b)

Figure 10.6

(a) A sample realization of the random telegraph signal and (b) the estimate of the autocorrelation function based on that realization. The dotted line is the true autocorrelation function. ■

In order to improve the quality of the autocorrelation estimate, it is common to introduce a windowing function to suppress the erratic behavior of the estimate at large values of τ. This is particularly important when estimating the PSD since the wild behavior at large values of $|\tau|$ will distort the estimate of the PSD at all frequencies once the Fourier transform of the autocorrelation estimate is taken.

> **Definition 10.6:** For a WSS random process $X(t)$, the windowed estimate of the autocorrelation function using a windowing function $w(t)$ is given by

$$\hat{R}_{XX}^{(w)}(\tau) = \frac{w(t)}{2t_o - |\tau|} \int_{-t_o + \frac{|\tau|}{2}}^{t_o - \frac{|\tau|}{2}} X\left(t - \frac{\tau}{2}\right) X\left(t + \frac{\tau}{2}\right) dt. \tag{10.27}$$

There are many possible windowing functions that can be used. The previous autocorrelation estimate (without the windowing function) can be viewed as a windowed estimate with a rectangular window,

$$w(t) = \text{rect}\left(\frac{t}{4t_o}\right). \tag{10.28}$$

Another option would be to use a triangular window,

$$w(t) = \text{tri}\left(\frac{t}{2t_o}\right) = \begin{cases} 1 - \dfrac{|t|}{2t_o}, & |t| < t_o, \\[2mm] 0, & |t| > t_o. \end{cases} \tag{10.29}$$

This would lead to the autocorrelation estimate,

$$\hat{R}_{XX}^{(\text{tri})}(\tau) = \frac{1}{2t_o} \int_{-t_o + \frac{|\tau|}{2}}^{t_o - \frac{|\tau|}{2}} X\left(t - \frac{\tau}{2}\right) X\left(t + \frac{\tau}{2}\right) dt. \tag{10.30}$$

While this estimator is biased, the mean-squared error in the estimate will generally be smaller than when the rectangular window is used. Much of the classical spectral estimation theory focuses on how to choose an appropriate window function to satisfy various criteria.

■ Example 10.8:

The autocorrelation function of the random telegraph signal is once again estimated, this time with the windowed autocorrelation estimator using the triangular window. The sample realization as well as the autocorrelation estimate are shown in Figure 10.7. Note this time that the behavior of the estimate for large values of τ is more controlled.

(Continued)

Figure 10.7
(a) A sample realization of the random telegraph signal and (b) the windowed estimate of the autocorrelation function (using a triangular window) based on that realization.

Once an estimate of the autocorrelation can be found, the estimate of the PSD is obtained through Fourier transformation.

Definition 10.7: For a WSS random process $X(t)$, the correlation-based estimate (with windowing function $w(t)$) of the PSD is given by

$$\hat{S}_{XX}^{(w)}(f) = \mathscr{F}[\hat{R}_{XX}^{(w)}(\tau)] = \int_{-\infty}^{\infty} \hat{R}_{XX}^{(w)}(\tau)e^{-j2\pi f\tau}d\tau = \int_{-2t_o}^{2t_o} \hat{R}_{XX}^{(w)}(\tau)e^{-j2\pi f\tau}d\tau. \qquad (10.31)$$

■ Example 10.9:

The PSD estimates corresponding to the autocorrelation estimates of the previous example are illustrated in Figure 10.8. There the correlation-based PSD estimates are plotted and compared with the true PSD. Note that when no windowing is used, the PSD estimate tends to overestimate the true PSD. Another observation is that it appears from these results that the PSD estimates could be improved by smoothing. We will elaborate on that shortly.

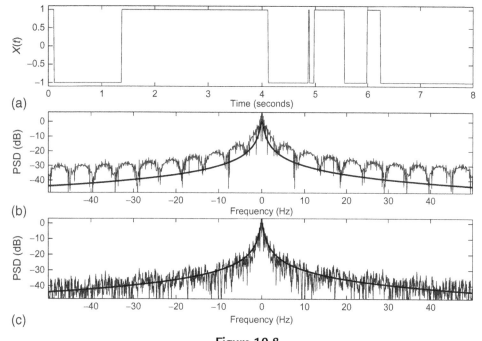

Figure 10.8
(a) A sample realization of the random telegraph signal and (b, c) the estimate of the PSD function. Plot (b) is for the unwindowed estimator while the plot (c) is for the triangular windowed estimator. For both PSD plots, the smooth thick line is the true PSD.

■

The next approach we consider for estimating the PSD of a random process is to directly use the definition of PSD in Equation (10.11). This approach is referred to as the periodogram estimate.

Definition 10.8: Given an observation of the process $X(t)$ over an interval $(-t_o, t_o)$, $X_{t_o}(t)$, the *periodogram* estimate of the PSD is

$$\hat{S}_{XX}^{(p)}(f) = \frac{1}{2t_o}|X_{t_o}(f)|^2. \tag{10.32}$$

Theorem 10.2: The periodogram estimate of the PSD is equivalent to the autocorrelation-based estimate with a triangular window. That is,

$$\hat{S}_{XX}^{(p)}(f) = \hat{S}_{XX}^{(\text{tri})}(f). \tag{10.33}$$

Proof: The proof of this theorem is a fairly elementary exercise in manipulating the properties of Fourier transforms. Recall that for any two signals, $x(t)$ and $y(t)$, the product of their spectra form a transform pair with the convolution of the two signals. That is, $\mathcal{F}[x(t)*y(t)] = X(f)Y(f)$. Applying this to Equation (10.32) results in

$$\left|X_{t_o}(f)\right|^2 = \mathcal{F}[X_{t_o}(\tau)*X_{t_o}(-\tau)] = \mathcal{F}\left[\int_{-\infty}^{\infty} X_{t_o}(u)X_{t_o}(u-\tau)\mathrm{d}u\right] = \mathcal{F}\left[\int_{-t_o}^{t_o} X_{t_o}(t)X_{t_o}(t+\tau)\mathrm{d}t\right],$$

$$\frac{1}{2t_o}\left|X_{t_o}(f)\right|^2 = \mathcal{F}\left[\frac{1}{2t_o}\int_{-t_o}^{t_o} X_{t_o}(t)X_{t_o}(t+\tau)\mathrm{d}t\right] = \mathcal{F}[\hat{R}_{XX}^{(\mathrm{tri})}(\tau)] = \hat{S}_{XX}^{(\mathrm{tri})}(f). \quad \square \qquad (10.34)$$

An example of the periodogram was given in Figure 10.8c. At the time, it was referred to as the correlation-based estimate with a triangular windowing function. Now, it is clear that the two are the same. It was mentioned in Example 10.9, that the quality of the periodogram might be improved by smoothing the PSD estimate. This can be accomplished by convolving $\hat{S}_{XX}^{(p)}(f)$ with some smoothing function, $\tilde{w}(f)$.

Definition 10.9: The smoothed periodogram with smoothing function $\tilde{w}(f)$ is given by

$$\hat{S}_{XX}^{(wp)}(f) = \tilde{w}(f)*\hat{S}_{XX}^{(p)}(f). \qquad (10.35)$$

The smoothed periodogram can be viewed in terms of the correlation-based estimate as well. Note that if $w(\tau) = F^{-1}[\tilde{w}(f)]$, then

$$\hat{S}_{XX}^{(wp)}(f) = \mathcal{F}[w(\tau)]*\mathcal{F}[\hat{R}_{XX}^{(\mathrm{tri})}(\tau)] = \mathcal{F}[w(\tau)\hat{R}_{XX}^{(\mathrm{tri})}(\tau)] \quad . \qquad (10.36)$$

Therefore, the smoothed periodogram is nothing more than the windowed correlation-based estimate with a window that is the product of $w(t)$ and the triangular window. This seems to indicate that there would be some potential benefit to using windowing functions other than what has been presented here. The reader is referred to the many books on spectral estimation for discussions of other possibilities.

In all of the spectral estimators presented thus far, an ensemble average was estimated using a single realization. A better estimate could be obtained if several independent realizations of the random process were observed and a sample average were used to replace the ensemble average. Even though we may be able to observe only a single realization, it may still be possible to achieve the same effect. This is done by breaking the observed time waveform into segments and treating each segment as an independent realization of the random process. The periodogram is computed on each segment and then the resulting estimates are averaged.

■ Example 10.10:

Figure 10.9 compares the periodogram estimate of the PSD of the random telegraph signal with and without segmentation. In plot (b), no segmentation is used, while in plot (c), the data are segmented into $M = 8$ frames. A periodogram is computed for each frame and the results are then averaged. Note the improvement in the PSD estimate when the segmentation is used. Also note that there is a slight bias appearing in the segmented estimate. This is most noticeable at the higher frequencies. This bias will get worse as more segments are used. There is a trade-off in wanting to use a large value of M to reduce the "jitter" in the estimate and wanting to use a small value of M to keep the bias to a minimum. The following MATLAB functions were used to implement the periodogram estimates with and without segmentation. This same functions can be used to estimate the PSD of any input signal.

```
function [Shat, f]=Periodogram(x,dx)
% This function computes the periodogram estimate of the PSD of the
% input signal. The vector x contains the samples of the input while
% dx indicates the time interval between samples.
Nx=length(x);
Rhat=conv(x,fliplr(x))/Nx;
Nr=length(Rhat);
Shat=fft(Rhat);
Shat=fftshift(dx*abs(Shat));
Nf=(Nr-1)/2;  df=1/(dx*Nr);
f=[-Nf:Nf]*df;

function [S,f]=EPrdgm(x,dx,M)
% This function computes the periodogram estimate of the PSD of the
% input signal by breaking the signal into M frames and performing a
% periodogram estimate on each frame then averaging the results.  The
% vector x contains the samples of the signal while dx is the sampling
% interval.  S is the estimated PSD and f is a vector of frequency
% samples which gives the frequency scale to be used when plotting S.
Nx=length(x);
```

(Continued)

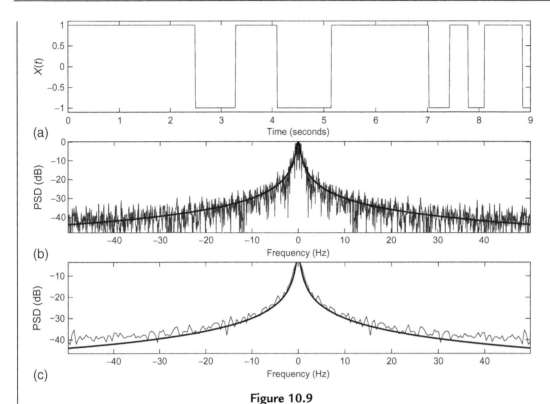

Figure 10.9

(a) A sample realization of the random telegraph signal and (b,c) the periodogram estimate of the PSD function. Plot (b) is for the unsegmented data while plot (c) is for when the data are segmented into $M = 8$ frames. For both PSD plots, the smooth thick line is the true PSD.

```
Nframe=floor(Nx/M);                  % frame length
S=zeros(1,2*Nframe-1);
for m=1:M
    xm=x(((m-1)*Nframe+1):(m*Nframe));
    [Stemp,f]=Periodogram(xm,dx);
    S=S+Stemp;
end
S=S/M;
```

∎

10.4.2 Parametric Spectral Estimation

In parametric spectral estimation, a general model of the data is assumed, which usually contains one or more unknown parameters. Given the general model, the PSD can be calculated analytically. The problem then becomes one of estimating the unknown parameters and plugging the result into the analytic form of the PSD. To provide an example of how this general approach works, we present a specific class of random process models.

Definition 10.10: Given a process $X[n]$ with known statistics, a new process, $Y[n]$, is formed according to the difference equation

$$a_0 Y[n] = \sum_{i=1}^{p} a_i Y[n-i] + \sum_{i=0}^{q} b_i X[n-i]. \tag{10.37}$$

This process is referred to as an *autoregressive moving average process* (ARMA). As special cases, if all of the a_i are equal to zero (except a_0, which is usually set equal to unity), then Equation (10.37) simplifies to

$$Y[n] = \sum_{i=0}^{q} b_i X[n-i]. \tag{10.38}$$

and the process is referred to as a *moving average* (MA) process. The notation MA(q) is used to refer to a qth order MA process. If all of the b_i are equal to zero except for b_0, then the difference equation becomes

$$a_0 Y[n] = \sum_{i=0}^{p} a_i Y[n-i] + b_0 X[n]. \tag{10.39}$$

and the process is referred to as an *autoregressive* (AR) process. The notation AR(p) is used to refer to a pth order AR process. For the general case, the notation ARMA(p,q) is used.

To demonstrate the basic principles of parametric estimation, suppose it was determined that a certain random process, $Y[n]$, is well modeled by an AR(1) model,

$$Y[n] = a_1 Y[n-1] + X[n], \tag{10.40}$$

where $X[n]$ is an IID random process with zero-mean and a variance of σ_X^2. It is noted that

$$Y[n+1] = a_1 Y[n] + X[n+1], \tag{10.41}$$

$$Y[n+2] = a_1^2 Y[n] + a_1 X[n+1] + X[n+2], \tag{10.42}$$

$$Y[n+3] = a_1^3 Y[n] + a_1^2 X[n] + a_1 X[n+1] + X[n+2], \tag{10.43}$$

and in general,

$$Y[n+k] = a_1^k Y[n] + \sum_{i=0}^{k-1} a_1^{k-1-i} X[n+i]. \tag{10.44}$$

Using this expression, the autocorrelation function of the AR(1) process can be computed.

$$R_{YY}[n, n+k] = E[Y[n]Y[n+k]] \tag{10.45}$$

$$= a_1^k E[Y^2[n]] + \sum_{i=0}^{k-1} a_1^{k-1-i} E[Y[n]X[n+i]] = a_1^k R_{YY}[n, n].$$

The last step is accomplished using the fact that $Y[n]$ is independent of $X[n+i]$ for $i > 0$. The expression $R_{YY}[n, n]$ is calculated according to

$$R_{YY}[n, n] = E[(a_1 Y[n-1] + X[n])^2] = a_1^2 R_{YY}[n-1, n-1] + \sigma_X^2. \tag{10.46}$$

Assuming that the process $Y[n]$ is WSS[3], this recursion becomes

$$R_{YY}[0] = a_1^2 R_{YY}[0] + \sigma_X^2 \Rightarrow R_{YY}[0] = \frac{\sigma_X^2}{1 - a_1^2}. \tag{10.47}$$

Therefore, the autocorrelation function of the AR(1) process is

$$R_{YY}[k] = \frac{\sigma_X^2}{1 - a_1^2} a_1^{|k|}. \tag{10.48}$$

Assuming that the samples of this discrete-time process are taken at a sampling interval of Δt, the PSD of this process works out to be

$$S_{YY}(f) = \frac{\Delta t \sigma_X^2}{\left|1 - a_1 e^{-j2\pi f \Delta t}\right|^2}. \tag{10.49}$$

For this simple AR(1) model, the PSD can be expressed as a function of two unknown parameters, a_1 and σ_X^2. The problem of estimating the PSD then becomes one of estimating the two parameters and then plugging the result into the general expression for the PSD. In many cases, the total power in the process may be known, which eliminates the need to

---

[3] It will be shown in the next chapter that this is the case provided that $X[n]$ is WSS.

estimate σ_X^2. Even if that is not the case, the value of σ_X^2 is just a multiplicative factor in the expression for PSD and does not change the shape of the curve. Hence, in the following, we focus attention on estimating the parameter, a_1.

Since we know the AR(1) model satisfies the recursion of Equation (10.40), the next value of the process can be predicted from the current value according to

$$\hat{Y}[n+1] = \hat{a}_1 Y[n]. \tag{10.50}$$

This is known as *linear prediction* since the predictor of the next value is a linear function of the current value. The error in this estimate is

$$\varepsilon = Y[n+1] - \hat{Y}[n+1] = Y[n+1] - \hat{a}_1 Y[n]. \tag{10.51}$$

Typically, we choose as an estimate of a_1 the value of $\hat{a}_1$ that makes the linear predictor as good as possible. Usually, "good" is interpreted as minimizing the mean-square error, which is given by

$$E[\varepsilon^2] = E[(Y[n+1] - \hat{a}_1 Y[n])^2] = R_{YY}(0)(1 + \hat{a}_1^2) - 2\hat{a}_1 R_{YY}(1). \tag{10.52}$$

Differentiating the mean-square error with respect to $\hat{a}_1$ and setting equal to zero results in

$$2\hat{a}_1 R_{YY}(0) - 2R_{YY}(1) = 0 \Rightarrow \hat{a}_1 = \frac{R_{YY}(1)}{R_{YY}(0)}. \tag{10.53}$$

Of course, we do not know what the autocorrelation function is. If we did, we would not need to estimate the PSD. So, the preceding ensemble averages must be replaced with sample averages, and the minimum mean-square error (MMSE) linear prediction coefficient is given by

$$\hat{a}_1 = \frac{\hat{R}_{YY}(1)}{\hat{R}_{YY}(0)} = \frac{\displaystyle\sum_{n=-n_0}^{n_0-1} y[n]y[n+1]}{\displaystyle\sum_{n=-n_0}^{n_0} y^2[n]}. \tag{10.54}$$

Note that we have used a lower case $y[n]$ in Equation (10.54) since we are dealing with a single realization rather than the ensemble $Y[n]$.

Example 10.11:

In this example, we use the AR(1) model to estimate the PSD of the random telegraph process. Clearly, the AR(1) model does not describe the random telegraph process; however, the autocorrelation function of the random telegraph signal is a two-sided

(Continued)

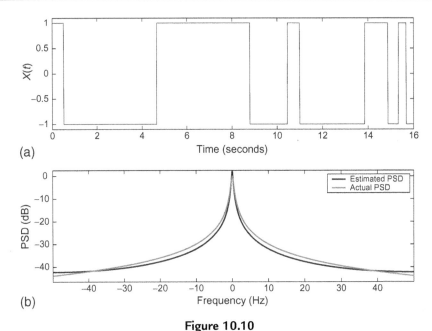

Figure 10.10
(a) A sample realization of the random telegraph signal and (b) the parametric estimate of the PSD function based on the AR(1) model.

exponential, as is the autocorrelation function of the AR(1) process. As a consequence, we expect this model to give good results. The results are shown in Figure 10.10. Notice how nicely the estimated PSD matches the actual PSD.

In the previous example, the results were quite good because the functional form of the PSD of the AR(1) process nicely matched the functional form of the true PSD. If the fit had not been so good, it might have been necessary to move to a higher order AR(p) model. In the exercises at the end of the chapter (see Exercises 10.19 and 10.20), the reader is led through the problem of finding the MMSE linear prediction coefficients for a general AR(p) model. The problem of analytically finding the PSD of the AR(p) process is dealt with in the next chapter (also see Exercise 10.18).

10.5 Thermal Noise

The most commonly encountered source of noise in electronic systems is that caused by thermal agitation of electrons in any conductive material, which is commonly referred to as *thermal noise*. Unlike shot noise, thermal noise does not require the presence of a direct current and hence is always present. We will not delve into the underlying thermodynamics to derive a model for this type of noise, but rather will just summarize some of the important results. Nyquist's theorem states that for a resistive element with an impedance of r ohms, at a temperature of t_k (measured in Kelvin), the mean-square voltage of the thermal noise measured in a incremental frequency band of width Δf centered at frequency f is found to be

$$E[V^2(t)] = v_{rms}^2 = 4kt_k r\Delta f\left[\frac{h|f|/kt_k}{\exp(h|f|/kt_k)-1}\right]\text{volts}^2,\qquad(10.55)$$

where

$$h = \text{Planck's constant} = 6.2\times10^{-34}\,J\text{-s}\,;$$

$$k = \text{Boltzman's constant} = 1.38\times10^{-23}\,J/^\circ K\,;$$

$$t_k = \text{absolute temperature} = 273+{}^\circ C\,.$$

Typically, a practical resistor is modeled as a Thevenin equivalent circuit, as illustrated in Figure 10.11, consisting of a noiseless resistor in series with a noise source with a mean-square value as specified in the previous equation. If this noisy resistor was connected to a resistive load of impedance r_L, the average power delivered to the load would be

$$\overline{P_L} = \frac{v_{rms}^2 r_L}{(r+r_L)^2}.\qquad(10.56)$$

The power delivered to the load is maximized when the source and the load impedance are matched (i.e., $r_L = r$). It is common to refer to the maximum power that can be delivered to a load as the *available power*. For a noisy resistor, the available power (in a bandwidth of Δf) is

$$\overline{P} = \frac{v_{rms}^2}{4r} = kt_k\Delta f\left[\frac{h|f|/kt_k}{\exp(h|f|/kt_k)-1}\right]\text{watts}.\qquad(10.57)$$

The PSD of the thermal noise in the resistor is then

$$S_{NN}(f) = \frac{1}{2}kt_k\left[\frac{h|f|/kt_k}{\exp(h|f|/kt_k)-1}\right].\qquad(10.58)$$

Figure 10.11
A Thevenin equivalent circuit for a noisy resistor.

The extra factor of $1/2$ is due to the fact that our PSD function is a two-sided function of frequency, and so the actual power in a given frequency band is evenly split between the positive and negative frequencies. Note that the power available to a load and the resulting PSD are independent of the impedance of the resistor, r.

This PSD function is plotted in Figure 10.12 for several different temperatures. Note that for frequencies that are of interest in most applications (except optical, infrared, etc.), the PSD function is essentially constant. It is straightforward (and left as an exercise to the reader) to show that this constant is given by

$$S_{NN}(f) = \frac{1}{2}kt_k = \frac{N_o}{2}, \tag{10.59}$$

where we have defined the constant $N_o = kt_f$. At $t_k = 298°K$ [4], the parameter N_o takes on a value of $= 4.11 \times 10^{-21}$ W/Hz $= -173.86$ dBm/Hz.

It is common to use this simpler function as a model for the PSD of thermal noise. Because the PSD contains equal power at all frequencies, this noise model is referred to as *white noise* (analogous to white light, which contains all frequencies). The corresponding autocorrelation function is

$$R_{NN}(\tau) = \frac{N_o}{2}\delta(\tau). \tag{10.60}$$

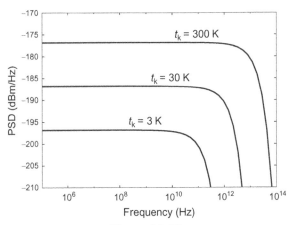

Figure 10.12
PSD of thermal noise in a resistor.

---

[4] Most texts use $t_k = 290°K$ as "room temperature"; however, this corresponds to a fairly chilly room ($17°C \approx 63°F$). On the other hand, $t_k = 298°K$ is a more balmy ($25°C \approx 77°F$) environment. These differences would only change the value of N_o by a small fraction of a dB.

It should be pointed out that the noise model of Equation (10.59) is a mathematical approximation to the actual PSD. There is no such thing as truly white noise, since such a process (if it existed) would have infinite power and would destroy any device we tried to measure it with. However, this mathematical model is simple, easy to work with, and serves as a good approximation to thermal noise for most applications.

In addition to modeling thermal noise as having a flat PSD, it can also be shown that the first-order characteristics of thermal noise can be well approximated with a zero-mean Gaussian process. We say that thermal noise is zero-mean white Gaussian noise (WGN) with a (two-sided) PSD of $N_{\mathrm{o}}/2$. While thermal noise is the most common source of noise in electronic devices, there are other sources as well. Shot noise was discussed at the end of Chapter 8. In addition, one may encounter flicker noise, which occurs primarily in active devices; burst or popcorn noise, which is found in integrated circuits and some discrete transistors; avalanche noise, which is produced by avalanche breakdown in a *p–n* junction; as well as several other types of noise. For the purposes of this text, we will stick with the white Gaussian model of thermal noise.

10.6 Engineering Application: PSDs of Digital Modulation Formats

In this section, we evaluate the PSD of a class of signals that might be used in a digital communication system. Suppose we have a sequence of data symbols $\{B_k\}$ that we wish to convey across some communication medium. We can use the data symbols to determine the amplitude of a sequence of pulses which we would then transmit across the medium (e.g., a twisted copper pair, or an optical fiber). This is know as pulse amplitude modulation (PAM). If the pulse amplitudes are represented by the sequence of random variables $\{..., A_{-2}, A_{-1}, A_0, A_1, A_2, ...\}$ and the basic pulse shape is given by the waveform $p(t)$, then the transmitted signal might be of the form

$$S(t) = \sum_{k = -\infty}^{\infty} A_k p(t - kt_{\mathrm{s}} - \Theta), \qquad (10.61)$$

where t_{s} is the symbol interval (that is, one pulse is launched every t_{s} seconds) and Θ is a random delay, which we take to be uniformly distributed over $[0, t_{\mathrm{s}})$ and independent of the pulse amplitudes. The addition of the random delay in the model makes the process $S(t)$ WSS. This is not necessary and the result we will obtain would not change if we did not add this delay, but it does simplify slightly the derivation.

If the data symbols are drawn from an alphabet of size 2^n symbols, then each symbol can be represented by an n-bit word, and thus the data rate of the digital communication system is $r = n/t_{\mathrm{s}}$ bits/second. The random process $S(t)$ used to represent this data has a certain spectral content, and thus requires a communications channel with a bandwidth adequate to carry that spectral content. It would be interesting to see how the required bandwidth relates to the data rate. Toward that end, we seek to determine the PSD of the PAM signal $S(t)$. We will find the PSD by first computing the autocorrelation function of $S(t)$ and then converting this to PSD via the Wiener–Khintchine–Einstein Theorem.

Using the definition of autocorrelation, the autocorrelation function of the PAM signal is given by

$$R_{SS}(t, t + \tau) = E[S(t)S(t + \tau)] = E\left[\sum_{k = -\infty}^{\infty} \sum_{m = -\infty}^{\infty} A_k A_m p(t - kt_s - \Theta)(p(t + \tau - mt_s - \Theta)) \right]$$

$$= \sum_{k = -\infty}^{\infty} \sum_{m = -\infty}^{\infty} E[A_k A_m] E[p(t - kt_s - \Theta)p(t + \tau - mt_s - \Theta)]$$

$$= \frac{1}{t_s} \sum_{k = -\infty}^{\infty} \sum_{m = -\infty}^{\infty} E[A_k A_m] \int_0^{t_s} p(t - kt_s - \theta)p(t + \tau - mt_s - \theta)d\theta. \tag{10.62}$$

To simplify notation, we define $R_{AA}[n]$ to be the autocorrelation function of the sequence of pulse amplitudes. Note that we are assuming the sequence is stationary (at least in the wide sense). Going through a simple change of variables ($v = t - kt_s - \theta$) then results in

$$R_{SS}(t, t + \tau) = \frac{1}{t_s} \sum_{k = -\infty}^{\infty} \sum_{m = -\infty}^{\infty} R_{AA}[m - k] \int_{t - (k + 1)t_s}^{t - kt_s} p(v)p(v + \tau - (m - k)t_s)dv. \tag{10.63}$$

Finally, we go through one last change of variables ($n = m - k$) to produce

$$R_{SS}(t, t + \tau) = \frac{1}{t_s} \sum_{k = -\infty}^{\infty} \sum_{n = -\infty}^{\infty} R_{AA}[n] \int_{t - (k + 1)t_s}^{t - kt_s} p(v)p(v + \tau - nt_s)dv$$

$$= \frac{1}{t_s} \sum_{n = -\infty}^{\infty} R_{AA}[n] \sum_{k = -\infty}^{\infty} \int_{t - (k + 1)t_s}^{t - kt_s} p(v)p(v + \tau - nt_s)dv$$

$$= \frac{1}{t_s} \sum_{n = -\infty}^{\infty} R_{AA}[n] \int_{-\infty}^{\infty} p(v)p(v + \tau - nt_s)dv. \tag{10.64}$$

To aid in taking the Fourier transform of this expression, we note that the integral in the previous equation can be written as a convolution of $p(t)$ with $p(-t)$,

$$\int_{-\infty}^{\infty} p(v)p(v + \tau - nt_{\text{s}})dv = p(t)*p(-t)\big|_{t = \tau - nt_{\text{s}}}. \tag{10.65}$$

Using the fact that convolution in the time domain becomes multiplication in the frequency domain along with the time reversal and time shifting properties of Fourier transforms (see Appendix C), the transform of this convolution works out to be

$$\mathscr{F}\left[\int_{-\infty}^{\infty} p(v)p(v + \tau - nt_{\text{s}})dv\right] = |P(f)|^2 e^{-j2\pi nft_{\text{s}}}, \tag{10.66}$$

where $P(f) = \mathscr{F}[p(t)]$ is the Fourier transform of the pulse shape used. With this result, the PSD of the PAM signal is found by taking the transform of Equation (10.64), resulting in

$$S_{SS}(f) = \frac{|P(f)|^2}{t_{\text{s}}} \sum_{n=-\infty}^{\infty} R_{AA}[n]e^{-j2\pi nft_{\text{s}}}. \tag{10.67}$$

It is seen from the previous equation that the PSD of a PAM signal is the product of two terms, the first of which is the magnitude squared of the pulse shape's spectrum, while the second term is essentially the PSD of the discrete sequence of amplitudes. As a result, we can control the spectral content of our PAM signal by carefully designing a pulse shape with a compact spectrum and also by introducing memory into the sequence of pulse amplitudes.

■ Example 10.12:

To start with, suppose the pulse amplitudes are an IID sequence of random variables which are equally likely to be +1 or −1. In that case, $R_{AA}[n] = \delta[n]$ and the PSD of the sequence of amplitudes is

$$\sum_{n=-\infty}^{\infty} R_{AA}[n]e^{-j2\pi nft_{\text{s}}} = 1.$$

In this case, $S_{SS}(f) = |P(f)|^2/t_{\text{s}}$ and the spectral shape of the PAM signal is completely determined by the spectral content of the pulse shape. Suppose we use as a pulse shape a square pulse of height a and width t_{s},

(Continued)

$$p(t) = a\operatorname{rect}(t/t_{s}) \leftrightarrow P(f) = at_{s}\operatorname{sinc}(ft_{s}).$$

The PSD of the resulting PAM signal is then $S_{SS}(f) = a^2 t_s \operatorname{sinc}^2(ft_s)$. Note that the factor $a^2 t_s$ is the energy in each pulse sent, E_p. A sample realization of this PAM process along with a plot of the PSD is given in Figure 10.13. Most of the power in the process is contained in the main lobe, which has a bandwidth of $1/t_s$ (equal to the data rate), but there is also a non-trivial amount of power in the sidelobes, which die off very slowly. This high-frequency content can be attributed to the instantaneous jumps in the process. These frequency sidelobes can be suppressed by using a pulse with a smoother shape. Suppose, for example, we used a pulse which was a half-cycle of a sinusoid of height a,

$$p(t) = a\cos\left(\frac{\pi t}{t_s}\right)\operatorname{rect}\left(\frac{t}{t_s}\right) \leftrightarrow P(f) = \frac{at_s}{2}\left[\operatorname{sinc}\left(ft_s - \frac{1}{2}\right) + \operatorname{sinc}\left(ft_s + \frac{1}{2}\right)\right] = \frac{at_s}{2\pi}\frac{\cos(\pi ft_s)}{\frac{1}{4} - (ft_s)^2}.$$

The resulting PSD of the PAM signal with half-sinusoidal pulse shapes is then

$$S_{SS}(f) = \frac{a^2 t_s}{4\pi^2}\frac{\cos^2(\pi ft_s)}{\left[\frac{1}{4} - (ft_s)^2\right]^2}.$$

In this case, the energy in each pulse is $E_p = a^2 t_s/2$. As shown in Figure 10.14, the main lobe is now 50 % wider than it was with square pulses, but the sidelobes decay much more rapidly.

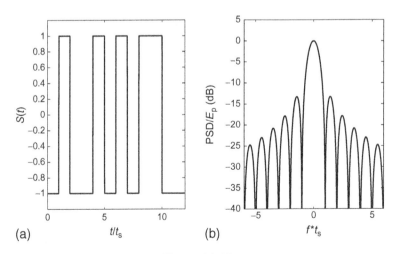

Figure 10.13
(a) A sample realization and (b) the PSD of a PAM signal with square pulses.

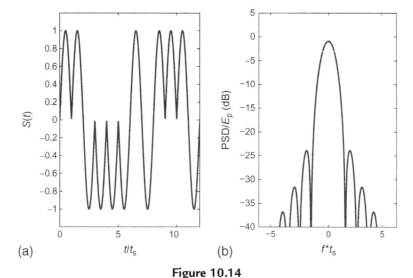

Figure 10.14
(a) A sample realization and (b) the PSD of a PAM signal with half-sinusoidal pulses.

Example 10.13:

In this example, we show how the spectrum of the PAM signal can also be manipulated by adding memory to the sequence of pulse amplitudes. Suppose the data to be transmitted $\{..., B_{-2}, B_{-1}, B_0, B_1, B_2, ...\}$ is an IID sequence of Bernoulli random variables, $B_k \in \{+1, -1\}$. In the previous example, we formed the pulse amplitudes according to $A_k = B_k$. Suppose instead we formed these amplitudes according to $A_k = B_k + B_{k-1}$. Now the pulse amplitudes can take on three values (even though each pulse still only carries one bit of information). This is known as *duobinary* precoding. The resulting autocorrelation function for the sequence of pulse amplitudes is

$$R_{AA}[n] = E[A_k A_{k+n}] = E[(B_k + B_{k-1})(B_{k+n} + B_{k+n-1})] = \begin{cases} 2, & n = 0, \\ 1, & n = \pm 1, \\ 0, & \text{otherwise.} \end{cases}$$

The PSD of this sequence of pulse amplitudes is then

$$\sum_{n=-\infty}^{\infty} R_{AA}[n] e^{-j2\pi n f t_s} = 2 + e^{j2\pi f t_s} + e^{-j2\pi f t_s} = 2 + 2\cos(2\pi f t_s) = 4\cos^2(\pi f t_s).$$

This expression then multiplies whatever spectral shape results from the pulse shape chosen. The PSD of duobinary PAM with square pulses is illustrated in Figure 10.15. In this case, the duobinary precoding has the benefit of suppressing the frequency sidelobes without broadening the main lobe.

(Continued)

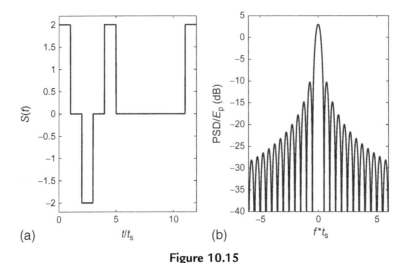

Figure 10.15
(a) A sample realization and (b) the PSD of a PAM signal with duobinary precoding and square pulses.

∎

Example 10.14:

The following MATLAB code creates a realization of a binary PAM signal where the pulse amplitudes are either +1 or −1. In this example, a half-sinusoidal pulse shape is used, but the code is written so that it is easy to change the pulse shape (just change the 6th line where the pulse shape is assigned to the variable p). After a realization of the PAM signal is created, the PSD of the resulting signal is estimated using the segmented periodogram technique given in Example 10.10. The resulting PSD estimate is shown in Figure 10.16. Note the agreement between the estimate and the actual PSD shown in Figure 10.14. The reader is encouraged to try running this program with different pulse shapes to see how the pulse shape changes the spectrum of the PAM signal.

```
N=100;                              % No. of bit intervals in realization.
Ns=19;                             % No. of time samples per bit.
Rb=9600;                           % bit rate (bits/sec)
dt=1/(Rb*Ns);                      % time between samples
t=([1:Ns]-0.5)/Ns;                 % time axis for pulse shape
p=sin(pi*t);                       % pulse shape
Ep=p*p'*dt;                        % energy per pulse
X(1:Ns:(Ns*(N-1)+1))=sign(rand(1,N)-0.5);% random data bits
X=conv(X,p);                       % PAM signal with pulse shape added
```

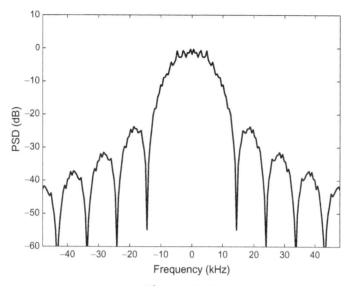

Figure 10.16
An estimate of the PSD of a PAM signal with half-sinusoidal pulse shapes.

```
M=10;                              % No. of segments
[S,f]=EPrdgm(X,dt,M);              % (Normalized) PSD estimate
plot(f/1000,10*log10(abs(S/Ep)))  % plot results
axis([-5*Rb/1000 5*Rb/1000 -60 10])
xlabel('frequency (kHz)')
ylabel('PSD (dB)')
```

Exercises

Section 10.1: Definition of PSD

10.1 Find the PSD of the process described in Exercise 8.1.

10.2 Find the PSD of the process described in Exercise 8.2.

10.3 Consider a constant random process, $X(t) = A$, where A is a random variable. Use Definition 10.1 to calculate the PSD of $X(t)$.

Section 10.2: Wiener–Khintchine–Einstein Theorem

10.4 Consider a random process of the form

$$X(t) = b\cos(2\pi\Psi t + \Theta),$$

where b is a constant, Θ is a uniform random variable over $[0, 2\pi)$, and Ψ is a random variable which is independent of Θ and has a PDF, $f_\Psi(\psi)$. Find the PSD, $S_{XX}(f)$ in terms of $f_\Psi(\psi)$. In so doing, prove that for any $S(f)$ which is a valid PSD function, we can always construct a random process with PSD equal to $S(f)$.

10.5 Let $X(t) = A\cos(\omega t) + B\sin(\omega t)$ where A and B are independent, zero-mean, identically distributed, non-Gaussian random variables.
   (a) Show that $X(t)$ is WSS, but not strict sense stationary. *Hint:* For the latter case consider $E[X^3(t)]$. *Note:* If A and B are Gaussian, then $X(t)$ is also stationary in the strict sense.
   (b) Find the PSD of this process.

10.6 Let $X(t) = \sum_{n=1}^{N} a_n\cos(\omega_n t + \theta_n)$ where all of the ω_n are non-zero constants, the a_n are constants, and the θ_n are IID random variables, each uniformly distributed over $[0, 2\pi)$.
   (a) Determine the autocorrelation function of $X(t)$.
   (b) Determine the PSD of $X(t)$.

10.7   Let $X(t) = \sum_{n=1}^{\infty} [A_n \cos(n\omega t) + B_n \sin(n\omega t)]$ be a random process, where A_n and B_n are random variables such that $E[A_n] = E[B_n] = 0$, $E[A_n B_m] = 0$, $E[A_n A_m] = \delta_{n,m} E[A_n^2]$, and $E[B_n B_m] = \delta_{n,m} E[B_n^2]$ for all m and n, where $\delta_{n,m}$ is the Kronecker delta function. This process is sometimes used as a model for random noise.

(a) Find the time-varying autocorrelation function $R_{XX}(t, t + \tau)$.

(b) If $E[B_n^2] = E[A_n^2]$, is this process WSS?

(c) Find the PSD of this process.

10.8   Find the PSD for a process for which $R_{XX}(\tau) = 1$ for all τ.

10.9   Suppose $X(t)$ is a stationary zero-mean Gaussian random process with PSD, $S_{XX}(f)$.

(a) Find the PSD of $Y(t) = X^2(t)$ in terms of $S_{XX}(f)$.

(b) Sketch the resulting PSD if $S_{XX}(f) = \text{rect}\left(\dfrac{f}{2B}\right)$.

(c) Is $Y(t)$ WSS?

10.10   Consider a random sinusoidal process of the form $X(t) = b\cos(2\pi f t + \Theta)$, where Θ has an arbitrary PDF, $f_\Theta(\theta)$. Analytically determine how the PSD of $X(t)$ depends on $f_\Theta(\theta)$. Give an intuitive explanation for your result.

10.11   Let $s(t)$ be a deterministic periodic waveform with period t_0. A random process is constructed according to $X(t) = s(t - T)$ where T is a random variable uniformly distributed over $[0, t_0)$. Show that the random process $X(t)$ has a line spectrum and write the PSD of $X(t)$ in terms of the Fourier Series coefficients of the periodic signal $s(t)$.

10.12   A sinusoidal signal of the form $X(t) = b\cos(2\pi f_0 t + \Theta)$ is transmitted from a fixed platform. The signal is received by an antenna which is on a mobile platform that is in motion relative to the transmitter, with a velocity of V relative to the direction of signal propagation between the transmitter and receiver. Therefore, the received signal experiences a Doppler shift and (ignoring noise in the receiver) is of the form

$$Y(t) = b\cos\left(2\pi f_0\left(1 + \frac{V}{c}\right)t + \Theta\right),$$

where c is the speed of light. Find the PSD of the received signal if V is uniformly distributed over $(-v_0, v_0)$. Qualitatively, what does the Doppler effect do to the PSD of the sinusoidal signal?

10.13 Two zero-mean discrete random processes, $X[n]$ and $Y[n]$, are statistically independent and have autocorrelation functions given by $R_{XX}[k] = (1/2)^k$ and $R_{YY}[k] = (1/3)^k$. Let a new random process be $Z[n] = X[n] + Y[n]$.
   (a) Find $R_{ZZ}[k]$. Plot all three autocorrelation functions.
   (b) Determine all three PSD functions analytically and plot the PSDs.

10.14 Let $S_{XX}(f)$ be the PSD function of a WSS discrete-time process $X[n]$. Recall that one way to obtain this PSD function is to compute $R_{XX}[n] = E[X[k]X[k+n]]$ and then take the DFT of the resulting autocorrelation function. Determine how to find the average power in a discrete-time random process directly from the PSD function, $S_{XX}(f)$.

10.15 A binary phase shift keying signal is defined according to
$$X(t) = \cos\left(2\pi f_c t + B[n]\frac{\pi}{2}\right) \quad \text{for } nT \le t < (n+1)T,$$
for all n, and $B[n]$ is a discrete-time Bernoulli random process that has values of $+1$ or -1.
   (a) Determine the autocorrelation function for the random process $X(t)$. Is the process WSS?
   (b) Determine the PSD of $X(t)$.

10.16 Let $X(t)$ be a random process whose PSD is shown in the accompanying figure. A new process is formed by multiplying $X(t)$ by a carrier to produce
$$Y(t) = X(t)\cos(\omega_0 t + \Theta),$$
where Θ is uniform over $[0, 2\pi)$ and independent of $X(t)$. Find and sketch the PSD of the process $Y(t)$.

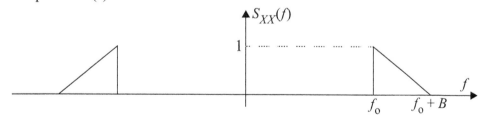

10.17 Consider a random process $Z(t) = X(t) + Y(t)$.
   (a) Find an expression for $S_{ZZ}(f)$ in terms of $S_{XX}(f)$, $S_{YY}(f)$, and $S_{XY}(f)$.
   (b) Under what conditions does $S_{ZZ}(f) = S_{XX}(f) + S_{YY}(f)$?

Section 10.3: Bandwidth of a Random Process

10.18 Develop a formula to compute the RMS bandwidth of a random process, $X(t)$, directly from its autocorrelation function, $R_{XX}(\tau)$.

10.19 A random process has a PSD function given by

$$S(f) = \frac{1}{\left(1 + \left(\frac{f}{B}\right)^2\right)^3}.$$

(a) Find the absolute bandwidth.
(b) Find the 3 dB bandwidth.
(c) Find the RMS bandwidth.

Can you generalize your result to a spectrum of the form

$$S(f) = \frac{1}{\left(1 + \left(\frac{f}{B}\right)^2\right)^N},$$

where N is an integer greater than 1?

10.20 A random process has a PSD function given by

$$S(f) = \frac{f^2}{\left(1 + \left(\frac{f}{B}\right)^2\right)^3}.$$

(a) Find the absolute bandwidth.
(b) Find the 3 dB bandwidth.
(c) Find the RMS bandwidth.

Can you generalize your result to a spectrum of the form

$$S(f) = \frac{f^2}{\left(1 + \left(\frac{f}{B}\right)^2\right)^N},$$

where N is an integer greater than 2?

Section 10.4: Spectral Estimation

10.21 Consider the linear prediction random process $X[n] = (1/2)X[n-1] + E[n]$, $n = 1, 2, 3, \ldots$, where $X[0] = 0$ and $E[n]$ is a zero-mean, IID random process.
(a) Find the mean and autocorrelation functions for $X[n]$. Is $X[n]$ WSS?
(b) Find the PSD of $X[n]$.

10.22 Consider an AR(2) process which is described by the recursion

$$Y[n] = a_1 Y[n-1] + a_2 Y[n-2] + X[n]$$

where $X[n]$ is an IID random process with zero-mean and variance σ_X^2.
(a) Show that the autocorrelation function of the AR(2) process satisfies the difference equation,

$$R_{YY}[k] = a_1 R_{YY}[k-1] + a_2 R_{YY}[k-2], \quad k = 2, 3, 4, \ldots.$$

(b) Show that the first two terms in the autocorrelation function satisfy

$$(1 - a_1^2 - a_2^2)R_{YY}[0] - 2a_1a_2R_{YY}[1] = \sigma_X^2,$$

$$\text{and } (1 - a_2)R_{YY}[1] = a_1R_{YY}[0].$$

From these two equations, solve for $R_{YY}[0]$ and $R_{YY}[1]$ in terms of a_1, a_2, and σ_X^2.

(c) Using the difference equation in part (a) together with the initial conditions in part (b), find a general expression for the autocorrelation function of an AR(2) process.

(d) Use your result in part (c) to find the PSD of an AR(2) process.

10.23 Suppose we use an AR(2) model to predict the next value of a random process based on observations of the two most recent samples. That is, we form

$$\hat{Y}[n+1] = a_1 Y[n] + a_2 Y[n-1].$$

(a) Derive an expression for the mean-square estimation error,

$$E[\varepsilon^2] = E[(Y[n+1] - \hat{Y}[n+1])^2].$$

(b) Find the values of the prediction coefficients, a_1 and a_2, that minimize the mean-square error.

10.24 Extend the results of Exercise 10.23 to a general AR(p) model. That is, suppose we wish to predict the next value of a random process by forming a linear combination of the p most recent samples:

$$\hat{Y}[n+1] = \sum_{k=1}^{p} a_k Y[n-k+1].$$

Find an expression for the values of the prediction coefficients which minimize the mean-square prediction error.

10.25 Show that the estimator for the autocorrelation function, $\hat{R}_{XX}(\tau)$, described in Equation (10.26) is unbiased. That is, show that $E[\hat{R}_{XX}(\tau)] = R_{XX}(\tau)$.

10.26 Suppose $X(t)$ is a zero-mean, WSS, Gaussian random process. Find an expression for the variance of the estimate of the autocorrelation function, $\hat{R}_{XX}(\tau)$, given in Equation (10.26). That is, find $\text{Var}(\hat{R}_{XX}(\tau))$. *Hint:* Remember $\hat{R}_{XX}(\tau)$ is unbiased (see Exercise 10.25) and you might find the Gaussian moment factoring theorem (see Exercise 6.18) useful.

10.27   Using the expression for $\mathrm{Var}(\hat{R}_{XX}(\tau))$ found in Exercise 10.26, show that as $|\tau| \to 2t_o$, $\mathrm{Var}(\hat{R}_{XX}(\tau)) > \mathrm{Var}(X(t))$ and therefore, the estimate of the autocorrelation function is at least as noisy as the process itself as $|\tau| \to 2t_o$.

10.28   Determine whether or not the periodogram is an unbiased estimate of the PSD.

10.29   Suppose we form a smoothed periodogram of the PSD, $\hat{S}_{XX}^{(wp)}(f)$, as defined in Equation (10.35), using a rectangular smoothing function,

$$\tilde{w}(f) \;=\; \frac{1}{f_\Delta}\mathrm{rect}\!\left(\frac{f}{f_\Delta}\right),$$

where f_Δ is the width of the rectangle. If we want to form the same estimator using a windowed correlation-based estimate, what window function (in the time domain) should we use?

Section 10.5: Thermal Noise

10.30   (a)   Prove that the expression for the PSD of thermal noise in a resistor converges to the constant $N_o/2 = kt_k/2$ as $f \to 0$.

(b)   Assuming a temperature of $298°\mathrm{K}$, find the range of frequencies over which thermal noise has a PSD which is within 99% of its value at $f = 0$.

(c)   Suppose we had a very sensitive piece of equipment which was able to accurately measure the thermal noise across a resistive element. Furthermore, suppose our equipment could respond to a range of frequencies which spanned 50 MHz. Find the power (in watts) and the RMS voltage (in volts) that we would measure across a $75\ \Omega$ resistor. Assume the equipment had a load impedance matched to the resistor.

10.31   Suppose two resistors of impedance r_1 and r_2 are placed in series and held at different physical temperatures, t_1 and t_2. We would like to model this series combination of noisy resistors as a single noiseless resistor with an impedance of $r = r_1 + r_2$, together with a noise source with an effective temperature of t_e. In short, we want the two models shown in the accompanying figure to be equivalent. Assuming the noise produced by the two resistors is independent, what should t_e, the effective noise temperature of the series combination of resistors, be? If the two resistors are held at the same physical temperature, is the effective temperature equal to the true common temperature of the resistors?

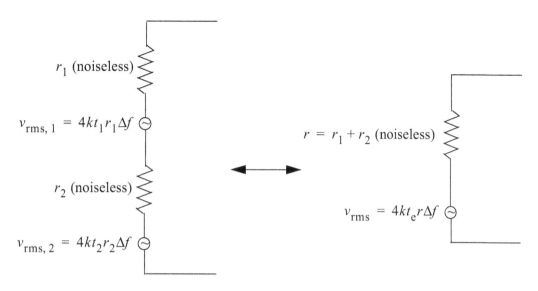

10.32   Repeat Exercise 10.31 for a parallel combination of resistors.

MATLAB Exercises

10.33

(a) Create a random process $X[n]$ where each sample of the random process is an IID, Bernoulli random variable equally likely to be ± 1. Form a new process according to the MA(2) model $Y[n] = X[n] - \frac{1}{2}X[n-1] + \frac{1}{4}X[n-2]$. Assume $X[n] = 0$ for $n < 0$.

(b) Compute the time-average autocorrelation function $\langle Y[n]Y[n+k] \rangle$ from a single realization of this process.

(c) Compute the ensemble average autocorrelation function $E[Y[n]Y[n+k]]$ from several realizations of this process. Does the process appear to be ergodic in the autocorrelation?

(d) Estimate the PSD of this process using the periodogram method.

10.34

(a) Create a random process $X[n]$ where each sample of the random process is an IID, Bernoulli random variable equally likely to be ± 1. Form a new process according to the AR(2) model $Y[n] = \frac{1}{2}Y[n-1] - \frac{1}{4}Y[n-2] + X[n]$. Assume $Y[n] = 0$ for $n < 0$.

(b) Compute the time-average autocorrelation function $\langle Y[n]Y[n+k] \rangle$ from a single realization of this process.

(c) Compute the ensemble average autocorrelation function $E[Y[n]Y[n+k]]$ from several realizations of this process. Does the process appear to be ergodic in the autocorrelation?

(d) Estimate the PSD of this process using the periodogram method.

10.35

(a) For the process in Exercise 10.34, find a parametric estimate of the PSD by using an AR(1) model. Compare the resulting PSD estimate with the non-parametric estimate found in Exercise 10.34(d). Explain any differences you see.

(b) Again, referring to the process in Exercise 10.34, find a parametric estimate of the PSD this time using an AR(2) model. Compare the resulting PSD estimate with the non-parametric estimate found in Exercise 10.34(d). Explain any differences you see.

10.36

(a) Write a MATLAB program to create a realization of a binary PAM signal with square pulses. You can accomplish this with a simple modification to the program given in Example 10.14. Call this signal $x(t)$.

(b) We can create a frequency shift keying (FSK) signal according to

$$y(t) = \cos\left(2\pi f_c t + \frac{\pi}{2t_s}\int_0^t x(u)\mathrm{d}u\right),$$

where t_s is the duration of the square pulses in $x(t)$ and f_c is the carrier frequency. Write a MATLAB program to create a 10 ms realization of this FSK signal assuming $t_s = 100\,\mu s$ and $f_c = 20\,kHz$.

(c) Using the segmented periodogram, estimate the PSD of the FSK signal you created in part (b).

(d) Estimate the 30 dB bandwidth of the FSK signal. That is, find the bandwidth where the PSD is down 30 dB from its peak value.

10.37 Construct a signal plus noise random sequence using 10 samples of the following:

$$X[n] = \cos(2\pi n f_0 t_s) + N[n],$$

where $N[n]$ is a sequence of zero-mean, unit variance, IID Gaussian random variables, $f_0 = 0.1/t_s = 100\,kHz$, and $t_s = 1\,\mu s$ is the time between samples of the process.

(a) Calculate the periodogram estimate of the PSD, $S_{XX}(f)$.

(b) Calculate a parametric estimate of the PSD using AR models with $p = 1, 2, 3,$ and 5. Compare the parametric estimates with the periodogram. In your opinion, which order AR model is the best fit.

(c) Repeat parts (a) and (b) using 100 samples instead of 10.

10.38 Construct a signal plus noise random sequence using 10 samples of:

$$X[n] = \cos(2\pi n f_1 t_s) + \cos(2\pi n f_2 t_s) + N[n],$$

where $N[n]$ is a sequence of zero-mean, unit variance, IID Gaussian random variables, and $f_1 = 0.1/t_s = 100$ kHz, $f_2 = 0.4/t_s = 400$ kHz, and $t_s = 1\mu s$ is the time between samples of the process.

(a) Calculate the periodogram estimate of the PSD, $S_{XX}(f)$.

(b) Calculate a parametric estimate of the PSD using AR models with $p = 3, 4, 5, 6,$ and 7. Compare the parametric estimates with the periodogram. In your opinion, which order AR model is the best fit.

(c) Repeat parts (a) and (b) using 100 samples instead of 10.

Random Processes in Linear Systems

In this chapter, we consider the response of both continuous time and discrete-time linear systems to random processes, such as a signal plus noise. We develop statistical descriptions of the output of linear systems with random inputs by viewing the systems in both the time domain and the frequency domain. An engineering application section at the end of this chapter demonstrates how filters can be optimized for the purpose of enhancing signal-to-noise ratios. It is assumed that the reader is familiar with the study of linear time-invariant (LTI) systems. A brief overview is provided in Appendix C for those needing a refresher.

11.1 Continuous Time Linear Systems

Consider a linear time-invariant (LTI) system described by an impulse response $h(t)$ or a transfer function $H(f)$. If a random process, $X(t)$, is the input to this system, the output will also be random and is given by the convolution integral

$$Y(t) = \int_{-\infty}^{\infty} X(u)h(t-u)\mathrm{d}u. \tag{11.1}$$

We would like to statistically describe the output of the system. Ultimately, the joint PDF of any number of samples of the output would be nice. In general, this is a very difficult problem and hence we have to be satisfied with a simpler description. However, if the input process is Gaussian, then the output process will also be Gaussian, since any linear processing of Gaussian random variables (processes) produces new Gaussian random variables (processes). In that case, to completely describe the output of the system, we need merely to compute the mean and the autocovariance (or autocorrelation) function of the output. Even if the processes involved are not Gaussian, the mean and autocorrelation functions will serve as a good start toward describing the process. Therefore, our first goal will be to specify the mean and autocorrelation functions of the output of an LTI system with a random input.

Probability and Random Processes. DOI: 10.1016/B978-0-12-386981-4.00011-5
© 2012 by Elsevier Inc. All rights reserved.

To start with, consider the mean function of the output:

$$\mu_Y(t) = E[Y(t)] = E\left[\int_{-\infty}^{\infty} X(u)h(t-u)du\right]$$

$$= \int_{-\infty}^{\infty} E[X(u)]h(t-u)du = \int_{-\infty}^{\infty} \mu_X(u)h(t-u)du. \tag{11.2}$$

Therefore, the output mean is the convolution of the input mean process with the impulse response of the system. For the special case when the input is wide sense stationary (WSS) and the input mean function is thereby constant, the output mean function becomes

$$\mu_Y(t) = \mu_X\int_{-\infty}^{\infty} h(t-u)du = \mu_X\int_{-\infty}^{\infty} h(s)ds = \mu_X H(0). \tag{11.3}$$

Note that the mean function of the output is also constant provided the input mean is constant.

The autocorrelation function of the output is calculated in a similar manner.

$$R_{YY}(t_1, t_2) = E[Y(t_1)Y(t_2)] = E\left[\left(\int_{-\infty}^{\infty} X(u)h(t_1-u)du\right)\left(\int_{-\infty}^{\infty} X(v)h(t_2-v)dv\right)\right] \tag{11.4}$$

$$= \int_{-\infty}^{\infty}\int_{-\infty}^{\infty} E[X(u)X(v)]h(t_1-u)h(t_2-v)dudv$$

$$= \int_{-\infty}^{\infty}\int_{-\infty}^{\infty} R_{XX}(u, v)h(t_1-u)h(t_2-v)dudv .$$

For WSS inputs, this expression can be simplified a little by using the fact that $R_{XX}(u, v) = R_{XX}(v-u)$. The output autocorrelation function is then

$$R_{YY}(t_1, t_2) = \int_{-\infty}^{\infty}\int_{-\infty}^{\infty} R_{XX}(v-u)h(t_1-u)h(t_2-v)dudv . \tag{11.5}$$

Although it may not appear like it from this expression, here, the output autocorrelation function is also a function of time difference only. To see this, perform the change of variables $s = t_1 - u$ and $w = t_2 - v$. Then Equation (11.5) becomes

$$R_{YY}(t_1, t_2) = \int_{-\infty}^{\infty}\int_{-\infty}^{\infty} R_{XX}(t_2-t_1+s-w)h(s)h(w)dwds. \tag{11.6}$$

Now it is clear that $R_{YY}(t_1, t_2) = R_{YY}(t_2-t_1)$. To write this result in a more compact form, note that the inner integral in Equation (11.6) can be expressed as

$$\int_{-\infty}^{\infty} R_{XX}(t_2 - t_1 + s - w)h(w)\mathrm{d}w = R_{XX}(t)*h(t)\Big|_{t = t_2 - t_1 + s}, \tag{11.7}$$

where $*$ denotes convolution. Let $g(t) = R_{XX}(t)*h(t)$, then the output autocorrelation can be expressed as

$$R_{YY}(t_2 - t_1) = \int_{-\infty}^{\infty} g(t_2 - t_1 + s)h(s)\mathrm{d}s = g(t)*h(-t)\Big|_{t = t_2 - t_1}. \tag{11.8}$$

Putting all these results together, we get

$$R_{YY}(\tau) = R_{XX}(\tau)*h(\tau)*h(-\tau). \tag{11.9}$$

Thus, the output autocorrelation function is found by a double convolution. The presence of the double convolution in the previous equation begs for an equivalent frequency domain representation. Taking Fourier transforms of both sides gives an expression for the power spectral density (PSD) of the output of the filter in terms of the input PSD:

$$S_{YY}(f) = S_{XX}(f)H(f)H^*(f) = S_{XX}(f)|H(f)|^2. \tag{11.10}$$

The term $|H(f)|^2$ is sometimes referred to as the *power transfer function* because it describes how the power is transferred from the input to the output of the system. In summary, we have shown the following results.

Theorem 11.1: Given an LTI system with impulse response $h(t)$ or transfer function $H(f)$ and a random input process $X(t)$, the mean and autocorrelation functions of the output process, $Y(t)$, can be described by

$$\mu_Y(t) = \mu_X(t)*h(t), \tag{11.11a}$$

$$R_{YY}(t_1, t_2) = \int_{-\infty}^{\infty} \int_{-\infty}^{\infty} R_{XX}(u, v)h(t_1 - u)h(t_2 - v)\mathrm{d}u\mathrm{d}v. \tag{11.11b}$$

Furthermore, if $X(t)$ is WSS, then $Y(t)$ is also WSS with

$$\mu_Y = \mu_X H(0), \tag{11.12a}$$

$$R_{YY}(\tau) = R_{XX}(\tau)*h(\tau)*h(-\tau), \tag{11.12b}$$

$$S_{YY}(f) = S_{XX}(f)|H(f)|^2. \tag{11.12c}$$

At times, it is desirable to specify the relationship between the input and output of a filter. Toward that end, we can calculate the cross-correlation function between the input and output.

$$R_{XY}(t_1, t_2) = E[X(t_1)Y(t_2)] = E\left[X(t_1)\int_{-\infty}^{\infty} X(u)h(t_2-u)du\right] \quad (11.13)$$

$$= \int_{-\infty}^{\infty} E[X(t_1)X(u)]h(t_2-u)du = \int_{-\infty}^{\infty} R_{XX}(t_1, u)h(t_2-u)du$$

$$= \int_{-\infty}^{\infty} R_{XX}(t_1, t_2-v)h(v)dv.$$

If $X(t)$ is WSS, then this simplifies to

$$R_{XY}(\tau) = \int_{-\infty}^{\infty} R_{XX}(\tau-v)h(v)dv = R_{XX}(\tau)*h(\tau). \quad (11.14)$$

In a similar manner, it can be shown that

$$R_{YX}(\tau) = R_{XX}(\tau)*h(-\tau). \quad (11.15)$$

In terms of cross-spectral densities, these equations can be written as

$$S_{XY}(f) = S_{XX}(f)H(f) \text{ and } S_{YX}(f) = S_{XX}(f)H^*(f). \quad (11.16)$$

■ Example 11.1:

White Gaussian noise, $N(t)$, with a PSD of $S_{NN}(f) = N_0/2$ is input to an RC lowpass filter (LPF). Such a filter will have a transfer function and impulse response given by

$$H(f) = \frac{1}{1+j2\pi fRC} \text{ and } h(t) = \frac{1}{RC}\exp\left(-\frac{t}{RC}\right)u(t),$$

respectively. If the input noise is zero-mean, $\mu_N = 0$, then the output process will also be zero-mean, $\mu_Y = 0$. Also

$$S_{YY}(f) = S_{NN}(f)|H(f)|^2 = \frac{N_0/2}{1+(2\pi fRC)^2}.$$

Using inverse Fourier transforms, the output autocorrelation is found to be

$$R_{YY}(\tau) = \frac{N_0}{4RC}\exp\left(-\frac{|\tau|}{RC}\right).$$

■

■ Example 11.2:

Suppose we wish to convert a white noise process from continuous time to discrete time using a sampler. Since white noise has infinite power, it cannot be sampled directly and must be filtered first. Suppose for simplicity we use an ideal LPF of bandwidth B to

Figure 11.1
Block diagram of a sampling system to convert white noise from continuous
time to discrete time.

perform the sampling so that the system is as illustrated in Figure 11.1. Let $N_f(t)$ be the
random process at the output of the LPF. This process has a PSD of

$$S_{N_f N_f}(f) = \frac{N_0}{2}\mathrm{rect}\!\left(\frac{f}{2B}\right) = \begin{cases} N_0/2, & |f| < B, \\ 0, & \text{otherwise.} \end{cases}$$

The corresponding autocorrelation function is

$$R_{N_f N_f}(\tau) = N_0 B\,\mathrm{sinc}(2B\tau).$$

If the output of the filter is sampled every t_0 seconds, the discrete-time noise process will
have an autocorrelation of $R_{NN}[k] = N_0 B\,\mathrm{sinc}(2kBt_0)$. If the discrete-time output process
$N[n]$ is to be white, then we want all samples to be uncorrelated. That is, we want
$R_{NN}[k] = 0$ for all $k \neq 0$. Recall that the sinc function has nulls whenever its argument is
an integer. Thus, the discrete-time process will be white if (and only if) $2Bt_0$ is an integer.
In terms of the sampling rate, $f_0 = 1/t_0$, for the discrete-time process to be white, the
sampling rate must be $f_0 = 2B/m$ for some integer m. ∎

11.2 Discrete-Time Linear Systems

The response of a discrete-time linear system to a (discrete-time) random process is found
using virtually identical techniques to those used with continuous time systems. As such, we
do not repeat the derivations here, but rather summarize the relevant results. We start with a
linear system described by the difference equation

$$\sum_{i=0}^{p} a_i Y[n-i] = \sum_{k=0}^{q} b_k X[n-k], \tag{11.17}$$

where $X[n]$ is the input to the system and $Y[n]$ is the output. The reader might recognize this
system as producing an autoregressive moving average process as described in Section 10.4.
This system can be described by a transfer function expressed using either z-transforms or
discrete-time Fourier transforms (DTFT) as

$$H(z) = \frac{\sum_{k=0}^{q} b_k z^{-k}}{\sum_{i=0}^{p} a_i z^{-i}} \quad \text{or} \quad H(f) = \frac{\sum_{k=0}^{q} b_k e^{-j2\pi kf}}{\sum_{i=0}^{p} a_i e^{-j2\pi if}} . \tag{11.18}$$

If the DTFT is used, it is understood that the frequency variable f is actually a normalized frequency (normalized by the sampling rate). The system can also be described in terms of a discrete-time impulse response, $h[n]$, which can be found through either an inverse z-transform or an inverse DTFT. The following results apply to any discrete-time system described by an impulse response, $h[n]$, and transfer function, $H(f)$.

Theorem 11.2: Given a discrete-time LTI system with impulse response $h[n]$ and transfer function $H(f)$, and a random input process $X[n]$, the mean and autocorrelation functions of the output process, $Y[n]$, can be described by

$$\mu_Y[n] = \mu_X[n] * h[n], \tag{11.19a}$$

$$R_{YY}[n_1, n_2] = \sum_{k_1 = -\infty}^{\infty} \sum_{k_2 = -\infty}^{\infty} R_{XX}[k_1, k_2] h[n_1 - k_1] h[n_2 - k_2]. \tag{11.19b}$$

Furthermore, if $X[n]$ is WSS, then $Y[n]$ is also WSS with

$$\mu_Y = \mu_X H(0), \tag{11.20a}$$

$$R_{YY}[n] = R_{XX}[n] * h[n] * h[-n], \tag{11.20b}$$

$$S_{YY}(f) = S_{XX}(f)|H(f)|^2. \tag{11.20c}$$

Again, it is emphasized that the frequency variable in the PSD of a discrete-time process is to be interpreted as frequency normalized by the sampling rate.

■ Example 11.3:

A discrete-time Gaussian white noise process has zero-mean and an autocorrelation function of $R_{XX}[n] = \sigma^2 \delta[n]$. This process is input to a system described by the difference equation

$$Y[n] = aY[n-1] + bX[n].$$

Note that this produces an AR(1) process at the output. The transfer function and impulse response of this system are

$$H(f) = \frac{b}{1 - ae^{-j2\pi f}} \quad \text{and} \quad h[n] = ba^n u[n],$$

respectively, assuming that $|a| < 1$. The autocorrelation and PSD functions of the output process are

$$R_{YY}[n] = \frac{b^2 a^{|n|}}{1 - a^2} \quad \text{and} \quad S_{YY}(f) = \frac{\sigma^2 b^2}{|1 - ae^{-j2\pi f}|^2} = \frac{\sigma^2 b^2}{1 + a^2 - 2a\cos(2\pi f)},$$

respectively.

11.3 Noise Equivalent Bandwidth

Consider an ideal LPF with a bandwidth B whose transfer function is shown in Figure 11.2. Suppose white Gaussian noise with PSD $N_0/2$ is passed through this filter. The total output power would be $P_o = N_o B$. For an arbitrary LPF, the output noise power would be

$$P_o = \frac{N_0}{2} \int_{-\infty}^{\infty} |H(f)|^2 df. \tag{11.21}$$

One way to define the bandwidth of an arbitrary filter is to construct an ideal LPF that produces the same output power as the actual filter. This results in the following definition of bandwidth known as *noise equivalent bandwidth*.

Definition 11.1: The *noise equivalent bandwidth* of a LPF with transfer function $H(f)$ is

$$B_{\text{neq}} = \frac{1}{2|H(0)|^2} \int_{-\infty}^{\infty} |H(f)|^2 df = \frac{1}{|H(0)|^2} \int_{0}^{\infty} |H(f)|^2 df. \tag{11.22}$$

Figure 11.2
Power transfer function of an arbitrary and ideal LPF. $B = B_{\text{neq}}$ if areas under the two curves are equal.

This definition needs to be slightly adjusted for bandpass filters (BPFs). If the center of the passband is taken to be at some frequency, f_o, then the noise equivalent bandwidth is

$$B_{neq} = \frac{1}{2|H(f_o)|^2} \int_{-\infty}^{\infty} |H(f)|^2 df = \frac{1}{|H(f_o)|^2} \int_{0}^{\infty} |H(f)|^2 df. \tag{11.23}$$

■ Example 11.4:

Consider the RC LPF whose transfer function is

$$H(f) = \frac{1}{1 + j2\pi fRC}.$$

The noise equivalent bandwidth of this filter is

$$B_{neq} = \int_{0}^{\infty} \frac{1}{1 + (2\pi fRC)^2} df = \frac{1}{2\pi RC} \tan^{-1}(u) \Big|_{0}^{\infty} = \frac{1}{4RC}.$$

In addition to using the noise equivalent bandwidth, the definitions in Section 10.3 presented for calculating the bandwidth of a random process can also be applied to find the bandwidth of a filter. For example, the absolute bandwidth and the RMS bandwidth of this filter are both infinite while the 3 dB (half power) bandwidth of this filter is

$$B_{3dB} = \frac{1}{2\pi RC},$$

which for this example is slightly smaller than the noise equivalent bandwidth.
■

11.4 Signal-to-Noise Ratios

Often the input to a linear system will consist of signal plus noise, namely,

$$X(t) = S(t) + N(t), \tag{11.24}$$

where the signal part can be deterministic or a random process. We can invoke linearity to show that the mean process of the output can be viewed as a sum of the mean due to the signal input alone plus the mean due to the noise input alone. That is,

$$\mu_Y(t) = \mu_S(t)*h(t) + \mu_N(t)*h(t). \tag{11.25}$$

In most cases, the noise is taken to be zero-mean, in which case the mean at the output is due to the signal part alone.

When calculating the autocorrelation function of the output, we cannot invoke superposition since autocorrelation is not a linear operation. First, we calculate the autocorrelation function of the signal plus noise input.

$$R_{XX}(t_1, t_2) = E[(S(t_1) + N(t_1))(S(t_2) + N(t_2))]. \tag{11.26}$$

If the signal and noise part are independent, which is generally a reasonable assumption, and the noise is zero-mean, then this autocorrelation becomes

$$R_{XX}(t_1, t_2) = R_{SS}(t_1, t_2) + R_{NN}(t_1, t_2), \tag{11.27}$$

or, assuming all processes involved are WSS,

$$R_{XX}(\tau) = R_{SS}(\tau) + R_{NN}(\tau). \tag{11.28}$$

As a result, the PSD of the output can be written as

$$S_{YY}(f) = S_{XX}(f)|H(f)|^2 = S_{SS}(f)|H(f)|^2 + S_{NN}(f)|H(f)|^2, \tag{11.29}$$

which is composed of two terms, namely that due to the signal and that due to the noise. We can then calculate the output power due to the signal part and the output power due to the noise part.

Definition 11.2: The *signal-to-noise ratio* (SNR) for a signal comprises the sum of a desired (signal) part and a noise part is defined as the ratio of the power of the signal part to the power (variance) of the noise part. That is, for $X(t) = S(t) + N(t)$,

$$\text{SNR} = \frac{E[S^2(t)]}{E[N^2(t)]} = \frac{R_{SS}(0)}{R_{NN}(0)} = \frac{\int_{-\infty}^{\infty} S_{SS}(f)\,\mathrm{d}f}{\int_{-\infty}^{\infty} S_{NN}(f)\,\mathrm{d}f}. \tag{11.30}$$

■ Example 11.5:

Suppose the input to the RC LPF of the previous example consists of a sinusoidal signal plus white noise. That is, let the input be $X(t) = S(t) + N(t)$, where $N(t)$ is white Gaussian noise as in the previous example and $S(t) = a\cos(\omega_0 t + \Theta)$, where Θ is a uniform random variable over $[0, 2\pi)$ that is independent of the noise. The output can be written as $Y(t) = S_o(t) + N_o(t)$, where $S_o(t)$ is the output due to the sinusoidal signal input and $N_o(t)$ is the output due to the noise. The signal output can be expressed as

$$S_o(t) = a|H(f_o)|\cos(\omega_0 t + \angle H(f_o) + \Theta),$$

and the power in this sinusoidal signal is

$$R_{S_o S_o}(0) = \frac{a^2 |H(f_o)|^2}{2}.$$

(Continued)

From the results of Example 11.1, the noise power at the output is

$$R_{N_{o}N_{o}}(0) = \frac{N_{o}}{4RC}.$$

Therefore, the SNR of the output of the RC LPF is

$$\text{SNR} = \frac{2a^{2}RC|H(f_{o})|^{2}}{N_{o}} = \frac{2a^{2}RC}{N_{o}(1 + (2\pi f_{o}RC)^{2})}.$$

Suppose we desire to adjust the RC time constant (or, equivalently, adjust the bandwidth) of the filter so that the output SNR is optimized. Differentiating with respect to the quantity RC, setting equal to zero and solving the resulting equation produces the optimum time constant

$$RC_{\text{opt}} = \frac{1}{2\pi f_{o}}.$$

Stated another way, the 3 dB frequency of the RC filter is set equal to the frequency of the input sinusoid in order to optimize output SNR. The resulting optimum SNR is

$$\text{SNR}_{\text{opt}} = \frac{a^{2}}{2\pi N_{o}f_{o}}.$$

∎

11.5 The Matched Filter

Suppose we are given an input process consisting of a (known, deterministic) signal plus an independent white noise process (with a PSD of $N_{o}/2$). It is desired to filter out as much of the noise as possible while retaining the desired signal. The general system is shown in Figure 11.3. The input process $X(t) = s(t) + N(t)$ is to be passed through a filter with impulse response $h(t)$ that produces an output process $Y(t)$. The goal here is to design the filter to maximize the SNR at the filter output. Due to the fact that the input process is not necessarily stationary, the output process may not be stationary and therefore the output SNR may be time varying. Hence, we must specify at what point in time we want the SNR to be maximized. Picking an arbitrary sampling time, t_{o}, for the output process, we desire to design the filter such that the SNR is maximized at time $t = t_{o}$.

Let Y_{o} be the value of the output of the filter at time t_{o}. This random variable can be expressed as

$$Y_{o} = Y(t_{o}) = s(t)*h(t)\big|_{t = t_{o}} + N_{Y}(t_{o}), \qquad (11.31)$$

Figure 11.3
Linear system for filtering noise from a desired signal.

where $N_Y(t)$ is the noise process out of the filter. The powers in the signal and noise parts, respectively, are given by

$$\text{signal power} = [s(t)*h(t)|_{t=t_0}]^2 = \left[\int_{-\infty}^{\infty} h(u)s(t_0-u)du\right]^2, \qquad (11.32)$$

$$\text{noise power} = \frac{N_0}{2}\int_{-\infty}^{\infty}|H(f)|^2 df = \frac{N_0}{2}\int_{-\infty}^{\infty} h^2(t)dt. \qquad (11.33)$$

The SNR is then expressed as the ratio of these two quantities,

$$\text{SNR} = \frac{2}{N_0}\frac{\left[\displaystyle\int_{-\infty}^{\infty} h(u)s(t_0-u)du\right]^2}{\displaystyle\int_{-\infty}^{\infty} h^2(t)dt}. \qquad (11.34)$$

We seek the impulse response (or equivalently the transfer function) of the filter that maximizes the SNR as given in Equation (11.34). To simplify this optimization problem, we use Schwarz's inequality, which states that

$$\left|\int_{-\infty}^{\infty} x(t)y(t)dt\right|^2 \leq \int_{-\infty}^{\infty}|x(t)|^2 dt \int_{-\infty}^{\infty}|y(t)|^2 dt, \qquad (11.35)$$

where equality holds if and only if $x(t) \propto y(t)$ [1]. Applying this result to the expression for SNR produces an upper bound on the SNR:

$$\text{SNR} \leq \frac{2}{N_0}\frac{\displaystyle\int_{-\infty}^{\infty}|h(t)|^2 dt \int_{-\infty}^{\infty}|s(t_0-t)|^2 dt}{\displaystyle\int_{-\infty}^{\infty} h^2(t)dt} = \frac{2}{N_0}\int_{-\infty}^{\infty}|s(t_0-t)|^2 dt$$

$$= \frac{2}{N_0}\int_{-\infty}^{\infty}|s(t)|^2 dt = \frac{2E_s}{N_0}, \qquad (11.36)$$

where E_s is the energy in the signal $s(t)$. Furthermore, this maximum SNR is achieved when $h(t) \propto s(t_0-t)$. In terms of the transfer function, this relationship is expressed as $H(f) \propto S^*(f)e^{-j2\pi f t_0}$. The filter that maximizes the SNR is referred to as a *matched filter*

---

[1] The notation $x(t) \propto y(t)$ means that $x(t)$ is proportional to $y(t)$.

since the impulse response is matched to that of the desired signal. These results are summarized in the following theorem.

Theorem 11.3: If an input to an LTI system characterized by an impulse response, $h(t)$, is given by $X(t) = s(t) + N(t)$ where $N(t)$ is a white noise process, then a matched filter will maximize the output SNR at time t_o. The impulse response and transfer function of the matched filter are given by

$$h(t) = s(t_o - t) \text{ and } H(f) = S^*(f)e^{-j2\pi f t_o}. \tag{11.37}$$

Furthermore, if the white noise at the input has a PSD of $S_{NN}(f) = N_o/2$, then the optimum SNR produced by the matched filter is

$$\text{SNR}_{\text{max}} = \frac{2E_s}{N_o}, \tag{11.38}$$

where E_s is the energy in the signal $s(t)$.

■ Example 11.6:

A certain communication system transmits a square pulse given by

$$s(t) = \begin{cases} 1, & 0 \le t < t_1, \\ 0, & \text{otherwise.} \end{cases}$$

This signal is received in the presence of white noise at a receiver producing the received process $R(t) = s(t) + N(t)$. The matched filter that produces the optimum SNR at time t_o for this signal has an impulse response of the form

$$h(t) = s(t_o - t) = \begin{cases} 1, & t_o - t_1 < t \le t_o, \\ 0, & \text{otherwise.} \end{cases}$$

The output of the matched filter is then given by

$$Y(t) = h(t) * R(t) = \int_{-\infty}^{\infty} h(t-u)R(u)\,du = \int_{-\infty}^{\infty} s(u+t_o-t)R(u)\,du = \int_{t-t_o}^{t-t_o+t_1} R(u)\,du.$$

Therefore, the matched filter for a square pulse is just a finite time integrator. The matched filter simply integrates the received signal for a period of time equal to the width of the pulse. When sampled at the correct point in time, the output of this integrator will produce the maximum SNR. The operation of this filter is illustrated in Figure 11.4.

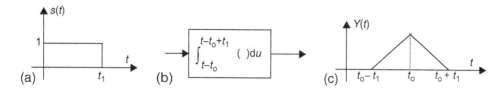

Figure 11.4
(a) A square pulse, (b) the corresponding matched filter, and (c) the response of the matched filter to the square pulse.

Example 11.7:

In this example, we expand on the results of the previous example and consider a sequence of square pulses with random (binary) amplitudes as might be transmitted in a typical communication system. Suppose this signal is corrupted by white Gaussian noise and we must detect the transmitted bits. That is, we must determine whether each pulse sent has a positive or negative amplitude. Figure 11.5a shows both the square pulse train and the same signal corrupted by noise. Note that by visually observing the signals, it is very difficult to make out the original signal from the noisy version. We attempt to clean up this signal by passing it through the matched filter from Example 11.6. In the absence of noise, we would expect to see a sequence of overlapping triangular pulses. The matched filter output both with and without noise is illustrated in Figure 11.5b. Notice that a great deal of noise has been eliminated by the matched filter. To detect the data bits, we would sample the matched filter output at the end of each bit interval (shown by circles in the plot) and use the sign of the sample to be the estimate of the transmitted data bit. In this example, all of our decisions would be correct. The MATLAB code used to generate these signals follows. This is just a modified version of the code used to generate the PAM signal in Example 10.14.

```
N=10;                       % No. of bit intervals.
Ns=25;                      % No. of time samples per bit.
t=([1:Ns]-0.5)/Ns;          % time axis for pulse shape
p=ones(size(t));            % square pulse shape
d=sign(rand(1,N)-0.5);      % random data bits
X(1:Ns:(Ns*(N-1)+1))=d;
X=conv(X,p);                % PAM signal with pulse shape.
sigma=sqrt(Ns/10);          % noise strength
noise=sigma*randn(size(X)); % Gaussian noise
```

(Continued)

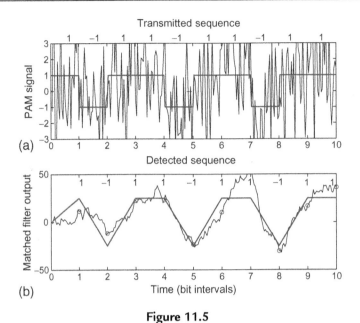

Figure 11.5
(a) A binary PAM signal with and without additive noise along with (b) the result of passing both signals through a matched filter.

```
R=X+noise;
subplot(2,1,1)                          % plot clean and noisy signal
x_axis=[1:length(R)]/Ns;
plot(x_axis,R, x_axis,X)
axis([0 N -3 3])
h=fliplr(p);                            %  matched  filter  impulse
response
z=conv(R,h);                            % noisy output of matched fil-
ter
z2=conv(X,h);                           % noise free MF output
zs=z(Ns*[1:N]);                         % sample matched filter out-
puts
subplot(2,1,2)                          % plot matched filter output
x_axis=[1:length(z)]/Ns;
plot(x_axis,z,'-',x_axis,z2,'-',[1:N],zs,'o')
```

■

11.6 The Wiener Filter

In this section, we consider another filter design problem that involves removing the noise from a sum of a desired signal plus noise. In this case, the desired signal is also a random process (rather than a known, deterministic signal as in the last section) and the goal here is to estimate the desired part of the signal plus noise. In its most general form, the problem is stated as follows. Given a random process $X(t)$, we want to form an estimate $Y(t)$ of some other zero-mean process $Z(t)$ based on observation of some portion of $X(t)$. We require the estimator to be linear. That is, we will obtain $Y(t)$ by filtering $X(t)$. Hence,

$$Y(t) = \int_{-\infty}^{\infty} h(t-u)X(u)\mathrm{d}u. \qquad (11.39)$$

We want to design the filter to minimize the mean square error

$$E[\varepsilon^2(t)] = E[(Z(t) - Y(t))^2]. \qquad (11.40)$$

In this section, we will consider the special case where the observation consists of the process we are trying to estimate plus independent noise. That is $X(t) = Z(t) + N(t)$. We observe $X(t)$ for some time interval $t \in (t_1, t_2)$ and based on that observation, we will form an estimate of $Z(t)$. Consider a few special cases:

- *Case I:* If $(t_1, t_2) = (-\infty, t)$, then we have a filtering problem in which we must estimate the present based on the entire past. We may also have $(t_1, t_2) = (t - t_o, t)$ in which case we have a filtering problem where we must estimate the present based on the most recent past.
- *Case II:* If $(t_1, t_2) = (-\infty, \infty)$, then we have a smoothing problem where we must estimate the present based on a noisy version of the past, present, and future.
- *Case III:* If $(t_1, t_2) = (-\infty, t - t_o)$, then we have a prediction problem where we must estimate the future based on the past and present.

All of these cases can be cast in the same general framework, and a single result will describe the optimal filter for all cases. In order to derive the optimal filter, it is easier to view the problem in discrete time and then ultimately pass to the limit of continuous time. Thus, we reformulate the problem in discrete time. Given an observation of the discrete-time process $X[n] = Z[n] + N[n]$ over some time interval $n \in [n_1, n_2]$, we wish to design a filter $h[n]$ such that the linear estimate

$$Y[n] = \sum_{k=n_1}^{n_2} h[n-k]X[k] \qquad (11.41)$$

minimizes the mean square error $E[\varepsilon^2[n]] = E[(Z[n] - Y[n])^2]$.

The filter $h[n]$ can be viewed as a sequence of variables. We seek to jointly optimize with respect to each variable in that sequence. This can be done by differentiating with respect to each variable and setting the resulting equations equal to zero:

$$\frac{\mathrm{d}}{\mathrm{d}h[m]}E[\varepsilon^2[n]] = 2E\left[\varepsilon[n]\frac{\mathrm{d}}{\mathrm{d}h[m]}\varepsilon[n]\right] = 0, \quad m \in [n - n_2, n - n_1]. \qquad (11.42)$$

Noting that

$$\frac{d}{dh[m]}\varepsilon[n] = \frac{d}{dh[m]}(Z[n] - Y[n]) = -\frac{d}{dh[m]}Y[n] = -X[n-m], \quad (11.43)$$

the system of equations to solve becomes

$$E[\varepsilon[n]X[n-m]] = 0, \quad \text{for } m \in [n-n_1, n-n_2]. \quad (11.44)$$

Equivalently, this can be rewritten as

$$E[\varepsilon[n]X[m]] = 0, \quad \text{for } m \in [n_1, n_2]. \quad (11.45)$$

In summary, the filter that minimizes the mean square error will cause the observed data to be orthogonal to the error. This is the *orthogonality principle* that was previously developed in Chapter 6, Section 6.5.3. Applying the orthogonality principle, we have

$$E[\varepsilon[n]X[m]] = E\left[X[m]\left(Z[n] - \sum_{k=n_1}^{n_2} h[n-k]X[k]\right)\right] = 0. \quad (11.46)$$

Assuming all the processes involved are jointly WSS, these expectations can be written in terms of autocorrelation and cross-correlation functions as

$$\sum_{k=n_1}^{n_2} h[n-k]R_{XX}[k-m] = R_{XZ}[n-m], \quad m \in [n_1, n_2], \quad (11.47)$$

or equivalently (making the change of variables $i = n-k$ and $j = n-m$),

$$\sum_{i=n-n_2}^{n-n_1} h[i]R_{XX}[j-i] = R_{XZ}[j], \quad j \in [n-n_2, n-n_1]. \quad (11.48)$$

These equations are known as the *Wiener–Hopf equations*, the *normal equations*, or the *Yule–Walker* equations. The resulting filter found by solving this system of equations is known as the *Wiener filter*.

A similar result can be found for continuous time systems by applying the orthogonality principle in continuous time. Given an observation of $X(t)$ over the time interval (t_1, t_2), the orthogonality principle states that the filter which minimizes the mean square prediction error will satisfy

$$E[\varepsilon(t)X(s)] = 0, \quad \text{for } s \in (t_1, t_2). \quad (11.49)$$

This produces the continuous time version of the Wiener–Hopf equation

$$\int_{t-t_2}^{t-t_1} h(v)R_{XX}(\tau - v)dv = R_{XZ}(\tau), \quad \tau \in (t - t_2, t - t_1). \tag{11.50}$$

The techniques used to solve the Wiener–Hopf equation depend on the nature of the observation interval. For example, consider the smoothing problem where the observation interval is $(t_1, t_2) = (-\infty, \infty)$. In that case, the Wiener–Hopf equation becomes

$$\int_{-\infty}^{\infty} h(v)R_{XX}(\tau - v)dv = R_{XZ}(\tau). \tag{11.51}$$

The left hand side of the equation is a convolution and the integral equation can easily be solved using Fourier transforms. Taking a Fourier transform of both sides of the equation results in

$$H(f)S_{XX}(f) = S_{XZ}(f) \Rightarrow H(f) = \frac{S_{XZ}(f)}{S_{XX}(f)}. \tag{11.52}$$

Note also that if the noise is zero-mean and independent of $Z(t)$, then $R_{XZ}(\tau) = R_{ZZ}(\tau)$ and $R_{XX}(\tau) = R_{ZZ}(\tau) + R_{NN}(\tau)$. The transfer function of the Wiener filter for the smoothing problem then becomes

$$H(f) = \frac{S_{ZZ}(f)}{S_{ZZ}(f) + S_{NN}(f)}. \tag{11.53}$$

■ Example 11.8:

Suppose the desired signal $Z(t)$ has a spectral density of

$$S_{ZZ}(f) = \frac{1}{1 + f^2}$$

and the noise is white with a PSD of $S_{NN}(f) = 1$. Then the Wiener filter for the smoothing problem has the form

$$H(f) = \frac{\dfrac{1}{1 + f^2}}{\dfrac{1}{1 + f^2} + 1} = \frac{1}{2 + f^2},$$

The corresponding impulse response is

$$h(t) = \frac{\pi}{\sqrt{2}}\exp(-\sqrt{8}\,\pi|t|).$$

Note that this filter is not causal. This is due to the nature of the smoothing problem, whereby we estimate the present based on past, present, and future.

■

Next, consider the filtering problem where the observation interval is $(t_1, t_2) = (-\infty, t)$. In this case, the Wiener–Hopf equation becomes

$$\int_0^\infty h(v)R_{XX}(\tau - v)dv = R_{XZ}(\tau), \qquad \tau \in (0, \infty). \tag{11.54}$$

It is emphasized now that the left hand side of the equation is not a convolution since the lower limit of the integral is not $-\infty$. The resulting integral equation is much trickier to solve than in the case of the smoothing problem. In order to develop a procedure for solving this general equation, consider the special case when $R_{XX}(\tau) = \delta(\tau)$. In that case, the above integral equation becomes

$$h(\tau) = R_{XZ}(\tau), \quad \text{for } \tau > 0. \tag{11.55}$$

Because we are estimating the present based on observing the past, the filter must be causal and thus its impulse response must be zero for negative time. Therefore, for the special case when $R_{XX}(\tau) = \delta(\tau)$, the Wiener filter is $h(\tau) = R_{XZ}(\tau)u(\tau)$.

This example in itself is not very interesting since we would not expect $X(t)$ to be white, but it does help to find the general solution to the Wiener–Hopf equation. First, before estimating $Z(t)$, suppose we pass the input $X(t)$ through a causal filter with a transfer function $1/G(f)$. Call the output $\tilde{X}(t)$. If $G(f)$ is chosen such that $|G(f)|^2 = S_{XX}(f)$, then the process $\tilde{X}(t)$ will be a white process and the filter is called a *whitening filter*. We can then use the result of the previous special case to estimate $Z(t)$ based on the white process $\tilde{X}(t)$. Hence, we are designing the Wiener filter in two stages as illustrated in Figure 11.6.

To find the impulse response of the second filter, we start with the result that $h_2(\tau) = R_{\tilde{X}Z}(\tau)u(\tau)$. Also, since $\tilde{X}(t)$ can be written as $\tilde{X}(t) = X(t)*h_1(t)$, then

$$S_{\tilde{X}Z}(f) = S_{XZ}(f)H_1^*(f). \tag{11.56}$$

The resulting quantities needed to form the second filter are then

$$S_{\tilde{X}Z}(f) = \frac{S_{XZ}(f)}{G^*(f)} \leftrightarrow R_{\tilde{X}Z}(\tau) = \mathscr{F}^{-1}\left\{\frac{S_{XZ}(f)}{G^*(f)}\right\}. \tag{11.57}$$

Whitening filter   Wiener filter for white input

Figure 11.6
Constructing the Wiener filter for the filtering problem as a cascade of two filters.

To construct the whitening filter, we need to find a $G(f)$ such that (1) $H_1(f) = 1/G(f)$ is causal and (2) $|G(f)|^2 = S_{XX}(f)$. The procedure for doing this is known as *spectral factorization*. Since $S_{XX}(f)$ is a PSD and thus is an even function of f, it will factor in the form

$$S_{XX}(f) = G(f)G^*(f),\tag{11.58}$$

where half of the poles and zeros are assigned to $G(f)$ and the other half are assigned to $G^*(f)$. In order to be sure that $H_1(f)$ is causal, we assign to $G(f)$ those zeros in the upper half plane. As will be shown in the next example, it is also important to assign poles from the upper half plane to $G(f)$ as well.

■ Example 11.9:

For this example, let $X(t) = Z(t) + N(t)$, where $N(t)$ is white noise with a spectral density of $S_{NN}(f) = 1$ and independent of $Z(t)$, which has a spectral density of

$$S_{ZZ}(f) = \frac{3}{1 + (2\pi f)^2}.$$

Note also that

$$S_{XZ}(f) = S_{ZZ}(f) \text{ and } S_{XX}(f) = S_{ZZ}(f) + S_{NN}(f) = \frac{3}{1 + (2\pi f)^2} + 1 = \frac{4 + (2\pi f)^2}{1 + (2\pi f)^2} = \frac{(2 + j2\pi f)(2 - j2\pi f)}{(1 + j2\pi f)(1 - j2\pi f)}.$$

A pole–zero plot for the PSD function $S_{XX}(f)$ is shown in Figure 11.7. We assign the poles and zeros in the upper–half plane to the function $G(f)$ (and hence the poles and zeros in the lower half plane go to $G^*(f)$). This results in

$$G(f) = \frac{2 + j2\pi f}{1 + j2\pi f} \Rightarrow H_1(f) = \frac{1 + j2\pi f}{2 + j2\pi f} = 1 - \frac{1}{2 + j2\pi f}.$$

The corresponding impulse response is

$$h_1(t) = \delta(t) - e^{-2t}u(t).$$

As desired, the whitening filter is causal. To find the form of the second filter, we first calculate the cross-spectral density between $\tilde{X}(t)$ and $Z(t)$,

Figure 11.7
Pole–zero plot for the PSD function of Example 11.9.

(Continued)

$$S_{\tilde{X}Z}(f) = S_{XZ}(f)H_1^*(f) = \frac{3}{1+(2\pi f)^2}\frac{1-j2\pi f}{2-j2\pi f} = \frac{3}{(1+j2\pi f)(2-j2\pi f)} = \frac{1}{1+j2\pi f}+\frac{1}{2-j2\pi f}.$$

Taking an inverse Fourier transform, the cross-correlation function is

$$R_{\tilde{X}Z}(\tau) = e^{-\tau}u(\tau)+e^{2\tau}u(-\tau).$$

The impulse response of the second filter is then given by

$$h_2(\tau) = R_{\tilde{X}Z}(\tau)u(\tau) = e^{-\tau}u(\tau).$$

When the actual Wiener filter is implemented, there is no reason the filter has to be implemented as a cascade of two filters. We did this for ease of determining the filter. Combining these two filters into one produces

$$H(f) = H_1(f)H_2(f) = \left(\frac{1+j2\pi f}{2+j2\pi f}\right)\left(\frac{1}{1+j2\pi f}\right) = \frac{1}{2+j2\pi f} \leftrightarrow h(t) = e^{-2t}u(t).$$

It can be easily verified that this filter does indeed satisfy the Wiener–Hopf equation. Note, we could have also chosen $G(f) = (2+j2\pi f)/(1-j2\pi f)$, and $H_1(f) = 1/G(f)$ would still have been causal:

$$H_1(f) = \frac{1-j2\pi f}{2+j2\pi f} = -1+\frac{3}{2+j2\pi f} \leftrightarrow h_1(t) = -\delta(t)+3e^{-2t}u(t).$$

In this case,

$$S_{\tilde{X}Z}(f) = S_{XZ}(f)H_1^*(f) = \frac{3}{1+(2\pi f)^2}\frac{1+j2\pi f}{2-j2\pi f} = \frac{3}{(1-j2\pi f)(2-j2\pi f)} = \frac{3}{1-j2\pi f}+\frac{-3}{2-j2\pi f} .$$
$$\leftrightarrow R_{\tilde{X}Z}(\tau) = 3e^{\tau}u(-\tau)-3e^{2\tau}u(-\tau)$$

This leads to $h_2(t) = 0$!! Thus, we see it is important to assign both poles and zeros from the upper half plane to $G(f)$.

■

Finally, we consider the prediction problem where we wish to estimate the value of $Z(t)$ based on observing $X(t)$ over the time interval $(-\infty, t-t_o)$. Applying the orthogonality principle, the appropriate form of the Wiener–Hopf equation for the prediction problem becomes

$$\int_{t_o}^{\infty} h(v)R_{XX}(\tau-v)dv = R_{XZ}(\tau), \qquad \tau \in (t_o, \infty). \tag{11.59}$$

This equation is solved using the same technique as with the filtering problem. First, the input is passed through a whitening filter and then Equation (11.59) for the case when the input process is white. The procedure for finding the whitening filter is exactly the same as before. The solution to Equation (11.59) when $R_{XX}(\tau) = \delta(\tau)$ is

$$h(\tau) = R_{XZ}(\tau)u(\tau - t_o). \tag{11.60}$$

In summary, the solution to the prediction problem is found by following these steps:

- Step 1. Factor the input PSD according to $S_{XX}(f) = \dfrac{1}{G(f)G^*(f)}$, where $G(f)$ contains all poles and zeros of $S_{XX}(f)$ that are in the upper half plane. The whitening filter is then specified by $H_1(f) = 1/G(f)$. Call $\tilde{X}(t)$ the output of the whitening filter when $X(t)$ is input.
- Step 2. Calculate the cross-correlation function, $R_{\tilde{X}Z}(\tau)$. The second stage of the Wiener filter is then specified by $h_2(\tau) = R_{\tilde{X}Z}(\tau)u(\tau - t_o)$.
- Step 3. The overall Wiener filter is found by combining these two filters, $H(f) = H_1(f)H_2(f)$.

It should be noted that the filtering problem can be viewed as a special case of the prediction problem when $t_o = 0$ and so this summary applies to the filtering problem as well.

■ Example 11.10:

In this example, we repeat the filter design of Example 11.9 for the case of the prediction problem. As before, we pick the whitening filter to be of the form

$$H_1(f) = \frac{1 + j2\pi f}{2 + j2\pi f} \quad \leftrightarrow \quad h_1(t) = \delta(t) - e^{-2t}u(t).$$

Again, the resulting cross-correlation function is then

$$R_{\tilde{X}Z}(\tau) = e^{-\tau}u(\tau) + e^{2\tau}u(-\tau).$$

Assuming that $t_o > 0$ so that the problem is indeed one of prediction, the impulse response of the second filter is

$$h_2(t) = R_{\tilde{X}Z}(t)u(t - t_o) = e^{-t}u(t - t_o).$$

(Continued)

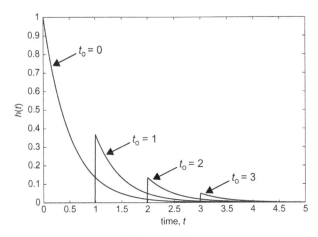

Figure 11.8
Impulse response of the Wiener prediction filter for Example 11.10.

The impulse response of the Wiener prediction filter is then

$$h(t) = h_1(t) * h_2(t) = [\delta(t) - e^{-2t}u(t)] * [e^{-t}u(t - t_0)] = e^{-2t + t_0}u(t - t_0).$$

This agrees with the result of the previous example when $t_0 = 0$. Figure 11.8 illustrates the impulse response of the Wiener prediction filter for several values of t_0.

∎

11.7 Bandlimited and Narrowband Random Processes

A random processes is said to be *bandlimited* if all of its frequency components are limited to some bandwidth, B. Specifically, if a random process $X(t)$ has a PSD function with an absolute bandwidth of B, then the process is said to be bandlimited to B Hz. For many bandlimited random processes, the frequency components are clustered at or near direct current (d.c.) Such a process is referred to as a *lowpass* random process. If, on the other hand, the frequency components of a random process are removed from d.c. and reside in some non-zero frequency band, the process is called a *bandpass* process. These distinctions are illustrated in Figure 11.9. For a bandpass process, in addition to the bandwidth, the location of the frequency band where the PSD is non-zero must also be specified. In the figure, the parameter f_0 describes that location. While often f_0 is taken to be the center of the band as illustrated in the figure, this does not have to be the case. The parameter f_0 can be chosen to be any convenient frequency within the band. In any event, f_0 is referred to as the *center frequency* of the band (even though it may not really be in the center). Finally, if a bandpass random process has a center frequency that is large compared to its bandwidth, $f_0 \gg B$, then the process is said to be *narrowband*.

Figure 11.9
The PSD functions of (a) a lowpass and (b) a bandpass random process.

Narrowband random processes frequently are found in the study of communication systems. For example, a commercial FM radio broadcast system uses channels with bandwidths of 200 kHz which are located near 100 MHz. Thus, the center frequency of an FM radio signal is about 500 times greater than its bandwidth. In the US digital cellular system, 30 kHz channels are used at frequencies near 900 MHz. In that case, the center frequencies are on the order of 30,000 times the bandwidth.

From studies of deterministic signals, the reader is probably aware that working with bandpass signals can be rather cumbersome. Trigonometric functions pop up everywhere and lead to seemingly endless usage of various identities. On the other hand, working with lowpass signals is often much simpler. To ease the complexity of working with bandpass signals, various representations have been formulated that allow bandpass signals to be decomposed into combinations of related lowpass signals. In this section, we focus on the most common of those decompositions, which is valid for narrowband signals. Generalizations of the following results are available for signals that are bandpass but not necessarily narrowband but will not be covered here.

Suppose a random process $Z(t)$ is narrowband. Then $Z(t)$ can be expressed in terms of two lowpass processes $X(t)$ and $Y(t)$ according to

$$Z(t) = X(t)\cos(\omega_0 t) - Y(t)\sin(\omega_0 t). \tag{11.61}$$

The two processes $X(t)$ and $Y(t)$ are referred to as the *inphase* (I) and *quadrature* (Q) components of $Z(t)$. Although it is not proven here, the equality in the previous equation is in the mean-square sense. That is,

$$E[\{Z(t) - (X(t)\cos(\omega_0 t) - Y(t)\sin(\omega_0 t))\}^2] = 0. \tag{11.62}$$

This Cartesian representation of the narrowband random process can also be replaced by a polar representation of the form

$$Z(t) = R(t)\cos(\omega_0 t + \Theta(t)), \qquad (11.63)$$

where $R(t)$ is called the *real envelope* of $Z(t)$ and $\Theta(t)$ is the *excess phase*. We next describe the relationship between the statistics of the I and Q components and the statistics of the original random process.

The I and Q components of a signal can be extracted using the system shown in Figure 11.10. The passbands of the LPF need to be large enough to pass the desired components (i.e., $>B/2$ Hz) but not so large as to pass the double frequency components produced by the mixers (at and around $2f_0$ Hz). For narrowband signals where $f_0 \gg B$, the filters can be very loosely designed and hence we do not need to worry too much about the particular forms of these filters. To see how this network functions, consider the output of the upper mixer.

$$2Z(t)\cos(\omega_0 t) = 2\{X(t)\cos(\omega_0 t) - Y(t)\sin(\omega_0 t)\}\cos(\omega_0 t)$$

$$= X(t)\{1 + \cos(2\omega_0 t)\} - Y(t)\sin(2\omega_0 t). \qquad (11.64)$$

After passing this through the LPF, the terms involving the double frequencies will be attenuated and the output of the upper LPF is indeed $X(t)$. Similar calculations reveal that $Y(t)$ is indeed the output of the lower branch.

Next, we calculate PSDs involving the I and Q components. Consider first, multiplication of the process $Z(t)$ by a sinusoid. Let $A(t) = 2Z(t)\cos(\omega_0 t)$. The autocorrelation function of $A(t)$ is easily calculated to be

$$R_{AA}(t, t + \tau) = 2R_{ZZ}(\tau)\{\cos(\omega_0 \tau) + \cos(2\omega_0 t + \omega_0 \tau)\}. \qquad (11.65)$$

Note that the process $A(t)$ is not WSS. In order to compute the PSD of a process which is not stationary, we must first take the time average of the autocorrelation function (with respect to t)

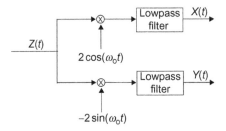

Figure 11.10
Network for decomposing a narrowband process into its I and Q components.

so that the result will be a function of τ only. This time-averaged autocorrelation function works out to be

$$\langle R_{AA}(t, t + \tau) \rangle = 2R_{ZZ}(\tau)\{ \cos(\omega_0 \tau) + \langle \cos(2\omega_0 t + \omega_0 \tau) \rangle \} = 2R_{ZZ}(\tau)\cos(\omega_0 \tau). \quad (11.66)$$

At this point, the PSD of $A(t)$ can then be found through Fourier transformation to be

$$S_{AA}(f) = S_{ZZ}(f - f_0) + S_{ZZ}(f + f_0). \quad (11.67)$$

Recall that the process $Z(t)$ was assumed to be narrowband. That is, its PSD has components near f_0 and $-f_0$. After shifting by f_0, the term $S_{ZZ}(f - f_0)$ has components near d.c. and also near $2f_0$. The components near d.c. will pass through the filter, while those at and around $2f_0$ will be attenuated. Similarly, $S_{ZZ}(f + f_0)$ will have terms near d.c. that will pass through the filter and terms near $-2f_0$ which will not. This is illustrated in Figure 11.11. For notational convenience, let L.P.{ } denote the lowpass part of a quantity. Then the PSD of the inphase component of a narrowband process can be written in terms of the PSD of the original process as

$$S_{XX}(f) = \text{L.P.}\{ S_{ZZ}(f - f_0) + S_{ZZ}(f + f_0) \}. \quad (11.68)$$

Following a similar set of steps, the PSD of the Q component is found to be identical to the I component. That is,

$$S_{YY}(f) = S_{XX}(f) = \text{L.P.}\{ S_{ZZ}(f - f_0) + S_{ZZ}(f + f_0) \}. \quad (11.69)$$

The cross-spectral density can also be calculated in a manner similar to the PSDs found above. The result is

$$S_{XY}(f) = j\text{L.P.}\{ S_{ZZ}(f - f_0) - S_{ZZ}(f + f_0) \}. \quad (11.70)$$

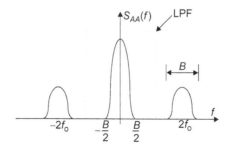

Figure 11.11
PSD of the input to the LPF in the I branch.

It is noted that if the PSD function $S_{ZZ}(f)$ is symmetric about $f = f_o$, then the cross-spectral density works out to be zero. In that case, the I and Q components are orthogonal (since $R_{XY}(\tau) = 0$). Furthermore, if the process $Z(t)$ is zero-mean, then the I and Q components will also be zero-mean. In this case, the I and Q components are then uncorrelated. Finally, if in addition, $Z(t)$ is a Gaussian random process, then the I and Q components are also statistically independent. In summary, we have proven the results of the following theorem.

Theorem 11.4: For a narrowband process $Z(t)$, the PSDs involving the I and Q components $X(t)$ and $Y(t)$ are given by

$$S_{YY}(f) = S_{XX}(f) = \text{L.P.}\{S_{ZZ}(f-f_o) + S_{ZZ}(f+f_o)\}, \qquad (11.71a)$$

$$S_{XY}(f) = j\text{L.P.}\{S_{ZZ}(f-f_o) - S_{ZZ}(f+f_o)\}. \qquad (11.71b)$$

If $Z(t)$ is a zero-mean Gaussian random process and its PSD is symmetric about $f = f_o$, then the I and Q components are statistically independent.

■ Example 11.11:

Suppose zero-mean white Gaussian noise with a PSD of $N_0/2$ is passed through an ideal BPF with a bandwidth of B Hz to produce the narrowband noise process $Z(t)$ as shown in Figure 11.12. The I and Q components will then have a PSD which is given by

$$S_{YY}(f) = S_{XX}(f) = N_0\text{rect}(f/B).$$

The corresponding autocorrelation functions are

$$R_{XX}(\tau) = R_{YY}(\tau) = N_0 B\text{sinc}(B\tau).$$

Since the PSD of $Z(t)$ is symmetric about $f = f_o$, the cross PSD is zero and therefore the I and Q components are independent.

Figure 11.12
(a) Generation of a narrowband noise process, (b) its PSD function, and (c) the PSD function of the I and Q components.

■

11.8 Complex Envelopes

When working with narrowband random processes, it is convenient to combine the I and Q components into a single lowpass random process whose real part is the I component and whose imaginary part is the Q component. The resulting random process is a complex lowpass random process.

> **Definition 11.3:** For a narrowband process, $Z(t)$, with I and Q components, $X(t)$ and $Y(t)$, respectively, the *complex envelope*, $G_z(t)$, is defined as
>
> $$G_z(t) = X(t) + jY(t). \qquad (11.72)$$

With this definition, the random process is written in terms of its complex envelope according to

$$Z(t) = \mathrm{Re}[G_z(t)e^{j\omega_o t}]. \qquad (11.73)$$

The properties developed in the previous section for the I and Q components of a narrowband random process can be used to determine equivalent properties for the complex envelope. To be consistent with the definitions for complex random variables given in Chapter 5, we define the autocorrelation function of a complex random process as follows.

> **Definition 11.4:** For any complex random process $G(t)$, the autocorrelation function is defined as[2]
>
> $$R_{GG}(t, t+\tau) = \frac{1}{2}E[G(t)G^*(t+\tau)]. \qquad (11.74)$$

If $G(t)$ represents the complex envelope of a narrowband random process and the I and Q components are jointly WSS, then this autocorrelation function will be a function only of τ. Also, the corresponding PSD can be found through Fourier transformation:

$$S_{GG}(f) = \mathscr{F}[R_{GG}(\tau)]. \qquad (11.75)$$

Using this definition together with the properties developed in the previous section, the autocorrelation function for a complex envelope is found to be

---

[2] The same disclaimer must be made here as in Definition 5.13. Many texts do not include the factor of $1/2$ in the definition of the autocorrelation function for complex random processes, and therefore the reader should be aware that there are two different definitions prevalent in the literature.

$$R_{GG}(\tau) = \frac{1}{2}E[(X(t)+jY(t))(X(t+\tau)-jY(t+\tau))]$$

$$= \frac{1}{2}R_{XX}(\tau) + \frac{1}{2}R_{YY}(\tau) + \frac{j}{2}R_{YX}(\tau) - \frac{j}{2}R_{XY}(\tau). \tag{11.76}$$

For the case where the I and Q components are orthogonal, this reduces to

$$R_{GG}(\tau) = \frac{1}{2}R_{XX}(\tau) + \frac{1}{2}R_{YY}(\tau) = R_{XX}(\tau) = R_{YY}(\tau). \tag{11.77}$$

The corresponding PSD is then

$$S_{GG}(f) = S_{XX}(f) = S_{YY}(f) = \text{L.P.}\{S_{ZZ}(f-f_o) + S_{ZZ}(f+f_o)\}. \tag{11.78}$$

Hence, the complex envelope has the same PSD and autocorrelation function as the I and Q components. It is left as an exercise for the reader to show that the autocorrelation and PSD functions of the original narrowband process can be found from those of the complex envelope according to

$$S_{ZZ}(f) = \frac{1}{2}S_{GG}(f-f_o) + \frac{1}{2}S_{GG}(-f-f_o), \tag{11.79}$$

$$R_{ZZ}(\tau) = \text{Re}[R_{GG}(\tau)e^{j\omega_o\tau}]. \tag{11.80}$$

11.9 Engineering Application: An Analog Communication System

A block diagram of a simple amplitude modulation (AM) analog communication system is shown in Figure 11.13. A message (usually voice or music) is represented as a random process $X(t)$ with some bandwidth B. This message is modulated onto a carrier using AM. The resulting AM signal is of the form

$$S_{AM}(t) = [A_o + X(t)]\cos(\omega_c t + \Theta). \tag{11.81}$$

Figure 11.13
A block diagram of an amplitude modulation (AM) communication system.

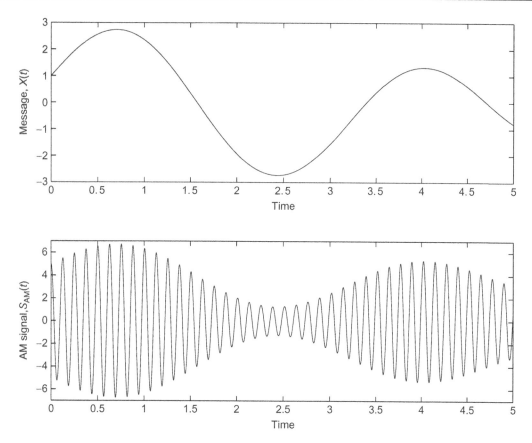

Figure 11.14
A sample message and the corresponding AM signal.

In AM, a bias, A_o, is added to the message and the result forms the time-varying amplitude of a radio frequency (RF) carrier. In order to allow for envelope detection at the receiver, the bias must satisfy $A_o > \max|X(t)|$. Some example waveforms are shown in Figure 11.14. Note that due to the process of modulation, the AM signal now occupies a bandwidth of $2B$. The modulated RF signal is transmitted using an antenna and propagates through the environment to the receiver where it is picked up with a receiving antenna.

To study the effects of noise on the AM system, we ignore other sources of corruption (e.g., interference, distortion) and model the received signal as simply a noisy version of the transmitted signal,

$$R(t) = S_{AM}(t) + N_w(t), \qquad (11.82)$$

where $N_w(t)$ is a Gaussian white noise process with a PSD of $N_o/2$. To limit the effects of noise, the received signal is passed through a BPF whose passband is chosen to allow the AM signal to pass, while removing as much of the noise as possible. This receiver filter is taken to be an ideal BPF whose bandwidth is $2B$. The output of the filter is then modeled as an AM signal plus narrowband noise:

$$Z(t) = S_{AM}(t) + N_x(t)\cos(\omega_c t + \Theta) - N_y(t)\sin(\omega_c t + \Theta)$$

$$= [A_o + X(t) + N_x(t)]\cos(\omega_c t + \Theta) - N_y(t)\sin(\omega_c t + \Theta). \tag{11.83}$$

The envelope detector is a device that outputs the real envelope of the input. Hence, for the preceding input, the demodulated output will be

$$X_d(t) = \sqrt{(A_o + X(t) + N_x(t))^2 + (N_y(t))^2}. \tag{11.84}$$

In its normal operating mode, the desired portion of the filter output will generally be much stronger than the noise. In that case, we observe that most of the time $A_o + X(t) + N_x(t) \gg N_y(t)$ so that the demodulated output can be well approximated by

$$X_d(t) \approx A_o + X(t) + N_x(t). \tag{11.85}$$

The d.c. component can easily be removed and hence $X(t)$ represents the desired component of the demodulated output and $N_x(t)$ is the undesired (noise) component. The power in the desired component depends on the message and we simply write it as $E[X^2(t)]$. The I component of the narrowband noise has a spectral density which is equal to N_o for $|f| < B$ (and zero otherwise). Therefore, the noise power at the demodulator output is $2N_oB$. The resulting SNR at the output of the AM system is

$$SNR = \frac{E[X^2(t)]}{2N_oB}. \tag{11.86}$$

It is common to express this SNR in terms of the transmitted power required to support the AM modulator. For an AM signal of the form in Equation (11.81), the transmitted power is

$$P_T = E[(A_o + X(t))^2\cos^2(\omega_c t + \Theta)]$$

$$= E[(A_o + X(t))^2]E[\cos^2(\omega_c t + \Theta)], \tag{11.87}$$

assuming the carrier phase is independent of the message. If the message is a zero-mean random process (which is usually the case), this expression simplifies to

$$P_T = \frac{A_o^2 + E[X^2(t)]}{2}. \tag{11.88}$$

Using this relationship, the SNR of the AM system can then be expressed as

$$\text{SNR} = \frac{E[X^2(t)]}{A_o^2 + E[X^2(t)]}\frac{P_T}{N_o B}. \tag{11.89}$$

The factor,

$$\eta_{\text{AM}} = \frac{E[X^2(t)]}{A_o^2 + E[X^2(t)]}, \tag{11.90}$$

is known as the *modulation efficiency* of AM and is usually expressed as a percentage. Note that due to the requirement that $A_o > \max|X(t)|$, the modulation efficiency of AM must be less than 50% (which would occur for square wave messages). For a sinusoidal message, the modulation efficiency would be no more than 33%, while for a typical voice or music signal, the modulation efficiency might be much smaller.

■ Example 11.12:

The MATLAB code that follows simulates the modulation and demodulation of an AM signal. The message signal is the sum of two sinusoids (one at 100 Hz and one at 250 Hz). For this example, the carrier frequency is taken to be $f_c = 100$ kHz. We have added noise to the AM signal to produce a typical received signal as shown in Figure 11.15a. To demodulate the signal, we decompose the received signal into its I and Q components using the technique illustrated in Figure 11.9. The LPF we used was a second-order Butterworth filter with a cutoff frequency of 400 Hz, but the particular form of the filter is not crucial. Once these components are found, the real envelope is computed according to $R(t) = \sqrt{X^2(t) + Y^2(t)}$. Our estimate of the message is then a scaled version of this envelope with the d.c. component removed. Figure 11.15b shows the original message (dashed) and the recovered message (solid).

```
fc=100000;                                   % carrier freq.
dt=1/(4*fc);                                 % sampling interval.
t=[0:dt:0.025];
f1=100; f2=250;
m=sin(2*pi*f1*t)+2*cos(2*pi*f2*t);           % message signal (two tones)
s=(m+1+max(abs(m))).*cos(2*pi*(fc*t+rand(1)));% AM signal
s=s+randn(1,length(s));                      % Noisy AM signal
subplot(2,1,1)
plot(t,s); ylabel('AM signal');

f0=400;
a=2*pi*f0/sqrt(2);
h=exp(-a*t).*sin(a*t);                       % Second order lowpass filter
                                             % with cutoff freq at f0.
temp=s.*cos(2*pi*fc*t);
```

(Continued)

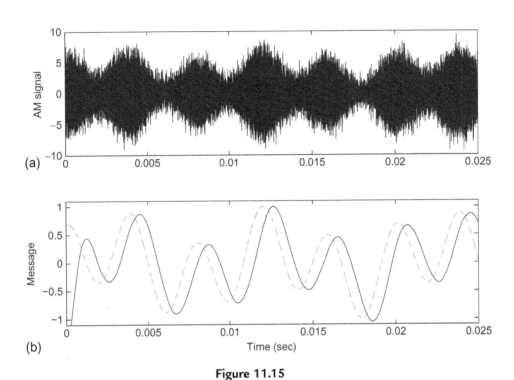

Figure 11.15
(a) An AM signal corrupted by noise along with (b) a comparison of the original message
and the demodulated message.

```
x=conv(temp,h)*dt;                              % I component
temp=s.*sin(2*pi*fc*t);
y=conv(temp,h)*dt;                              % Q component
r=sqrt(x.^2+y.^2);                              % real envelope
r=r(1:length(t));
mhat=r-sum(r)/length(r);
subplot(2,1,2)
plot(t,mhat/max(mhat),'b',t,m/max(m),'g--')
axis([0 max(t) -1.1 1.1])
ylabel('Message'); xlabel('time (sec)')
```

Section 11.1: Continuous Time Linear Systems

11.1 A white noise process, $X(t)$, with a PSD of $S_{XX}(f) = N_0/2$ is passed through a finite time integrator whose output is given by

$$Y(t) = \int_{t-t_o}^{t} X(u)du.$$

Find the following:
(a) the PSD of the output process,
(b) the total power in the output process,
(c) the noise equivalent bandwidth of the integrator (filter).

11.2 A certain LTI system has an input/output relationship given by

$$Y(t) = \frac{X(t) - X(t - t_o)}{t_o}.$$

(a) Find the output autocorrelation, $R_{YY}(\tau)$, in terms of the input autocorrelation, $R_{XX}(\tau)$.
(b) Find the output PSD, $S_{YY}(f)$, in terms of the input PSD, $S_{XX}(f)$.
(c) Does your answer to part (b) make sense in the limit as $t_o \to 0$?

11.3 The output $Y(t)$ of a linear filter is c times the input $X(t)$. Show that $R_{YY}(\tau) = c^2 R_{XX}(\tau)$.

11.4 The output $Y(t)$ of a filter is given in terms of its input $X(t)$ by

$$Y(t) = X(t) + X(t - t_o) + X(t - 2t_o).$$

(a) Determine $R_{YY}(\tau)$ as a function of $R_{XX}(\tau)$.
(b) Find $E[Y^2(t)]$.

11.5 Consider a non-linear device such that the output is $Y(t) = aX^2(t)$, where the input $X(t)$ consists of a signal plus a noise component, $X(t) = S(t) + N(t)$. Determine the mean and autocorrelation function for $Y(t)$ when the signal $S(t)$ and the noise $N(t)$ are both Gaussian random processes and wide sense stationary (WSS) with zero-mean, and $S(t)$ is independent of $N(t)$.

11.6 Calculate the spectrum for $Y(t)$ in Exercise 11.5 if

$$S_{SS}(f) = \frac{A^2}{4}[\delta(f+f_c) + \delta(f-f_c)]$$

and

$$S_{NN}(f) = \begin{cases} \dfrac{N_o}{2}, & f_c - \dfrac{B}{2} < |f| < f_c + \dfrac{B}{2}, \\ 0, & \text{otherwise.} \end{cases}$$

11.7 If the input to a linear filter is a random telegraph process with c zero-crossings per second and an amplitude A, determine the output PSD. The filter impulse response is $h(t) = b\exp(-at)u(t)$.

11.8 The input to a linear filter is a random process with the following autocorrelation function:

$$R_{XX}(\tau) = \frac{A\omega_o}{\pi}\frac{\sin(\omega_o\tau)}{\omega_o\tau}.$$

The impulse response of the filter is of the same form and is

$$h(t) = \frac{\omega_1}{\pi}\frac{\sin(\omega_1 t)}{\omega_1 t}.$$

Determine the autocorrelation function of the filter output for $\omega_o \geq \omega_1$ and for $\omega_o < \omega_1$.

11.9 The power spectrum at the input to an ideal BPF is

$$S_{XX}(f) = \frac{a}{1 + (f/f_o)^2}$$

Let the transfer function for the ideal BPF be

$$H(f) = \begin{cases} b, & \text{for } f_1 < |f| < f_2, \\ 0, & \text{otherwise.} \end{cases}$$

Determine the autocorrelation function of the output. You may have to make a reasonable approximation to obtain a simplified form. Assume that $f_2 - f_1 \ll 1$.

11.10 A random process with a PSD of $S_{XX}(f) = \dfrac{1}{1+f^2}$ is input to a filter. The filter is to be designed such that the output process is white (constant PSD). This filter is called a *whitening filter*.

(a) Find the transfer function of the whitening filter for this input process. Be sure that the filter is causal.

(b) Sketch a circuit which will realize this transfer function.

11.11 White Gaussian noise is input to an RC LPF.

(a) At what sampling instants is the output independent of the input at time $t = t_1$?

(b) At what sampling instants is the output independent of the output at time $t = t_1$?

(c) Repeat parts (a) and (b) if the filter is replaced by the finite time integrator of Exercise 11.1.

11.12

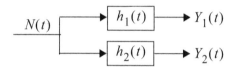

A white Gaussian noise process, $N(t)$, is input to two filters with impulse responses, $h_1(t)$ and $h_2(t)$, as shown in the accompanying figure. The corresponding outputs are $Y_1(t)$ and $Y_2(t)$, respectively.

(a) Derive an expression for the cross-correlation function of the two outputs, $R_{Y_1 Y_2}(\tau)$.

(b) Derive an expression for the cross-spectral density of the two outputs, $S_{Y_1 Y_2}(\tau)$.

(c) Under what conditions (on the filters) are the two outputs independent when sampled at the same instants in time? That is, when are $Y_1(t_o)$ and $Y_2(t_o)$ independent? Express your constraints in terms of the impulse responses of the filters and also in terms of their transfer functions.

(d) Under what conditions (on the filters) are the two outputs independent when sampled at different instants in time. That is, when are $Y_1(t_1)$ and $Y_2(t_2)$ independent for arbitrary t_1 and t_2? Express your constraints in terms of the impulse responses of the filters and also in terms of their transfer functions.

11.13 If the inputs to two linear filters $h_1(t)$ and $h_2(t)$ are $X_1(t)$ and $X_2(t)$, respectively, show that the cross-correlation between the outputs $Y_1(t)$ and $Y_2(t)$ of the two filters is

$$R_{Y_1 Y_2}(\tau) = \int_{-\infty}^{\infty} \int_{-\infty}^{\infty} h_1(\alpha) h_2(\beta) R_{X_1 X_2}(\tau + \alpha - \beta) d\alpha d\beta.$$

Section 11.2: Discrete-Time Systems

11.14 Is the following function a valid discrete-time autocorrelation function? Justify your answer.

$$R_{XX}[k] = \begin{cases} 1, & k = 1, \\ \dfrac{3}{4}, & k = \pm 3, \\ 0, & \text{otherwise.} \end{cases}$$

11.15 A discrete random sequence $X[n]$ is the input to a discrete linear filter $h[n]$. The output is $Y[n]$. Let $Z[n] = X[n+i] - Y[n]$. Find $E[Z^2[n]]$ in terms of the autocorrelation functions for $X[n]$ and $Y[n]$ and the cross-correlation function between $X[n]$ and $Y[n]$.

11.16 The unit impulse response of a discrete linear filter is $h[n] = a^n u[n]$, where $|a| < 1$. The autocorrelation function for the input random sequence is

$$R_{XX}[k] = \begin{cases} 1, & k = 0, \\ 0, & \text{otherwise.} \end{cases}$$

Determine the cross-correlation function between the input and output random sequences.

11.17 Find the PSD of a discrete random sequence with the following autocorrelation function: $R_{XX}[k] = a(b^{|k|})$, where $|b| < 1$.

11.18 A discrete-time linear filter has a unit pulse response $h[n]$. The input to this filter is a random sequence with uncorrelated samples. Show that the output PSD is real and non-negative.

11.19 The input, $X[k]$, to a filter is a discrete-time zero-mean random process whose autocorrelation function is $R_{XX}[n] = \delta[n]$. The input/output relationship of the filter is given by

$$Y[k] = \frac{1}{n} \sum_{m=0}^{n-1} X[k-m].$$

(a) Find the autocorrelation function of the output, $R_{YY}[n]$.
(b) Find the PSD of the output, $S_{YY}(f)$.

11.20 The input to a filter is a discrete-time, zero-mean, random process whose autocorrelation function is

$$R_{XX}[n] = |a|^n,$$

for some constant a such that $|a| < 1$. We wish to filter this process so that the output of the filter is white. That is, we want the output, $Y[k]$, to have an autocorrelation function, $R_{YY}[n] = \delta[n]$.

(a) Find the transfer function of the filter. You may express your answer in terms of either a DTFT or a z-transform.

(b) Find the impulse response of the filter.

For both parts, make sure your answer results in a causal filter.

Section 11.3: Noise Equivalent Bandwidth

11.21 Determine the noise equivalent bandwidth for a filter with impulse response

$$h(t) = b\exp(-at)u(t).$$

11.22 A filter has the following transfer function:

$$H(f) = \frac{4}{10 + j2\pi f}.$$

Determine the ratio of the noise equivalent bandwidth for this filter to its 3-dB bandwidth.

11.23 Suppose you want to learn the characteristics of a certain filter. A white noise source with an amplitude of 15 watts/Hz is connected to the input of the filter. The power spectrum of the filter output is measured and found to be

$$S_{YY}(f) = \frac{30}{(2\pi f)^2 + 10^2}.$$

(a) What is the bandwidth (3 dB) of the filter?

(b) What is the attenuation (or gain) at zero frequency?

(c) Show one possible (i.e., real, causal) filter that could have produced this output PSD.

11.24 A filter has an impulse response of $h(t) = te^{-t}u(t)$. Find the noise equivalent bandwidth of the filter.

11.25   A filter has a transfer function given by

$$H(f) = \frac{1 + j2\pi f}{(1 + j2\pi f)^2 + (2000\pi)^2}.$$

(a)   Is this filter, lowpass, highpass, or bandpass?

(b)   Find the noise equivalent bandwidth of this filter.

11.26   The definition for the noise equivalent bandwidth of a filter was given in terms of the transfer function of the filter in Definition 11.1. Derive an equivalent expression for the noise equivalent bandwidth of a filter in terms of the impulse response of the filter, $h(t)$.

11.27   Suppose a filter has a transfer function given by $H(f) = \text{sinc}^2(f)$. Find the noise equivalent bandwidth of the filter. *Hint*: You might want to use the result of Exercise 11.26.

Section 11.4: Signal-to-Noise Ratios

11.28   For the high-pass RC network shown, let
$X(t) = A\sin(\omega_c t + \Theta) + N(t)$, where $N(t)$ is white, WSS, Gaussian noise and Θ is a random variable uniformly distributed over $[0, 2\pi)$. Assuming zero initial conditions:

(a)   Find the output mean and variance.

(b)   Find and plot the autocorrelation function of the output.

(c)   Find and plot the output PSD.

(d)   Find the output SNR.

11.29   A parallel RLC network is driven by an input current source of
$X(t) = A\sin(\omega_c t + \Theta) + N(t)$, where $N(t)$ is white, WSS noise with zero-mean. The output is the voltage across the network. The phase Θ is a random variable uniformly distributed over $[0, 2\pi)$.

(a)   Find the output power spectrum by first computing the output autocorrelation function and then transforming.

(b)   Check the result of part (a) by using (11.12c).

(c)   Determine the output SNR and optimize the bandwidth to maximize the SNR. Assume ω_c differs from the center frequency of the RLC filter.

Hints: You may have to calculate the autocorrelation function as a function of t and τ and then let t go to infinity to find the steady-state output. There are several conditions you may want to consider; for example, the filter may be overdamped, critically damped, or under damped. It may also have an initial voltage on the capacitor and a current through the inductor. State your assumption about these conditions.

11.30 A one-sided exponential pulse, $s(t) = \exp(-t)u(t)$, plus white noise is input to the finite time integrator of Exercise 11.1. Adjust the width of the integrator, t_0, so that the output SNR is maximized at $t = t_0$.

11.31 A square pulse of width $t_0 = 1\,\mu s$ plus zero-mean white Gaussian noise is input to a filter with impulse response, $h(t) = \exp(-t/t_1)u(t)$.
   (a) Find the value of the constant t_1 such that the SNR at the output of the filter will be maximum.
   (b) Assuming the square pulse is non-zero over the time interval $(0, t_0)$, at what sampling time will the SNR at the output of the filter be maximized?

11.32 The input to a filter consists of a half-sinusoidal pulse

$$s(t) = \begin{cases} \sin(\pi t), & 0 \le t < 1, \\ 0, & \text{otherwise}, \end{cases}$$

plus zero-mean white Gaussian noise.
   (a) Suppose the impulse response of the filter is rectangular, $h(t) = \text{rect}(t/t_a)$. What should the width of the rectangle, t_a, be in order to maximize the SNR at the output? At what point in time does that maximum occur?
   (b) Suppose the impulse response of the filter is triangular, $h(t) = \text{tri}(t/t_b)$. What should the width of the triangle, t_b, be in order to maximize the SNR at the output? At what point in time does that maximum occur?
   (c) Which filter (rectangle or triangle) produces the larger SNR and by how much? Specify your answer in decibels (dBs).

Section 11.5: The Matched Filter

11.33
   (a) Determine the impulse response of the filter matched to the pulse shape shown in the accompanying figure. Assume that the filter is designed to maximize the output SNR at time $t = 3t_0$.
   (b) Sketch the output of the matched filter designed in part (a) when the signal $s(t)$ is at the input.

11.34 Find the impulse response and transfer function of a filter
matched to a triangular waveform as shown in the
accompanying figure when the noise is stationary and
white with a power spectrum of $N_0/2$.

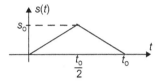

11.35 A known deterministic signal, $s(t)$, plus colored (not white) noise, $N(t)$, with a PSD
of $S_{NN}(f)$ is input to a filter. Derive the form of the filter that maximizes the SNR at
the output of the filter at time $t = t_0$. To make this problem simpler, you do not need
to insist that the filter is causal.

11.36 Consider the system described in Example 11.6 where a communication system
transmits a square pulse of width t_1. In that example, it was found that the
matched filter can be implemented as a finite time integrator which integrates over
a time duration to t_1 seconds. For the sake of this problem, suppose $t_1 = 100 \,\mu s$.
Suppose that due to imperfections in the construction of this filter, we actually
implement a filter which integrates over a time interval of t_2 seconds. Determine
what range of values of t_2 would produce an output SNR which is within 0.25 dB
of the optimal value (when $t_2 = t_1$). The answer to this question will give you an
idea of how precise the design of the match filter needs to be (or how imperise it
can be).

Section 11.6: The Wiener Filter

11.37 Suppose we observe a random process $Z(t)$ (without any noise) over a time interval
$(-\infty, t)$. Based on this observation, we wish to predict the value of the same random
process at time $t + t_0$. That is, we desire to design a filter with impulse response,
$h(t)$, whose output will be an estimate of $Z(t + t_0)$:

$$Y(t) = \hat{Z}(t + t_0) = \int_0^\infty h(u)Z(t - u)\mathrm{d}u.$$

(a) Find the Wiener–Hopf equation for the optimum (in the minimum mean square
error (MMSE) sense) filter.
(b) Find the form of the Wiener filter if $R_{ZZ}(\tau) = \exp(-|\tau|)$.
(c) Find an expression for the mean square error $E[\varepsilon^2(t)] = E[(Z(t + t_0) - Y(t))^2]$.

11.38 Suppose we are allowed to observe a random process $Z(t)$ at two points in time, t_0
and t_2. Based on those observations we would like to estimate $Z(t)$ at time $t = t_1$
where $t_0 < t_1 < t_2$. We can view this as an interpolation problem. Let our estimator be
a linear combination of the two observations,

$$Y(t_1) = \hat{Z}(t_1) = aZ(t_0) + bZ(t_2) \ .$$

(a) Use the orthogonality principle to find the MMSE estimator.

(b) Find an expression for the mean square error of the MMSE estimator.

11.39 Suppose we are allowed to observe a random process $Z(t)$ at two points in time, t_0 and t_1. Based on those observations we would like to estimate $Z(t)$ at time $t = t_2$ where $t_0 < t_1 < t_2$. We can view this as a prediction problem. Let our estimator be a linear combination of the two observations,

$$Y(t_2) = \hat{Z}(t_2) = aZ(t_0) + bZ(t_1).$$

(a) Use the orthogonality principle to find the MMSE estimator.

(b) Suppose that $R_{ZZ}(\tau) = c\exp(-b|\tau|)$ for positive constants b and c. Show that in this case, the sample at time $t = t_0$ is not useful for predicting the value of the process at time $t = t_2$ (given we have observed the process at time $t = t_1 > t_0$). In other words, show that $a = 0$.

11.40 The sum of two independent random processes with PSDs

$$S_{SS}(f) = \frac{2}{50 + (2\pi f)^2} \text{ and } S_{NN}(f) = 40$$

are input to a LTI filter.

(a) Determine the Wiener smoothing filter. That is, find the impulse response of the filter that produces an output that is an MMSE estimate of $S(t)$.

(b) Find the PSD of the filtered signal plus noise.

11.41 The sum of two independent random sequences with autocorrelation functions

$$R_{SS}[k] = \frac{10}{1 - \left(\frac{1}{10}\right)^2} \left(\frac{1}{10}\right)^{|k|} \text{ and } R_{NN}[k] = \frac{100}{1 - \left(\frac{1}{4}\right)^2} \left(\frac{1}{4}\right)^{|k|}$$

is input to an LTI filter.

(a) Determine the Wiener smoothing filter. That is, find the impulse response of the filter which produces an output which is an MMSE estimate of $S[n]$.

(b) Find the PSD of the filtered signal plus noise.

(c) Find the input SNR and an estimate of the output SNR. Discuss whether or not the Wiener filter improves the SNR.

Hint: Compare the spectra of the Wiener filter, the signal, and the noise by plotting each on the same graph.

Sections 11.7 and 11.8: Narrowband Random Processes and Complex Envelopes

11.42 The PSD of a narrowband Gaussian noise process, $N(t)$, is as shown in the accompanying figure.
   (a) Find and sketch the PSD of the I and Q components of the narrowband noise process.
   (b) Find and sketch the cross-spectral density of the I and Q components.

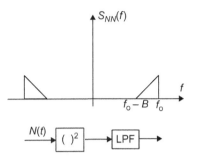

11.43 The narrowband noise process of Exercise 11.42 is passed through the system shown in the accompanying figure. In the system, the LPF has a bandwidth which is large compared to B but small compared to f_o.
   (a) Show that the output of the system is proportional to $|G_N(t)|^2$, where $G_N(t)$ is the complex envelope of $N(t)$.
   (b) Find and sketch the autocorrelation function of the output of the system. (*Hint*: You might find the Gaussian moment factoring theorem useful, see Exercise 6.18.)

11.44 Let $Z(t)$ be a zero-mean, stationary, narrowband process whose I and Q components are $X(t)$ and $Y(t)$, respectively. Show that the complex envelope, $G_Z(t) = X(t) + jY(t)$, satisfies $E[G_Z(t)G_Z(t + \tau)] = 0$.

MATLAB Exercises

11.45
   (a) Construct a signal plus noise random sequence using 10 samples of

$$X[n] = \cos(2\pi f_o nt_s) + N[n],$$

   where $N[n]$ is generated using randn in MATLAB and $f_o = 0.1/t_s$. Design a discrete-time matched filter for the cosine signal. Filter the signal plus noise sequence with the matched filter. At what sample value does the filter output peak.
   (b) Construct a signal plus noise random sequence using 10 samples of the following

$$X[n] = \cos(2\pi f_1 nt_s) + 10\cos(2\pi f_2 nt_s) + N[n],$$

   where $N[n]$ is generated using randn in MATLAB, $f_1 = 0.1/t_s$, and $f_2 = 0.4/t_s$. Design a discrete-time matched filter for the f_2 cosine signal. Filter the signal plus noise sequence with the matched filter. At what sample value does the filter output peak.

11.46   If a random sequence has an autocorrelation function $R_{XX}[k] = 10(0.8^{|k|})$, find the discrete-time, pure prediction Wiener filter. That is, find the filter $h[n]$ such that when $X[n]$ is input the output, $Y[n]$, will be an MMSE estimate of $X[n+1]$. Determine and plot the PSD for the random sequence and the power transfer function, $|H(f)|^2$, of the filter.

11.47   You have a random process with the following correlation matrix,

$$
\begin{bmatrix}
1.0 & 0.3 & 0.09 & 0.027 \\
0.3 & 1.0 & 0.3 & 0.09 \\
0.09 & 0.3 & 1.0 & 0.3 \\
0.027 & 0.09 & 0.3 & 1.0
\end{bmatrix}.
$$

Determine the pure prediction Wiener filter. That is, find the coefficients of the impulse response, $h = [h_0, h_1, h_2, 0, 0, 0, \ldots]$. Then determine the power spectrum of the Wiener filter. Are the results what you expected?

11.48   Generate a 100-point random sequence using `randn(1,100)` in MATLAB. Use a first-order AR filter to filter this random process. That is, the filter is

$$
A(z) = \frac{1}{1 + a_1 z^{-1}}.
$$

Let $a_1 = -0.1$. Use the filtered data to obtain an estimate for the first-order prediction Wiener filter. Compare the estimated filter coefficient with the true value.

Simulation Techniques

With the increasing computational power of very inexpensive computers, simulation of various systems is becoming very common. Even when a problem is analytically tractable, sometimes it is easier to write a quick program to simulate the desired results. However, there is a certain art to building good simulations, and many times avoidable mistakes have led to incorrect simulation results. This chapter aims at helping the reader to build a basic knowledge of some common techniques used for simulation purposes. Most of the results presented in this chapter are just applications of material covered in previous chapters, so there is nothing fundamentally new here. Nevertheless, armed with some of the basic simulation principles presented in this chapter, the reader should be able to develop simulation tools with confidence.

12.1 Computer Generation of Random Variables

In this section, we study techniques used to generate random numbers. However, we must start with a disclaimer. Most of the techniques used in practice to generate so-called random numbers will actually generate a completely deterministic sequence of numbers. So, what is actually random about these random number generators? Strictly speaking, nothing! Rather, when we speak of *computer-generated* random numbers, we are usually creating a sequence of numbers that have certain statistical properties that make them behave like random numbers, but in fact they are not really random at all. Such sequences of numbers are more appropriately referred to as *pseudorandom* numbers.

12.1.1 Binary Pseudorandom Number Generators

To start with, suppose we would like to simulate a sequence of independent identically distributed (IID) Bernoulli random variables, $X_1, X_2, X_3, \ldots$. One way to do this would be to grab a coin and start flipping it and observe the sequence of heads and tails, which could then be mapped to a sequence of 1s and 0 s. One drawback to this approach is that it is very time consuming. If our application demanded a sequence of length 1 million, not many of us would have the patience to flip the coin that many times. Therefore, we seek to assign this task to a computer. So, to simulate an IID sequence of random variables, essentially we would like to create a computer program that will output a binary sequence with the desired statistical

Probability and Random Processes. DOI: 10.1016\B978-0-12-386981-4-00012-7

Figure 12.1
A four-stage, binary linear feedback shift register.

properties. For example, in addition to having the various bits in the sequence be statistically independent, we might also want 0 s and 1s to be equally likely.

Consider the binary sequence generated by the linear feedback shift register (LFSR) structure illustrated in Figure 12.1. In that figure, the square boxes represent binary storage elements (i.e., flip-flops) while the adder is modulo-2 (i.e., an exclusive OR gate). Suppose the shift register is initially loaded with the sequence 1, 1, 1, 1. It is not difficult to show that the shift register will output the sequence

$$1, 1, 1, 1, 0, 0, 0, 1, 0, 0, 1, 1, 0, 1, 0, \ldots . \tag{12.1}$$

If the shift register were clocked longer, it would become apparent that the output sequence would be periodic with a period of 15 bits. While periodicity is not a desirable property for a pseudorandom number generator, if we are interested in generating short sequences (of length less than 15 bits), then the periodicity of this sequence generator would not come into play. If we are interested in generating longer sequences, we could construct a shift register with more stages so that the period of the resulting sequence would be sufficiently long so that its periodicity would not be of concern.

The sequence generated by our LFSR does possess several desirable properties. First, the number of 1s and 0 s is almost equal (eight 1s and seven 0 s is as close as we can get to equally likely with a sequence of length 15). Second, the autocorrelation function of this sequence is nearly equal to that of a truly random binary IID sequence (again, it is as close as we can possibly get with a sequence of period 15; see Exercise 12.1). Practically speaking, the sequence output by this completely deterministic device does a pretty good job of mimicking the behavior of an IID binary sequence. It also has the advantage of being repeatable. That is, if we load the shift register with the same initial contents, we can always reproduce the exact same sequence.

It should be noted that not all LFSRs will serve as good pseudorandom number generators. Consider for example the four-stage LFSR in Figure 12.2. This shift register is only a slightly modified version of the one in Figure 12.1, yet when loaded with the sequence 1, 1, 1, 1, this shift register outputs a repeating sequence of all 1s (i.e., the period of the output sequence is 1). The only difference between the two shift registers is in the placement of the feedback tap

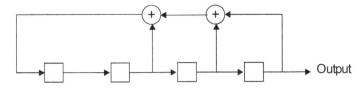

Figure 12.2
Another four-stage, binary linear feedback shift register.

Figure 12.3
A general *N*-stage, binary linear feedback shift register.

connections. Clearly, the placement of these tap connections is crucial in creating a good pseudorandom sequence generator.

A general *N*-stage LFSR is shown in Figure 12.3. The feedback tap gains, g_i, $i = 0, 1, 2, ..., N$, are each either 0 or 1. A 1 represents the presence of a tap connection, while a 0 represents the absence of a tap connection. It is also understood that $g_0 = g_N = 1$. That is, there must always be connections at the first and last position. A specific LFSR can then be described by the sequence of tap connections. For example, the four stage LFSR in Figure 12.1 can be described by the sequence of tap connections $[g_0, g_1, g_2, g_3, g_4] = [1, 0, 0, 1, 1]$. It is also common to further simplify this shorthand description of the LFSR by converting the binary sequence of tap connections to an octal number. For example, the sequence $[1, 0, 0, 1, 1]$ becomes the octal number 23. Likewise, the LFSR in Figure 12.2 is described by $[g_0, g_1, g_2, g_3, g_4] = [1, 0, 1, 1, 1] \rightarrow 27$.

An *N*-stage LFSR must necessarily generate a periodic sequence of numbers. At any point in time, the *N*-bit contents of the shift register can take on only one of 2^N possibilities. Given any starting state, the LFSR could at best cycle through all possible states, but sooner or later must return to its initial state (or some other state it has already been in). At that point, the output will then begin to repeat itself. Also, note that if the LFSR ever gets into the all zero state, it will remain in that state and output a sequence of all 0 s from that point on. Therefore, to get the maximum period out of a LFSR, we would like the shift register to cycle through all possible non-zero states exactly once before returning to its initial state. This will produce a

Table 12.1 LFSR feedback connections for m-sequences

SR Length, N	Feedback Connections (in Octal Format)
2	7
3	13
4	23
5	45, 67, 75
6	103, 147, 155
7	203, 211, 217, 235, 277, 313, 325, 345, 367
8	435, 453, 537, 543, 545, 551, 703, 747

periodic output sequence with period of $2^N - 1$. Such a sequence is referred to as a *maximal length linear feedback shift register* (MLLFSR) sequence, or an *m-sequence,* for short.

To study the question of how to connect the feedback taps of an LFSR in order to produce an m-sequence requires a background in abstract algebra beyond the scope of this book. Instead, we include a short list in Table 12.1 describing a few feedback connections, in the octal format described, which will produce m-sequences. This list is not exhaustive in that it does not list all possible feedback connections for a given shift register length that lead to m-sequences.

As mentioned, m-sequences have several desirable properties in that they mimic those properties exhibited by truly random IID binary sequences. Some of the properties of m-sequences generated by an N-stage LFSR are summarized as follows:

- m-sequences are periodic with a period of $2^N - 1$.
- In one period, an m-sequence contains $2^{N/2}$ 1s and $2^{N/2} - 1$ 0 s. Hence, 0 s and 1s are almost equally likely.
- The autocorrelation function of m-sequences is almost identical to that of an IID sequence.
- Define a run of length n to be a sequence of either n consecutive 1s or n consecutive 0 s. An m-sequence will have one run of length N, one run of length $N - 1$, two runs of length $N - 2$, four runs of length $N - 3$, eight runs of length $N - 4$, ..., and 2^{N-2} runs of length 1.

m-sequences possess many other interesting properties that are not as relevant to their use as random number generators.

■ Example 12.1:

Suppose we wish to construct an m-sequence of length 31. Since the period of an m-sequence is $2^N - 1$, we will need an $N = 5$ stage shift register to do the job. From Table 12.1, there are three different feedback connections listed, any of which will work. Using the first entry in the table, the octal number 45 translates to the feedback tap connections (1, 0, 0, 1, 0, 1). This describes the LFSR shown in Figure 12.4.

Figure 12.4
A five-stage LFSR with feedback tap connections specified by the octal number 45.

Assuming this LFSR is initially loaded with the sequence (1, 0, 0, 0, 0), the resulting m-sequence will be

(0 0 0 0 1 0 0 1 0 1 1 0 0 1 1 1 1 1 0 0 0 1 1 0 1 1 1 0 1 0 1). ∎

12.1.2 Nonbinary Pseudorandom Number Generators

Next suppose that it is desired to generate a sequence of pseudorandom numbers drawn from a nonbinary alphabet. One simple way to do this is to modify the output of the binary LFSRs discussed previously. For example, suppose we want a pseudorandom number generator that produces octal numbers (i.e, from an alphabet of size 8) that simulates a sequence of IID octal random variables where the eight possible outcomes are equally probable. One possible approach would be to take the five-stage LFSR of Example 12.1, group the output bits in triplets (i.e., three at a time), and then convert each triplet to an octal number. Doing so (assuming the LFSR is loaded initially as in Example 12.1), one period of the resulting octal sequence is

$$(0\ 2\ 2\ 6\ 3\ 7\ 0\ 6\ 7\ 2\ 4\ 1\ 1\ 3\ 1\ 7\ 4\ 3\ 3\ 5\ 2\ 0\ 4\ 5\ 4\ 7\ 6\ 1\ 5\ 6\ 5).\quad (12.2)$$

Note that each of the octal numbers 0, 1, 2, ..., 7 occurs exactly four times in this sequence except the number 0, which occurs three times. This is as close as we can get to equally likely octal numbers with a sequence of length 31. The number of runs of various lengths in this sequence also matches what we might expect from a truly random IID sequence of equally likely octal numbers. For example, the probability of a run of length two occurring in a random sequence of octal numbers is 1/8. Given a random sequence of length 31, the expected number of runs of length 2 is 31/8 = 3.875. This pseudorandom sequence has three runs of length 2. The expected number of runs of length 3 is 31/64 = 0.4844. The pseudorandom sequence has no runs of length 3.

It should be noted that since the octal sequence in Equation (12.2) is generated by the underlying feedback structure of Example 12.1, there should be a recursive formula that can be used to generate the sequence. In this case, the recursion is fairly complicated and not very instructive, but this leads us to the idea of generating pseudorandom sequences of nonbinary numbers using some recursive formula. One commonly used technique is the *power residue method* whereby a pseudorandom sequence is generated through a recursion of the form

$$x_k = ax_{k-1} \bmod q\ ,\quad k = 1, 2, 3, \dots,\qquad (12.3)$$

where a and q are suitably chosen constants. Due to the modulo-q operation, the elements of the sequence are from the set $\{0, 1, 2, ..., q-1\}$. The first element of the sequence x_0 is called the seed and given the set $\{a, q, x_0\}$, the resulting sequence is completely deterministic. Note also that the resulting sequence must be periodic since once any element appears a second time in the sequence, the sequence will then repeat itself. Furthermore, since the elements of the sequence are from a finite alphabet with q symbols, the maximum period of the sequence is $q-1$ (the period is not q because the element 0 must never appear or the sequence will produce all 0 s after that and therefore be of period 1). The next example shows that the desireable statistical properties of the sequence are dependent upon a careful selection of the numbers a and q.

■ Example 12.2:

First suppose that $\{a, q\} = \{4, 7\}$. Then the sequence produced has a period of 3 and assuming that the seed is $x_0 = 1$, the sequence is given by

$$(x_0, x_1, x_2, ...) = (1, 4, 2, 1, 4, 2, 1, 4, 2, ...).$$

This is not a particularly good result since with the selection of $q = 7$, we would hope for a period of $q - 1 = 6$. However, if we make a slight change so that $\{a, q\} = \{3, 7\}$, with the seed of $x_0 = 1$, the sequence becomes

$$(x_0, x_1, x_2, ...) = (1, 3, 2, 6, 4, 5, 1, 3, 2, 6, 4, 5, ...).$$

Now, as desired, the sequence has the maximal period of 6 and cycles through each of the integers from 1 through 6 exactly once each. As another example of a choice of $\{a, q\}$ which leads to a bad sequence, suppose we selected $\{a, q\} = \{4, 8\}$. Then the sequence produced (assuming $x_0 = 1$) would be

$$(x_0, x_1, x_2, ...) = (1, 4, 0, 0, 0, ...).$$

■

Clearly, we can get pseudorandom sequences with very different statistical properties depending on how we choose $\{a, q\}$. By choosing the number q to be very large, the resulting period of the pseudorandom sequence will also be very large and thus the periodicity of the sequence will not become an issue. Most math packages and high-level programming languages have built in random number generators that use this method. Commonly, the parameters $\{a, q\} = \{7^5, 2^{31} - 1\}$ are used. This produces a sequence of length $2^{31} - 2$, which is over 2 billion. Furthermore, by normalizing the elements of the sequence by q, the resulting pseudorandom sequence has elements from the set $\{1/q, 2/q, ..., (q-1)/q\}$. With a very large choice for the value of q, for almost all practical purposes, the elements will appear to be drawn from the continuous interval $(0, 1)$. Therefore, we have constructed a simple method to simulate random variables drawn from a uniform distribution.

12.1.3 Generation of Random Numbers from a Specified Distribution

Quite often, we are interested in generating random variables that obey some distribution other than a uniform distribution. In this case, it is generally a fairly simple task to transform a uniform random number generator into one that follows some other distribution. Consider forming a monotonic increasing transformation $g(\ )$ on a random variable X to form a new random variable Y. From the results of Chapter 4, the PDFs of the random variables involved are related by

$$f_Y(y) = \frac{f_X(x)}{\mathrm{d}g/\mathrm{d}x}. \tag{12.4}$$

Given an arbitrary PDF, $f_X(x)$, the transformation $Y = g(X)$ will produce a uniform random variable Y if $\mathrm{d}g/\mathrm{d}x = f_X(x)$ or equivalently $g(x) = F_X(x)$. Viewing this result in reverse, if X is uniformly distributed over $(0, 1)$ and we want to create a new random variable, Y with a specified distribution, $F_Y(y)$, the transformation $Y = F_Y^{-1}(X)$ will do the job.

■ Example 12.3:

Suppose we want to transform a uniform random variable into an exponential random variable with a PDF of the form

$$f_Y(y) = a\exp(-ay)u(y).$$

The corresponding CDF is

$$F_Y(y) = [1 - \exp(-ay)]u(y).$$

Therefore, to transform a uniform random variable into an exponential random variable, we can use the transformation

$$Y = F_Y^{-1}(X) = -\frac{\ln(1-X)}{a}.$$

Note that if X is uniformly distributed over $(0, 1)$, then $1 - X$ will be uniformly distributed as well so that the slightly simpler transformation

$$Y = -\frac{\ln(X)}{a}$$

will also work.   ■

This approach for generation of random variables works well provided that the CDF of the desired distribution is invertible. One notable exception where this approach will be difficult is the Gaussian random variable. Suppose, for example, we wanted to transform a uniform random variable, X, into a standard normal random variable, Y. The CDF in this case is the complement of a Q-function, $F_Y(y) = 1 - Q(y)$. The inverse of this function would then provide the appropriate transformation, $y = Q^{-1}(1-x)$, or as with the previous example, we could simplify this to $y = Q^{-1}(x)$. The problem here lies with the inverse Q-function which can not be expressed in a closed form. One could devise efficient numerical routines to compute the inverse Q-function, but fortunately there is an easier approach.

An efficient method to generate Gaussian random variables from uniform random variables is based on the following 2×2 transformation. Let X_1 and X_2 be two independent uniform random variables (over the interval $(0, 1)$). Then if two new random variables, Y_1 and Y_2 are created according to

$$Y_1 = \sqrt{-2\ln(X_1)}\cos(2\pi X_2), \tag{12.5a}$$

$$Y_2 = \sqrt{-2\ln(X_1)}\sin(2\pi X_2), \tag{12.5b}$$

then Y_1 and Y_2 will be independent standard normal random variables (see Example 5.24). This famous result is known as the *Box–Muller transformation* and is commonly used to generate Gaussian random variables. If a pair of Gaussian random variables is not needed, one of the two can be discarded. This method is particularly convenient for generating complex Gaussian random variables since it naturally generates pairs of independent Gaussian random variables. Note that if Gaussian random variables are needed with different means or variances, this can easily be accomplished through an appropriate linear transformation. That is, if $Y \sim N(0, 1)$, then $Z = \sigma Y + \mu$ will produce $Z \sim N(\mu, \sigma^2)$.

12.1.4 Generation of Correlated Random Variables

Quite often, it is desirable to create a sequence of random variables that are not independent, but rather have some specified correlation. Suppose we have a Gaussian random number generator that generates a sequence of IID standard normal random variables, $X = (X_1, X_2, \ldots, X_N)^{\mathrm{T}}$ and it is desired to transform this set of random variables into a set of Gaussian random variables, $Y = (Y_1, Y_2, \ldots, Y_N)^{\mathrm{T}}$ with some specified covariance matrix, C_Y. By using a linear transformation of the form $Y = AX$, the joint Gaussian distribution will be preserved. The problem of how to choose the transformation to produce the desired covariance matrix was covered in Chapter 6, Section 6.4.1. Recall that to specify this transformation, an eigendecomposition of the covariance matrix is performed to produce $C_Y = Q\Lambda Q^{\mathrm{T}}$, where Λ is the diagonal matrix of eigenvalues of C_Y and Q is the corresponding matrix of eigenvectors. Then the matrix $A = Q\sqrt{\Lambda}$ will produce the vector Y with the correct covariance matrix C_Y. Once again, if a random vector with a nonzero mean vector is desired, the above approach can be augmented as $Y = AX + B$, where B is the vector of appropriate means.

12.2 Generation of Random Processes

Next consider the problem of simulating a random process, $X(t)$, with a desired PSD, $S_{XX}(f)$. It is not feasible to create a continuous time random process with a computer. Fortunately, we can invoke the sampling theorem to represent the continuous time random process by its samples. Let $X_k = X(kT_s)$ be the kth sample of the random process taken at a sampling rate of $R_s = 1/T_s$. Then provided the sampling rate is chosen to be at least twice the absolute bandwidth of $X(t)$ (i.e., twice the largest nonzero frequency component of $S_{XX}(f)$), the random process can be reproduced from its samples. Thus, the problem of generating a random process can be translated into one of creating a sequence of random variables. The question is how should the random variables be correlated in order to produce the correct PSD? Of course, the autocorrelation function, $R_{XX}(\tau)$, provides the answer to this question. If the random process is sampled at a rate of $R_s = 1/T_s$, then the kth and mth sample will have a correlation specified by $E[X_k X_m] = R_{XX}((k-m)T_s)$. Hence, if $X = (X_1, X_2, \ldots, X_N)$ is a sequence of samples of the random process $X(t)$, the correlation matrix of these samples will have a Toeplitz structure (assuming $X(t)$ is stationary) of the form

$$
\boldsymbol{R}_{XX} = \begin{bmatrix}
R_{XX}(0) & R_{XX}(T_s) & R_{XX}(2T_s) & \cdots & R_{XX}(NT_s) \\
R_{XX}(-T_s) & R_{XX}(0) & R_{XX}(T_s) & \cdots & R_{XX}((N-1)T_s) \\
R_{XX}(-2T_s) & R_{XX}(-T_s) & R_{XX}(0) & \cdots & R_{XX}((N-2)T_s) \\
\cdots & \cdots & \cdots & & \cdots \\
R_{XX}(-NT_s) & R_{XX}(-(N-1)T_s) & R_{XX}(-(N-2)T_s) & \cdots & R_{XX}(0)
\end{bmatrix} . \quad (12.6)
$$

Once the appropriate correlation matrix is specified, the procedure developed in the last section can be used to generate the samples with the appropriate correlation matrix.

This approach will work fine provided that the number of samples desired is not too large. However in many cases, we need to simulate a random process for a large time duration. In that case, the number of samples, N, needed becomes large and hence the matrix $\boldsymbol{R}_{XX}$ is also large. Performing the necessary eigendecomposition on this matrix then becomes a computationally intensive problem. The following subsections look at some alternative approaches to this general problem that offer some computational advantages.

12.2.1 Frequency Domain Approach

If the random process to be simulated is a Gaussian random process, we can approach the problem by creating samples of the random process in the frequency domain. Suppose we wish to create a realization of the random process, $X(t)$, of time duration T_d, say over the interval $(0, T_d)$, and that we do not much care what happens to the process outside this interval.

Figure 12.5

A realization of the random process $X(t)$ along with its periodic extension $\tilde{X}(t)$.

To start with we produce a periodic signal $\tilde{X}(t)$ by repeating $X(t)$ every T_d seconds as illustrated in Figure 12.5. Since $\tilde{X}(t)$ is periodic it has a Fourier series representation

$$\tilde{X}(t) = \sum_k X_k e^{j2\pi k f_o t} \ , \quad f_o = \frac{1}{T_d} . \tag{12.7}$$

Note that due to the linearity of the Fourier series construction, if the X_k are zero mean Gaussian random variables, then the resulting process $\tilde{X}(t)$ will be a zero mean Gaussian random process. Furthermore, the periodic random process $\tilde{X}(t)$ has a line spectrum given by

$$S_{\tilde{X},\tilde{X}}(f) = \sum_k s_k^2 \delta(f - k f_o) \ , \quad s_k^2 = E[|X_k|^2] . \tag{12.8}$$

The s_k can be chosen to shape the spectrum to any desired form. If the desired PSD of the random process is $S_{XX}(f)$ then we could pick $s_k^2 \propto S_{xx}(k f_o)$. The constant of proportionality can be chosen so that the total power in the process $\tilde{X}(t)$ matches that of $X(t)$. In particular, suppose that $X(t)$ is bandlimited so that $S_{XX}(f) = 0$ for $|f| > W$. Then the number of terms in the Fourier series in Equation (12.7) is finite. Let

$$M = \left\lfloor \frac{W}{f_o} \right\rfloor = \lfloor W T_d \rfloor . \tag{12.9}$$

Then s_k will be nonzero only for $|k| \leq M$. Hence, we need to generate a total of $2M + 1$ random variables, $X_{-M}, X_{-M+1}, ..., X_{-1}, X_0, X_1, ..., X_M$. The variances of these random variables are chosen such that

$$s_k^2 = E[|X_k|^2] = \beta S_{XX}(k f_o), \quad \beta = \frac{\displaystyle\int_{-W}^{W} S_{XX}(f) df}{\displaystyle\sum_{k=-M}^{M} S_{XX}(k f_o)} . \tag{12.10}$$

In summary, the random process can be simulated by first generating a sequence of $2M+1$ zero-mean complex Gaussian random variables. Each random variable should be scaled so that the variances are as specified in Equation (12.10). Samples of the random process in the time domain can be constructed for any desired time resolution, Δt, according to

$$X[i] = X(i\Delta t) = \sum_{k=-M}^{M} X_k (e^{j2\pi f_0 \Delta t})^{ik} . \tag{12.11}$$

If the random process is real, it is sufficient to generate the $M+1$ random variables $X_0, X_1, ..., X_M$ independently and then form the remaining random variables $X_{-1}, X_{-2}, ..., X_{-M}$ using the conjugate symmetry relationship $X_{-k} = X_k^*$. In this case, X_0 must also be real so that a total of $2M+1$ real Gaussian random variables are needed (one for X_0 and two each for X_k, $k = 1, 2, ..., M$) to construct the random process, $X(t)$.

■ Example 12.4:

Suppose we wish to generate a 5-ms segment of a real zero-mean Gaussian random process with a PSD given by

$$S_{XX}(f) = \frac{1}{1 + (f/f_3)^4},$$

where $f_3 = 1 kHz$ is the 3 dB frequency of the PSD. Strictly speaking, this process has an infinite absolute bandwidth. However, for sufficiently high frequencies there is minimal power present. For the purposes of this example, we (somewhat arbitrarily) take the bandwidth to be $W = 6f_3$ so that approximately 99.9% of the total power in the process is contained within $|f| < W$. Since we want to simulate a time duration of $T_d = 5$ ms, the number of Fourier series coefficients needed is given by $M = \lfloor WT_d \rfloor = 30$. Figure 12.6 shows a comparison of the actual PSD, $S_{XX}(f)$, with the discrete line spectrum approximation. Also, one realization of the random process generated by this method is shown in Figure 12.7. The MATLAB code used to create these plots follows.

```
% Set parameters
I=sqrt(-1);
Td=5e-3; fo=1/Td; f3=1e3; dt=1e-5;
M=floor(6*f3*Td); m=[-M:M];

% Construct discrete samples of PSD
x=[0:0.01:10]; psd=1./(1+x.^4);
power=2*f3*sum(psd)*0.01;
s=1./(1+((m*fo)/f3).^4);
beta=power/sum(s);
s=beta*s;
% Construct "continuous" PSD
f=[-8:0.01:8]*f3;
```

(Continued)

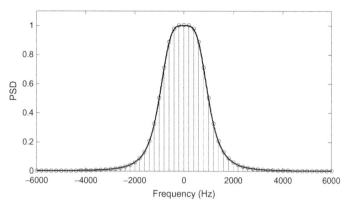

Figure 12.6
A comparison of the exact PSD along with the discrete approximation for Example 12.4.
(For color verssion of this figure, the reader is referred to web verssion of this chapter.)

Figure 12.7
A realization of the random process of Example 12.4.

```
psd=1./(1+(f/f3).^4);
% Plot results
subplot(2,1,1)
stem(m*fo,s/fo); hold on
plot(f,psd,'g'); hold off
axis([-8*f3 8*f3 0 1.2])
xlabel('frequency (Hz)'); ylabel('PSD');
% Generate Frequency domain samples
z0=randn(1); z0=z0*sqrt(s(M+1));
zplus=sqrt(s(M+2:2*M+1)/2).*(randn(1,M)+I*randn(1,M));
zminus=conj(fliplr(zplus));
z=[zminus z0 zplus];
% Create time domain process
t=[0:dt:Td];
rp=zeros(1,length(t));
for m=-M:M
    rp=rp+z(m+M+1)*exp(I*2*pi*m*fo*t);
```

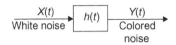

Figure 12.8
Time domain filtering to create a colored Gaussian random process.

12.2.2 Time Domain Approach

A simple alternative to the previous frequency domain approach is to perform time domain filtering on a white Gaussian noise process as illustrated in Figure 12.8. Suppose white Gaussian noise with PSD, $S_{XX}(f) = 1$ is input to an LTI filter that can be described by the transfer function, $H(f)$. Then, from the results in Chapter 11, it is known that the PSD of the output process is $S_{YY}(f) = |H(f)|^2$. Therefore, in order to create a Gaussian random process, $Y(t)$, with a prescribed PSD, $S_{YY}(f)$, we can construct a white process and pass this process through an appropriately designed filter. The filter should have a magnitude response which satisfies

$$|H(f)| = \sqrt{S_{YY}(f)}. \tag{12.12}$$

The phase response of the filter is irrelevant, and hence, any convenient phase response can be given to the filter.

This technique is particularly convenient when the prescribed PSD, $S_{YY}(f)$, can be written as the ratio of two polynomials in f. Then the appropriate transfer function can be found through spectral factorization techniques. If the desired PSD is a more complicated function of f, then designing and/or implementing a filter to produce that PSD may be difficult. In that case, it may be necessary to use an approximate filter design.

■ Example 12.5:

In this example, we design the filter needed to generate the random process specified in Example 12.4 using the time domain method. The PSD is factored as follows:

$$S(f) = \frac{1}{1 + (f/f_3)^4} = \frac{f_3^4}{(f - f_3 e^{j\pi/4})(f - f_3 e^{j3\pi/4})(f - f_3 e^{j5\pi/4})(f - f_3 e^{j7\pi/4})}.$$

If the first two poles are associated with $H(f)$ (and the last two with $H^*(f)$), then the filter has a transfer function of

(Continued)

$$H(f) = \frac{f_3^2}{(f - f_3 e^{j\pi/4})(f - f_3 e^{j3\pi/4})} = \frac{f_3^2}{f^2 - j\sqrt{2}f_3 f - f_3^2}$$

which can be represented in the time domain as

$$h(t) = -2\omega_o e^{-\omega_o t} \sin(\omega_o t) u(t) ,$$

where $\omega_o = \sqrt{2}\pi f_3$. For the purposes of creating a random process with the desired spectrum, the negative sign in front of this impulse response is irrelevant and can be ignored. Therefore, to produce the desired random process, we start by generating a white Gaussian random process and then convolve the input with the impulse response specified above. ∎

Once an appropriate analog filter has been designed, the filter must be converted to a discrete time form. If the sampling rate is taken to be sufficiently high, then the impulse response of the discrete time filter can be found by simply sampling the impulse response of the continuous time filter. This is the so-called impulse invariance transformation. However, because of aliasing that occurs in the process of sampling the impulse response, the frequency response of the digital filter will not be identical to that of the original analog filter unless the analog filter is absolutely bandlimited. Of course, if the analog filter is approximately bandlimited and if the sampling rate is sufficiently large, this aliasing can be kept to a minimum and the resulting digital filter will produce a very good approximation to the desired frequency response.

An alternative popular approach for producing a digital filter from an analog design is to use a bilinear transformation. That is, suppose we have an analog filter with transfer function $H_a(s)$. A digital approximation to this filter, $H_d(z)$, can be obtained (assuming a sampling frequency of f_s) according to

$$H_d(z) = H_a(s)\Big|_{s = 2f_s\left(\frac{1 - z^{-1}}{1 + z^{-1}}\right)} . \tag{12.13}$$

One advantage of the bilinear transformation is that if the analog filter is stable, then the digital filter will be stable as well. Note also that if the analog filter is an nth order filter, then the order of the digital filter will be no more than n as well.

■ Example 12.6:

In this example, we find the digital approximation to the analog filter designed in Example 12.5 using the bilinear approximation. From the results of that example, the analog filter was a second-order Butterworth filter whose transfer function (in terms of s) was given by

$$H_a(s) = \frac{\omega_3^2}{s^2 + \sqrt{2}\,\omega_3 s + \omega_3^2},$$

where $\omega_3 = 2\pi f_3$ is the 3-dB frequency of the filter in radians per second. After a little bit of algebraic manipulation, application of the bilinear transformation in Equation (12.13) results in

$$H_d(z) = \frac{b_0 + b_1 z^{-1} + b_2 z^{-2}}{a_0 + a_1 z^{-1} + a_2 z^{-2}},$$

where $b_0 = 1$, $b_1 = 2$, $b_2 = 1$, $a_0 = 1 + \sqrt{2}\gamma + \gamma^2$, $a_1 = 2 - 2\gamma^2$, $a_2 = 1 - \sqrt{2}\gamma + \gamma^2$, and $\gamma = f_s/(\pi f_3)$. Figure 12.9 shows a plot of the impulse response of this filter as well as one realization of the random process created by passing white Gaussian noise through this filter. Note that for this example, the impulse response of the filter lasts for about 1 ms (this makes sense since the bandwidth was 1 kHz). Therefore, when creating the filtered process, at least the first millisecond of output data should be discarded since the contents of the digital filter have not reached a statistical steady state until that point. The relevant MATLAB code follows

(a)

(b)

Figure 12.9
(a) Impulse response and (b) a single realization of the output of the filter designed in Example 12.6.

(Continued)

```
I=sqrt(-1);
imp_len=150;                              % length of impulse response in
                                            samples
sim_len=150;                              % length of simulation in samples
f3=1000;                                  % 3dB frequency of desired PSD
fs=30000;                                 % sampling rate
g=fs/(pi*f3);                             % compute filter coefficients
b0=1; b1=2; b2=1; b=[b0 b1 b2];
a0=1+sqrt(2)*g+g^2; a1=2-2*g^2;
a2=1-sqrt(2)*g+g^2; a=[a0 a1 a2];
x=zeros(1,imp_len); x(1)=1;               % impulse
y=filter(b,a,x);                          % impulse response of filter
time_scale=[1:length(y)]/fs;
subplot(2,1,1)                            % plot impulse response
stem(time_scale*1000,y,'o')
xlabel('time(msec)=n/f_s')
ylabel('h[n]')
x=randn(1,sim_len);                       % white Gaussian random process
y=filter(b,a,x);                          % filtered process
time_scale=[1:length(y)]/fs;
subplot(2,1,2)                            % plot realization
plot(time_scale*1000,y);
xlabel('time (msec)'); ylabel('X(t)');
```

■

One advantage of the time domain approach is that it is convenient for generating very long realizations of the random process. Once the filter is designed, the output process is created by performing a convolution of the impulse response of the filter with a white input sequence. The complexity of this operation is linear in the length of the sequence. Furthermore, the process of creating a long sequence can be broken into several smaller sequences. The smaller sequences can then be concatenated together to create a long sequence. There will be no discontinuities at the points of concatenation if the contents of the filter are stored after each realization and used as the starting contents for the next realization.

Some care must be taken when using the time domain method if the desired sampling rate of the process is much larger than the bandwidth of the filter. In this case, the poles of the digital filter will be close to the unit circle and the filter might exhibit stability problems. This is illustrated in Figure 12.10 where the magnitude of the poles of the digital filter from Example 12.6 is plotted as a function of the sampling rate. Note that when the rate at which the process is sampled becomes a few hundred times the bandwidth of the filter, the poles of the digital filter become very close to the unit circle. This problem can easily be avoided by creating a digital filter to create samples of the process at a lower rate (perhaps at several times the bandwidth of the filter so as to avoid aliasing) and then upsampling (through interpolation) the resulting process to any desired rate.

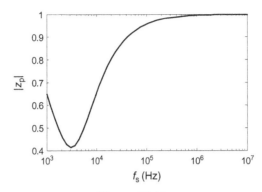

Figure 12.10
Magnitude of the filter poles as a function of sampling
frequency for the filter designed in Example 12.6.

12.2.3 Generation of Gaussian White Noise

Generation of a white noise process is exceedingly common and also is very simple. However, it is often the source of frequent mistakes among engineers, so we felt it worth making a few comments about computer generation of white Gaussian noise processes. The source of confusion in the generation of white noise is that one cannot represent white noise from its time domain samples. White noise has infinite power; therefore, samples of a white noise process would require infinite variance. Alternatively, white noise has infinite bandwidth, so the Nyquist rate for recovering white noise from its samples would be infinite. In order to represent a "white" process in discrete time, we must invoke some form of prefiltering before sampling. This will limit the bandwidth so that a finite sampling rate can be used, and at the same time, it will limit the power of the process so that the resulting samples of the filtered white process will have a finite variance.

Strictly speaking, once we filter the white noise it is no longer white, but this should not be of too much concern. In practice, there is no such thing as truly white noise. Recall that the white noise model was an approximation to a noise process which had a constant PSD over a very large (but not infinite) bandwidth. Furthermore, any equipment we use to measure or receive the noise will automatically filter the process. With this in mind, we imagine a prefilter that has a bandwidth that is much larger than any bandwidth we are concerned with in the specific system we are simulating. The noise we create, although not truly white, will behave as though it were white for all practical purposes.

In order to simplify the process of creating the samples of the prefiltered white noise, it is common to employ an ideal lowpass prefilter with bandwidth W as illustrated in Figure 12.11. Now that the process is bandlimited, it can be represented by samples at any rate that satisfies

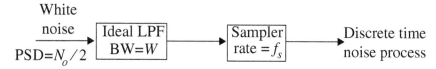

Figure 12.11
A/D conversion process for white noise.

$f_s \geq 2W$. Since the prefilter is ideal, the autocorrelation function of the filter output is easily calculated to be

$$R(\tau) = N_o W \text{sinc}(2W\tau). \tag{12.14}$$

Note that since this sinc function has nulls at multiples of $1/2W$, the samples of the filtered process will be uncorrelated provided that the samples are spaced by any integer multiple of $1/2W$. In other words, if the sampling rate satisfies $f_s = 2W/n$ for any integer n, then the samples of the prefiltered white noise will be uncorrelated. By choosing $n = 1$ so that the sampling rate is exactly the Nyquist rate, $f_s = 2W$, the process can be recovered from the discrete samples and the samples are uncorrelated. For Gaussian noise, this implies that the filtered white noise can be represented by a sequence of independent, zero-mean, Gaussian random variables with variance of $\sigma^2 = N_o W$. Note that the variance of the samples and the rate at which they are taken are related by $\sigma^2 = N_o f_s/2$.

The lesson to be learned here is that if we wish to represent Gaussian white noise as a sequence of *independent* Gaussian random variables, then there is an implicit assumption about the nature of the prefiltering. Furthermore, to be consistent with this assumption, the variance of the samples must be adjusted when the sampling rate is changed. The variance and sampling rate cannot be selected independently.

12.3 Simulation of Rare Events

Quite often, we are interested in estimating the probability of some event, A. If analytically calculating this probability is too cumbersome, we can design a computer program to simulate the system in question and then observe whether or not the event occurs. By repeating this procedure many times, we can observe how often the event A occurs and hence get an estimate of its probability through a relative frequency approach. The event A could be a bit error in a digital communications system—in which case we are interested in calculating bit error probability—or it could be a buffer overflow in a computer network or even something as extravagant as breaking the bank at the blackjack table.

12.3.1 Monte Carlo Simulations

In general, suppose we have the ability to recreate (simulate) the experiment an arbitrary number of times and define a sequence of Bernoulli random variables, X_i, that are defined according to

$$X_i = \begin{cases} 1, & \text{if } A \text{ occurs during the } i\text{th experiment,} \\ 0, & \text{otherwise.} \end{cases} \tag{12.15}$$

Hence, X_i is simply an indicator function for the event A. If the experiments are independent, then the probability of the event A, p_A, can be estimated according to

$$\hat{p}_A = \frac{1}{n} \sum_{i=1}^{n} X_i . \tag{12.16}$$

This is nothing more than estimating the mean of an IID sequence of random variables. From the development of Chapter 7, we know that this estimator is unbiased and that as $n \to \infty$ the estimate converges (almost everywhere via the strong law of large numbers) to the true probability.

In practice, we do not have the patience to run our simulation an infinite number of times nor do we need to. At some point, the accuracy of our estimate should be "good enough," but how many trials is enough? Some very concrete answers to this question can be obtained using the theory developed in Chapter 7. If the event A is fairly probable, then it will not take too many trials to get a good estimate of the probability, in which case runtime of the simulation is not really too much of an issue. However, if the event A is rare, then we will need to run many trials to get a good estimate of p_A. In the case when n gets large, we want to be sure not to make it any larger than necessary so that our simulation runtimes do not get excessive. Thus, the question of how many trials to run becomes important when simulating rare events.

Assuming n is large, the random variable $\hat{p}_A$ can be approximated as a Gaussian random variable via the central limit theorem. The mean and variance are $E[\hat{p}_A] = p_A$ and $\sigma_{\hat{p}_A}^2 = n^{-1} p_A (1 - p_A)$, respectively. One can then set up a confidence interval based on the desired accuracy. For example, suppose we wish to obtain an estimate that is within 1% of the true value with 90% probability. That is, we want to run enough trials to insure that

$$\Pr(|\hat{p}_A - p_A| < 0.01 p_A) = 0.9 = 1 - \alpha . \tag{12.17}$$

From the results of Chapter 7, Section 7.5, we get

$$\varepsilon_{0.1} = 0.01 p_A = \frac{\sigma_X}{\sqrt{n}} c_{0.1} = \sqrt{\frac{p_A(1 - p_A)}{n}} c_{0.1}, \tag{12.18}$$

where the value of $c_{0.1}$ is taken from Table 7.1 as $c_{0.1} = 1.64$. Solving for n gives us an answer for how long to run the simulation:

$$n = \frac{(100c_{0.1})^2(1-p_A)}{p_A} = \frac{(164)^2(1-p_A)}{p_A} \approx \frac{(164)^2}{p_A}.$$ (12.19)

Or in general, if we want the estimate, $\hat{p}_A$, to be within β percent of the true value (i.e., $|\hat{p}_A - p_A| < \beta p_A / 100$) with probability α, then the number of trials in the simulation should be chosen according to

$$n = \frac{\left(\frac{100c_\alpha}{\beta}\right)^2 (1-p_A)}{p_A} \approx \frac{\left(\frac{100c_\alpha}{\beta}\right)^2}{p_A}.$$ (12.20)

This result is somewhat unfortunate because in order to know how long to run the simulation, we have to know the value of the probability we are trying to estimate in the first place. In practice, we may have a crude idea of what to expect for p_A which we could then use to guide us in selecting the number of trials in the simulation. However, we can use this result to give us very specific guidance in how to choose the number of trials to run, even when p_A is completely unknown to us. Define the random variable N_A to be the number of occurrences of the event A in n trials, that is,

$$N_A = \sum_{i=1}^{n} X_i .$$ (12.21)

Note that $E[N_A] = np_A$. That is, the quantity np_A can be interpreted as the expected number of occurrences of the event A in n trials. Multiplying both sides of Equation (12.20) by p_A then produces

$$E[N_A] = \left(\frac{100c_\alpha}{\beta}\right)^2 (1-p_A) \approx \left(\frac{100c_\alpha}{\beta}\right)^2 .$$ (12.22)

Hence, one possible procedure to determine how many trials to run is to repeat the experiment for a random number of trials until the event A occurs some fixed number of times as specified by Equation (12.22). Let M_k be the random variable which represents the trial number of the kth occurrence of the event A. Then, one could form an estimate of p_A according to

$$\hat{p}_A = \frac{k}{M_k}.$$ (12.23)

It turns out that this produces a biased estimate; however, a slightly modified form,

$$\hat{p}_A = \frac{k-1}{M_k - 1},$$ (12.24)

produces an unbiased estimate (see Exercise 7.12).

■ Example 12.7:

Suppose we wish to estimate the probability of an event that we expect to be roughly on the order of $p \sim 10^{-4}$. Assuming we want 1% accuracy with a 90 % confidence level, the number of trials needed will be

$$n = \frac{1}{p}\left(\frac{100c_{\alpha}}{\beta}\right)^2 = 10^4\left(\frac{100*1.64}{1}\right)^2 = 268,960,000 \; .$$

Alternatively, we need to repeat the simulation experiment until we observe the event

$$N_A = \left(\frac{100c_{\alpha}}{\beta}\right)^2 = 26,896$$

times. Assuming we do not have enough time available to repeat our simulation over 1/4 of a billion times, we would have to accept less accuracy. Suppose that due to time limitations we decide that we can only repeat our experiment 1 million times, then we can be sure that with 90% confidence, the estimate will be within the interval $(p - \varepsilon, p + \varepsilon)$, if ε is chosen according to

$$\varepsilon = \frac{\sqrt{p(1-p)}}{\sqrt{n}}c_{\alpha} \approx \frac{\sqrt{p}}{\sqrt{n}}c_{\alpha} = 1.64\frac{10^{-2}}{10^3} = 1.64 \times 10^{-5} = 0.164p \; .$$

With 1 million trials we can only be 90% sure that the estimate is within 16.4% of the true value.

■

The preceding example demonstrates that using the Monte Carlo approach to simulating rare events can be very time consuming in that we may need to repeat our simulation experiments many times to get a high degree of accuracy. If our simulations are complicated, this may put a serious strain on our computational resources. The next subsection presents a novel technique, which when applied intelligently can substantially reduce the number of trials we may need to run.

12.3.2 Importance Sampling

The general idea behind importance sampling is fairly simple. In the Monte Carlo approach, we spend a large amount of time with many repetitions of an experiment while we are waiting for an event to occur which may happen only very rarely. In importance sampling, we skew the distribution of the underlying randomness in our experiment so that the "important" events happen more frequently. We can then use analytical tools to convert our distorted simulation results into an unbiased estimate of the probability of the event in which we are interested. To help present this technique, we first generalize the problem treated in Section 12.3.1. Suppose the simulation experiment consisted of creating a sequence of random variables, $X = (X_1, X_2, ..., X_m)$ according to some density, $f_X(x)$, and then observing whether or not some event A occurred which could be defined in terms of the X_i. For example, suppose X_i represents the number of messages that arrive at a certain node in a computer communications network at time instant i. Furthermore, suppose that it is known that the node requires a fixed amount of time to forward each message along to its intended destination and that the node has some finite buffer capacity for storing messages. The event A might represent the event that the node's buffer overflows, and thus a message is lost during the time interval $i = 1, 2, ..., m$. Ideally, the communication network has been designed so that this event is fairly uncommon, but it is desirable to quantify how often this overflow will occur. While it may be fairly straightforward to determine whether or not a buffer overflow has occurred given a specific sequence of arrivals, $X = x$, determining analytically the probability of buffer overflow may be difficult, so we decide to approach this via simulation. Let $\eta_A(X)$ be an indicator function for the event A. That is, let $\eta_A(x) = 1$ if x is such that the event A occurs and $\eta_A(x) = 0$ otherwise. Also, let $x^{(i)}$ be the realization of the random vector X that occurs on the ith trial of the simulation experiment. Then the Monte Carlo approach to estimating p_A is

$$\hat{p}_{A, \text{MC}} = \frac{1}{n} \sum_{i=1}^{n} \eta_A(x^{(i)}). \tag{12.25}$$

Now suppose instead we generate a sequence of random variables $Y = (Y_1, Y_2, ..., Y_m)$ according to a different distribution $f_Y(y)$ and form the estimate

$$\hat{p}_{A, \text{IS}} = \frac{1}{n} \sum_{i=1}^{n} \frac{f_X(y^{(i)})}{f_Y(y^{(i)})} \eta_A(y^{(i)}) \quad . \tag{12.26}$$

It is pretty straightforward (see Exercise 12.11) to establish that this estimator is also unbiased. By carefully choosing the density function, $f_Y(y)$, we may be able to drastically speed up the convergence of the series in Equation (12.26) relative to that in Equation (12.25).

The important step here is to decide how to choose the distribution of Y. In general, the idea is to choose a distribution of Y so that the event $\{ \eta_A(Y) = 1 \}$ occurs more frequently than the event $\{ \eta_A(X) = 1 \}$. In other words, we want to choose a distribution so that the "important" event is sampled more often. It is common to employ the so-called twisted distribution, which calls on concepts taken from large deviation theory. But using these techniques is beyond the scope of this book. Instead, we take an admittedly ad hoc approach here and on a case-by-case basis we try to find a good (but not necessarily optimal) distribution. An example of using importance sampling is provided in the following application section.

12.4 Engineering Application: Simulation of a Coded Digital Communication System

In this section, we demonstrate use of the importance sampling technique outlined in the previous section in the simulation of a digital communication system with convolutional coding. A basic block diagram of the system is illustrated in Figure 12.12. A source outputs binary data, X_i, which is input to an encoder that adds redundancy to the data stream for the purposes of error protection. For this particular example, an encoder with a code rate of 1/2 is used. Simply put, this means that for every one bit input, the convolutional encoder outputs two coded bits, $\{Y_{2i-1}, Y_{2i}\}$. To keep this example simple, the channel is modeled as one which randomly inverts bits with some probability p in a memoryless fashion. That is, what happens to one bit on the channel is independent of any of the other bits. Given a vector of bits input to the channel, $\boldsymbol{Y} = (Y_1, Y_2, ..., Y_n)$, the probability of observing a certain output of the channel $\boldsymbol{R} = (R_1, R_2, ..., R_n)$ is given by

$$\Pr(\boldsymbol{R}|\boldsymbol{Y}) = \prod_{i=1}^{n} \Pr(R_i|Y_i) \quad , \tag{12.27}$$

where

$$\Pr(R_i|Y_i) = \begin{cases} 1-p, & \text{if } R_i = Y_i, \\ p, & \text{if } R_i = \overline{Y}_i. \end{cases} \tag{12.28}$$

The decoder then takes the received sequence output from the channel and determines what was the most likely data sequence.

For this example, it is not necessary to understand the workings of the encoder and decoder. We will just assume the existence of some computer subroutines which simulate their functions. Each trial of our simulation experiment will consist of randomly generating an IID sequence of equally likely data bits, passing them through the encoder, randomly corrupting some of the encoded bits according to the channel model we have developed, and then

Figure 12.12
Block diagram of a digital communication system.

decoding the received sequence. The decoded bits are then compared with the original bits to measure the decoded bit error probability. For the purposes of this example, it is assumed that data are transmitted in blocks of 50 information bits (which are encoded into blocks of 100 coded bits). We refer to each of these blocks as a frame. The channel is simulated by creating an error pattern $\boldsymbol{E} = (E_1, E_2, ..., E_{100})$ where the E_i are a sequence of IID random variables with $\Pr(E_i=1) = p$ and $\Pr(E_i=0) = 1-p$. Then

$$\boldsymbol{R} = \boldsymbol{Y} \oplus \boldsymbol{E}. \tag{12.29}$$

The event $\{E_i=1\}$ implies that the ith bit is inverted in the process of going through the channel while $\{E_i=0\}$ means that the ith bit is received correctly.

Using the standard Monte Carlo approach, the decoded bit error rate is estimated according to

$$\hat{P}_{e, MC} = \frac{1}{mn}\sum_{j=1}^{m} \eta(\boldsymbol{x}^{(j)}, \hat{\boldsymbol{x}}^{(j)}, \boldsymbol{e}) \quad , \tag{12.30}$$

where m is the number of packets transmitted in the simulation; n is the number of bits per packet; and the function $\eta(\boldsymbol{x}^{(j)}, \hat{\boldsymbol{x}}^{(j)}, \boldsymbol{e})$ counts the number of bit errors that occurred in the jth packet. If the channel error rate, p, is fairly high (e.g., a noisy channel), then the Monte Carlo approach will work quite nicely. However, if $p \ll 1$, then channel errors will be infrequent and the decoder will usually correct them. Thus, the decoded error probability will be very small and the Monte Carlo approach will require us to simulate an astronomical number of packets.

Alternatively, consider a simple importance sampling approach. To avoid simulating endless packets which ultimately end up error free, we now create IID error patterns with $\Pr(E_i = 1) = q$, where q is some suitably chosen value which is larger than p. Note that any pattern $\boldsymbol{e} = (e_1, e_2, ..., e_{100})$ that contains exactly w 1s and $100 - w$ zeros will be generated with probability $\Pr(\boldsymbol{e}) = q^w(1-q)^{100-w}$. Let $w(\boldsymbol{e})$ be the number of 1s in a particular error pattern, $\boldsymbol{e}$. Then, our importance sampling estimate of the decoded error probability will be

$$\hat{P}_{e, IS} = \frac{1}{mn}\sum_{j=1}^{m} \frac{p^{w(e)}(1-p)^{100-w(e)}}{q^{w(e)}(1-q)^{100-w(e)}} \eta(\boldsymbol{x}^{(j)}, \hat{\boldsymbol{x}}^{(j)}, \boldsymbol{e}) \quad . \tag{12.31}$$

Simulation results are shown in Figure 12.13 where the channel bit error probability is $p = 0.01$. Note that there are theoretical bounds that tell us that for this example, the actual probability of decoded bit error should be bounded by $1.97 \times 10^{-5} \le P_e \le 6.84 \times 10^{-5}$. To get fairly accurate results via the Monte Carlo approach, we would expect to have to simulate on the order of several hundred thousand packets. It is seen in Figure 12.13 that indeed, even after simulating 10,000 packets, the estimated error probability has still not converged well. For the importance sampling results, we used $q = 0.05$ so that important error events

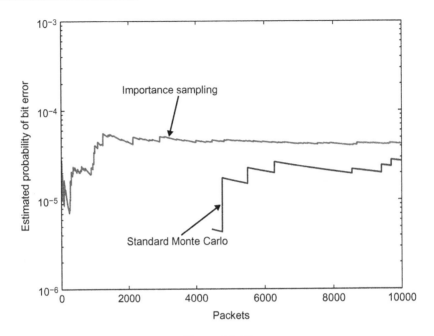

Figure 12.13
Simulation results for a coded digital communication system
using standard Monte Carlo and importance sampling techniques.

occurred much more frequently. As seen in the figure, the estimated error probability has converged rather nicely after simulating only a few thousand packets. Therefore, for this example, using the importance sampling method has sped up the simulation time by a few orders of magnitude.

Exercises

Section 12.1: Computer Generation of Random Variables

12.1 Consider the periodic sequence generated by the four-stage shift register in Figure 12.1. Suppose the sequence is converted to ± 1-valued sequence by mapping all 1's to -1's and all 0's to 1's. One period of the resulting sequence is:

$$-1, -1, -1, -1, 1, 1, 1, -1, 1, 1, -1, -1, 1, -1, 1 .$$

Calculate the autocorrelation function of this periodic sequence. *Note:* Do not just treat this sequence as having finite length. The sequence is infinite in both directions. The finite sequence shown above is just one period of the infinite periodic sequence.

12.2 Sketch the shift register described by the octal number 75. Find the sequence output by this shift register assuming that the shift register is initially loaded with all ones.

12.3 A certain N-stage shift register has its tap connections configured so that it produces an m-sequence. The shift register is initially loaded with the contents $c = (c_0, c_1, ..., c_{N-1})$ resulting in a periodic sequence of period $2^N - 1$. Prove that if the shift register is loaded with some other contents $c' \neq 0$, the new output sequence will be a cyclic shift of the original output sequence. *Hint:* Recall that for an m-sequence, the shift register must cycle through all nonzero states.

12.4 Suppose we create a binary ± 1-valued sequence of length N by drawing N independent realizations of a Bernoulli random variable to form one period of the sequence. Compute the autocorrelation function of this random Bernoulli sequence.

12.5 Suppose a pseudorandom sequence is constructed using the power residue method as described by

$$x_k = ax_{k-1} \bmod q , \ k = 1, 2, 3, ...$$

Find the period and the sequence which results for the following values of (a, q). For each case, assume the seed is $x_0 = 1$.
 (a) $a = 4, q = 9$,
 (b) $a = 5, q = 9$,
 (c) $a = 2, q = 5$,
 (d) $a = 5, q = 11$.

12.6 Suppose a pseudorandom sequence is constructed using the power residue method as discussed in Exercise 12.5. If $q = 11$, find a value of a that leads to a sequence with maximum possible period.

12.7 Find a transformation which will change a uniform random variable into each of the following distributions (see Appendix D for the definitions of these distributions if necessary):
  (a)   arcsine,
  (b)   Cauchy,
  (c)   Rayleigh,
  (d)   Geometric.

Section 12.2: Generation of Random Processes

12.8 Suppose we wish to generate a 10-ms realization of a zero-mean Gaussian random process with a PSD of

$$S(f) = \frac{1}{1 + (f/f_3)^2}.$$

  (a)   Find the bandwidth that contains 99% of the total power in the random process.
  (b)   Determine how many frequency samples are needed for the frequency domain method described in Section 12.2.1.

12.9 Suppose we wanted to generate the random process whose PSD is given in Exercise 12.8 using the time domain method discussed in Section 12.2.2.
  (a)   Find the transfer function of the analog filter which will produce the desired output PSD when the input is a zero-mean, white Gaussian random process.
  (b)   Use a bilinear transformation to convert the analog filter to a digital filter.
  (c)   What is the approximate duration of the impulse response of the digital filter if the 3-dB frequency of the random process is $f_3 = 1\,\text{kHz}$.

12.10 Suppose a zero-mean Gaussian random process, $X(t)$, has a PSD given by

$$S_{XX}(f) = \text{rect}\left(\frac{f}{f_o}\right),$$

where $f_o = 1$ kHz. We desire to simulate a 5-ms segment of this process sampled at a rate of 2 kHz. Therefore, 10 samples will need to be created in our simulation. We will create the samples of the process by first generating 10 uncorrelated Gaussian random variables and then pass those variables through an appropriate transformation matrix, A.
  (a) Find the correlation matrix of the 10 samples.
  (b) Find the form of the matrix A that will produce samples with the desired correlation matrix.

Section 12.3: Simulation of Rare Events

12.11 Suppose we use a Monte Carlo simulation approach to simulate the probability of some rare event A. It is decided that we will repeat the simulation until the event A occurs 35 times. With what accuracy will we estimate p_A to within a 90% confidence level?

12.12 Prove that the importance sampling (IS) estimator of Equation (12.26) is unbiased. That is, show that

$$E[\hat{p}_{A, \text{IS}}] = p_A.$$

12.13 Show that the variance of the IS estimator of Equation (12.26) is given by

$$\text{Var}(\hat{p}_{A, \text{IS}}) = \frac{1}{n} \left\{ E_X \left[\eta_A(X) \frac{f_X(X)}{f_Y(X)} \right] - p_A^2 \right\}.$$

12.14 Suppose the random variable X has an exponential distribution, $f_X(x) = \exp(-x) u(x)$. We wish to estimate the probability of the event $A = \{X > x_0\}$ via simulation. We will compare the standard Monte Carlo estimate,

$$\hat{p}_{A, \text{MC}} = \frac{1}{n} \sum_{k=1}^{n} u(X_i - x_0),$$

where the random variables X_i are chosen according to the exponential distribution specifed by PDF $f_X(x)$, with an importance sampling estimate,

$$\hat{p}_{A, \text{IS}} = \frac{1}{n} \sum_{k=1}^{n} u(Y_i - x_0) \frac{f_X(Y_i)}{f_Y(Y_i)},$$

where the random variables Y_i are chosen from a suitable distribution specified by its PDF, $f_Y(y)$. Note that both estimators are unbiased, so we will compare these estimators by examining their variances.

(a) Find the variance of the Monte Carlo estimate.

(b) Find the variance of the IS estimator assuming that the random variables Y_i are chosen from a scaled exponential distribution, $f_Y(y) = a \exp(-ay) u(y)$.

(c) Assuming that $x_0 = 20$, find the value of a that minimizes the variance of the IS estimator using the scale exponential distribution.

(d) How much faster do you expect the IS simulation to run as compared to the MC simulation?

12.15   Repeat Exercise 12.14 using random variables X that follow a Gaussian distribution $f_X(x) = \exp(-x^2/2)$. Also, for parts (b)–(d) use a shifted distribution for the importance sampling estimator of the form $f_Y(y) = \exp(-(x-a)^2/2)$. Also, for this problem, use $x_o = 6$.

MATLAB Exercises

12.16   We wish to generate a periodic sequence of numbers that cycles through the integers from 1 to 100 in a pseudorandom fashion. Choose a pair of integers (a, q) that can be used in the power residue method to produce a sequence of the desired form. Write a MATLAB program to verify that the sequence produced does in fact cycle through each of the integers 1–100 exactly once each before the sequence repeats.

12.17   Let $X \sim N(2, 1)$, $Y \sim N(0, 1)$, $Z \sim N(0, 1)$, and $W = \sqrt{X^2 + Y^2 + Z^2}$. We desire to find $\Pr(W > 3)$. Write a MATLAB program to estimate this probability through Monte Carlo simulation techniques. If we want to be 90% sure that our estimate is within 5% of the true value, about how many times should we observe the event $\{W > 3\}$? Run your program and provide the estimate of the desired probability. Can you find the probability analytically?

12.18   Write a MATLAB program to generate a realization of the random process from Exercise 12.8 (using the frequency domain method). Use a periodogram to estimate the PSD of the process using the realization of the process you generated. Does your PSD estimate agree with the PSD that the process is designed to possess?

12.19   Write a MATLAB program to generate a realization of the random process from Exercise 12.9 (using the time domain method). Use a periodogram to estimate the PSD of the process using the realization of the process you generated. Does your PSD estimate agree with the PSD that the process is designed to possess?

Review of Set Theory

The purpose for reviewing set theory is to provide a mathematical structure for organizing methods for counting and grouping objects. Set theory may be used to define the probabilities of possible outcomes of experiments. There are two common methods for defining a set. The first method, known as the roster method, is to list the elements of a set. Synonyms for sets include class, aggregate, and collection. We will denote sets by capital letters, A, B, C, etc. The elements or objects of a set will be indicated by lowercase letters, such as a, b, c, etc. If a is an element (or object or member or point) of A, then we denote this as $a \in A$. If a is not an element of A, this is denoted as $a \notin A$. A second way of defining a set is called the property method, which describes some property held by all elements of the set, but is not held by objects that do not belong to the set.

> **Definition A.1:** A set A is said to be a *subset* of another set B if all elements of A are also in B. In which case we write $A \subseteq B$. With this definition, it is possible that the two sets are equal (i.e., they have all the same elements) in which case $A \subseteq B$ and at the same time $B \subseteq A$. If on the other hand, A is a subset of B and there are some elements of B which are not in A, then we say that A is a *proper subset* of B and we write $A \subset B$.

> **Definition A.2:** The *universal set*, S, is the set of all objects under consideration in a given problem, while the *empty set*, $\varnothing$, is the set that contains no elements.

> **Definition A.3:** The *complement* of a set A, written $\overline{A}$, is the set of all elements in S which are not in A. For two sets A and B which satisfy $A \subset B$, the *difference set*, written $B - A$, is the set of elements in B which are not in A.

Note that for any set A, $\varnothing \subseteq A \subseteq S$ and $A \subseteq A$. Also, if $A \subseteq B$ and $B \subseteq C$, then $A \subseteq C$. Finally, we also note the relationship $\overline{S} = \varnothing$.

> **Definition A.4:** For any two sets A and B, the *union* of the two sets, $A \cup B$, is the set of all elements which are contained in either A or B and the *intersection* of the two sets, $A \cap B$, is the set of all elements which are contained in both A and B. In the algebra of sets, the union operation plays the role of addition and so sometimes the

Probability and Random Processes. DOI: 10.1016/B978-0-12-386981-4.00014-0

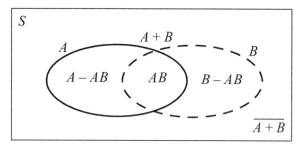

Figure A.1
A Venn diagram illustrating some of the concepts of sets.

notation $A + B$ is used while the intersection operation plays the role of multiplication and hence the alternative notations $A \bullet B$ or AB are common.

Some of the concepts just presented are illustrated using a Venn diagram in Figure A.1. The set A is contained within the thick solid line, the set B within the dashed line, the set $A + B$ is the set of points inside either line and the set AB is the set of points inside both. The set $A - AB$ is the set of points inside the solid line but not inside the dashed line, while the set $B - AB$ is the set of points inside the dashed line but not inside the solid line. The set $\overline{A + B}$ is the set of all points outside of both lines.

Definition A.5: Two sets A and B are said to be *mutually exclusive*, or *disjoint*, if and only if they have no common elements, in which case $A \cap B = \varnothing$. A collection of sets $A_1, A_2, \ldots, A_n$ are said to be *exhaustive* if each element in the universal set is contained in at least one of the sets in the collection. For exhaustive sets, $A_1 \cup A_2 \cup \ldots \cup A_n = S$.

The following laws are consequences of the definitions we have just introduced. The reader should verify these laws to gain familiarity with the algebra of sets.

- Idempotent: $A \cup A = A$, $A \cap A = A$, for all sets A.
- Commutative: $A \cup B = B \cup A$, $A \cap B = B \cap A$, for all sets A and B.
- Associative: $A \cup (B \cup C) = (A \cup R) \cup C = A \cup B \cup C$,
  $A \cap (B \cap C) = (A \cap B) \cap C = A \cap B \cap C$ for all sets A, B, and C.
- Distributive: $A \cap (B \cup C) = (A \cap B) \cup (A \cap C)$,
  $A \cup (B \cap C) = (A \cup B) \cap (A \cup C)$ for all sets A, B, and C.
- Consistency: The three conditions $A \subseteq B$, $A \cap B = A$, and $A \cup B = B$ are all consistent or mutually equivalent.
- Universal bounds: $\varnothing \subseteq A \subseteq S$ for all sets A.
- Product: $\varnothing \cap A = \varnothing$, $S \cap A = A$ for all sets A.
- Sum: $\varnothing \cup A = A$, $S \cup A = S$ for all sets A.

- Involution: $\overline{(\overline{A})} = A$ for all sets A.
- Complementarity: $A \cup \overline{A} = S$, $A \cap \overline{A} = \varnothing$ for all sets A.
- De Morgan's first law: $\overline{A \cup B} = \overline{A} \cap \overline{B}$ for all sets A and B.
- De Morgan's second law: $\overline{A \cap B} = \overline{A} \cup \overline{B}$ for all sets A and B.

De Morgan's laws can be stated in the following way. To find the complement of an expression replace each set by its complement and interchange additions with multiplications and multiplications with additions.

Review of Linear Algebra

The study of probability and random processes draws heavily upon concepts and results from elementary linear algebra. In the text, we assume that the reader has a working knowledge of undergraduate level linear algebra. The aim of this appendix is to provide a review for those who need to brush up on these concepts. This review is not intended to be an exhaustive treatment of the topic, but rather is a summary of selected concepts that are used within the text.

Definition B.1: A matrix is a two-dimensional array of numbers. We say that a matrix has size $m \times n$ if the array has m rows and n columns. If the matrix has only one row, then we refer to it as a *row vector* while if there is only one column it is a *column vector*. A matrix with one row and one column is called a *scalar*. The elements of a matrix, B, are written as $b_{i,j}$, where the first subscript represents the row number and the second subscript gives the column number.

Definition B.2: The *transpose* of a matrix is obtained by exchanging the row and column number of each element. That is, if matrix B has elements $b_{i,j}$, then the element in the ith row and jth column of B^T is $b_{j,i}$. Hence if a matrix has size $m \times n$, then its transpose has size $n \times m$. Also, the transpose of a column vector is a row vector and the transpose of a row vector is a column vector. A matrix B is said to be *symmetric* if $B^T = B$.

Definition B.3: The Hermitian transpose (or just Hermitian) of a matrix B is written B^H and is formed by taking the complex conjugate of the transpose. That is, if matrix B has elements $b_{i,j}$, then the element in the ith row and jth column of B^H is $b_{j,i}^*$. A matrix B is said to be *Hermitian symmetric* if $B^H = B$. Sometimes such a matrix is simply called a *Hermitian* matrix.

Definition B.4: Addition and multiplication of matrices is defined as follows. If two matrices A and B have the same size (same number of rows and columns), then their sum is defined as $C = A + B$ where $c_{i,j} = a_{i,j} + b_{i,j}$. That is, matrix addition is done on an element-by-element basis. If the two matrices do not have the same dimensions, they cannot be added. If A has size $m \times k$ and B has size $k \times n$, then their product $C = AB$ will be an $m \times n$ matrix whose elements are given by

Probability and Random Processes. DOI: 10.1016/B978-0-12-386981-4.00015-2

$$c_{i,j} = \sum_{l=1}^{k} a_{i,l} b_{l,j} . \tag{B.1}$$

In order for the matrix product AB to be defined, the number of columns in A must equal the number of rows in B.

It is also common to define two different products involving vectors, the so-called scalar (or dot) product and the matrix (or cross) product. We have no occasion to use the cross product in this text and so we do not consider it here. For two column vectors a and b (both with the same number of elements), the dot product is defined as $a \bullet b = a^H b$ where the standard definition of matrix multiplication as it applies to vectors is used. Two vectors are said to be *orthogonal* if their dot product is zero. Finally, the norm of a vector is $\|b\| = \sqrt{b^H b}$.

With these definitions in place, the reader should be able to verify the following properties of matrix arithmetic:

- Commutative: $A + B = B + A$ for matrices for which the addition is defined. However, the same property does not usually hold for multiplication. That is, AB does not necessarily equal BA. In fact, BA may not even be defined.
- Associative: $A + (B + C) = (A + B) + C$ and $A(BC) = (AB)C$.
- Distributive: $A(B + C) = AB + AC$.
- Transposes of Sums: $(A + B)^T = A^T + B^T$ and $(A + B)^H = A^H + B^H$.
- Transposes of Products: $(AB)^T = B^T A^T$ and $(AB)^H = B^H A^H$.

In much of this text, many of the matrices we deal with are square. The following definition identifies some characteristics which can be applied to square matrices.

Definition B.5: A matrix B is *diagonal* if its elements satisfy $b_{i,j} = 0$ for all $i \neq j$. A matrix is *upper triangular* if $b_{i,j} = 0$ for all $i > j$ and *lower triangular* if $b_{i,j} = 0$ for all $i < j$. Note that a diagonal matrix is simultaneously upper and lower triangular. Finally, a matrix is an $m \times m$ *identity* matrix if it is a diagonal matrix whose diagonal entries are all equal to one. The letter I is reserved to represent an identity matrix.

An identity matrix has the form

$$I = \begin{bmatrix} 1 & 0 & 0 & \dots & 0 \\ 0 & 1 & 0 & \dots & 0 \\ 0 & 0 & 1 & \dots & 0 \\ \dots & \dots & \dots & & \dots \\ 0 & 0 & 0 & \dots & 1 \end{bmatrix} . \tag{B.2}$$

Sometimes we use a subscript to indicate the dimensions of the identity matrix. For example, $I_{m \times m}$ would represent an identity matrix with m rows and columns. Note that the identity matrix is the identity with respect to matrix multiplication. That is, for any $m \times n$ matrix B, $I_{m \times m} B = B I_{n \times n} = B$. The identity with respect to matrix addition would be a matrix of all zeros.

> **Definition B.6:** The *inverse* of a square matrix B, written B^{-1} (if it exists), is a matrix which satisfies $BB^{-1} = B^{-1}B = I$. If the matrix B is not square, then it may have a left inverse which satisfies $B^{-1}B = I$ and a different right inverse which satisfies $BB^{-1} = I$. In fact, for non-square matrices, the left inverse and right inverse will not have the same dimensions.

The inverse of a square matrix need not exist, but if it does, it is unique and the left and right inverses are identical. If the inverse of a matrix does not exist, then the matrix is said to be *singular*, while if the inverse exists, the matrix is *non-singular* or *invertible*.

The inverse of a matrix plays an important role in the solutions of simultaneous linear equations. Consider the following system of n linear equations in n unknowns, $x_1, x_2, \ldots, x_n$.

$$a_{1,1} x_1 + a_{1,2} x_2 + \ldots + a_{1,n} x_n = b_1,$$

$$a_{2,1} x_1 + a_{2,2} x_2 + \ldots + a_{2,n} x_n = b_2,$$

$$\ldots$$

$$a_{n,1} x_1 + a_{n,2} x_2 + \ldots + a_{n,n} x_n = b_n. \tag{B.3}$$

By defining A as the $n \times n$ matrix of coefficients whose elements are $a_{i,j}$ and the column vectors $x = (x_1, x_2, \ldots, x_n)^T$ and $b = (b_1, b_2, \ldots, b_n)^T$, this system of equations can be written in matrix form as

$$Ax = b. \tag{B.4}$$

Multiplying both sides by the inverse of A leads us to the solution

$$x = A^{-1}b. \tag{B.5}$$

Hence if the coefficient matrix is invertible, the system has a unique solution. On the other hand, if the coefficient matrix is singular, then A^{-1} does not exist and the set of equations does not have a unique solution. This would be the case if some of the equations in (B.3) were linearly dependent (redundant), in which case the system would have more than one solution, or if some of the equations were inconsistent which would lead to no solution. In either case of redundant or inconsistent equations, the rows of the A matrix will be linearly dependent and the inverse will not exist. Conversely, if the rows of A are linearly independent, the matrix will be invertible.

A few properties of matrices that are related to the inverse are listed below:

*   Inverse of Transposes: $(A^T)^{-1} = (A^{-1})^T$.
*   Inverse of Hermitian Transposes: $(A^H)^{-1} = (A^{-1})^H$.
*   Inverse of Products: $(AB)^{-1} = B^{-1}A^{-1}$.
*   Inverse of Identities: $I^{-1} = I$.

A single quantity that is extremely useful in characterizing the invertibility of a matrix is its determinant. The determinant can be rather difficult to define in a simple manner, but it has many useful properties. We use a recursive definition which may seem rather cryptic, but is probably the simplest definition and is also consistent with how determinants are often calculated.

> **Definition B.7:** Let B be an $n \times n$ matrix with elements $b_{i,j}$. Define $B_{(i,j)}$ to be the $(n-1) \times (n-1)$ matrix obtained by removing the ith row and jth column from B. Then, the determinant of B is defined recursively according to

$$\det(B) = \sum_{j=1}^{n} (-1)^{i+j} \det(B_{(i,j)}), \quad \text{for any } i = 1, 2, ..., n. \tag{B.6}$$

This recursion, together with the definition that for a 1×1 matrix $B = [b]$, $\det(B) = b$, is sufficient to calculate the determinant of any $n \times n$ matrix.

To see how this works, consider a 2×2 matrix.

$$\det\left(\begin{bmatrix} a & b \\ c & d \end{bmatrix}\right) = a\det([d]) - b\det([c]) = ad - bc. \tag{B.7}$$

This was obtained using $i = 1$ in (B.6). We could have also used $i = 2$ and achieved the same result

$$\det\left(\begin{bmatrix} a & b \\ c & d \end{bmatrix}\right) = -c\det([b]) + d\det([a]) = -cb + da. \tag{B.8}$$

For a general 3×3 matrix, the determinant works out to be

$$\det\left(\begin{bmatrix} a & b & c \\ d & e & f \\ g & h & i \end{bmatrix}\right) = a\det\left(\begin{bmatrix} e & f \\ h & i \end{bmatrix}\right) - b\det\left(\begin{bmatrix} d & f \\ g & i \end{bmatrix}\right) + c\det\left(\begin{bmatrix} d & e \\ g & h \end{bmatrix}\right)$$

$$= a(ei - fh) - b(di - fg) + c(dh - eg). \tag{B.9}$$

Probably the most important property of determinants is that if $\det(B) = 0$ then the matrix B is singular and conversely if $\det(B) \neq 0$, then B is invertible. This, along with some other important properties, are listed below. We will not prove any of these properties in this review.

- Invertibility: $\{\det(B) = 0\} \Leftrightarrow \{B \text{ is singular}\}$.
- Row Exchange: If the matrix A is formed by exchanging any two rows in the matrix B, then $\det(A) = -\det(B)$.
- Identity Matrices: For any identity matrix, $\det(I) = 1$.
- Triangular Matrices: If B is a triangular matrix,

$$\det(B) = \prod_{i=1}^{n} b_{i,i}.$$

  That is, the determinant is the product of the diagonal elements. Note that diagonal matrices are a special case and this property applies to them as well.
- Products of Matrices: $\det(AB) = \det(A)\det(B)$.
- Inverses: $\det(B^{-1}) = 1/\det(B)$.
- Transposes: $\det(B^T) = \det(B)$.
- Hermitian Transposes: $\det(B^H) = (\det(B))^*$.

In addition to computing determinants of matrices, we will also find need throughout the text to compute eigenvalues and eigenvectors of square matrices.

Definition B.8: For a square matrix B, the scalar λ is an eigenvalue and the vector x is a corresponding eigenvector if

$$Bx = \lambda x. \tag{B.10}$$

Note that the previous equation can be written in the slightly different form $(B - \lambda I)x = 0$. The eigenvector x will be non-zero only if the matrix $B - \lambda I$ is singular. If it were non-singular, we could multiply both sides by its inverse to obtain the trivial solution $x = 0$. Hence, the eigenvalues of the matrix B must be solutions to the equation

$$\det(\boldsymbol{B} - \lambda \boldsymbol{I}) = 0. \tag{B.11}$$

This is the so-called characteristic equation for the matrix $\boldsymbol{B}$. For an $n \times n$ matrix, $\boldsymbol{B}$, the characteristic equation will be an nth order polynomial equation in λ and hence an $n \times n$ matrix will have n eigenvalues (although some of them may be repeated). Corresponding to each eigenvalue, λ_k, is an eigenvector, $\boldsymbol{x}_k$. Note that the eigenvector is not unique since if $\boldsymbol{x}_k$ satisfies (B.10) then any multiple of $\boldsymbol{x}_k$ will also satisfy the same equation and hence will also be an eigenvector. In order to resolve this ambiguity, it is common to normalize the eigenvectors so that $\|\boldsymbol{x}\|^2 = 1$, but the vector is still an eigenvector even if it is not normalized. In the case of repeated eigenvalues, there may also be corresponding eigenvectors which differ by more than just a scale constant.

Before listing the important properties of eigenvalues and eigenvectors, it is necessary to include one more definition.

> **Definition B.9 :** A matrix $\boldsymbol{B}$ is *positive definite* if $\boldsymbol{z}^H \boldsymbol{B} \boldsymbol{z} > 0$ for any vector $\boldsymbol{z}$ and it is *negative definite* if $\boldsymbol{z}^H \boldsymbol{B} \boldsymbol{z} < 0$ for all $\boldsymbol{z}$. If $\boldsymbol{z}^H \boldsymbol{B} \boldsymbol{z} \geq 0$, then the matrix is referred to as *positive semi-definite* and if $\boldsymbol{z}^H \boldsymbol{B} \boldsymbol{z} \leq 0$, the matrix is *negative semi-definite*.

With this definition in place, we now list the following properties of eigenvalues and eigenvectors.

- Trace of a matrix: $\text{trace}(\boldsymbol{B}) = \sum_{k=1}^{n} b_{k,k} = \sum_{k=1}^{n} \lambda_k$. That is, the sum of the eigenvalues is equal to the sum of the diagonal elements of a matrix, also known as its *trace*.
- Determinant of a matrix:

$$\det(\boldsymbol{B}) = \prod_{k=1}^{n} \lambda_k.$$

  That is, the product of the eigenvalues is the determinant of a matrix. As a result, any singular matrix must have at least one eigenvalue which is zero.
- Triangular matrices: If a matrix $\boldsymbol{B}$ is triangular (or diagonal), the eigenvalues are just the diagonal entries, $\lambda_k = b_{k,k}$.
- Positive and Negative Definite Matrices: If $\boldsymbol{B}$ is positive definite, then its eigenvalues are all real and positive while if $\boldsymbol{B}$ is positive semi-definite, all its eigenvalues are non-negative. Similarly if $\boldsymbol{B}$ is negative definite than the eigenvalues are negative while if $\boldsymbol{B}$ is negative semi-definite, the eigenvalues are non-positive.
- Linear Independence: If the eigenvectors $\boldsymbol{x}_1, \boldsymbol{x}_2, ..., \boldsymbol{x}_n$ are non-zero and correspond to distinct (not repeated) eigenvalues $\lambda_1, \lambda_2, ..., \lambda_n$, then the eigenvectors are linearly independent.
- Diagonal Form: Suppose $\boldsymbol{B}$ is an $n \times n$ matrix and has n linearly independent eigenvectors. Construct a matrix $\boldsymbol{S}$ whose columns are the n eigenvectors and a

diagonal matrix Λ whose diagonal entries are the eigenvalues. Then the matrix $\boldsymbol{B}$ can be factored as $\boldsymbol{B} = \boldsymbol{S}\Lambda\boldsymbol{S}^{-1}$.

• Powers of a Matrix: A direct result of the previous property is that for matrices with linearly independent eigenvectors, $\boldsymbol{B}^k = \boldsymbol{S}\Lambda^k\boldsymbol{S}^{-1}$. Furthermore, since Λ is diagonal, Λ^k is computed by raising each diagonal entry to the kth power.

In many of the applications encountered in this text, the matrices we are dealing with are Hermitian. These matrices possess further properties which are not necessarily shared by all matrices. These additional properties make Hermitian matrices particularly convenient to work with.

• Positive Semi-definite: Any Hermitian matrix has all real eigenvalues and is at least positive semi-definite. Furthermore, if the matrix is also non-singular, then it will be positive definite.
• Orthogonal Eigenvectors: Eigenvectors of a Hermitian matrix which correspond to different eigenvalues are orthogonal.
• Spectral Decomposition: Any Hermitian matrix can be decomposed into the form

$$\boldsymbol{B} = \boldsymbol{U}\Lambda\boldsymbol{U}^H = \sum_{k=1}^{n} \lambda_k \boldsymbol{x}_k \boldsymbol{x}_k^H, \tag{B.12}$$

where $\boldsymbol{U}$ is the matrix whose columns are the eigenvectors (normalized).

Review of Signals and Systems

This appendix provides a summary of some important results in the area of signal representation and linear time invariant systems. Any engineering student embarking on a serious study of probability and random processes should be familiar with these concepts and hence a rigorous development is not attempted here. Rather, this review is intended as a brief refresher for those who need it and also as a quick reference for some important results that are used throughout the text. In this appendix, attention is focused on deterministic signals and systems in both continuous and discrete time.

Definition C.1: Consider a periodic signal $x(t)$ whose period is T_0. That is, $x(t) = x(t + T_0)$ for all t. The inverse of the period $f_0 = 1/T_0$ is called the *fundamental frequency* of $x(t)$ and any frequency, $f_n = nf_0$, which is a multiple of the fundamental frequency is called a *harmonic*.

Any periodic signal (subject to some mild constraints known as the Dirichlet conditions) can be represented as a linear combination of complex exponential signals, $\exp(j2\pi f_n t)$, whose frequencies are at the harmonics of the signal. That is, if $x(t)$ is periodic with period T_0, then

$$x(t) = \sum_{k=-\infty}^{\infty} x_n e^{j2\pi n f_0 t}. \tag{C.1}$$

This is known as the *Fourier Series* expansion and the series coefficients can be computed according to

$$x_n = \frac{1}{T_0} \int_{T_0} x(t) e^{-j2\pi n f_0 t} dt. \tag{C.2}$$

Since the signal is periodic, the integral in the previous expression can be taken over any convenient interval of length T_0. In general, the series coefficients are complex numbers, and it is common to express them in terms of their magnitude and phase, $x_n = |x_n| \exp(j\angle x_n)$. The Fourier series coefficients display the frequency content of periodic signals.

Probability and Random Processes. DOI: 10.1016/B978-0-12-386981-4.00016-4

For signals which are not periodic, the Fourier transform can be used to display the frequency content of a signal. The Fourier transform of a signal is given by

$$X(f) = \int_{-\infty}^{\infty} x(t)e^{-j2\pi ft}dt,$$ (C.3)

and the inverse Fourier Transform is

$$x(t) = \int_{-\infty}^{\infty} X(f)e^{j2\pi ft}df.$$ (C.4)

Sometimes we use the notation $x(t) \leftrightarrow X(f)$ to indicate that $x(t)$ and $X(f)$ are a Fourier Transform pair. A table of some common Fourier Transform pairs is provided in Table E.1. Some of the more important properties of Fourier Transforms are listed below.

- Linearity: If $x(t) \leftrightarrow X(f)$ and $y(t) \leftrightarrow Y(f)$, then $ax(t) + by(t) \leftrightarrow aX(f) + bY(f)$ for any constants a and b.
- Symmetry: If $x(t)$ is real valued, then $X(-f) = X^*(f)$. As a result $|X(f)|$ must then be an even function of f and $\angle X(f)$ must be an odd function of f. In addition, if $x(t)$ is both real and even, then $X(f)$ will be real and even.
- Time Shifting: If $x(t) \leftrightarrow X(f)$, then $x(t - t_o) \leftrightarrow e^{-j2\pi ft_o}X(f)$. As a consequence, shifting a signal in time does not alter the magnitude of its Fourier Transform.
- Differentiation: If $x(t) \leftrightarrow X(f)$, then $\dfrac{dx(t)}{dt} \leftrightarrow j2\pi f X(f)$.
- Integration: If $x(t) \leftrightarrow X(f)$, then $\int_{-\infty}^{t} x(u)du \leftrightarrow \dfrac{X(f)}{j2\pi f} + \dfrac{1}{2}X(0)\delta(f)$. The term $X(0)$ that appears in this expression is the *direct current* (d.c.) value of the signal.
- Time and Frequency Scaling: If $x(t) \leftrightarrow X(f)$, then $x(at) = \dfrac{1}{|a|}X\left(\dfrac{f}{a}\right)$ for any constant $a \neq 0$.
- Parseval's Relation: If $x(t) \leftrightarrow X(f)$, then $\int_{-\infty}^{\infty} |x(t)|^2 dt = \int_{-\infty}^{\infty} |X(f)|^2 df$. This is a statement of conservation of energy. That is, the energy in the time domain is equal to the energy in the frequency domain.
- Convolution: If $x(t) \leftrightarrow X(f)$ and $y(t) \leftrightarrow Y(f)$, then

$$x(t)*y(t) = \int_{-\infty}^{\infty} x(u)y(t-u)dt \leftrightarrow X(f)Y(f).$$

For signals in discrete time, $x[n]$, a Discrete-time Fourier Transform (DTFT) is defined according to:

$$X(f) = \sum_{n=-\infty}^{\infty} x[n]e^{-j2\pi fn},$$ (C.5)

and the inverse DTFT is

$$x[n] = \int_{-1/2}^{1/2} X(f)e^{j2\pi fn}df .$$

(C.6)

Since $X(f)$ is periodic with period of 1, the integral in the previous equation can be taken over any interval of length 1. It is common to view the DTFT using discrete frequency samples as well. In that case, the definition of the DTFT and its inverse is modified to give the N-point DTFT:

$$X[m] = \sum_{n=0}^{N-1} x[n]e^{-j2\pi mn/N},$$

(C.7)

$$x[n] = \frac{1}{N}\sum_{m=0}^{N-1} X[m]e^{j2\pi mn/N}.$$

(C.8)

Alternatively, by replacing $\exp(j2\pi f)$ with z in the definition of the DFT, we get the z-transform:

$$X(z) = \sum_{n=-\infty}^{\infty} x[n]z^{-n} .$$

(C.9)

The inverse z-transform is given by a complex contour integral,

$$x[n] = \frac{1}{2\pi j}\oint X(z)z^{n-1}dz ,$$

(C.10)

where the contour of integration is any closed contour which encircles the origin in the counterclockwise direction and is within the region of convergence of $X(z)$. Because of the complicated nature of the inverse transform, it is common to compute these inverse transforms via tables. A table of some common z-transform pairs is provided in Table E.2.

These various transform representations of signals are particularly useful when studying the passage of signals through linear time-invariant (LTI) systems.

> **Definition C.2:** Suppose when $x(t)$ is input to a system, the output is $y(t)$. The system is said to be *time-invariant* if the input $x(t-t_0)$ produces an output of $y(t-t_0)$. That is, a time delay in the input produces the same time delay on the output but no other changes to the output. Furthermore, suppose the two inputs $x_1(t)$ and $x_2(t)$ produce the two outputs $y_1(t)$ and $y_2(t)$, respectively. Then, the system is *linear* if the input $ax_1(t)+bx_2(t)$ produces the output $ay_1(t)+by_2(t)$ for any constants a and b. Identical definitions apply to discrete time systems as well.

A direct consequence of the linearity of a system is the concept of superposition which states that if the input can be written as a linear combination of several terms $x(t) = a_1 x_1(t) + a_2 x_2(t) + \ldots + a_n x_n(t)$, then the corresponding output can be written as the same linear combination of the corresponding outputs $y(t) = a_1 y_1(t) + a_2 y_2(t) + \ldots + a_n y_n(t)$.

Any LTI system can be described in terms of its impulse response, $h(t)$. If the input is a delta (impulse) function, $\delta(t)$, the output is then the impulse response $y(t) = h(t)$. For any LTI system, the input/output relationship is given in terms of the impulse response according to the convolution integral

$$y(t) = x(t)*h(t) = \int_{-\infty}^{\infty} x(u)h(t-u)du = \int_{-\infty}^{\infty} x(t-u)h(u)du. \qquad \text{(C.11)}$$

If the input is a complex exponential at some frequency f, i.e., $x(t) = \exp(j2\pi ft)$, then the corresponding output is then

$$y(t) = e^{j2\pi ft} \int_{-\infty}^{\infty} h(u)e^{-j2\pi fu}du = e^{j2\pi ft}H(f). \qquad \text{(C.12)}$$

That is, the output will also be a complex exponential whose magnitude and phase have been adjusted according to the complex number $H(f)$. This function of frequency is called the *transfer function* of the system and is the Fourier Transform of the impulse response. Since complex exponentials form eigenfunctions of any LTI system, when studying LTI systems, it is convenient to decompose signals into linear combinations of complex exponentials. If $x(t)$ is a periodic signal it can be written as a linear combination of complex exponentials through its Fourier Series representation,

$$x(t) = \sum_k x_k e^{j2\pi k f_o t}. \qquad \text{(C.13)}$$

Then using the concept of superposition together with the previous result, we find that the output of an LTI system, when $x(t)$ is input is

$$y(t) = \sum_k x_k H(k f_o)e^{j2\pi k f_o t} = \sum_k y_k e^{j2\pi k f_o t}. \qquad \text{(C.14)}$$

Hence the Fourier series coefficients of the input and output of an LTI system are related by the simple form

$$y_k = x_k H(k f_o). \qquad \text{(C.15)}$$

A similar relationship holds for the Fourier transforms of non-periodic signals. Taking Fourier Transforms of both sides of (C.11) and using the convolution property of Fourier Transforms results in

$$Y(f) = X(f)H(f).$$
(C.16)

Identical relationships hold for the DFT and z-transforms as well.

Summary of Common Random Variables

This appendix provides a quick reference of some of the most common random variables. Special functions that are used in this appendix are defined in the following list:

- Gamma function: $\Gamma(\alpha) = \int_0^\infty u^{\alpha-1}e^{-u}du$, $\mathrm{Re}[\alpha] > 0$.

- Incomplete Gamma function: $\gamma(\alpha, \beta) = \int_0^\beta u^{\alpha-1}e^{-u}du$, $\mathrm{Re}[\alpha] > 0$.

- Beta function: $B(a, b) = \int_0^1 u^{a-1}(1-u)^{b-1}du = \dfrac{\Gamma(a)\Gamma(b)}{\Gamma(a+b)}$.

- Incomplete Beta function: $\beta(a, b, x) = \int_0^x u^{a-1}(1-u)^{b-1}du$, $0 < x < 1$.

- Modified Bessel function of order m: $I_m(x) = \dfrac{1}{2\pi}\int_0^{2\pi} e^{x\cos(\theta)}\cos(m\theta)d\theta$.

- Q-function: $Q(x) = \int_x^\infty \dfrac{1}{\sqrt{2\pi}}\exp\left(-\dfrac{u^2}{2}\right)du$.

- Marcum's Q-function: $Q(\alpha, \beta) = \int_\beta^\infty u\exp\left(-\dfrac{\alpha^2+u^2}{2}\right)I_0(\alpha u)du$.

Continuous Random Variables

Arcsine

For any $b > 0$

$$f_X(x) = \frac{1}{\pi\sqrt{b^2-x^2}}, \ -b < x < b. \tag{D.1}$$

$$F_X(x) = \begin{cases} 0, & x < -b, \\ \dfrac{1}{2} + \dfrac{1}{\pi}\sin^{-1}\left(\dfrac{x}{b}\right), & -b \le x \le b, \\ 1, & x > b. \end{cases} \tag{D.2}$$

$$\mu_X = 0, \ \sigma_X^2 = \frac{b^2}{2}. \tag{D.3}$$

Probability and Random Processes. DOI: 10.1016/B978-0-12-386981-4.00017-6

Note:

1) Formed by a transformation $X = b\cos(2\pi U + \theta)$ where b and θ are constants and U is a uniform random variable over $[0, 1)$.

Beta

For any $a > 0$ and $b > 0$

$$f_X(x) = \frac{1}{B(a, b)}x^{a-1}(1-x)^{b-1}, \quad 0 < x < 1. \tag{D.4}$$

$$F_X(x) = \begin{cases} 0, & x < 0, \\ \dfrac{B(a, b, x)}{B(a, b)}, & 0 \leq x \leq 1, \\ 1, & x > 1. \end{cases} \tag{D.5}$$

$$\mu_X = \frac{a}{a+b}, \quad \sigma_X^2 = \frac{ab}{(a+b)^2(a+b+1)}. \tag{D.6}$$

Cauchy

For any $b > 0$

$$f_X(x) = \frac{b/\pi}{b^2 + x^2}. \tag{D.7}$$

$$F_X(x) = \frac{1}{2} + \frac{1}{\pi}\tan^{-1}\left(\frac{x}{b}\right). \tag{D.8}$$

$$\Phi_X(\omega) = e^{-b|\omega|}. \tag{D.9}$$

Notes:

1) Both the mean and variance are undefined.
2) Formed by a transformation of the form $X = b\tan(2\pi U)$ where U is uniform over $[0, 1)$.

Chi-Square

For integer $n > 0$

$$f_X(x) = \frac{x^{n/2-1}}{2^{n/2}\Gamma(n/2)}e^{-x/2}, \qquad x \geq 0. \tag{D.10}$$

$$F_X(x) = \begin{cases} 0, & x < 0, \\ \dfrac{\gamma(n/2, x/2)}{\Gamma(n/2)}, & x \geq 0. \end{cases} \tag{D.11}$$

$$\Phi_X(\omega) = \frac{1}{(1-2j\omega)^{n/2}}. \tag{D.12}$$

$$\mu_X = n, \ \sigma_X^2 = 2n. \tag{D.13}$$

Notes:

1) The Chi-Square random variable is a special case of the Gamma random variable.
2) The parameter n is referred to as the number of degrees of freedom of the chi-square random variable.
3) The Chi-Square random variable is formed by a transformation of the form $X = \sum_{k=1}^{n} Z_k^2$, where the Z_k are IID zero-mean, unit variance, Gaussian random variables.

Erlang

For any integer $n > 0$ and any $b > 0$

$$f_X(x) = \frac{b^n x^{n-1}e^{-bx}}{(n-1)!}, \qquad x \geq 0. \tag{D.14}$$

$$F_X(x) = \begin{cases} 0, & x < 0, \\ \dfrac{\gamma(n, bx)}{(n-1)!}, & x \geq 0. \end{cases} \tag{D.15}$$

$$\Phi_X(\omega) = \frac{1}{(1-j\omega/b)^n}. \tag{D.16}$$

$$\mu_X = n/b, \ \sigma_X^2 = n/b^2. \tag{D.17}$$

Notes:

1) The Erlang random variable is a special case of the Gamma random variable.
2) The Erlang random variable is formed by summing n IID exponential random variables.
3) The CDF can also be written as a finite series

$$\frac{\gamma(n, bx)}{(n-1)!} = 1 - e^{bx}\sum_{k=0}^{n-1}\frac{(bx)^k}{k!}, \quad x \geq 0. \tag{D.18}$$

Exponential

For any $b > 0$

$$f_X(x) = be^{-bx}, \quad x \geq 0. \tag{D.19}$$

$$F_X(x) = \begin{cases} 0, & x < 0, \\ 1 - e^{-bx}, & x \geq 0. \end{cases} \tag{D.20}$$

$$\Phi_X(\omega) = \frac{1}{1 - j\omega/b}. \tag{D.21}$$

$$\mu_X = 1/b, \ \sigma_X^2 = 1/b^2. \tag{D.22}$$

Notes:

1) The Exponential random variable is a special case of the Erlang and Gamma random variables.
2) The Exponential random variable possesses the memoryless property,

$$f_X(x|X > a) = f_X(x - a). \tag{D.23}$$

F

For any integers $n > 0$ and $m > 0$

$$f_X(x) = \frac{\left(\frac{n}{m}\right)^{n/2}}{B\left(\frac{n}{2}, \frac{m}{2}\right)}x^{\frac{n}{2}-1}\left(1 + \frac{n}{m}x\right)^{-\frac{n+m}{2}}, \quad x > 0. \tag{D.24}$$

$$\mu_X = \frac{m}{m-2} \text{ for } m > 2, \qquad \sigma_X^2 = \frac{m^2(2n+2m-4)}{n(m-2)^2(m-4)} \text{ for } m > 4. \tag{D.25}$$

Note:

1) If U and V are independent Chi-square random variables with n and m degrees of freedom respectively, then $F = (U/n)/(V/m)$ will be an F random variable with n and m degrees of freedom.

Gamma

For any $a > 0$ and $b > 0$

$$f_X(x) = \frac{b^a x^{a-1} e^{-bx}}{\Gamma(a)}, \qquad x \geq 0. \tag{D.26}$$

$$F_X(x) = \frac{\gamma(a, bx)}{\Gamma(a)}. \tag{D.27}$$

$$\Phi_X(\omega) = \frac{1}{(1 - j\omega/b)^a}. \tag{D.28}$$

$$\mu_X = a/b, \ \sigma_X^2 = a/b^2. \tag{D.29}$$

Note:

1) The Gamma random variable contains the Chi-Square, Erlang, and Exponential random variables as special cases.

Gaussian

For any μ and any $\sigma > 0$

$$f_X(x) = \frac{1}{\sqrt{2\pi\sigma^2}} \exp\left(-\frac{(x-\mu)^2}{2\sigma^2}\right). \tag{D.30}$$

$$F_X(x) = 1 - Q\left(\frac{x-\mu}{\sigma}\right). \tag{D.31}$$

$$\Phi_X(\omega) = \exp\left(j\omega\mu - \frac{1}{2}\omega^2\sigma^2\right). \tag{D.32}$$

$$\mu_X = \mu, \ \sigma_X^2 = \sigma^2. \tag{D.33}$$

Gaussian-Multivariate

For any n element column vector $\boldsymbol{\mu}$ and any valid $n \times n$ covariance matrix $\boldsymbol{C}$

$$f_X(x) = \frac{1}{(2\pi)^{n/2}\det(\boldsymbol{C})} \exp\left(-\frac{1}{2}(\boldsymbol{X}-\boldsymbol{\mu})^T\boldsymbol{C}^{-1}(\boldsymbol{X}-\boldsymbol{\mu})\right). \tag{D.34}$$

$$\Phi_X(\omega) = \exp\left(j\boldsymbol{\mu}^T\omega - \frac{1}{2}\omega^T\boldsymbol{C}\omega\right). \tag{D.35}$$

$$E[\boldsymbol{X}] = \boldsymbol{\mu}, \quad E[(\boldsymbol{X}-\boldsymbol{\mu})(\boldsymbol{X}-\boldsymbol{\mu})^T] = \boldsymbol{C}. \tag{D.36}$$

Laplace

For any $b > 0$

$$f_X(x) = \frac{b}{2}\exp(-b|x|). \tag{D.37}$$

$$F_X(x) = \begin{cases} \frac{1}{2}e^{bx}, & x < 0, \\ 1 + -\frac{1}{2}e^{-bx}, & x \geq 0. \end{cases} \tag{D.38}$$

$$\Phi_X(\omega) = \frac{1}{1 + (\omega/b)^2}. \tag{D.39}$$

$$\mu_X = 0, \quad \sigma_X^2 = 2/b^2. \tag{D.40}$$

Log-Normal

For any μ and any $\sigma > 0$

$$f_X(x) = \frac{1}{x\sqrt{2\pi\sigma^2}}\exp\left(-\frac{(\ln(x)-\mu)^2}{2\sigma^2}\right), \quad x > 0. \tag{D.41}$$

$$F_X(x) = \begin{cases} 0, & x < 0, \\ 1 - Q\left(\frac{\ln(x)-\mu}{\sigma}\right), & x \geq 0. \end{cases} \tag{D.42}$$

$$\mu_X = \exp\left(\mu + \frac{\sigma^2}{2}\right), \quad \sigma_X^2 = [\exp(\sigma^2) - 1]\exp(2\mu + \sigma^2). \tag{D.43}$$

Notes:

1) The log-normal random variable is formed by a transformation of the form $X = \exp(Z)$, where Z is a Gaussian random variable with mean μ and variance σ^2.
2) It is common to find instances in the literature where σ is referred to as the standard deviation of the log-normal random variable. This is a misnomer. The quantity σ is not the standard deviation of the log-normal random variable but rather is the standard deviation of the underlying Gaussian random variable.

Nakagami

For any $b > 0$ and $m > 0$

$$f_X(x) = \frac{2m^m}{\Gamma(m)b^m}x^{2m-1}\exp\left(-\frac{m}{b}x^2\right), \qquad x \geq 0. \tag{D.44}$$

$$F_X(x) = \begin{cases} 0, & x < 0, \\ \dfrac{\gamma\left(m, \dfrac{m}{b}x^2\right)}{\Gamma(m)}, & x \geq 0. \end{cases} \tag{D.45}$$

$$\mu_X = \frac{\Gamma(m+1/2)}{\Gamma(m)}\sqrt{\frac{b}{m}}, \qquad \sigma_X^2 = b - \mu_X^2. \tag{D.46}$$

Rayleigh

For any $\sigma > 0$

$$f_X(x) = \frac{x}{\sigma^2}\exp\left(-\frac{x^2}{2\sigma^2}\right), \qquad x \geq 0. \tag{D.47}$$

$$F_X(x) = \begin{cases} 0, & x < 0, \\ 1 - \exp\left(-\dfrac{x^2}{2\sigma^2}\right), & x \geq 0. \end{cases} \tag{D.48}$$

$$\mu_X = \sqrt{\frac{\pi\sigma^2}{2}}, \qquad \sigma_X^2 = \frac{(4-\pi)\sigma^2}{2}. \tag{D.49}$$

Notes:

1) The Rayleigh random variable arises when performing a Cartesian to Polar transformation of two independent zero-mean Gaussian random variables. That is, if Y_1 and Y_2 are independent zero-mean Gaussian random variables with variances of σ^2, then $X = \sqrt{Y_1^2 + Y_2^2}$ follows a Rayleigh distribution.

2) The Rayleigh random variable is a special case of the Rician random variable.

Rician

For any $a \geq 0$ and any $\sigma > 0$

$$f_X(x) = \frac{x}{\sigma^2} \exp\left(-\frac{x^2 + a^2}{2\sigma^2}\right) I_0\left(\frac{ax}{\sigma^2}\right), \qquad x \geq 0. \tag{D.50}$$

$$F_X(x) = \begin{cases} 0, & x < 0, \\ 1 - Q\left(\frac{a}{\sigma}, \frac{x}{\sigma}\right), & x \geq 0. \end{cases} \tag{D.51}$$

$$\mu_X = \sqrt{\frac{\pi\sigma^2}{2}} \exp\left(-\frac{a^2}{4\sigma^2}\right)\left[\left(1 + \frac{a^2}{2\sigma^2}\right) I_0\left(\frac{a^2}{4\sigma^2}\right) + \frac{a^2}{2\sigma^2} I_1\left(\frac{a^2}{4\sigma^2}\right)\right]. \tag{D.52}$$

$$\sigma_X^2 = 2\sigma^2 + a^2 - \mu_X^2. \tag{D.53}$$

Notes:

1) The Rician random variable arises when performing a Cartesian to Polar transformation of two independent Gaussian random variables. That is, if Y_1 and Y_2 are independent Gaussian random variables with means of μ_1 and μ_2, respectively and equal variances of σ^2, then $X = \sqrt{Y_1^2 + Y_2^2}$ follows a Rician distribution, with $a = \sqrt{\mu_1^2 + \mu_2^2}$.

2) The ratio a^2/σ^2 is often referred to as the Rician parameter or the Rice factor. As the Rice factor goes to zero, the Rician random variable becomes a Rayleigh random variable.

Student t

For any integer $n > 0$

$$f_X(x) = \frac{1}{B(n/2, 1/2)\sqrt{n}}\left(1 + \frac{x^2}{n}\right)^{-\frac{n+1}{2}}. \tag{D.54}$$

$$\mu_X = 0, \qquad \sigma_X^2 = \frac{n}{n-2} \text{ for } n > 2. \tag{D.55}$$

Notes:

1) This distribution was first published by W. S. Gosset in 1908 under the pseudonym "Student." Hence this distribution has come to be known as the Student's *t*-distribution.
2) The parameter *n* is referred to as the number of degrees of freedom.
3) If X_i, $i = 1, 2, ..., n$, is a sequence of IID Gaussian random variables and $\hat{\mu}$ and $\hat{s^2}$ are the sample mean and sample variance, respectively, then the ratio

   $T = (\hat{\mu} - \mu)/\sqrt{\hat{s^2}/n}$ will have a *t*-distribution with $n - 1$ degrees of freedom.

Uniform

For any $a < b$

$$f_X(x) = \frac{1}{b-a}, \quad a \le x < b. \tag{D.56}$$

$$F_X(x) = \begin{cases} 0, & x < a, \\ \dfrac{x-a}{b-a}, & a \le x \le b, \\ 1, & x > b. \end{cases} \tag{D.57}$$

$$\Phi_X(\omega) = \frac{e^{jb\omega} - e^{ja\omega}}{j\omega(b-a)}. \tag{D.58}$$

$$\mu_x = \frac{a+b}{2}, \quad \sigma_X^2 = \frac{(b-a)^2}{12}. \tag{D.59}$$

Weibull

For any $a > 0$ and any $b > 0$

$$f_X(x) = abx^{b-1}\exp(-ax^b), \quad x \ge 0. \tag{D.60}$$

$$F_X(x) = \begin{cases} 0, & x < 0, \\ 1 - \exp(-ax^b), & x \ge 0. \end{cases} \tag{D.61}$$

$$\mu_X = \frac{\Gamma\left(1 + \frac{1}{b}\right)}{a^{1/b}}, \quad \sigma_X^2 = \frac{\Gamma\left(1 + \frac{2}{b}\right) - \left[\Gamma\left(1 + \frac{1}{b}\right)\right]^2}{a^{2/b}}. \tag{D.62}$$

Note:

1) The Weibull random variable is a generalization of the Rayleigh random variable and reduces to a Rayleigh random variable when $b = 2$.

Discrete Random Variables

Bernoulli

For $0 < p < 1$

$$P_X(k) = \begin{cases} 1 - p, & k = 0, \\ p, & k = 1, \\ 0, & \text{otherwise.} \end{cases} \tag{D.63}$$

$$H_X(z) = 1 - p(1 - z) \quad \text{for all } z. \tag{D.64}$$

$$\mu_X = p, \; \sigma_X^2 = p(1 - p). \tag{D.65}$$

Binomial

For $0 < p < 1$ and any integer $n > 0$

$$P_X(k) = \begin{cases} \binom{n}{k} p^k (1 - p)^{n - k}, & k = 0, 1, 2, \ldots, n, \\ 0, & \text{otherwise.} \end{cases} \tag{D.66}$$

$$H_X(z) = (1 - p(1 - z))^n \quad \text{for any } z. \tag{D.67}$$

$$\mu_x = np, \quad \sigma_X^2 = np(1 - p). \tag{D.68}$$

Note:
1) The binomial random variable is formed as the sum of n independent Bernoulli random variables.

Geometric

For $0 < p < 1$

$$P_X(k) = \begin{cases} (1 - p)p^k, & k \geq 0, \\ 0, & k < 0. \end{cases} \tag{D.69}$$

$$H_X(z) = \frac{1-p}{1-pz}, \quad \text{for } |z| < 1/p. \tag{D.70}$$

$$\mu_X = \frac{p}{1-p}, \quad \sigma_X^2 = \frac{p}{(1-p)^2}. \tag{D.71}$$

Pascal (or Negative Binomial)

For $0 < q < 1$ and any integer $n > 0$

$$P_X(k) = \begin{cases} 0, & k < n, \\ \binom{k-1}{n-1}(1-q)^n q^{k-n}, & k = n, n+1, n+2, \dots. \end{cases} \tag{D.72}$$

$$H_X(z) = \left(\frac{(1-q)z}{1-qz}\right)^n, \quad \text{for } |z| < 1/q. \tag{D.73}$$

$$\mu_X = \frac{n}{1-q}, \quad \sigma_X^2 = \frac{nq}{(1-q)^2}. \tag{D.74}$$

Poisson

For any $b > 0$

$$P_X(k) = \begin{cases} \frac{b^k}{k!} e^{-b}, & k \geq 0, \\ 0, & k < 0. \end{cases} \tag{D.75}$$

$$H_X(z) = \exp(b(z-1)), \text{ for all } z. \tag{D.76}$$

$$\mu_X = b, \quad \sigma_X^2 = b. \tag{D.77}$$

Mathematical Tables

A. Trigonometric Identities

$$\sin^2(x) + \cos^2(x) = 1. \tag{E.1}$$

$$\cos(x \pm y) = \cos(x)\cos(y) \mp \sin(x)\sin(y). \tag{E.2}$$

$$\sin(x \pm y) = \sin(x)\cos(y) \pm \cos(x)\sin(y). \tag{E.3}$$

$$\cos(x)\cos(y) = \frac{1}{2}\cos(x+y) + \frac{1}{2}\cos(x-y). \tag{E.4}$$

$$\sin(x)\sin(y) = \frac{1}{2}\cos(x-y) - \frac{1}{2}\cos(x+y). \tag{E.5}$$

$$\sin(x)\cos(y) = \frac{1}{2}\sin(x+y) + \frac{1}{2}\sin(x-y). \tag{E.6}$$

$$\exp(jx) = \cos(x) + j\sin(x). \tag{E.7}$$

$$\cos(x) = \frac{e^{jx} + e^{-jx}}{2}. \tag{E.8}$$

$$\sin(x) = \frac{e^{jx} - e^{-jx}}{2j}. \tag{E.9}$$

B. Series Expansions

$$\frac{1}{1-x} = \sum_{k=0}^{\infty} x^k, \quad \text{for } |x| < 1. \tag{E.10}$$

$$\frac{1 - x^{n+1}}{1-x} = \sum_{k=0}^{n} x^k, \quad \text{for all } x. \tag{E.11}$$

Probability and Random Processes. DOI: 10.1016/B978-0-12-386981-4.00018-8

$$\frac{1}{(1-x)^{n+1}} = \sum_{k=n}^{\infty} \binom{k}{n} x^{k-n} = \sum_{k=0}^{\infty} \binom{k+n}{n} x^k, \quad \text{for } |x| < 1. \tag{E.12}$$

$$(x+y)^n = \sum_{k=0}^{n} \binom{n}{k} x^k y^{n-k}, \quad \text{for all } x, y. \tag{E.13}$$

$$\exp(x) = \sum_{k=0}^{\infty} \frac{1}{k!} x^k, \quad \text{for all } x. \tag{E.14}$$

$$\cos(x) = \sum_{k=0}^{\infty} \frac{(-1)^k}{(2k)!} x^{2k}, \quad \text{for all } x. \tag{E.15}$$

$$\sin(x) = \sum_{k=0}^{\infty} \frac{(-1)^k}{(2k+1)!} x^{2k+1}, \quad \text{for all } x. \tag{E.16}$$

$$\ln(1-x) = -\sum_{k=1}^{\infty} \frac{1}{k} x^k, \quad \text{for } |x| < 1. \tag{E.17}$$

$$Q(x) = \frac{1}{2} + \frac{1}{\sqrt{2\pi}} \sum_{k=0}^{\infty} \frac{(-1)^{k+1}}{k! 2^k (2k+1)} x^{2k+1}, \quad \text{for all } x. \tag{E.18}$$

$$I_m(x) = \sum_{k=0}^{\infty} \frac{1}{k!(k+m)!} \left(\frac{x}{2}\right)^{2k+m}, \quad \text{for all } x. \tag{E.19}$$

C. Some Common Indefinite Integrals

Note: For each of the indefinite integrals, an arbitrary constant may be added to the result.

$$\int x^n dx = \begin{cases} \dfrac{x^{n+1}}{n+1}, & n \neq -1, \\ \ln(x), & n = -1. \end{cases} \tag{E.20}$$

$$\int b^x dx = \frac{b^x}{\ln(b)}, \quad b \neq 1. \tag{E.21}$$

$$\int \ln(x) dx = x \ln(x) - x. \tag{E.22}$$

$$\int \sin(x) dx = -\cos(x). \tag{E.23}$$

$$\int \cos(x) dx = \sin(x). \tag{E.24}$$

$$\int \tan(x) dx = -\ln(|\cos(x)|). \tag{E.25}$$

$$\int \sinh(x) dx = \cosh(x). \tag{E.26}$$

$$\int \cosh(x) dx = \sinh(x). \tag{E.27}$$

$$\int \tanh(x) dx = \ln(|\cosh(x)|). \tag{E.28}$$

$$\int e^{ax} \sin(bx) dx = e^{ax} \left(\frac{a \sin(bx) - b \cos(bx)}{a^2 + b^2} \right). \tag{E.29}$$

$$\int e^{ax} \cos(bx) dx = e^{ax} \left(\frac{b \sin(bx) + a \cos(bx)}{a^2 + b^2} \right). \tag{E.30}$$

$$\int x^n e^{bx} dx = e^{bx} \sum_{k=0}^{n} \frac{(-1)^k}{b^{k+1}} \frac{n!}{(n-k)!} x^{n-k} \quad (n \geq 0). \tag{E.31}$$

$$\int x^n \ln(bx) dx = x^{n+1} \left(\frac{\ln(bx)}{n+1} - \frac{1}{(n+1)^2} \right) \quad (n \neq -1). \tag{E.32}$$

$$\int \frac{1}{x^2 + b^2} dx = \frac{1}{b} \tan^{-1}\left(\frac{x}{b} \right) \quad (b > 0). \tag{E.33}$$

$$\int \frac{1}{\sqrt{b^2 - x^2}} dx = \sin^{-1}\left(\frac{x}{b} \right) \quad (b > 0). \tag{E.34}$$

$$\int \frac{1}{\sqrt{x^2 + b^2}} dx = \log(x + \sqrt{x^2 + b^2}) = \sinh^{-1}\left(\frac{x}{b} \right) \quad (b > 0). \tag{E.35}$$

$$\int \frac{1}{\sqrt{x^2 - b^2}} dx = \log\left|x + \sqrt{x^2 - b^2}\right| = \cosh^{-1}\left(\frac{x}{b} \right) \quad (b > 0). \tag{E.36}$$

$$\int \frac{1}{ax^2 + bx + c}\,dx = \begin{cases} \dfrac{1}{\sqrt{b^2 - 4ac}} \ln\left|\dfrac{2ax + b - \sqrt{b^2 - 4ac}}{2ax + b + \sqrt{b^2 - 4ac}}\right|, & b^2 > 4ac, \\[3mm] \dfrac{2}{\sqrt{4ac - b^2}} \tan^{-1}\left(\dfrac{2ax + b}{\sqrt{4ac - b^2}}\right), & b^2 < 4ac. \end{cases} \tag{E.37}$$

$$\int \frac{1}{\sqrt{ax^2 + bx + c}}\,dx = \begin{cases} \dfrac{1}{\sqrt{a}} \ln\left|2ax + b + 2\sqrt{a(ax^2 + bx + c)}\right|, & a > 0, \\[3mm] \dfrac{1}{\sqrt{-a}} \sin^{-1}\left(\dfrac{-2ax - b}{\sqrt{b^2 - 4ac}}\right), & a < 0. \end{cases} \tag{E.38}$$

D. Some Common Definite Integrals

$$\int_0^\infty x^n e^{-x}\,dx = \Gamma(n + 1) = n! \quad \text{for integer } n \geq 0. \tag{E.39}$$

$$\int_{-\infty}^\infty e^{-x^2}\,dx = \int_0^\infty x^{-1/2} e^{-x}\,dx = \Gamma(1/2) = \sqrt{\pi}. \tag{E.40}$$

$$\int_0^\infty x^{n-1/2} e^{-x}\,dx = \Gamma(n + 1/2) = \frac{(2n)!}{2^{2n} n!}\sqrt{\pi}, \quad \text{for integer } n \geq 1. \tag{E.41}$$

$$\int_{-\infty}^\infty \operatorname{sinc}(x)\,dx = \int_{-\infty}^\infty \operatorname{sinc}^2(x)\,dx = 1. \tag{E.42}$$

$$\frac{1}{2\pi}\int_0^{2\pi} \cos^n(x)\,dx = \frac{1}{2\pi}\int_0^{2\pi} \sin^n(x)\,dx = \begin{cases} 0, & n \text{ odd}, \\[3mm] \dbinom{n}{n/2}\dfrac{1}{2^n}, & n \text{ even}. \end{cases} \tag{E.43}$$

$$\int_{-\infty}^\infty \frac{1}{x^2 + b^2}\,dx = 2\int_0^\infty \frac{1}{x^2 + b^2}\,dx = \frac{\pi}{b}, \quad b > 0. \tag{E.44}$$

$$\int_{-b}^b \frac{1}{\sqrt{b^2 - x^2}}\,dx = 2\int_0^b \frac{1}{\sqrt{b^2 - x^2}}\,dx = \pi, \quad b > 0. \tag{E.45}$$

E. Definitions of Some Common Continuous Time Signals

$$\text{Step function: } u(x) = \begin{cases} 1, & x > 0, \\ 0, & x < 0. \end{cases} \tag{E.46}$$

$$\text{Rectangle function: } \text{rect}(x) = \begin{cases} 1, & |x| < 1/2, \\ 0, & |x| > 1/2. \end{cases} \tag{E.47}$$

$$\text{Triangle function: } \text{tri}(x) = \begin{cases} 1 - |x|, & |x| \leq 1, \\ 0, & |x| > 1. \end{cases} \tag{E.48}$$

$$\text{Sinc function: } \text{sinc}(x) = \frac{\sin(\pi x)}{\pi x}. \tag{E.49}$$

F. Fourier Transforms

Table E.1: Common Fourier transform pairs

Signal (Time Domain)	Transform (Frequency Domain)
$\text{rect}(t/t_0)$	$t_0 \text{sinc}(ft_0)$
$\text{tri}(t/t_0)$	$t_0 \text{sinc}^2(ft_0)$
$\exp\left(-\dfrac{t}{t_0}\right)u(t)$	$\dfrac{t_0}{1 + j2\pi ft_0}$
$\exp\left(-\dfrac{|t|}{t_0}\right)$	$\dfrac{2t_0}{1 + (2\pi ft_0)^2}$
$\text{sinc}(t/t_0)$	$t_0 \text{rect}(ft_0)$
$\text{sinc}^2(t/t_0)$	$t_0 \text{tri}(ft_0)$

(Continued)

Table E.1: Common Fourier transform pairs — Cont'd

Signal (Time Domain)	Transform (Frequency Domain)
$\exp(j2\pi f_0 t)$	$\delta(f - f_0)$
$\cos(2\pi f_0 t + \theta)$	$\frac{1}{2}\delta(f - f_0)e^{j\theta} + \frac{1}{2}\delta(f + f_0)e^{-j\theta}$
$\delta(t - t_0)$	$\exp(-j2\pi f t_0)$
$\mathrm{sgn}(t)$	$\dfrac{1}{j\pi f}$
$u(t)$	$\frac{1}{2}\delta(f) + \dfrac{1}{j2\pi f}$
$\exp(-(t/t_0)^2)$	$\sqrt{\pi t_0^2}\exp(-(\pi f t_0)^2)$

G. z-Transforms

Table E.2: Common z-transform pairs

Signal	Transform	Region of Convergence
$\delta[n]$	1	All z
$u[n]$	$\dfrac{1}{1 - z^{-1}}$	$|z| > 1$
$nu[n]$	$\dfrac{z^{-1}}{(1 - z^{-1})^2}$	$|z| > 1$
$n^2 u[n]$	$\dfrac{z^{-1}(1 + z^{-1})}{(1 - z^{-1})^3}$	$|z| > 1$

Table E.2: Common z-transform pairs — Cont'd

Signal	Transform	Region of Convergence
$n^3 u[n]$	$\dfrac{z^{-1}(1 + 4z^{-1} + z^{-2})}{(1 - z^{-1})^4}$	$\|z\| > 1$
$b^n u[n]$	$\dfrac{1}{1 - bz^{-1}}$	$\|z\| > \|b\|$
$nb^n u[n]$	$\dfrac{bz^{-1}}{(1 - bz^{-1})^2}$	$\|z\| > \|b\|$
$n^2 b^n u[n]$	$\dfrac{bz^{-1}(1 + bz^{-1})}{(1 - bz^{-1})^3}$	$\|z\| > \|b\|$
$b^n \cos[\Omega_0 n] u[n]$	$\dfrac{1 - b\cos(\Omega_0)z^{-1}}{1 - 2b\cos(\Omega_0)z^{-1} + bz^{-2}}$	$\|z\| > \|b\|$
$b^n \sin[\Omega_0 n] u[n]$	$\dfrac{b\sin(\Omega_0)z^{-1}}{1 - 2b\cos(\Omega_0)z^{-1} + bz^{-2}}$	$\|z\| > \|b\|$
$\dfrac{u[n-1]}{n}$	$\ln\left(\dfrac{1}{1 - z^{-1}}\right)$	$\|z\| > 1$
$\dbinom{n + m}{m} b^n u[n]$	$\dfrac{1}{(1 - bz^{-1})^{m+1}}$	$\|z\| > \|b\|$
$\dfrac{b^n}{n!} u[n]$	$\exp(bz^{-1})$	All z

H. Laplace Transforms

Table E.3: Common Laplace transform pairs

Function	Transform	Region of Convergence
$u(t)$	$1/s$	$\text{Re}[s] > 0$
$\exp(-bt)u(t)$	$\dfrac{1}{s+b}$	$\text{Re}[s] > -b$
$\sin(bt)u(t)$	$\dfrac{b}{s^2 + b^2}$	$\text{Re}[s] > 0$
$\cos(bt)u(t)$	$\dfrac{s}{s^2 + b^2}$	$\text{Re}[s] > 0$
$e^{-at}\sin(bt)u(t)$	$\dfrac{b}{(s+a)^2 + b^2}$	$\text{Re}[s] > -a$
$e^{-at}\cos(bt)u(t)$	$\dfrac{s+a}{(s+a)^2 + b^2}$	$\text{Re}[s] > -a$
$\delta(t)$	1	All s
$\dfrac{d}{dt}\delta(t)$	s	All s
$t^n u(t), \; n \geq 0$	$\dfrac{n!}{s^{n+1}}$	$\text{Re}[s] > 0$
$t^n e^{-bt}u(t), \; n \geq 0$	$\dfrac{n!}{(s+b)^{n+1}}$	$\text{Re}[s] > -b$

I. Table of the Q-function

The following table lists values of the function $Q(x)$ for $0 \leq x < 4$ in increments of 0.05. To find the appropriate value of x, add the value at the beginning of the row to the value at the top of the column. For example, to find $Q(1.75)$, find the entry from the column headed by 1.00 and the row headed by 0.75 to get $Q(1.75) = 0.04005916$.

Table E.4: Values of $Q(x)$ for $0 \leq x < 4$ (in increamants of 0.05)

$Q(x)$	0.00	1.00	2.00	3.00
0.00	0.50000000	0.15865525	0.02275013	0.00134990
0.05	0.48006119	0.14685906	0.02018222	0.00114421
0.10	0.46017216	0.13566606	0.01786442	0.00096760
0.15	0.44038231	0.12507194	0.01577761	0.00081635
0.20	0.42074029	0.11506967	0.01390345	0.00068714
0.25	0.40129367	0.10564977	0.01222447	0.00057703
0.30	0.38208858	0.09680048	0.01072411	0.00048342
0.35	0.36316935	0.08850799	0.00938671	0.00040406
0.40	0.34457826	0.08075666	0.00819754	0.00033693
0.45	0.32635522	0.07352926	0.00714281	0.00028029
0.50	0.30853754	0.06680720	0.00620967	0.00023263
0.55	0.29115969	0.06057076	0.00538615	0.00019262
0.60	0.27425312	0.05479929	0.00466119	0.00015911
0.65	0.25784611	0.04947147	0.00402459	0.00013112
0.70	0.24196365	0.04456546	0.00346697	0.00010780
0.75	0.22662735	0.04005916	0.00297976	0.00008842
0.80	0.21185540	0.03593032	0.00255513	0.00007235
0.85	0.19766254	0.03215677	0.00218596	0.00005906
0.90	0.18406013	0.02871656	0.00186581	0.00004810
0.95	0.17105613	0.02558806	0.00158887	0.00003908

Numerical Methods for Evaluating the Q-Function

In this appendix, we give an overview of several methods available for numerically evaluating the CDF of a Gaussian random variable and related integrals. Recall that for a zero-mean unit variance Gaussian random variable, the CDF is given by the integral

$$F_X(x) = \int_{-\infty}^{x} \frac{1}{\sqrt{2\pi}} \exp\left(-\frac{t^2}{2}\right) dt. \tag{F.1}$$

The Q-function is the complement of this integral

$$Q(x) = \int_{x}^{\infty} \frac{1}{\sqrt{2\pi}} \exp\left(-\frac{t^2}{2}\right) dt. \tag{F.2}$$

Many math packages have internal routines for evaluating related integrals, usually the error function or complementary error function. Given the most common definitions of these functions,

$$\text{erf}(x) = \frac{2}{\sqrt{\pi}} \int_{0}^{x} \exp(-t^2) dt, \tag{F.3}$$

$$\text{erfc}(x) = \frac{2}{\sqrt{\pi}} \int_{x}^{\infty} \exp(-t^2) dt, \tag{F.4}$$

the Q-function can then be written in terms of these functions as

$$Q(x) = \frac{1}{2}\text{erfc}\left(\frac{x}{\sqrt{2}}\right) = \frac{1}{2} - \frac{1}{2}\text{erf}\left(\frac{x}{\sqrt{2}}\right). \tag{F.5}$$

For situations where internally defined functions are not available, several numerical techniques are available for efficiently evaluating the Q-function. To start with, recall the symmetry relationship

$$Q(x) = 1 - Q(-x). \tag{F.6}$$

Probability and Random Processes. DOI: 10.1016/B978-0-12-386981-4.00019-X

Hence any routine for evaluating the Q-function only needs to work on positive values of x. To start with, we consider the Taylor series expansion of the Q-function about the point $x = 0$,

$$Q(x) = \frac{1}{2} + \frac{1}{\sqrt{2\pi}} \sum_{k=0}^{\infty} \frac{(-1)^{k+1}}{k!2^k(2k+1)} x^{2k+1}. \tag{F.7}$$

This series is convergent for all $x \geq 0$ but will converge faster for smaller values of x. A good approximation can be obtained by truncating the series to a sufficient number of terms. Since the series is alternating, the truncation error is bounded by the first term neglected. Figure F.1 shows the Q-function along with its approximations using the Taylor series truncated to various numbers of terms. It is seen from this figure that for $x > 2$, a large number of terms may be needed for the Taylor series to converge.

For larger values of x, it is common to use the following asymptotic series expansion

$$Q(x) = \frac{1}{\sqrt{2\pi}x} \exp\left(-\frac{x^2}{2}\right) \left\{ 1 + \sum_{n=1}^{\infty} \frac{(-1)^n(2n)!}{2^n n!} x^{-2n} \right\}, \quad \text{for } x > 0. \tag{F.8}$$

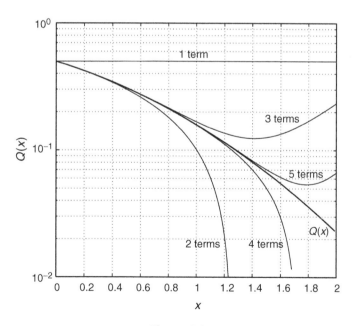

Figure F.1
The Q-function and its truncated Taylor series approximation.

Since the series has negative powers of x, the larger x is, the faster the series will converge. Also, as with the Taylor series expansion, since this is a convergent alternating series, the truncation error will be bounded by the first neglected term. Also, the sign of the error will be the same as that of the first neglected term. As a result, for large values of x, the Q-function can be upper and lower bounded by

$$\frac{1}{\sqrt{2\pi}x}\left(1-\frac{1}{x^2}\right)\exp\left(-\frac{x^2}{2}\right) \leq Q(x) \leq \frac{1}{\sqrt{2\pi}x}\exp\left(-\frac{x^2}{2}\right). \tag{F.9}$$

These two bounds are shown in Figure F.2. From the figure as well as from the expressions in the previous equation, it is clear that as $x \to \infty$ both bounds are asymptotically tight. Hence it is seen that

$$Q(x) \sim \frac{1}{\sqrt{2\pi}x}\exp\left(-\frac{x^2}{2}\right). \tag{F.10}$$

From the figure, it appears that this approximation is fairly accurate for $x > 4$. If more accuracy is desired, more terms can be included in the asymptotic expansion.

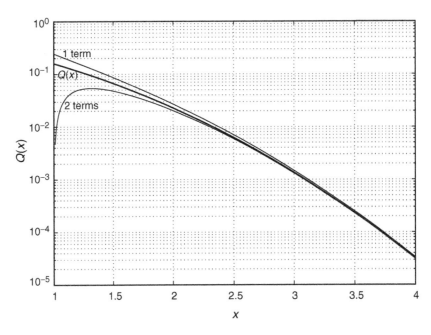

Figure F.2
Upper and lower bounds on the Q-function.

In addition to the Taylor series and asymptotic series expansions, there are also a few continued fraction expansions for the Q-function. These are listed below:

$$Q(x) = \frac{1}{\sqrt{2\pi}}\exp\left(-\frac{x^2}{2}\right)\left\{\cfrac{1}{x+\cfrac{1}{x+\cfrac{2}{x+\cfrac{3}{x+\cfrac{4}{x+\ldots}}}}}\right\}, \quad \text{for } x > 0, \tag{F.11}$$

$$Q(x) = \frac{1}{2} - \frac{1}{\sqrt{2\pi}}\exp\left(-\frac{x^2}{2}\right)\left\{\cfrac{x}{1-\cfrac{x^2}{3+\cfrac{2x^2}{5-\cfrac{3x^2}{7+\cfrac{4x^2}{9-\ldots}}}}}\right\}, \quad \text{for } x \geq 0. \tag{F.12}$$

A number of polynomial and rational approximations of the Q-function are available. Among them, the following seems to offer the best accuracy:

$$Q(x) = \frac{1}{\sqrt{2\pi}}\exp\left(-\frac{x^2}{2}\right)(b_1 t + b_2 t^2 + b_3 t^3 + b_4 t^4 + b_5 t^5), \quad t = \frac{1}{1+px}, \tag{F.13}$$

where,

$p = 0.2316419$,

$b_1 = 0.319381530$,

$b_2 = -0.35656378$,

$b_3 = 1.7814779$,

$b_4 = -1.821256$,

$b_5 = 1.3302744$.

The error in this approximation is less than 7.5×10^{-8} for all $0 \le x < \infty$.

Finally, we note that any desired accuracy can be obtained by computing the Q-function via numerical integration. Using the definition directly, the Q-function has an infinite limit which is inconvenient for performing numerical integration. For small to moderate values of x, this problem can be circumvented by rewriting the Q-function as

$$Q(x) = \frac{1}{2} - \frac{1}{\sqrt{2\pi}} \int_0^x \exp\left(-\frac{t^2}{2}\right) dt. \tag{F.14}$$

For large values of x it may be more efficient to work with the standard definition and truncate the upper limit to form the approximation

$$Q(x) \approx \frac{1}{\sqrt{2\pi}} \int_x^{x+c} \exp\left(-\frac{t^2}{2}\right) dt, \tag{F.15}$$

where the constant c is chosen to insure the desired accuracy. For $c > 2$ and $x > 1.5$ the relative error in this approximation can be shown to be bounded by

$$\frac{\varepsilon(x)}{Q(x)} = \frac{Q(x+c)}{Q(x)} \le \exp\left(-\left(\frac{c^2}{2} + 3\right)\right). \tag{F.16}$$

For example, choosing $c = 3.5$ will guarantee a relative accuracy of less than 10^{-4} (i.e., four digits of accuracy). Finally, we note an alternative form of the Q-function which has finite limits,

$$Q(x) = \frac{1}{\pi} \int_0^{\pi/2} \exp\left(-\frac{x^2}{2\sin^2\theta}\right) d\theta. \tag{F.17}$$

Since the integrand is fairly well behaved and the limits of integration are finite, numerical integration on this form is particularly convenient and can be performed to any desired accuracy.

Index

Note: Page numbers followed by *f* indicate figures, *t* indicate tables and *n* indicate notes.

A

a posteriori probability, 28, 28*n*2
  auditorium seat probability, 29
  in optical communication
    system, 39–40
  in statistical independence, 30
a priori probability, 28, 28*n*1
  auditorium seat probability, 29
  in optical communication system,
    40
  for radar system, 318
  in statistical independence, 30
Absorbing state, 385–386, 386*f*
Absorption probability, 407–409
Accessible state, 395
AM. *See* Amplitude modulation
Amplitude, of stochastic signals, 7
Amplitude modulation (AM),
    500–503
  MATLAB for, 503–504, 504*f*
Analog communication system,
    500–504, 500*f*, 501*f*, 504*f*
Aperiodic Markov chain, 396
AR. *See* Autoregressive process
Arcsine distribution, 124
Arcsine random variable, 565–566
ARMA. *See* Autoregressive moving
    average process
Arrival, 360
Arrival state, 403, 416
Atomic outcomes, 13–14
  of coin flipping, 14
  of dice rolling, 15
  of die rolling, 14
  exercises for, 46–47
  MATLAB rand command and,
    15
  probability, 14
Auditorium seat probability,
    Bayes's theorem and, 29

Autocorrelation function, 344–345
  of AR process, 450
  autocovariance function and, 347
  average normalized power, 356
  for complex envelope, 499–500
  of continuous time processes,
    344–345
  of discrete-time processes,
    344–347
  ergodic in, 351
    estimation of, 352–354, 353*f*
    influence of, 354–355
    sinusoidal random processes,
      352
  estimation of, 441–445, 442*f*,
    444*f*, 445*f*
    random telegraph process,
      442–444, 442*f*, 444*f*
    windowing function for, 443
  exercises for, 375
  Fourier transform pair of PSD
    and, 433
  of Gaussian random processes,
    358–359
  of LTI system, 473–475
  of PAM, 456–457
  of Poisson processes, 363
  properties of, 356–357
  for PSD spectral estimation,
    correlation method,
    441–445, 442*f*, 444*f*, 445*f*
  of shot noise processes,
    366–367
  of signal-to-noise ratios,
    480–481
  for sinusoidal random processes,
    346
  for thermal noise, 454
  wide sense stationary and,
    349–351, 356–357

Autocovariance function, 347
  of Poisson processes, 363
  of shot noise processes,
    366–367
Autoregressive moving average
    process (ARMA), for PSD
    estimation, 449
Autoregressive process (AR)
  autocorrelation function of, 450
  for PSD estimation, 449–450
  random telegraph process,
    451–452, 452*f*
Available power, 453
Average. *See* Expected values
Average information, 156
Average normalized power, 356
Average power, in random process,
    431
Axioms of probability, 10–13, 13*f*
  exercises for, 44–45

B

Backward Kolmogorov equations,
    403
Bandlimited random processes,
    494–498, 495*f*, 496*f*, 497*f*,
    498*f*
Bandpass filters (BPFs), noise
    equivalent bandwidth of,
    480
Bandpass process, 439
  bandwidth of, 439, 439*f*,
    440*f*
Bandpass random process, 494
  PSD of, 494–495, 495*f*
Bandwidth
  of random processes, 439–441,
    439*f*, 440*f*
    exercises for, 465–466
  of random telegraph process, 441

Bayes's theorem, 27–29, 28*f*
  a posteriori probability, 28, 28*n*2
  a priori probability, 28, 28*n*1
  auditorium seat probability, 29
  conditional, 90
  for estimation and detection
    problems, 266–267
  exercises for, 52–54
  in optical communication system,
    39–40
  theorem of total probability,
    27–28, 28*f*
Bernoulli, 8
Bernoulli random variable, 35, 574
  CDF of, 310, 310*f*
  mean of, 297
  PMF of, 310
  random telegraph signal, 340–341
Bernoulli trial
  binomial random variable and, 36
  Cartesian space for, 35
  exercises for, 56
Bernoulli trials, 35
Bessel function, 565
  Rician random variables and, 83
Best linear unbiased estimator
    (BLUE), 291
  for mean IID random variables,
    291–292
Beta function, 565
Beta random variable, 566
Bilinear transformation, 530
Binary data, in optical
    communication system,
    38–41, 38*f*
Binary entropy function, 156, 157*f*
Binary pseudorandom number
    generators, 517–521, 500*f*,
    519*f*, 520*t*, 521*f*
Binary symmetric channel (BSC),
    224, 224*f*
Binomial coefficient, 23–24. *See
    also* Combinations
  binomial random variable and, 36
  exercises for, 48–49
  MATLAB exercise for, 62
Binomial random variable, 35–37,
    574
  central moments of, 118
  in failure probability, 314
  moments of, 116

PMF of, 36
Poisson random variable for, 37
probability-generating functions
  of, 137–138
Binomial random variables,
    characteristic function of,
    132
Birth processes, 360–361
Birth–death processes, 401–411
  MATLAB exercise for,
    406–407, 406*f*, 412–413,
    412*f*
  M/M/∞ queue, 405
  M/M/1 queue, 405–407, 406*f*
  population modeling with,
    407–409
  telephone exchange described by,
    416–417, 417*f*
BLUE. *See* Best linear unbiased
    estimator
Bounds, on tail probabilities,
    140–147, 143*f*, 147*f*
Box-Muller transformation, 218, 524
BPFs. *See* Bandpass filters
Branching process, 388
Bridge, partitioning in, 25–26
BSC. *See* Binary symmetric
    channel

C

Cartesian coordinates
  transformation into polar
    coordinates, 216–217
  transformation into spherical
    coordinates, 262–263, 263*f*
Cartesian space, for Bernoulli trial,
    35
Cauchy distribution, 213
Cauchy random variables, 84, 566
CDF. *See* Cumulative distribution
    function
Central limit theorem, 306–310
  exercises for, 329
  Gaussian distribution and,
    308–310
  proof of, 307–308
Central moments
  of random variables, exercises
    for, 162–163
  of single random variables,
    117–121

MATLAB exercise for,
    120–121
Channel capacity, 221–225, 223*f*,
    224*f*, 225*f*
  exercises for, 239–240
Channel coding, 221–225, 223*f*,
    224*f*, 225*f*
  exercises for, 239–240
Channel coding theorem, 225
Chapman–Kolmogorov equation,
    389, 409–410
  continuous time version of, 402
Characteristic functions
  of chi-squared random variables,
    259
  joint, for pairs of random
    variables, 206–209,
    235–236
  of Laplace random variables, 260
  in quadratic transformations of
    Gaussian random vectors,
    258–260
  of single random variables,
    130–136
    exercises for, 167–168
    moments and, 133–135
    natural logarithm of,
      135–136
    PDFs and, 130–131
  transformations of pairs of
    random variables using,
    211–212, 214
Chebyshev's inequality, 142
  in law of large numbers, 304
Chernoff bound, 142–144, 143*f*
Chi-squared (χ^2) random variables,
    81–82, 567
  characteristic function of, 259
Circular Gaussian density function,
    220
Classical approach
  of assigning probabilities, 14
    exercises for, 46–47
    relative frequency approach to,
      16
    unsatisfactory results with,
      15–16
  for joint probability, 18
Coded digital communication
    system, simulation of,
    539–541, 539*f*, 541*f*

Coefficient of kurtosis, 119–120
    for Gaussian random variables,
        135
Coefficient of skewness, 119–120
    for Gaussian random variables,
        135
Coin flipping
    MATLAB simulation of, 15
    PMF for, 33
    probability assignment for, 14
    random process of, 336–338,
        336*f*, 338*f*
    sample space and outcomes of,
        8–9
Column vector, 551
Combinations, 23–24, 26*t*
    English *v.* mathematical usage
        of, 24
    in lottery game, 24
Combinatorics, 20–26, 26*t*
    combinations, 23–24
    exercises for, 48–52
    k-permutations, 23
    partitions, 24–26
    permutations, 22
    principle of counting, 21
Combined experiment, 21–22, 26*t*
Communicating states, 395
Communication system, 4–5, 5*f*.
        See also Digital modulation
        formats
    analog, 500–504, 500*f*, 501*f*, 504*f*
    Erlang random variable in, 82
    matched filter for SNR for,
        484–485, 485*f*
    optical, 38–41, 38*f*, 42*f*
    Rayleigh random variables in, 82
    Rician distribution in, 84
    simulation of coded digital,
        539–541, 539*f*, 541*f*
    sinusoid in, 79
    statistical independence and, 32
Complement, 545, 546*f*
Complementary error function
        integral, for CDF of Gaussian
        random variable, 74
Complex envelopes
    exercises for, 514
    for I and Q components, 499–500
Complex numbers, Rayleigh
        distribution and, 82

Complex random variables,
        219–221
    exercises for, 238–239
Computer communication network,
        Markov processes
        describing, 413–415, 415*f*
Computer generation
    of random processes, 525–534
        exercises for, 544
        frequency domain approach to,
            525–528, 526*f*, 528*f*
        of Gaussian white noise,
            533–534, 534*f*
        time domain approach,
            529–532, 529*f*, 531*f*, 533*f*
    of random variables, 517–524
        binary pseudorandom number
            generators, 517–521, 518*f*,
            519*f*, 520*t*, 521*f*
        correlated, 524
        exercises for, 543–544
        nonbinary pseudorandom
            number generators, 521–522
        from specified distribution,
            521–522
Conditional cumulative distribution
        function, 86–87
    exercises for, 104–106
    for pairs of random variables,
        188–191
        exercises for, 231
Conditional entropy, 221–222
Conditional expected values
    exercises for, 163–164
    of functions of single random
        variables, 122
    of pairs of random variables,
        196–197
    of single random variables,
        121–122
Conditional probability, 18–20
    deck of playing cards and, 19–20
    definition of, 18–19
    exercises for, 47–48
    independence and, 31
    joint probability compared with,
        19
    in optical communication system,
        40–41
Conditional probability density
        function, 87–89, 88*f*

exercises for, 104–106
    for multiple random variables,
        245–247
        exercises for, 277–278
    for pairs of random variables,
        188–191
        exercises for, 231
    properties of, 87–88
    transformations of pairs of
        random variables using,
        213–214
Conditional probability mass
        functions
    for multiple random variables,
        245–247
        exercises for, 277–278
    for pairs of random variables,
        188–191
        exercises for, 231
Confidence intervals, 310–315
    constants used to calculate, 311*t*
    exercises for, 330–331
    for failure probability, 314
    for IID random variables, 312
    sample mean and, 312
    Student's t-distribution, 313
Confidence level, 311, 311*t*
Continuous random variables,
        565–574
    CDF of, 65–67, 65*f*, 66*f*
        PDF for, 289
    conditional, 90
    PDF of, 69
    PMF and, 63
    staircase transformations of, 128,
        128*f*
Continuous sample space, 10
    events on, 10
    mix with discrete, 10
Continuous time and discrete
        amplitude signals, 335
    PDF of, 341
    random process of, 338–339, 339*f*
Continuous time linear systems
    exercises for, 505–507
    random processes in, 473–477
Continuous time Markov processes,
        401–411, 409–413
    exercises for, 426–427
    MATLAB exercise for, 406–407,
        406*f*, 412–413, 412*f*

Continuous time Markov processes
  (*continued*)
  M/M/∞ queue, 405
  M/M/1 queue, 405–407, 406*f*
  population modeling with,
    407–409
Continuous time processes
  autocorrelation function of,
    344–345
  mean function of, 343
  strict sense stationary,
    348–349
Continuous time signals, 335, 581
Convergence almost everywhere
  exercises for, 326–327
  of random sequences, 301, 303*t*
  strong law of large numbers and,
    305
Convergence everywhere
  exercises for, 326–327
  of random sequences, 300–301,
    303*t*
Convergence in distribution
  of central limit theorem, 308
  exercises for, 326–327
  of random sequences,
    302–303, 303*t*
Convergence in mean square sense
  exercises for, 326–327
  of random sequences, 302, 303*t*
Convergence in probability
  exercises for, 326–327
  of random sequences,
    301–302, 303*t*
  weak law of large numbers and,
    305
Convergence modes
  law of large numbers and,
    304–305
  of random sequences,
    298–303, 303*t* convergence
    almost everywhere, 301,
    303t convergence
    everywhere, 300–301, 303*t*
  convergence in distribution,
    302–303, 303*t*
  convergence in mean square
    sense, 302, 303*t*
  convergence in probability,
    301–302, 303*t*
  exercises for, 326–327

Convolution, transformations of
  pairs of random variables
  using, 211
Convolution function, MATLAB,
  352–354, 353*f*
Coordinate systems
  three-dimensional
    transformations of,
    262–263, 263*f*
  transformation of pairs of random
    variables for changes of,
    214–217
Corollary of probability, 11–12
Correlation
  of complex random variables,
    221
  independence and, 195, 198–199
    for Gaussian random variables
      in multiple dimensions, 251
    for jointly Gaussian random
      variables, 205–206, 206*f*
  of pairs of random variables,
    193–195
Correlation coefficient
  independence and, 198
  for jointly Gaussian random
    variables, 206
  of pairs of random variables,
    194–195
Correlation matrix
  for IID random variables,
    291–292
  of multiple random variables,
    247–249
Correlation method, for PSD
  spectral estimation,
    441–445, 442*f*, 444*f*, 445*f*
Counting processes, 362. *See also*
    Poisson counting processes
  PMF of, 361–362
  properties of, 360–361
Covariance
  of complex random variables,
    221
  independence and, 198
  of pairs of random variables,
    193–195
Covariance function, of random
  walk, 360
Covariance matrix, of multiple
  random variables, 247–249

Cross spectral density
  of I and Q components, 497–498
  between random processes,
    438–439
Cross-correlation function,
    347–348
  Fourier transform of, 438–439
  of LTI system, 475–476
Cross-covariance function, 348
Cumulants, 135–136
Cumulative distribution function
    (CDF), 64–68, 65*f*, 66*f*
  of Bernoulli random variable,
    310, 310*f*
  of Cauchy random variables, 84
  of chi-squared (χ^2) random
    variables, 81–82
  conditional, 86–87
    exercises for, 104–106
    for pairs of random variables,
      188–191, 231
  of continuous random variables,
    65–67, 65*f*, 66*f*
    PDF for, 289
  of discrete random variables, 68
    PMF for, 289
  of Erlang random variable, 81–82
  estimating of IID random
    variables, 297–298
  exercises for, 97–99
  of exponential random variables,
    79–80, 79*f*
  of gamma random variables, 81
  of Gaussian random variable, 69,
    73–76, 75*f*
    evaluation of, 75
    in Q-function terms, 75–76
    standard forms of, 74
    transformation of, 74–75
  joint
    exercises for, 227–228
    for multiple random variables,
      245–247, 277–278
    for pairs of random variables,
      178–180, 179*f*
  of Laplace random variables,
    80, 80*f*
  PDF compared with, 69–71, 69*f*
  PMF and, 68
  properties of, 64
  of random processes, 340, 342

Cumulative distribution function
(CDF) *(continued)*
of random variables, 65–66, 65*f*,
66*f*
conditional, 86–87
reliability function, 91
of Rayleigh random variables,
82, 83*f*
of Rician random variables,
83–84, 84*f*
of standard normal random
variables, 74
transformations of pairs of
random variables using,
210, 212–214
unconditional, 89–90
of uniform random variables,
78–79, 78*f*
Customer arrivals, exponential
random variables in, 80

D

Dark current, 41
De Morgan's laws, 549
Deck of playing cards
conditional probability and, 19–20
joint probability and, 18–20
Definite integrals, 580
Departure state, 403, 416
Detection, 264
exercises for, 283–284
MAP detection, 265–267
Detection probability
for radar system, 318–323
sequential detection and, 320
thresholds for, 319–320
Determinant of matrix, 554–555
Deterministic signal, 7
continuous, Fourier transform
for, 429
noise compared with, 7
Diagonal matrix, 552
Dice rolling
MATLAB exercise for, 62
probability assignment of, 15
relative frequency approach for,
16–17, 16*t*
sample space and outcomes of, 9
statistical independence and, 30–31
Die rolling
probability assignment of, 14

sample space and outcomes of, 9
Difference set, 545–546, 546*f*
Differential pulse code modulation
(DPCM), 272–275, 273*f*,
274*f*, 275*f*, 276*f*
Diffusion modeling, 411–413, 412*f*
Digital communication systems,
mutual information of,
222–225, 223*f*, 224*f*, 225*f*
Digital modulation formats, PSDs
of, 455–461, 458*f*, 459*f*,
460*f*, 461*f*
Discrete Fourier transform, for
discrete-time processes, 429
Discrete random variables, 32–38,
34*f*, 574–575
Bernoulli, 35
binomial, 35–37
CDF of, 68
PMF for, 289
central moments of, 117
characteristic function of,
131–132
for coin flipping, 33
conditional expected values of,
122
conditional PMF for, 188
exercises for, 55–59
expected value of, 112
expected value of functions of,
113
expected values for, 192
geometric, 37–38
joint PMFs for, 186–188
joint probability-generating
function for, 208–209
moments of, 115–116
PMF of, 33
Poisson, 37
probability-generating functions
of, 136–138
staircase transformations of
continuous random
variables into, 128, 128*f*
Discrete time and continuous
amplitude signals, 335
random process of, 339–340, 340*f*
Discrete time and discrete
amplitude signals, 335
Discrete-time Fourier transform
(DTFT), 560–561

for discrete-time linear systems,
477–478
Discrete-time linear systems
exercises for, 477–479, 508–509
Gaussian white noise in,
478–479
random processes in, 477–479
Discrete-time processes
autocorrelation function of,
344–347
autocovariance function of, 347
discrete Fourier transform for,
429
mean function of, 343
Discrete-valued Markov processes.
See Markov chains
Disjoint, 546
Distortion, signal quantization
causing, 148–153, 149*f*,
152*f*
Distribution
arcsine, 124
Cauchy, 213
Laplace, quantization of,
151–154, 152*f*
random sequence convergence
in, 302–303, 303*t*
Distribution of infinitesimal
increments, of Poisson
processes, 361
Domain, of functions, 32
DPCM. *See* Differential pulse code
modulation
Drift, of Markov chain, 414–415,
415*f*
DTFT. *See* Discrete-time Fourier
transform

E

Efficient estimator, 290
Eigenvalue, 555–557
Eigenvector, 555–557
Empty set, 545, 546*f*
Energy, of random process, 430
Ensemble averaged power, 430
Ensemble of functions, 336
Entropy
conditional, 221–222
of single random variables,
155–158, 157*f*, 158*t*
exercises for, 173

Envelope detector, for AM, 501–502
Ergodic in autocorrelation, sinusoidal random processes, 351
Ergodic in mean, 351
Ergodicity
  autocorrelation function
    and estimation of, 352–354, 353f
    influence of, 354–355
  of random processes, 351–356
  of sinusoidal random processes, 352
  strict sense stationary and, 351–352
  two limited forms of, 351
  of WSS, 351–356
Erlang random variable, 81–82, 567–568
  moment-generating functions of, 140
Erlang-B formula, 416, 417f
Error function integral, for CDF of Gaussian random variable, 74
Errors, in optical communication system, 40–41, 42f
Estimation, 264
  exercises for, 283–284
  MAP, 265–267
  ML, 267–268
  MMSE, 268–272
Events, 7–10
  of CDF, 64
  on continuous space, 10
  definition of, 8
  exercises for, 43–44
  intersection of, 17
  probability as function of, 11
  of random number generator, 10
Exam permutations, 22
Expected values. *See also* Mean
  conditional
    exercises for, 163–164
    of functions of single random variables, 122
    of pairs of random variables, 196–197
    of single random variables, 121–122

of functions of random variables, 113–114, 115t
  exercises for, 160–161
of multiple random variables, 247–249
  exercises for, 279–280
for pairs of random variables, 192–197
  conditional expected values, 196–197
  correlation, 193–195
  covariance, 193–195
  exercises for, 231–233
  (m, n)th joint moment, 195–196
of single random variables, 111–113
  exercises for, 159–160
Experiments, 7–10
  combined, 21–22, 26t
  definition of, 8
  exercises for, 43–44
  function domains of, 32
  sample spaces for, 10
Exponential random variables, 79–80, 79f, 568
  central moments of, 119
  characteristic function of, 131, 133–134
  expected value of, 112
  failure rate and, 93
  gamma random variable and, 81
  memoryless property of, 93
  transformation of, 124, 129
Extinction probability, 407–409

F

F random variable, 568–569
Factorial function, gamma function and, 81
Factorial moments, 137–138
  joint, 208
Fading
  Rayleigh random variables in, 82
  Rician distribution in, 84
Failure probability, MATLAB example for, 314–315, 314f
Failure rates, 91–95
  changes in, 92
Failure rate function, 92
  exponential distribution of, 93
  of Rayleigh distribution, 93

reliability function and, 92
of system
  parallel interconnection, 94–95
  series interconnection, 93–95
False alarm probability
  for radar system, 318–323
  sequential detection and, 320
  thresholds for, 319–320
Filtering problems, 264
First central moment, 118
First moment, 115–117. *See also* Expected values; Mean
First return probability, 396–397
Fokker–Planck equation, 411
Forward Kolmogorov equations, 403
Fourier Series, 559–560
Fourier Transform. *See also* Discrete Fourier transform
  of cross-correlation function, 438–439
  for deterministic continuous signal, 429
  for PAM, 457
  in PSD, 430
  review of, 560–561
Fourier Transform pairs, 560, 581–582t
  autocorrelation function and PSD as, 433
Fourth central moment, 118
Frequency domain techniques, 7
  for computer generation of random processes, 525–528, 526f, 528f
  MATLAB example for, 527–528, 528f
  for random processes, 429
Frequency shift keying modem, 337–338, 338f
Function(s). *See also specific* functions
  ensemble of, 336
  moment-generating
    of single random variables, 139–140, 169–170
    tail probability evaluations using, 142–147, 143f, 147f
  of outcomes, 32
  probability-generating, of single random variables, 136–138, 168–169

Function(s) *(continued)*
  of random sum of Gaussian
    random variables, 317
  of random sum of IID random
    variables, 316
  of random variables
    conditional expected value of,
      122
    exercises for, 160–161
    expected value of,
      113–114, 115*t*
  transformations with
    monotonically decreasing
      functions, 123*f*, 124–125
    monotonically increasing
      functions, 122–124, 123*f*
    nonmonotonic functions,
      125–129, 125*f*, 128*f*, 130*f*
    staircase functions, 128, 128*f*
Fundamental frequency, 559

G
Galileo, 7
Gambler's ruin problem, 386, 386*f*,
  392
Gambling, probability and, 7–8
Gamma distribution, random
  telegraph signal, 341
Gamma function, 81, 565
  exercise for, 104
Gamma random variables, 81, 127,
  569
Gaussian distribution
  central limit theorem and,
    308–310
  zero-mean, 82
Gaussian random processes, 342,
  357–360, 359*f*
  autocorrelation function of,
    346–347
  definition of, 357
  exercises for, 375–376
  PDF of, 357–359
  random walk, 359–360, 359*f*
  simulation of, 525–527, 526*f*
    MATLAB example for,
      527–528, 528*f*
Gaussian random variables, 71–78,
  73*f*, 75*f*, 569. *See also*
    Jointly Gaussian random
    variables

Box-Muller transformation for
  generation of, 218
CDF of, 69, 73–76, 75*f*
  evaluation of, 75
  in Q-function terms, 75–76
  standard forms of, 74
  transformation of, 74–75
characteristic function of, 132,
  135
continuous time and discrete
  amplitude signals as, 341
exercises for, 101–103
generation of, 523–524
  correlated, 524
likelihood ratio test for, 319–321
linear transformations of,
  253–257
MATLAB exercise for, 109
in multiple dimensions,
  249–251
  exercises for, 280–281
  MATLAB exercise for,
    251–252
pairs of, conditional PMF, PDF,
  and CDF for,
    190–191
PDF of, 72–73, 73*f*
  for radar system, 317–318
random sum of, 317
shorthand notation for, 73
transformation of, 124, 127
transformations of pairs of,
  212–213, 216–219
Gaussian random vectors, quadratic
  transformations of, 257–260
Gaussian white noise
  computer generation of,
    533–534, 534*f*
  in discrete-time linear systems,
    478–479
  in LTI system, 476
Gaussian-Multivariate random
  variables, 570
Genetic models, Markov chains in,
  388
Geometric random variable, 37–38,
  129, 574–575
  PMF of, 37
  probability-generating functions
    of, 138
Global balance equations, 403

H
Half-power bandwidth, 440, 440*f*
Harmonic frequency, 559
Heavy-tailed distribution, Cauchy
  PDF as, 84
Hermitian symmetry, 439, 551, 557
Histogram, for MATLAB random
  number generation, 71, 72*f*
Homogenous Markov process, 402

I
I components
  complex envelopes for, 499–500
  cross spectral density of,
    497–498
  extraction of, 496, 496*f*
  PSD of, 496–497, 497*f*
Identity matrix, 552
IID random variables. *See*
  Independent and identically
  distributed random variables
Importance sampling, 537–538
  for simulation of coded digital
    communication system,
    539–541, 539*f*, 541*f*
Impulse invariance transformation,
  530
Indefinite integrals, 578–580
Independence, 29–32
  communication network and, 32
  correlation and, 195, 198–199
  for Gaussian random variables
    in multiple dimensions, 251
  for jointly Gaussian random
    variables, 205–206, 206*f*
  dice rolling and, 30–31
  exercises for, 54–55
  mutual exclusiveness compared
    with, 31
  of random variables, 197–199
    exercises for, 233–234
    MATLAB exercise for,
      200–201, 201*f*
  of set of *N* random variables, 247
  test for, 30
Independent and identically
  distributed (IID) random
  variables, 289–298
  in central limit theorem, 306–308
  confidence intervals for, 312
  estimating CDF of, 297–298

Independent and identically
distributed (IID) random
variables (*continued*)
estimating mean of, 290–295
BLUE for, 291–292
confidence intervals for,
310–315, 311*t*
ML for, 292–294
estimating variance of, 295–297
exercises for, 325–326
likelihood ratio test for, 319
PAM with, 457–459, 458*f*, 459*f*
PDF of, 289
random sum of, 315–317
function of, 316
mean and variance of, 315–316
sample mean of, 299
Independent increments, of Poisson
processes, 361
Information
mutual, 221–225, 223*f*, 224*f*, 225*f*
exercises for, 239–240
quantitative definition of,
155–158, 157*f*, 158*t*
Integrals
definite, 580
indefinite, 578–580
Interference, accounting for, 7
Interpolation problems, 264
Intersection, 545, 546*f*
Interval. *See also* Confidence
intervals
CDF and random variable in,
64, 69
of exponential random variables,
79, 79*f*
of uniform random variables,
78–79, 78*f*
Inverse of matrix, 553–554
Inverse transform, for deterministic
continuous signal, 429
Irreducible Markov chain, 395

J

Joint CDFs
for multiple random variables,
245–247
exercises for, 277–278
for pairs of random variables,
178–180, 179*f*

exercises for, 227–228
Joint characteristic functions, for
pairs of random variables,
206–209
exercises for, 235–236
Joint factorial moments, 208
Joint moments, 192–197, 208–209
Joint moment-generating functions,
209
Joint PDFs
exercises for, 228–229
for Gaussian random variables in
multiple dimensions,
250–252
for multiple random variables,
245–247
exercises for, 277–278
for pairs of random variables,
180–185, 184*f*
MATLAB exercise for,
185–186
Joint PMFs
for multiple random variables,
245–247
exercises for, 277–278
for pairs of random variables,
186–188
exercises for, 229–230
Joint probability, 17–20
computation of, 17–18
conditional probability compared
with, 19
deck of playing cards and, 18–20
exercises for, 47–48
independence and, 31
of three events, 19
Joint probability-generating
functions, 208–209
Jointly Gaussian random variables,
202–206, 203–204*f*, 205*f*,
206*f*
exercises for, 234
joint characteristic functions for,
207–208
linear transformation of,
218–219, 254
MAP estimator for, 265
p-n Junction diode
operation of, 365
shot noise processes in, 365–370

with current pulse shape,
369–370, 370*f*
exercises for, 378–380
with square wave,
368–369, 368*f*, 369*f*

K

k-permutations, 23, 26*t*
of padlock combinations, 23
Kurtosis, 118. *See also* Coefficient
of kurtosis

L

Lagrange multiplier techniques, for
IID random variables, 291
Laplace, 7
Laplace distribution, signals with,
151–153, 152*f*
Laplace random variables, 80, 80*f*,
570
central moments of, 120
characteristic function of, 260
exercise for, 104
Laplace transforms, 584*t*
Law of large numbers, 304–306
exercises for, 327–328
strong, 304–305
weak, 304–305
Level of significance, 311, 311*t*
LFSR. *See* Linear feedback shift
register
License plates, principle of
counting and, 22
Likelihood ratio test
for IID Gaussian random
variables, 319–321
for radar system, 318, 321–322
Linear algebra, 551–557
Linear estimator, 290–292
Linear feedback shift register (LFSR),
for binary pseudorandom
number generators, 518–521,
518*f*, 519*f*, 520*t*
Linear minimum mean square error
(LMMSE) estimator,
270–272
Linear prediction
of PSD, 451
of speech, 272–275, 273*f*, 274*f*,
275*f*, 276*f*

Linear systems, random processes
in, 473–504
analog communication system,
500–504, 500*f*, 501*f*, 504*f*
bandlimited and narrowband,
494–498, 495*f*, 496*f*, 497*f*,
498*f*
complex envelopes, 499–500
exercises for, 514
continuous time, 473–477
exercises for, 505–507
discrete-time, 477–479
exercises for, 508–509
matched filter, 482–486, 482*f*,
485*f*, 486*f*
exercises for, 482–486, 482*f*,
485*f*, 486*f*, 511–512
MATLAB exercises for,
514–515
noise equivalent bandwidth,
479–480, 479*f*, 509–510
exercises for, 509–510
signal-to-noise ratios, 480–482
exercises for, 480–482,
510–511
Wiener filter, 486–494, 490*f*,
491*f*, 494*f*
exercises for, 486–494, 490*f*,
491*f*, 494*f*, 512–513
Linear time-invariant (LTI) system
autocorrelation function of,
473–475
cross-correlation function of,
475–476
mean function of, 473–475
passage of signals through,
561–563
random processes in, 473–477
white Gaussian noise in, 476
white noise conversion,
476–477, 477*f*
Linear transformations
of Gaussian random variables,
253–257
of jointly Gaussian random
variables, 218–219, 254
of multiple random variables,
253–257
of pairs of random variables,
218–219

of vector random variable,
248–249
LMMSE estimator. *See* Linear
minimum mean square error
estimator
Log-likelihood ratio test, for radar
system, 321–322
Log-normal random variables,
570–571
Lottery game, combinations in, 24
Lower triangular, 552
Lowpass filter (LPF)
Gaussian white noise in, 476
noise equivalent bandwidth of,
479–480, 479*f*
SNR and, 481–482
white noise conversion in,
476–477, 477*f*
Lowpass process, 439
bandwidth of, 439, 439*f*, 440*f*
Lowpass random process, 494
PSD of, 494–495, 495*f*
LPF. *See* Lowpass filter
LTI system. *See* Linear time-
invariant system

M

(*m*, *n*)th joint moment of two
random variables, 195–196
MA. *See* Moving average process
MAP. *See Maximum a posteriori*
marcumq, MATLAB, 110
Marcum's Q-function, 83–84, 84*f*,
565
Marginal CDFs, 178, 180, 246
Marginal PDFs, 182–183, 246
Marginal PMFs, 187–188, 246
Markov chains, 384
characterization of, 394–401
exercises for, 424–426
computer communication
network described by,
413–415, 415*f*
definitions and examples of,
384–385, 385*f*, 390–391
gambler's ruin problem, 386,
386*f*, 392
genetic models, 388
queuing system, 387, 393–394,
394*f*

random walks, 387, 396, 399–401
transition and state probability
calculations in, 388–393
Markov processes, 383
computer communication
network described by,
413–415, 415*f*
continuous time, 401–411
exercises for, 426–427
MATLAB exercise for,
406–407, 406*f*,
412–413, 412*f*
M/M/∞ queue, 405
M/M/1 queue, 405–407, 406*f*
population modeling with,
407–409
continuous time and continuous
amplitude, 409–411
definitions and examples of,
383–388
branching process, 388
exercises for, 419–421
gambler's ruin problem, 386,
386*f*, 392
genetic models, 388
Markov chains, 384–385, 385*f*,
390–391
Poisson counting process,
383–384
queuing system, 387, 393–394,
394*f*
random walks, 387, 396,
399–401
telephone exchange described by,
416–417, 417*f*
transition and state probability
calculations in, 388–393
exercises for, 421–424
MATLAB exercise for,
393–394, 394*f*
Markov's inequality, 141
Matched filter
exercises for, 482–486, 482*f*,
485*f*, 486*f*, 511–512
for SNR, 482–486, 482*f*, 485*f*,
486*f*
for binary PAM signal,
485–486, 486*f*
for communication system,
484–485, 485*f*

Mathematical tables, 577–585, 581–582*t*, 582–583*t*, 584*t*, 585*t*
Mathematical tools
  random telegraph process, 340–341
  for studying random processes, 340–348, 342*f*, 345*f*
    autocorrelation function, 344–345
    autocovariance function, 347
    cross-correlation function, 347–348
    cross-covariance function, 348
    exercises for, 371–372
MATLAB, 2
  for amplitude modulation, 503–504, 504*f*
  for bilinear approximation of random processes, 530–532, 531*f*
  binomial coefficient exercise for, 62
  central moment exercise for, 120–121
  continuous time Markov process exercise for, 406–407, 406*f*, 412–413, 412*f*
  convolution function, 352–354, 353*f*
  dice rolling exercise for, 62
  for ergodicity and autocorrelation function estimation, 352–354, 353*f*
  for estimating CDF of IID random variables, 297–298
  for failure probability, 314–315, 314*f*
  for Gaussian random processes simulation, 527–528, 528*f*
  for Gaussian random variable, 109
    in multiple dimensions, 251–252
  help, 15
  histogram of randn, 76–78, 77*f*
  independent random variable exercise for, 200–201, 201*f*
  joint PDF exercise for, 185–186

marcumq function, 110
Markov chain transition and state probability calculation exercise for, 393–394, 394*f*
Markov process exercises for, 427–428
mean function of sinusoidal random processes, 344, 345*f*
multiple random variables exercises for, 287–288
multiple random variables transformation exercise for, 256–257
pairs of random variables exercises for, 242–243
for periodogram estimate of PSD in random telegraph process, 447–448, 448*f*
for PMF, 34, 34*f*
PSD exercises, 469–471
  of PAM, 460–461, 461*f*
for Q-function, 110
random generation, 15
  exercises for, 61–62
  histogram construction for, 71, 72*f*
random process exercises, 381–382
  of continuous time and discrete amplitude signals, 338–339, 339*f*
  of discrete time and continuous amplitude signals, 339–340, 340*f*
  in linear systems, 514–515
relative frequency approach simulation, 16–17, 16*t*
for Rician random variable, 110
for sample mean convergence, 305–306, 305*f*
scalar quantization exercise for, 153–154, 154*f*
of shot noise processes, 369–370, 370*f*
simulation technique exercises, 546
sine of uniform random variables, 85, 85*f*
single random variable operation exercises for, 174–175

single random variable transformation exercise for, 129, 130*f*
  for uniform random variables, 109
  for variance of IID random variable estimation, 296–297
Matrices, 551–557
Maximal length linear feedback shift register (MLLFSR), 520–521, 521*f*
Maximum a posteriori (MAP), 39
  in estimation and detection problems, 265–267
Maximum likelihood (ML)
  for estimation problems, 267–268
  for mean of IID random variable estimation, 292–294
  for variance of IID random variable estimation, 295–296
Mean
  of Bernoulli random variable, 297
  of complex random variables, 220
  conditional
    exercises for, 163–164
    of functions of single random variables, 122
    of single random variables, 121–122
  ergodic in, 351
  estimating of IID random variables, 290–295
    BLUE for, 291–292
    confidence intervals for, 310–315, 311*t*
    ML for, 292–294
  of failure probability, 314
  of functions of random variables, 113–114, 115*t*
    exercises for, 160–161
  of Gaussian random variables, 72–73, 72*n*1, 73*f*, 317–318
  in law of large numbers, 304
  of multiple random variables, 247–249
  order statistics and, 261–262

Mean *(continued)*
 for pairs of random variables, 192–197
 probability-generating functions and, 137–138
 of random processes, 343
 of random sum of Gaussian random variables, 317
 of random sum of IID random variables, 315–316
 of random variables, estimation of, 304
 of Rician random variables, 296–297
 sample, 292
 of single random variables, 111–113
  characteristic functions and, 133
  exercises for, 159–160
  tail probability evaluations using, 141–142
 of WSS random processes, 351
Mean function
 autocovariance function and, 347
 of Gaussian random processes, 358–359
 of LTI system, 473–475
 of random processes, 343
 of random telegraph process, 343
 of shot noise processes, 365–366
 of sinusoidal random processes, 343–344
  MATLAB for, 344, 345*f*
 wide sense stationary and, 349–351
Mean square error
 for IID random variables, 291
 for PSD estimation, 451
 Wiener filter to minimize, 487–489
Mean square (MS) sense, random sequence convergence in, 302, 303*f*
Mean time to first return, 400–401
Mean vector, for IID random variables, 291
Median, order statistics and, 261–262
Memoryless property, of exponential random variables, 93

Minimum mean square error (MMSE), for estimation problems, 268–272
Mixed random variables, 67
 transformation of, 127
MLLFSR. *See* Maximal length linear feedback shift register
M/M/∞ queue, 405
M/M/1 queue, 405–407, 404*f*
MMSE. *See* Minimum mean square error
Modulation efficiency, 503
Moment-generating functions
 joint, 209
 of single random variables, 139–140
  exercises for, 169–170
 tail probability evaluations using, 142–147, 143*f*, 147*f*
Moments. *See also* Central moments
 joint, 192–197, 208–209
 of single random variables, 115–117
  characteristic functions and, 133–135
  exercises for, 161
  probability-generating functions and, 136–138
Monotonically decreasing function transformations, 123*f*, 124–125
Monotonically increasing function transformations, 122–124, 123*f*
Monte Carlo simulations, 535–537
 for simulation of coded digital communication system, 539–541, 539*f*, 541*f*
Monty Hall problem, 61–62
Moving average process (MA), for PSD estimation, 449
m-sequences. *See* Maximal length linear feedback shift register
Multinomial coefficient, 25. *See also* Partitions
 exercises for, 48
Multiple random variables, 245
 conditional PMF and PDF for, 245–247

 exercises for, 277–278
 in estimation and detection problems, 264
 exercises for, 283–284
 MAP estimation, 265–267
 ML estimation, 267–268
 MMSE estimation, 268–272
 expected values of, 247–249
 exercises for, 279–280
 Gaussian, 249–251
  exercises for, 280–281
  MATLAB exercise for, 251–252
 independence of, 247
 joint PMF, CDF, and PDF for, 245–247
  exercises for, 277–278
 in linear prediction of speech, 272–275, 273*f*, 274*f*, 275*f*, 276*f*
 transformations involving, 252–253
  coordinate systems in three dimensions, 262–263, 263*f*
  exercises for, 281–283
  linear, 253–257
  MATLAB exercise for, 256–257
  order statistics, 260–262
  quadratic, 257–260
Mutual exclusiveness, 31
 independence compared with, 31
 of PMF of binomial random variable, 36
Mutual information
 exercises for, 239–240
 of pairs of random variables, 221–225, 223*f*, 224*f*, 225*f*
Mutually exclusive, 546

N

Nakagami random variable, 571
Narrowband random processes, 494–498, 495*f*, 496*f*, 497*f*, 498*f*
 exercises for, 514
n-bit quantizer, 149–153, 149*f*, 152*f*
Negative binomial random variable. *See* Pascal random variable
Negative definite, 556–557

Noise, 7. *See also* Gaussian white
noise; Stochastic signals
accounting for, 7
autocovariance function and, 347
deterministic signal or function
compared with, 7
Poisson random variable for, 37
sensor voltage and, 289–290
thermal, 452–454
variance of, 290
white, 454
Noise equivalent bandwidth,
479–480, 479*f*
exercises for, 509–510
Nonbinary pseudorandom number
generators, 521–522
Noncoherent systems
Rayleigh random variables in, 82
Rician distribution in, 84
Nonmonotonic function
transformations, 125–129,
125*f*, 128*f*, 130*f*
Non-singular matrix, 553–554
Non-stationary
Wiener–Khintchine–Einstein
theorem for, 435
WSS and, 349
Normal equations, 488
Normal random variables, 73
n-step transition probability,
389–391, 396–397
*n*th central moment, 117
*n*th moment, of single random
variables, 115
*n*th order PDF, of random
processes, 342
Null recurrent, 400–401
Numerical routines, for CDF of
Gaussian random variable, 74
Nyquist's theorem, thermal noise
and, 453

O

Optical communication system,
38–41, 38*f*, 42*f*
Bayes's theorem for, 39–40
conditional probability for,
40–41
overview of, 38–39, 38*f*
PMFs of, 39
probability of error for, 41, 42*f*

Order statistics, 260–262
Orthogonal variables, 193
complex, 221
Orthogonal vectors, 552
Orthogonality principle, 270, 488
Outcomes. *See also* Atomic
outcomes
of coin flipping, 8
definition of, 8
of dice rolling, 9
of die rolling, 9
exercises for, 43–44
functions of, 32
of random number generator, 10
total number of possible, 21–22

P

Padlock combinations, *k*-
permutations of, 23
Pairs of random variables,
177–178
channel capacity and channel
coding, 221–225, 223*f*, 224*f*,
225*f*
exercises for, 239–240
complex random variables,
219–221
exercises for, 238–239
conditional CDFs for, 188–191
exercises for, 231
conditional PDFs for, 188–191
exercises for, 231
conditional PMFs for, 188–191
exercises for, 231
expected values for, 192–197
conditional expected values,
196–197
correlation, 193–195
covariance, 193–195
exercises for, 231–233
(*m*, *n*)th joint moment,
195–196
independence of, 197–199
exercises for, 233–234
MATLAB exercise for,
200–201, 201*f*
joint CDFs for, 178–180, 179*f*
exercises for, 227–228
joint characteristic functions for,
206–209
exercises for, 235–236

joint moment-generating
functions for, 209
joint PDFs for, 180–185, 184*f*
exercises for, 228–229
MATLAB exercise for,
185–186
joint PMFs for, 186–188
exercises for, 229–230
joint probability-generating
functions for, 208–209
jointly Gaussian, 202–206,
203–204*f*, 205*f*, 206*f*
exercises for, 234
joint characteristic functions
for, 207–208
linear transformation of,
218–219, 254
MAP estimator for, 265
mutual information of,
221–225, 223*f*, 224*f*, 225*f*
exercises for, 239–240
transformations of, 210–219
exercises for, 236–238
PAM. *See* Pulse amplitude
modulation
Parallel interconnection, system,
reliability function and
failure rate function of,
94–95
Parametric estimation, of PSD,
448–452, 452*f*
Parseval's theorem, in PSD, 430
Partitions, 24–26, 26*t*
of bridge, 25–26
three groups, 25
two groups, 24–25
Pascal, 8
Pascal random variable, 38, 575.
See also Geometric random
variable
PCM. *See* Pulse code modulation
PDF. *See* Probability density
function
Period of states, 396
Periodogram
for PSD spectral estimation, 441,
445–448, 448*f*
random telegraph process,
447–448, 448*f*
Permutations, 22, 26*t*
k-permutations, 23

Phase, of stochastic signals, 7
Φ-function
    for CDF of Gaussian random
        variable, 74
        evaluation of, 75
    Q-function relation with, 75, 75*f*
Photodetector, in optical
        communication system,
        38–39
Photoemissive surface, in optical
        communication system, 39
PMF. *See* Probability mass function
Points, 360
    of Poisson counting processes,
        364
Poisson counting processes,
        360–364, 364*f*, 383–384
    autocorrelation function of, 363
    PMF of, 360, 363
    points of, 364
Poisson distribution, random
        telegraph signal, 341
Poisson impulse processes, 364,
        364*f*
Poisson processes
    autocovariance function of, 363
    exercises for, 376–378
Poisson random variable, 37, 575
    exercises for, 56–58
    expected value of, 112
    in optical communication system,
        39
Polar coordinates, Cartesian
        coordinate transformation
        into, 216–217
Population modeling, with birth–
        death processes, 407–409
Positive definite, 556–557
Positive recurrent, 400–401
Power, in random process, 431
Power residue method, 521–522
Power spectral density (PSD), 430
    AR process for, 449–450
    random telegraph process,
        451–452, 452*f*
    of bandpass random process,
        494–495, 495*f*
    bandwidth of random processes,
        439–441, 439*f*, 440*f*
    exercises for, 465–466
    for complex envelope, 500

cross spectral density, 438–439
definition of, 430–433
    exercises for, 463
of digital modulation formats,
        455–461, 458*f*, 459*f*, 460*f*,
        461*f*
    autocorrelation of PAM,
        456–457
    Fourier transform for, 457
    with IID sequence of random
        variables, 457–459, 458*f*,
        459*f*
    MATLAB example for,
        460–461, 461*f*
    with memory, 459–460, 460*f*
Fourier transform in, 430
Fourier transform pair of
        autocorrelation function
        and, 433
    of I and Q components,
        496–497, 497*f*
    of lowpass random process,
        494–495, 495*f*
MATLAB exercises for, 469–471
Parseval's theorem in, 430
properties of, 431
of sinusoidal random processes,
        432
spectral estimation, 441–452
    exercises for, 466–468
    non-parametric, 441–448,
        442*f*, 444*f*, 445*f*, 448*f*
    parametric, 448–452, 452*f*
    thermal noise, 452–455, 453*f*, 454*f*
    exercises for, 468–469
Wiener–Khintchine–Einstein
        theorem, 433–439, 434*f*,
        436*f*, 437*f*
    exercises for, 463–465
    of WSS, 433
Power transfer function, 475
    of LPF, 479, 479*f*
Prediction problems, 264
Principle of counting, 21
Probability. *See also* Conditional
        probability; Joint probability
    a posteriori, 28, 28*n*2
    a priori, 28, 28*n*1
    applications of, 1–5, 3*f*, 4*f*, 5*f*
    assignment of, 13–17, 16*t*
        classical approach to, 14

of coin flipping, 14
of dice rolling, 15
of die rolling, 14
exercises for, 46–47
MATLAB rand command and,
        15
relative frequency approach
        for, 16
of atomic outcomes, 14
axioms of, 10–13, 13*f*
    exercises for, 44–45
definition of, 10–11
gambling and, 7–8
random sequence convergence
        in, 301–302, 303*t*
relative frequency interpretation
        of, 297
theorem of total probability,
        27–28, 28*f*
Probability densities, of randn
        function, 76–78, 77*f*
Probability density function (PDF),
        69–71, 69*f*
    of Cauchy random variables, 84
    CDF compared with, 69–71, 69*f*
    of chi-squared (χ^2) random
        variables, 81–82
    conditional, 87–89, 88*f*
        exercises for, 104–106
        for multiple random variables,
            245–247, 277–278
        for pairs of random variables,
            188–191, 231
        properties of, 87–88
        transformations of pairs of
            random variables using,
            213–214
    of continuous random variables,
        for CDF, 289
    of continuous time and discrete
        amplitude signals, 341
    of Erlang random variable, 81–82
    of exponential random variables,
        79–80, 79*f*
    of gamma random variables, 81
    of Gaussian random processes,
        357–359
    of Gaussian random variables,
        72–73, 73*f*
    histogram of randn function,
        76–78, 77*f*

Probability density function (PDF)
(*continued*)
of IID random variables, 289
joint
exercises for, 228–229
for Gaussian random variables
in multiple dimensions,
250–252
for multiple random variables,
245–247, 277–278
for pairs of random variables,
180–186, 184*f*
joint Gaussian, 202–206,
203–204*f*, 205*f*
of Laplace random variables, 80,
80*f*
properties of, 70
of random processes, 340, 342
*n*th order, 342
second-order, 342
of random variables, 69, 76
characteristic functions and,
130–131
conditional, 87–89, 88*f*
exercises for, 99–101
reliability function, 91
of Rayleigh random variables,
82, 83*f*
of Rician random variables,
83–84, 84*f*
of shot noise processes,
367–368
of uniform random variables,
78–79, 78*f*, 309, 309*f*
verify validity of, 70
Probability mass function (PMF),
33
of Bernoulli random variable, 35,
310, 310*f*
binomial random variable, 35–37
CDF and, 68
for coin flipping, 33
conditional
for multiple random variables,
245–247, 277–278
for pairs of random variables,
188–191, 231
of continuous random variables,
63
on counting processes, 361–362

of discrete random variables, for
CDF, 289
exercises for, 55–59
geometric random variable,
37–38
joint
exercises for, 229–230
for multiple random variables,
245–247, 277–278
for pairs of random variables,
186–188
MATLAB for, 34, 34*f*
in optical communication system,
39
of Poisson processes, 360, 363
Poisson random variable, 37
problems with, 63–64
of random processes, 340, 342,
342*f*
of random telegraph process,
342*f*
Probability of error, for optical
communication system,
40–41, 42*f*
Probability ratio test, for radar
system, 318
Probability theory, 7
assigning probabilities, 13–17, 16*t*
axioms of probability, 10–13, 13*f*
basic combinatorics, 20–26, 26*t*
Bayes's theorem, 27–29, 28*f*
conditional probability, 18–20
development of, 11
discrete random variables,
32–38, 34*f*
exercises for, 43–62
experiments, samples spaces, and
events, 7–10
joint probability, 17–20
statistical independence, 29–32
Probability-generating functions
joint, 208–209
of single random variables,
136–138
exercises for, 168–169
PSD. *See* Power spectral density
Pseudorandom number generators
binary, 517–521, 518*f*, 519*f*,
520*t*, 521*f*
nonbinary, 521–522

Pulse amplitude modulation
(PAM), 455–456
autocorrelation function of,
456–457
Fourier transform for, 457
with IID sequence of random
variables, 457–459, 458*f*,
459*f*
matched filter for SNR in,
485–486, 486*f*
MATLAB example for,
460–461, 461*f*
with memory, 459–460, 460*f*
Pulse code modulation (PCM),
differential, 272–275, 273*f*,
274*f*, 275*f*, 276*f*

Q

Q components
complex envelopes for, 499–500
cross spectral density of, 497–498
extraction of, 496, 496*f*
PSD of, 497
Q-function, 565
for CDF of Gaussian random
variable, 74
evaluation of, 74
expression of, 75–76
Marcum's, 83–84
MATLAB exercise for, 110
numerical methods for
evaluating, 587–591, 588*f*,
589*f*
Φ-function relation with, 75, 75*f*
probabilities in terms of, 76
values of, 584–585, 585*t*
Quadratic transformations, of
multiple random variables,
257–260
Quantization
of single random variables,
148–153, 149*f*, 152*f*
exercises for, 172–173
MATLAB exercise for,
153–154, 154*f*
of speech, 272–275, 273*f*, 274*f*,
275*f*, 276*f*
Queuing systems, 387, 393–394,
394*f*
conditional PDF in, 89

Queuing systems *(continued)*
exponential random variables in, 80
gamma random variables in, 81
M/M/∞ queue, 405
M/M/1 queue, 405–407, 406*f*
Queuing theory, 402–403

R

Radar system, 4, 4*f*, 317–323
approaches to, 318
false alarm and detection probabilities for, 318–319
probability ratio test for, 318
sequential detection, 320
signal-to-noise ratio for, 323
system performance for, 323
thresholds for, 319–320
Wald's inequalities for, 321–322
Radio frequency, phase of, 79
rand, MATLAB, probability assignment and, 15
exercises for, 61–62
histogram construction for, 71, 72*f*
randn, MATLAB, histogram of, 76–78, 77*f*
Random number generator
with MATLAB, 15
exercises for, 61–62
histogram creation for, 71, 72*f*
PMF and, 63
sample space, events, and outcomes of, 10
Random processes, 335–370
average normalized power, 356
bandpass, 494
bandwidth of, 439–441, 439*f*, 440*f*
exercises for, 465–466
CDF of, 340, 342
of coin flipping, 336–338, 336*f*, 338*f*
computer generation of, 525–534
exercises for, 544
frequency domain approach to, 525–528, 526*f*, 528*f*
of Gaussian white noise, 533–534, 534*f*
time domain approach, 529–532, 529*f*, 531*f*, 533*f*

of continuous time and discrete amplitude signals, 338–339, 339*f*
cross spectral density between, 438–439
definition of, 336
of discrete time and continuous amplitude signals, 339–340, 340*f*
ergodic in autocorrelation, 351
ergodic in mean, 351
ergodicity of, 351–356
frequency domain techniques for, 429
Gaussian, 342, 357–360, 359*f*
autocorrelation function of, 346–347
definition of, 357
exercises for, 375–376
PDF of, 357–359
in linear systems, 473–504
analog communication system, 500–504, 500*f*, 501*f*, 504*f*
bandlimited and narrowband, 494–498, 495*f*, 496*f*, 497*f*, 498*f*
complex envelopes, 499–500, 514
continuous time, 473–477, 505–507
discrete-time, 477–479, 508–509
matched filter, 482–486, 482*f*, 485*f*, 486*f*, 511–512
MATLAB exercises for, 514–515
noise equivalent bandwidth, 479–480, 479*f*
signal-to-noise ratios, 480–482, 510–511
Wiener filter, 486–494, 490*f*, 491*f*, 495*f*, 512–513
lowpass, 494
mathematical tools for studying, 340–348, 342*f*, 345*f*
autocorrelation function, 344–345
autocovariance function, 347
cross-correlation function, 347–348

cross-covariance function, 348
exercises for, 371–372
MATLAB exercises for, 381–382
mean function of, 343
mean of, 343
PDF of, 340, 342
*n*th order, 342–343
second-order, 342
PMF of, 340, 342, 342*f*
Poisson processes, 360–364, 364*f*
exercises for, 376–378
PSD for
definition of, 430–433
Wiener–Khintchine–Einstein theorem, 433–439, 434*f*, 436*f*, 437*f*
of random variable, 337, 337*f*
autocorrelation function for, 346
ergodicity of, 352
MATLAB for, 344, 345*f*
mean function of, 343–344
strict sense stationary, 349
realization of, 336
shot noise processes, 365–370, 368*f*, 369*f*, 370*f*
spectral characteristics of, 429–430
strict sense stationary, 348–349
exercises for, 372–375
wide sense stationary, 349–351
exercises for, 372–375
Wiener process, 360
Random sequences
convergence modes of, 298–303, 303*t*
convergence almost everywhere, 301, 303*t*
convergence everywhere, 300–301, 303*t*
convergence in distribution, 302–303, 303*t*
convergence in mean square sense, 302, 303*t*
convergence in probability, 301–302, 303*t*
exercises for, 326–327
IID random variables, 289–298
estimating CDF of, 297–298
estimating mean of, 290–295

Random sequences (continued)
   estimating variance of,
     295–297
   PDF of, 289
   sample mean of, 299
  radar system, 317–323
   approaches to, 318
   false alarm and detection
     probabilities for, 318–319
   probability ratio test for, 318
   sequential detection, 320
   system performance for, 323
   thresholds for, 319–320
   Wald's inequalities, 321–322
Random sums
  central limit theorem, 306–310
   Bernoulli random variables,
     310, 310f
   exercises for, 329
   proof of, 307–308
   uniform random variables,
     309, 309f
  of Gaussian random variables, 317
  law of large numbers, 304–306
   exercises for, 327–328
  of random variables, 315–317
   exercises for, 331–332
   function of, 316
   mean and variance of, 315–316
Random telegraph process
  autocorrelation estimate for, 442,
    442f
   with windowing function,
     443–445, 444f, 445f
  bandwidth of, 441
  mathematical study of,
    340–341
  MATLAB, 338–339, 339f
  mean function of, 343
  periodogram estimate of PSD for,
    447–448, 448f
  PMF of, 342f
  Wiener–Khintchine–Einstein
    theorem for, 436–438, 436f,
    437f
Random variables. See also specific
    types
  CDF of, 65–66, 65f, 66f
   conditional, 86–87
   exercises for, 97–99
   reliability function, 91

  central moments of, 117–121
   exercises for, 162–163
   MATLAB exercise for,
    120–121
  characteristic functions of,
    130–136
   exercises for, 167–168
   moments and, 133–135
   natural logarithm of, 135–136
   PDFs and, 130–131
  for coin flipping, 33
  computer generation of, 517–524
   binary pseudorandom number
    generators, 517–521, 518f,
    519f, 520t, 521f
   correlated, 524
   exercises for, 543–544
   nonbinary pseudorandom
    number generators, 521–522
   from specified distribution,
    521–522
  conditional expected values of,
    121–122
   exercises for, 163–164
  entropy associated with,
    155–158, 157f, 158t
   exercises for, 173
  expected value of, 111–113
   exercises for, 159–160
   functions of, 113–114, 115t
  functions of, 85, 85f
   conditional expected value of,
    122
   exercises for, 160–161
   expected value of,
    113–114, 115t
  independence of, 197–199
   exercises for, 233–234
   MATLAB exercise for,
    200–201, 201f
  in interval, CDF and, 64, 69
  list of common, 565–575
  MATLAB histogram of randn,
    76–78, 77f
  mean of, estimation of, 304
  moment-generating functions of,
    139–140
   exercises for, 169–170
  moments of, 115–117
   characteristic functions and,
    133–135

   exercises for, 161
   probability-generating
    functions and, 136–138
  PDF of, 69, 76
   characteristic functions and,
    130–131
   conditional, 87–89, 88f
   exercises for, 99–101
   reliability function, 91
  PMF of, 33
  probability-generating functions
    of, 136–138
   exercises for, 168–169
  random process of, 337, 337f
   autocorrelation function for, 346
   ergodicity of, 352
   MATLAB for, 344, 345f
   mean function of, 343–344
   strict sense stationary, 349
  random sums of, 315–317
   exercises for, 331–332
   function of, 316
   mean and variance of, 315–316
  scalar quantization of,
    148–153, 149f, 152f
   exercises for, 172–173
   MATLAB exercise for,
    153–154, 154f
  tail probability evaluations,
    140–147, 143f, 147f
   exercises for, 170–172
  tangent of, 84
  transformations of, 122
   exercises for, 164–167
   MATLAB exercise for, 129,
    130f
   monotonically decreasing
    functions, 123f, 124–125
   monotonically increasing
    functions, 122–124, 123f
   nonmonotonic functions,
    125–129, 125f, 128f, 130f
Random walks, 359–360, 359f, 387,
    396, 399–401
Rare events, simulation of, 534–538
  exercises for, 545–546
  importance sampling, 537–538
  Monte Carlo simulations, 535–537
Rayleigh distribution, 82
  reliability function and failure
    rate function of, 93

Rayleigh random variables, 82, 83*f*, 571–572
  exercise for, 104
  expected value of, 113
  Rician random variables and, 83
Realization, of random process, 336
Recurrent state, 398–401
Relative frequency approach
  to assigning probabilities, 16
  for dice rolling, 16–17, 16*t*
  for joint probability, 18
Relative frequency interpretation of
  probability, 297
Reliability function, 91
  derivative of, 91
  exponential distribution of, 93
  failure rate function and, 92
  of Rayleigh distribution, 93
  of system
    parallel interconnection, 94–95
    series interconnection, 93–95
Reliability rates, 91–95
Rician distribution, 84
Rician random variables, 83–84, 84*f*, 572
  central moments of, 120–121
  MATLAB exercise for, 110
  mean of, 296–297
  variance of, 296–297
RMS bandwidth. *See* Root-mean-square
Root-mean-square (RMS)
  bandwidth, 440
Row vector, 551

S

Saddle point approximation,
  145–147, 147*f*
Sample mean, 292
  confidence intervals and, 312
  of IID random variables, 299
  in law of large numbers, 304
  MATLAB example for,
    305–306, 305*f*
  of Rician random variables,
    296–297
Sample spaces, 7–10
  choice of, 10
  of coin flipping, 8–9
  continuous, 10
    mix with discrete, 10

definition of, 8
  of dice rolling, 9
  of die rolling, 9
  exercises for, 43–44
  of random number generator, 10
  random variables of, 32
Sample variance, 296
  MATLAB example for,
    305–306, 305*f*
  of Rician random variables,
    296–297
Satellite communication channels,
  Rician distribution in, 84
Scalar, 551
Scalar quantization, of single
  random variables, 148–153,
  149*f*, 152*f*
  exercises for, 172–173
  MATLAB exercise for, 153–154,
    154*f*
Second central moment, 118. *See
  also* Variance
Second moment, 115–117
Second-order PDF, of random
  processes, 342
Sensor, voltage of, noise and,
  289–290
Sequential detection
  performance of, 320–322
  for radar system, 320
Series expansions, 577–578
Series interconnection, system,
  reliability function and
  failure rate function of,
  93–95
Set theory, 545–549, 548*f*
Shannon entropy, 156
Shot noise processes, 365–370,
  368*f*, 369*f*, 370*f*
  autocorrelation function of,
    366–367
  in *p-n* junction diode, 365–370
    with current pulse shape,
      369–370, 370*f*
    exercises for, 378–380
    with square wave,
      368–369, 368*f*, 369*f*
  mean function of, 365–366
  PDF of, 367–368
Poisson counting processes, 365
Poisson impulse processes, 365

as strict sense stationary, 368
as WSS, 367
Signals
  review of, 559–563
  scalar quantization of, 148–153,
    149*f*, 152*f*
  sinusoidal, 219
  types of, 335
Signal-to-noise ratio, of radar
  system, 323
Signal-to-noise ratios (SNR)
  for AM, 502–503
  exercises for, 480–482, 510–511
  in linear systems, 480–482
  LPF and, 481–482
  matched filter for, 482–486, 482*f*,
    485*f*, 486*f*
    for binary PAM signal,
      485–486, 486*f*
    for communication system,
      484–485, 485*f*
  of radar system, 323
Signal-to-quantization-noise power
  ratio (SQNR), 148–150,
  153–154, 154*f*
  for DPCM speech coders, 275,
    276*f*
Simulation techniques, 517–541
  for coded digital communication
    system, 539–541, 539*f*, 541*f*
  computer generation of random
    processes, 525–534
    exercises for, 544
    frequency domain approach to,
      525–528, 526*f*, 528*f*
    of Gaussian white noise,
      533–534, 534*f*
    time domain approach,
      529–532, 529*f*, 531*f*, 533*f*
  computer generation of random
    variables, 517–524
    binary pseudorandom number
      generators, 517–521, 518*f*,
      519*f*, 520*t*, 521*f*
    correlated, 524
    exercises for, 543–544
    nonbinary pseudorandom
      number generators,
      521–522
    from specified distribution,
      521–522

Simulation techniques *(continued)*
  MATLAB exercises for, 546
  of rare events, 534–538
    exercises for, 545–546
    importance sampling, 537–538
    Monte Carlo simulations,
      535–537
Single random variables. *See*
  Random variables
Singular matrix, 553–554
Sinusoidal random processes, 337,
  337*f*
  autocorrelation function for, 346
  ergodicity of, 352
  mean function of, 343–344
    MATLAB for, 344, 345*f*
  PSD of, 432
  strict sense stationary, 349
  Wiener–Khintchine–Einstein
    theorem for, 435–436
Sinusoidal signals, 219
6 dB rule, 150
Skewness, 118. *See also* Coefficient
  of skewness
Slotted Aloha, 413–415, 415*f*
Smoothing function, for
  periodogram, 446
SNR. *See* Signal-to-noise ratios
Source coding, entropy and,
  155–158, 157*f*, 158*t*
  exercises for, 173
Spectral estimation
  exercises for, 466–468
  of PSD, 441–452
    non-parametric, 441–448,
      442*f*, 444*f*, 445*f*, 448*f*
    parametric, 448–452, 452*f*
Speech, linear prediction of,
  272–275, 273*f*, 274*f*, 275*f*,
  276*f*
Speech recognition system, 2–4, 3*f*
Speech signal amplitude, Laplace
  random variables in, 80
Spherical coordinates, Cartesian
  coordinate transformation
  into, 262–263, 263*f*
SQNR. *See* Signal-to-quantization-
  noise power ratio
Staircase function transformation,
  128, 128*f*

Standard deviation
  estimating of IID random
    variables, 295–297
  of Gaussian random variables,
    72–73, 72*n*1, 73*f*
  second central moment of
    random variables and, 118
Standard normal random variables,
  73
  CDF of, 74
  central limit theorem and, 307–308
State transition probability matrix,
  calculation of, 388–393
  exercises for, 421–424
  MATLAB exercise for, 393–394,
    394*f*
Stationary
  strict sense, random processes,
    348–349
  wide sense, random processes,
    349–351
Stationary increments, of Poisson
  processes, 361
Statistical independence. *See*
  Independence
Steady-state distribution, 391, 393,
  401
  for continuous time Markov
    processes, 403–405
Stochastic signals, 7
Strict sense stationary
  ergodicity and, 351–352
  exercises for, 372–375
  of random processes, 348–349
  shot noise processes as, 368
  sinusoidal random processes, 349
Strong law of large numbers,
  304–305
  convergence almost everywhere
    and, 305
Student's t-distribution, 313, 315*n*5,
  572–573
Subset, 545, 548*f*
Symmetric matrix, 551
Systems, review of, 559–563

T

Tail probabilities, evaluation of,
  140–147, 143*f*, 147*f*
  exercises for, 170–172

Taylor series expansion of
  Q-function, 588–589, 588*f*
Telephone exchange, Markov
  processes describing,
  416–417, 417*f*
Telephone traffic, Erlang random
  variable in, 82
Theorem of Total Probability,
  27–28, 28*f*
  unconditional cumulative
    distribution function and,
    89–90
Thermal noise, 452–454
  exercises for, 468–469
  PSD of, 452–455, 453*f*, 454*f*
Thevenin equivalent circuit, 453,
  453*f*
Third central moment, 118
Three dimensions. *See* Multiple
  random variables
3 dB bandwidth, 440, 440*f*
Time, continuous function of, 335
Time difference variable,
  autocorrelation function
  and, 345
Time domain techniques, 7
  for computer generation of
    random processes, 529–532,
    529*f*, 531*f*, 533*f*
    MATLAB example for,
      530–532, 531*f*
Time-averaged power, of random
  process, 430
Total probability, theorem of,
  27–28, 28*f*
Transfer function, 562
Transformations
  of multiple random variables,
    252–253
    coordinate systems in three
      dimensions, 262–263, 263*f*
    exercises for, 281–283
    linear, 253–257
    MATLAB exercise for,
      256–257
    order statistics, 260–262
    quadratic, 257–260
  of pairs of random variables,
    210–219
    exercises for, 236–238

Transformations *(continued)*
  of single random variables, 122
    exercises for, 164–167
    MATLAB exercise for, 129,
      130*f*
    monotonically decreasing
      functions, 123*f*, 124–125
    monotonically increasing
      functions, 122–124, 123*f*
    nonmonotonic functions,
      125–129, 125*f*, 128*f*, 130*f*
    on uniform random variables, 79
  of vector random variable,
    248–249
Transient state, 398–401
Transition probability matrix,
  384–385
  calculation of, 388–393
    exercises for, 421–424
    MATLAB exercise for,
      393–394, 394*f*
  for continuous time Markov
    processes, 402
Transpose, 551
Trigonometric identities, 577
Two dimensions. *See* Pairs of
  random variables

U

Unbiased estimate, 262, 290
Unconditional cumulative
  distribution function, 89–90
Uncorrelated random variables, 193
  complex, 221
  independence and, 195, 199, 251
Uniform quantizer, 149–150, 149*f*
Uniform random variables, 78–79,
  78*f*, 573
  central moments of, 118
  histogram of MATLAB
    generation of, 71, 72*f*
  MATLAB exercise for, 109
  moments of, 117
  PDF of, 309, 309*f*
  sine of, 85, 85*f*
  transformation on, 79
  WSS and, 350
Union, 545, 546*f*
Universal set, 545, 546*f*
Upper triangular, 552

V

Variability, accounting for, 7
Variance, 118
  of complex random variables,
    220–221
  estimating of IID random
    variables, 295–297
    MATLAB example for,
      296–297
    ML for, 295–296
  of failure probability, 314
  of Gaussian random variables,
    72–73, 72*n*1, 317–318
  of IID random variables, 292
  in law of large numbers, 304
  of noise, 290
  probability-generating functions
    and, 137–138
  of random sum of Gaussian
    random variables, 317
  of random sum of IID random
    variables, 315–316
  of Rician random variables,
    296–297
  sample, 296
  tail probability evaluations using,
    142
Vector random variable. *See*
    Multiple random variables
Voice conversation duration,
  exponential random
  variables in, 80

W

Wald's inequalities, for radar
  system, 321–323
Weak law of large numbers, 304–305
  convergence in probability and,
    305
Weibull random variable, 573–574
White noise, 454
  LTI system conversion of,
    476–477, 477*f*
Whitening, 256
Whitening filter, 490–491, 490*f*
Wide sense stationary (WSS)
  autocorrelation function and,
    349–351, 356–357
  ergodicity of, 351–356
  exercises for, 372–375

Gaussian random processes,
  358–359
  PSD of, 433
  random processes, 349–351
  shot noise processes as, 367
Wiener filter, 486–494, 490f, 491*f*,
  494*f*
  exercises for, 486–494, 490*f*,
    491*f*, 494*f*, 512–513
  impulse response with, 489
  to minimize mean square error,
    487–489
  for prediction, 492–494, 494*f*
  with two filters, 490–492
Wiener process, 360
Wiener–Hopf equations, 488–489
  for prediction, 492–494, 494*f*
  for two filters, 490–491
Wiener–Khintchine–Einstein
    theorem, 433–439, 434*f*,
    436*f*, 437*f*
  exercises for, 463–465
  for random telegraph process,
    436–438, 436*f*, 437*f*
  for sinusoidal random processes,
    435–436
Windowing function, for
    autocorrelation function
    estimation, 443
  random telegraph process,
    443–445, 444*f*, 445*f*
Wireless communication channels,
    Rayleigh random variables
    in, 82
Wireline telecommunication
    networks, Erlang random
    variable in, 82
WSS. *See* Wide sense stationary

Y

Yule-Walker equations, 488

Z

Zero-mean Gaussian distribution, 82
Zeroth central moment, 118
Zeroth moment, 115
Z-transform, 561, 582–583*t*
  for discrete-time linear systems,
    477–478
  for discrete-time processes, 429

Printed and bound by CPI Group (UK) Ltd, Croydon, CR0 4YY

03/10/2024

01040314-0004